Essentials of Atmospheric Sciences

Essentials of Atmospheric Sciences

Edited by **Smith Paul**

R CALLISTO REFERENCE

New York

Published by Callisto Reference,
106 Park Avenue, Suite 200,
New York, NY 10016, USA
www.callistoreference.com

Essentials of Atmospheric Sciences
Edited by Smith Paul

International Standard Book Number: 978-1-63239-328-9 (Hardback)

Printed in the United States of America.

Contents

Preface

From thunderstorms and cyclones to wave propagation and cloud propagation, everything broadly falls under the category of atmospheric sciences. This is a branch of science that contains many branches within itself. In other words, it is the study of the atmosphere and its phenomena which are further split into multiple disciplines like atmospheric physics, atmospheric chemistry, atmospheric dynamics, meteorology, climatology and volcanology.

Given human beings' interest in understanding, predicting and controlling the atmosphere, this is a branch of science that has seen some stellar work being done over the past few decades. Atmospheric models have been designed. Climatological simulations have been undertaken. Natural disasters have been predicted and massive strides have been made in understanding every phenomenon that the atmosphere undergoes.

Technological advances and recent discoveries have further enhanced the development of this branch of science. Now, the understanding of atmospheric chemistry, atmospheric physics, climatology, gravity waves and atmospheric dynamics has evolved to a very large extent.

This text brings together all the information related to atmospheric sciences. It pertains to understanding and documenting all the latest developments in the field of atmospheric sciences and sheds light on the current research being undertaken around the world for a better grasp of natural occurrences to be prepared for any eventuality. This book includes contributions from some renowned experts and researchers in this field.

Given its broad coverage and up-to-date information, this book should serve as a crucial source of information for students, professionals and enthusiasts, helping them understand the natural phenomena a lot better.

I would like to thank my publisher and my family for their faith and support.

Editor

Concerning the Lower Atmosphere Responses to Magnetospheric Storms and Substorms

P. A. Sedykh and I. Yu. Lobycheva

Institute of Solar-Terrestrial Physics SB RAS, Lermontov Street, 126 a, P.O. Box 291, Irkutsk 664033, Russia

Correspondence should be addressed to P. A. Sedykh; pvlsd@iszf.irk.ru

Academic Editor: Helena A. Flocas

The issue of existence and physical mechanism for solar-terrestrial couplings has rather a long history. Investigations into the solar activity effect on meteorological processes in the lower atmosphere have become especially topical recently. The aim of this study is to investigate the effect of geomagnetic activity on meteorological processes in the atmosphere. We analyze the data on magnetic storms and tropical cyclones that were observed in the North Atlantic, East Pacific, and West Pacific to understand the mechanism for magnetospheric disturbance effects on complicated nonlinear system of atmospheric processes.

1. Introduction

There exist statistical correlations among geomagnetic activity, atmospheric pressure, and temperature [1–3]. Authors in [4] suggested that the observed climate response to solar variability is caused by a dynamical response in the troposphere to heating predominantly in the stratosphere.

According to [5], a tropical cyclogenesis may be "a mechanism for effective discharge of the surplus heat in the atmosphere under the conditions when the routine mechanism effect becomes insufficient." Between the solar-terrestrial disturbance parameters, on the one hand, and the cyclogenesis characteristics, on the other, various researchers endeavor to trace hard-to-detect statistical associations. In [6], the correlation between tropical cyclones and the Cycle-23 storms was investigated. The revealed coincidence between the time of origin and evolution of the 2005 August 23-24 Hurricane Katrina with the powerful geomagnetic storm main phase [7] also boosted the research in this area.

The issue of physical mechanism for solar-terrestrial couplings has long interested researchers. Many geophysicists were almost prepared to reject the idea about a solar activity effect on the lower atmosphere condition as absolutely unacceptable. And, first of all, the matter was that the atmospheric process power enormously exceeds the solar-wind input energy flux into the near-Earth space. Due to this, it seems most unlikely that solar activity could significantly affect the lower atmosphere condition. However, the research done over the last years allowed us to find a clue to overcome this inconsistency. The main objection to a possibility of the solar activity effective influence on the lower atmosphere condition and on weather, based on insufficient power of the solar wind, appears quite surmountable; see, for example, [8]. Also, like the computations in [9] show, the energy necessary to create the atmospheric optical screen (shield) is incomparably lower than the variation amplitude in the screen-induced solar energy flux arriving at the lower atmosphere.

According to [8], variations in the atmospheric ionizing radiation flux observed during geomagnetospheric disturbances cause a noticeable variation in the chemical composition and contents of small components, as well as in the atmosphere transparency. The main types of such variations are [8] (1) galactic cosmic ray intensity short-term depressions (observed during geomagnetic disturbances (Forbush decreases)) caused by dispersion of energetic charged particles by the magnetic fields transported from the solar atmosphere by the solar wind high-velocity streams; (2) solar cosmic ray flux bursts caused by solar flares.

This paper aims to investigate geomagnetic activity effects on meteorological processes and a possible effect of magnetospheric disturbances on the tropical cyclogenesis evolution character. Usually, researchers considered magnetospheric

(a)

(b)

(c)

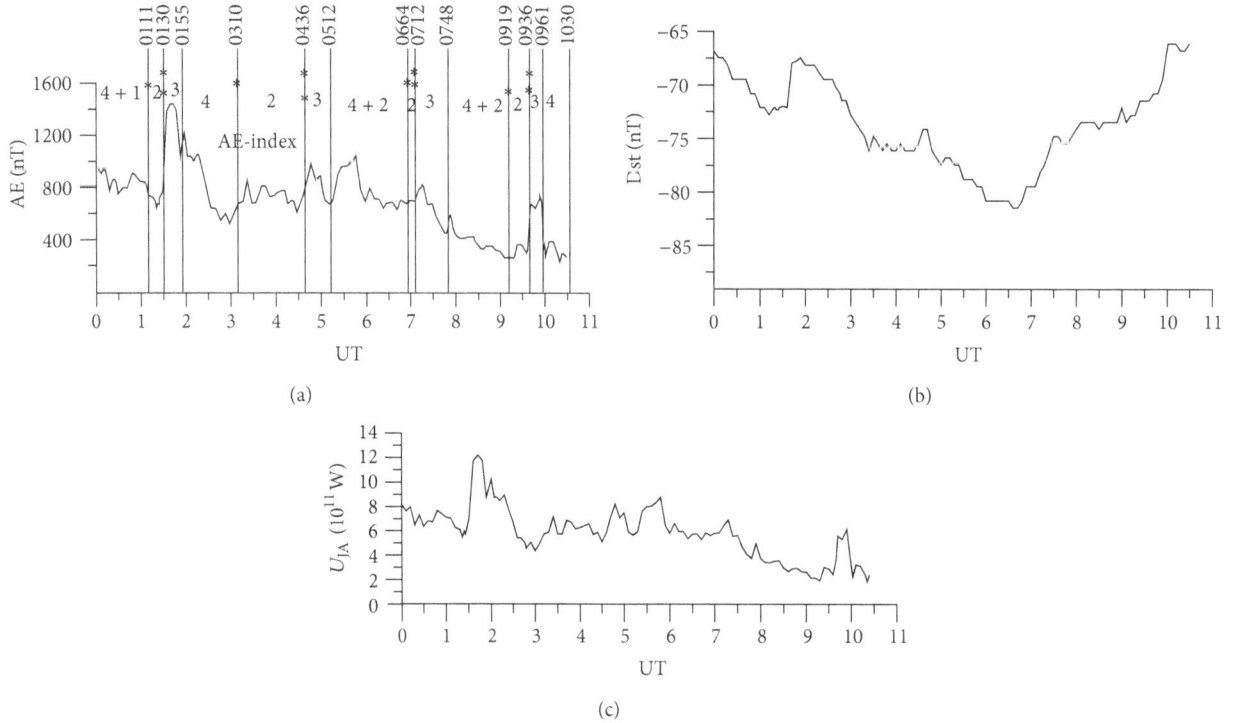

FIGURE 1: (a) AE-index for the chain of substorms, May 3, 1986. Timing of the chain of substorms CDAW9C [13]. Single and double asterisks indicate the actively convective phase and the expansion (explosive) phase of the substorm, respectively. (b) Dst-index for the chain of CDAW9C substorms [13]. (c) Estimations of power of Joule heating of the ionosphere and power of particles precipitation into the ionosphere for the chain of CDAW9C substorms.

and atmospheric problems in separate papers. This paper is an attempt to combine consistent parts taken from the meteorology ionospheric and magnetospheric physics.

2. Principal Results

A magnetospheric storm is a 1–3 day long phenomenon spanning all the magnetosphere regions, and it features sharp depressions in the magnetic field. During storms and substorms, the ionosphere undergoes rather significant Joule heating with a great power of precipitating energetic particles. Huge energy increases the ionosphere temperature and causes large-scale ion drifts and neutral winds.

We used magnetograms from ground-based stations, AE-indices, Dst-indices for each magnetospheric storm, and calculated some parameters characterizing energetic aspects of magnetospheric disturbances. The aim was to investigate a possible effect of powerful magnetospheric storms on the evolution character of meteorological processes in the atmosphere to study the correlation between magnetospheric disturbances and meteorological background variations. We also address the meteorological data for 10-day periods after these disturbances.

The power of the ionosphere's Joule heating $U_j = \int_S \Sigma_P E^2 dS$ is either calculated through the Assimilative Mapping of Ionospheric Electrodynamics (AMIE) [10] or estimated based on empirical formulas as the AE-index

function [11]: $U_j = 0.33 \times AE$, where U_j is in GWatt and the AE-index is in nT. The power of the energetic particles precipitating into the ionosphere U_a can be also estimated based on empirical formulas $U_a = 1.75 \times (AE/100 + 1.6) \times 10^{10}$ (W) [12]. We also used the databases for hurricanes at http://www.nhc.noaa.gov/pastall.shtml, http://www.csc.noaa .gov/hurricane_tracks, and http://www.aoml.noaa.gov/hrd/ hurdat/ushurrlist18512007.txt. To study the surface pressure, the temperature, and other characteristics, we used the data at ftp://ftp.cdc.noaa.gov/. The information on tropical cyclones was obtained from http://russian.wunderground.com/ hurricane/hurrarchive.asp?region=at.

The 1986 May 3 magnetospheric storm over the (0000-1100) UT interval was studied within the international CDAW9C (Coordinated Data Analysis Workshop) Project and represented a 4-substorm chain [13]. This event refers to magnetospheric storms that occurred during solar Cycle-21. It featured the AE-index reaching the values of about 1400–1500 nT, whereas the Dst-index reached only −70–80 nT (see Figures 1 and 2). Because this storm comprised 4 substorms, we detected the substorm phases by timing each of the 4 substorms. Following [14, 15], we will term these phases the growth phase (1), the actively-convective phase (2), the expansion (explosive) phase (3), and the recovery phase (4).

In summary, the data on the 1986 May 3 event (Figures 1 and 2) could be used to trace each substorm phase contribution (energy loading-unloading) to the composite set of processes in the lower atmosphere.

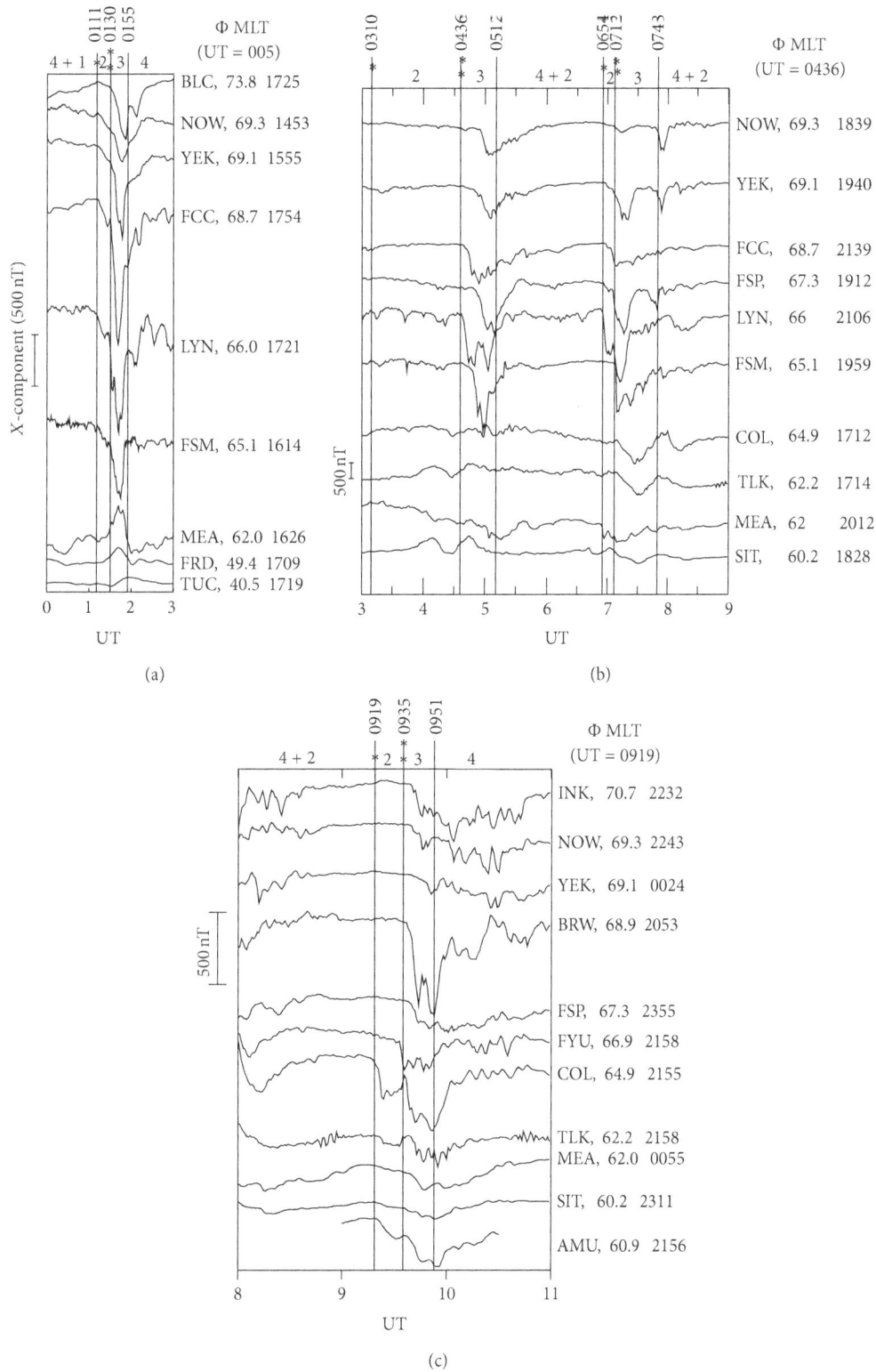

FIGURE 2: ((a), (b), and (c)) Timing of the chain of substorms CDAW9C. Magnetograms of the ground-based magnetometers for the chain of substorms; May 3, 1986. Single and double asterisks indicate the actively convective phase and the expansion (explosive) phase of the substorm, respectively [13].

TABLE 1

Region	1986	1989	2003
Northern Atlantic	June, August, September, November	June, July, August, September, October, November, December	April, June, July, August, September, October, November, December
Eastern Pacific	May, June, July, August, September, October	January, May, June, July, August, September, October	January, May, June, July, August, September, October
Western Pacific	Throughout the year	January, April, May, June, July, August, September, October, November, December	April, May, June, July, August, September, October, November, December

We indicate the months when storms were observed in those years. As seen from the table, the storm was observed in the Eastern Pacific in May 1986. It had existed May 22–29 which cannot be associated with the May 3, 1986 magnetic storm. Yet, it corroborates the climatology of tropical cyclone (tropical cyclones originate in the Pacific Ocean, May through November).

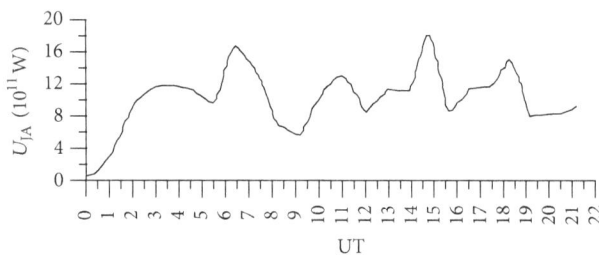

FIGURE 3: Estimations of power Joule heating and of power of particles precipitation into the ionosphere for the magnetospheric storm, March 13, 1989.

The 1989 March 13-14 magnetospheric storm refers to one of the strongest magnetospheric storms observed since the space era onset and occurred during solar Cycle-22, see, for example, [16, 17]. The geomagnetic storm reached the maximal intensity on 1989 March 13 when the Dst-index reached a record value of about −600 nT [16, 17]. To save space, we do not give the illustrations for the 1989 March 13-14 event here. There are many papers describing this storm.

The 2003 October-November solar activity sharp burst led to strong magnetospheric storms with the 2003 October 28–31 high AE-index values and Dst-index of about −400 nT [18]. For the reasons above, we do not give the illustrations for this event here.

3. Discussion and Conclusions

As one can see (Figures 1, 3, and 4), the ionosphere underwent rather a significant Joule heating with a great power of precipitating energetic particles. Nevertheless, there were no abnormal variations in the lower atmosphere meteoparameters. The variations in physical characteristics of hurricanes (wind velocity, temperature, and pressure distribution) were not associated with high magnetic activity both in the high-latitude regions and in the mid-latitude ones (Figures 5, 6, and 7). We have used a lot of data for the analysis; some examples are illustrated in figures. There is no principal difference in plots for the periods of magnetospheric disturbances in comparison to nonstorm days. The atmospheric process power incomparably exceeds energy flux from the solar wind into the geomagnetosphere and the power of extremely

strong magnetospheric disturbances. The energy flux from the magnetosphere into the atmosphere during the strong storm was about 1.5×10^{19} (erg/s) $\times 24 \times 3600 = 1.2 \times 10^{24}$ ergs/day, which is by 2-3 orders of magnitude less than the atmospheric process power whose values are in [19]. Investigations into the solar activity effect on atmospheric processes have become especially topical recently [7, 20]. But, if there is a mechanism for the magnetospheric disturbance effect on meteorological processes in the atmosphere, it supposes a more complicated series of many intermediates and is not associated directly with the energy that arrives at the ionosphere during storms.

The main result of our investigation (see Table 1) is that the 1989 March magnetic storm did not directly affect the tropical cyclogenesis evolution character (Figures 8, 9, and 10). As for the 2003 October-November storms, one cannot answer unambiguously, either, because 5 plus intense and long-lived cyclones were observed then. The 13 October–1 November strong storm in the North Atlantic had existed extremely long (19 days), taking into account that the cyclone life average duration in Atlantic is 9 days. In 2003 October, 3 tropical cyclones were observed in the East Pacific: Hurricane Nora (October 1–9, the maximal wind being 41 m/s); Tropical Storm Olaf (October 3–8, 30 m/s); Tropical Storm Patricia (October 20–26, 31 m/s). However, all of the strong storms in the East Pacific were observed prior to the magnetic storm onset.

Cloud layers play an important role in Earth's radiation balance [21–23], affecting the amount of heat from the Sun that reaches the surface and the heat radiated back from the surface that escapes out into space. One should pay special attention to the effect of the solar wind electric field sharp increase (via the global electric circuit during magnetospheric disturbances) on the cloud layer formation. It is necessary to test the assumptions that this layer may function as a screen decelerating radiative cooling of the air located *on the Central Antarctic ice dome* (as a result, there would be warming in the ground atmospheric layer and cooling above the cloud layer [24]). Authors in the paper [24] suggested that the interplanetary electric field influences the katabatic system of atmospheric circulation (*typical of the winter in the Antarctic*), via the global electric circuit affecting clouds and hence the radiation dynamics of the troposphere.

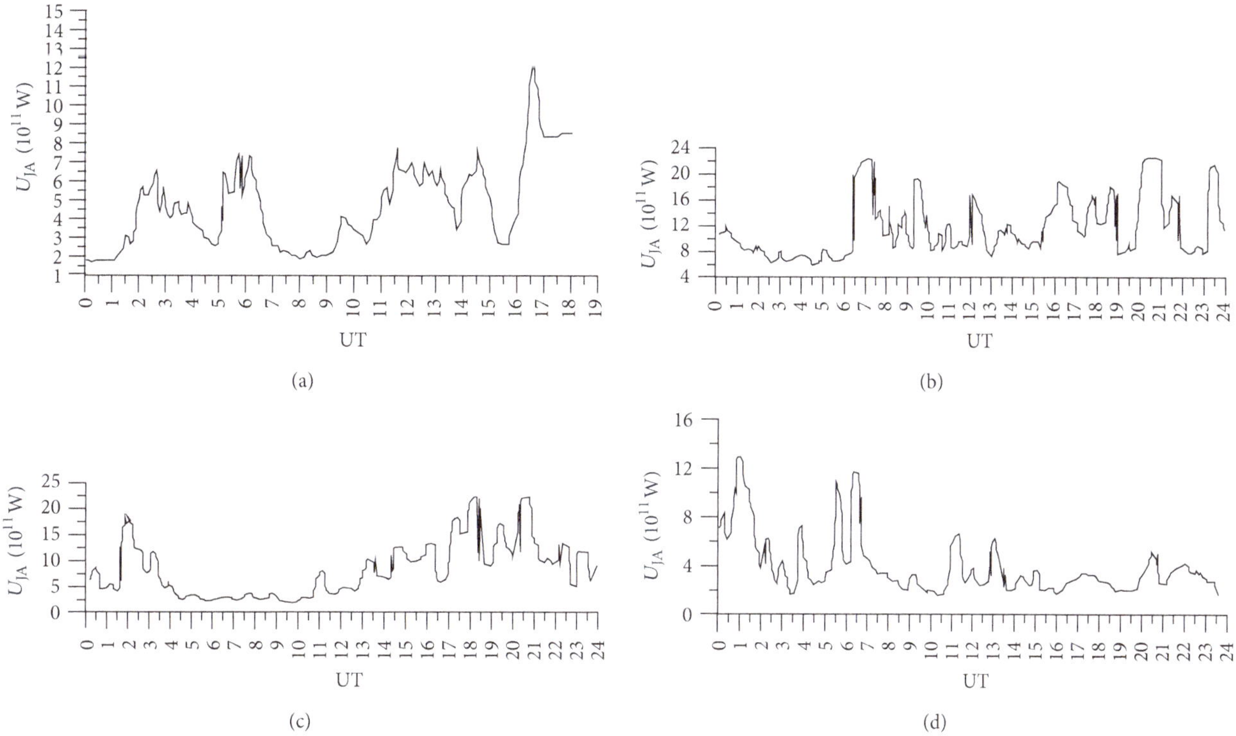

FIGURE 4: Estimations of power of Joule heating and power of particles precipitation into the ionosphere: (a) October 28, 2003; (b) October 29, 2003; (c) October 30, 2003; (d) October 31, 2003.

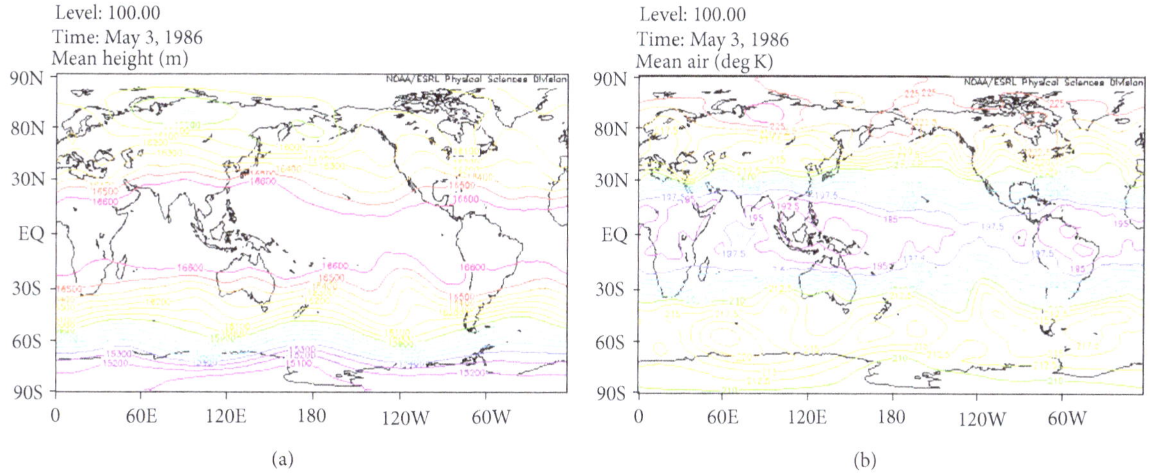

FIGURE 5: The NCEP/NCAR reanalysis example for the event, May 3, 1986: (a) height of an isobaric surface at the level 100 hPa (average per day); (b) the average temperature at the level 100 hPa.

Let us briefly address the problem of the extraneous electric field penetration into the Earth's magnetosphere. More than five decades ago, Dungey [25] suggested the following model for the magnetospheric electric field generation. The solar wind magnetic field partially permeates inside the magnetosphere therefore, the polar cap field lines leave for the solar wind. In the solar wind, there is the electric field $E = -(1/c)[VB]$, where V is the solar wind velocity and B is the interplanetary magnetic field. Because the conductance along field lines is very high, the electrical potential associated

with the field in the solar wind is transported into the polar cap ionosphere. In the polar cap, a Sun-away convection originates, and, for that convection to be closed, reverse motions on closed field lines of the inner geomagnetosphere are necessary. Two-vortex convection was assumed to be obtained. Through multiplying the solar wind electric field value by the geomagnetosphere size towards dawn-dusk being about 25–30 Re (Re~6371 km) and through assuming the solar wind velocity as ~450 km/s, we will obtain the potential difference between the dawn and the dusk sides of

Level: 100.00
Time: March 14, 1989
Mean height (m)

Level: 100.00
Time: March 14, 1989
Mean air (deg K)

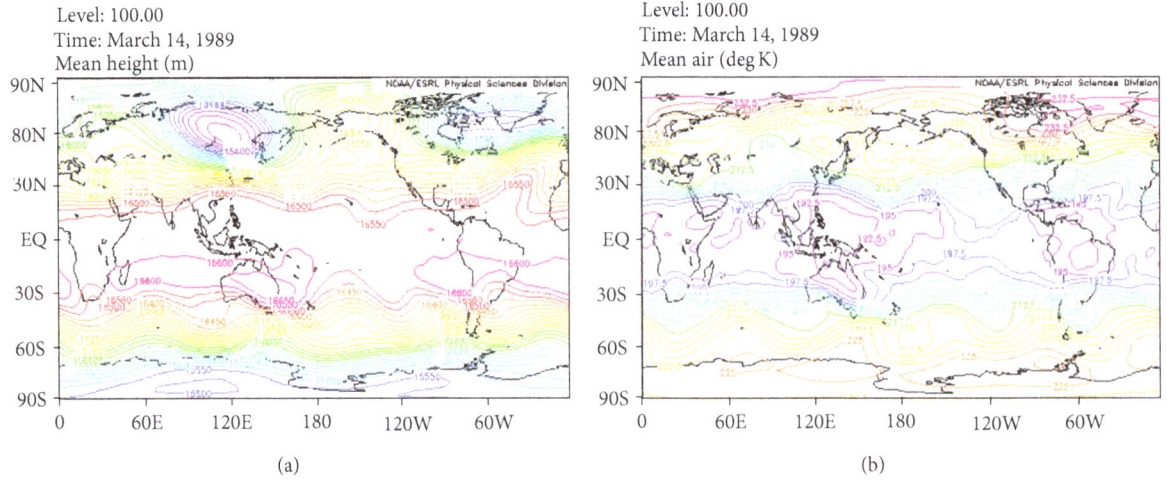

(a)

(b)

FIGURE 6: The NCEP/NCAR reanalysis example for the event, March 13-14, 1989: (a) height of an isobaric surface at the level 100 hPa (average per day); (b) the average temperature at the level 100 hPa.

Level: 100.00
Time: October 30, 2003
Mean height (m)

Level: 100.00
Time: October 30, 2003
Mean air (deg K)

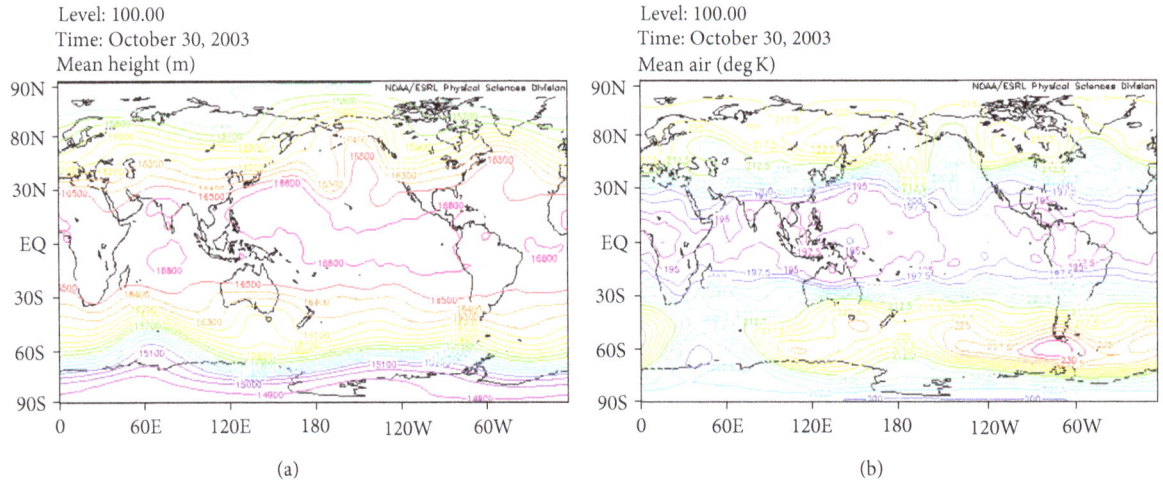

(a)

(b)

FIGURE 7: The NCEP/NCAR reanalysis example for the event, October 30, 2003: (a) height of an isobaric surface at the level 100 hPa (average per day); (b) the average temperature at the level 100 hPa.

the magnetosphere $\Delta\Psi \approx -80$ Bz, (kV) which is manifold less than the well-known experimental formula [26] $\Delta\Psi$(kV) = -11 Bz (nT) provides.

The penetration of the electric field and the current into the geomagnetosphere is a two-stage process and may be presented as follows (see Figure 11). Let an electric current component towards the magnetosphere appear at instant T. A potential value will be established at the magnetopause segment. In the thin near-side layer of the thickness $d \sim 2\pi c/\omega_{pp}$ (where ω_{pp} is proton plasma frequency and c is the speed of light), the charge division process will start, and the displacement current $j^* = (\varepsilon/4\pi) \times \partial E/\partial t$ will appear. Also, there will appear Ampere force $F = [j^* \times B]/c$ that will start accelerating plasma. The only force that withstands the Ampere one is the inertia force. Under the conditions of a homogeneous medium, the inertia force is $\rho \partial v/\partial t$:

$$\frac{\rho \partial v}{\partial t} = \frac{[j^* \times B]}{c} = \left(\frac{\varepsilon}{4\pi c}\right) \times \left[\frac{\partial E}{\partial t} \times B\right]. \quad (1)$$

Taking into account that $\varepsilon = c^2/V_A^2$, where V_A is the Alfven velocity, upon integrating we will have $v = c[E \times B]/B^2$ that is, the classic equation for the electric drift velocity (it is important for us to express the dynamic process in this case).

When the plasma is accelerated in the layer d up to the $V \times B$ drift velocity (and it happens during the gyroperiod), then there will be no field in the plasma coordinate system, and it appears at the boundary between the moving and stable plasmas in a stable coordinate system (see Figure 11). The boundary moving velocity separating the moving plasma from the stable one will be, consequently, $V_\phi \sim d\omega_B/2\pi$, where ω_B is the proton gyrofrequency. Taking the values d and ω_B into the equation for the phase velocity, we see that it is the Alfven velocity like we expected. Thus, the external electric field penetrates into the magnetosphere without any limitations of the Alfven-wave type, and the electric current is only *in a form of* the displacement current. The electric current flows through the system under consideration only when there is a transitive process. In the stationary regime, there is

(a)

(b)

(c)

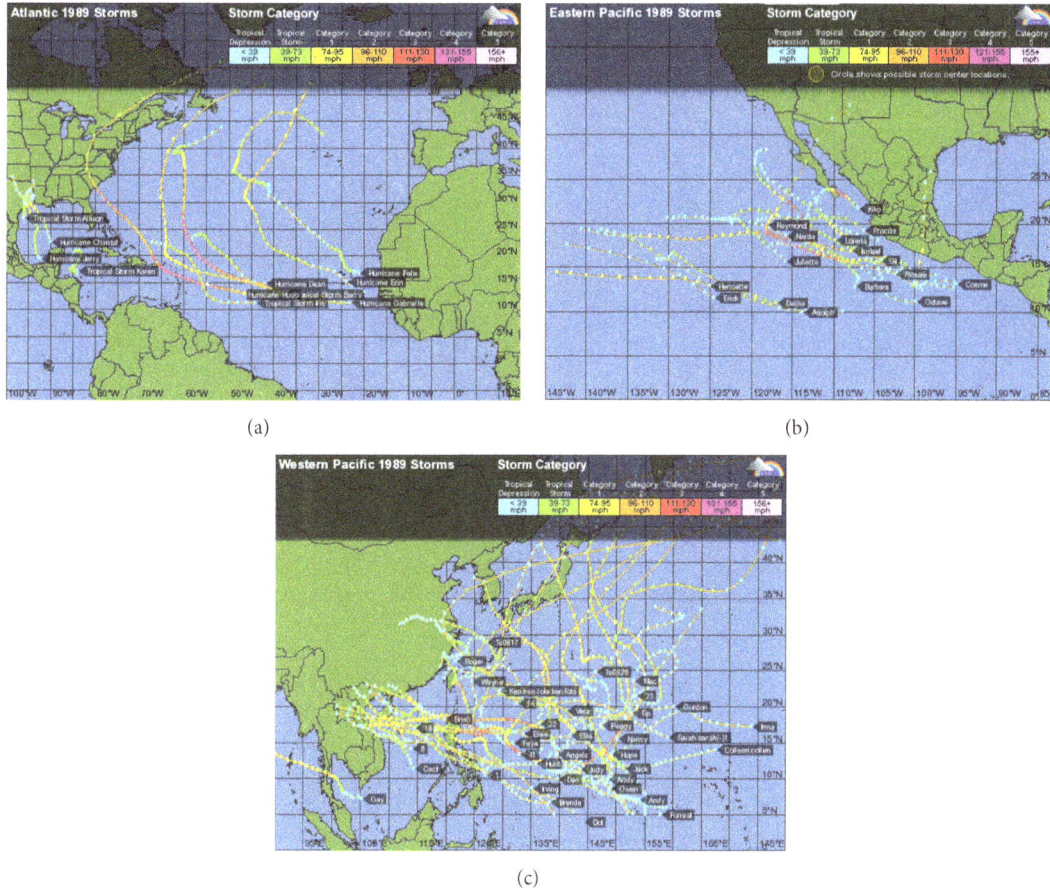

FIGURE 8: (a) The example of the trajectory of the cyclone observed in the North Atlantic, for events in 1989. (b) The example of the trajectory of the cyclone observed in Eastern Pacific, for events in 1989. (c) The example of the trajectory of the cyclone observed in Western Pacific, for events in 1989.

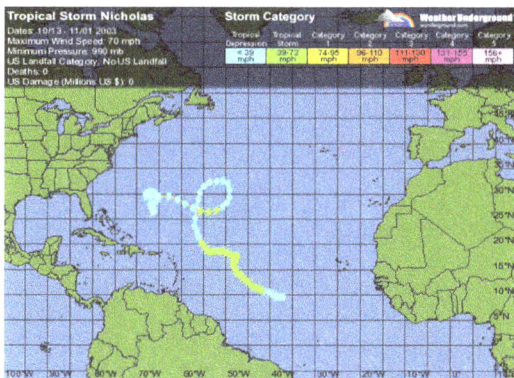

FIGURE 9: The trajectory of the cyclone observed in the North Atlantic 13 October–1 November, 2003. This storm was observed during the magnetic storm and has existed maximum for a long time (19 days) at average life time of a cyclone, which is 9 days, in the North Atlantic.

no electric current. If the magnetic field is inhomogeneous according to an axis X, then the gas pressure gradient will be originated independently due to the flow nonuniformity. The electric field establishment time in the system here is $\tau_E = L/V_\phi$, and the current establishment time is $\tau_I = L^*/V_c$, where L is the system size, V_ϕ is the phase velocity for the electromagnetic signal propagation across the system, $L^* = (B/\nabla B)$, and V_c is the plasma convection velocity. An approximate estimate applied to the magnetosphere gives the time of the electric field establishment to be hundreds of seconds and the electric current establishment time to be about an hour. Thus, the electric current penetration into plasma is a two-stage process. Initially, the polarization field that penetrates into plasma "layer by layer" is produced. Or, to be more exact, the momentum corresponding to this field penetrates into plasma. Here, if the system is inhomogeneous, the flow can redistribute pressure so that an electric current arises in plasma because of the appearance of gradients. This electric current is necessary to maintain plasma convection in the magnetosphere [27, 28].

Indeed, there is no simple global electric circuit via which a sharp increase in the solar wind electric field during magnetospheric disturbances would be possible. The solar wind electric field penetration process is complex and nonlinear. Field-aligned currents connect the magnetosphere and the ionosphere in the united electric circuit. The plasma convection generation in the geomagnetosphere is associated with the processes at the bow shock front [27]. The combined

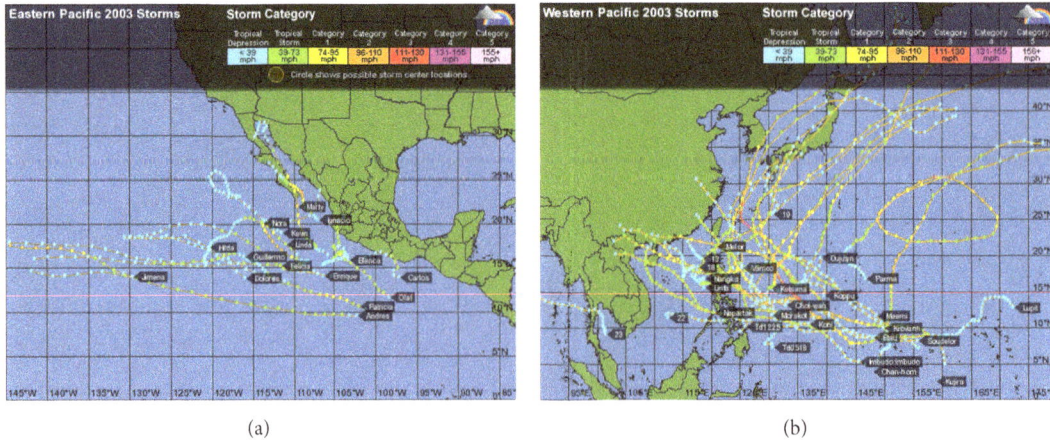

(a)   (b)

FIGURE 10: (a) The example of trajectories of the cyclone observed in Eastern Pacific. In October, 2003, in East part of Pacific ocean 3 outputs of a tropical cyclone were observed: Hurricane Nora, October, 1–9, 2003, with the maximal speed of wind ~41 m/s; Tropical Storm Olaf, October, 3–8, 2003, with maximal wind speed of 30 m/s; Tropical Storm Patricia, October, 20–26, 2003, with the maximal wind speed of 31 m/s. All storms were observed prior to the beginning of action of the magnetospheric storm. (b) The example of trajectories of the cyclone observed in Western Pacific for events in 2003. In October-November, 2003, more than 5 intensive and long cyclones were observed.

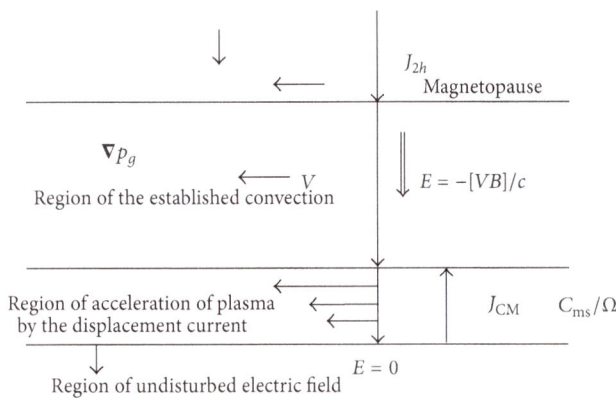

FIGURE 11: The scheme of the penetration of the external electric current and electric field in the magnetospheric plasma and generation of magnetospheric convection. In the figure, the part of magnetopause, through which the electric current and the electric field will penetrate, is shown.

action of plasma convection and pitch-angle diffusion of electrons and protons leads to the formation of plasma pressure distribution in the magnetosphere. As known, steady electric bulk currents are associated with the gas pressure distribution. The divergence of these bulk currents causes a spatial distribution of field-aligned currents, that is, magnetospheric sources of the ionospheric electric current systems [29]. The atmospheric conductivity sharply declines between the polar ionosphere and the layer at $h \sim 10$ km.

The geomagnetospheric disturbance effect on the troposphere is weak compared with a multitude of other factors affecting it. However, the existing works on a high correlation between tropical cyclones and magnetic storms may evidence either the existence of another mechanism for the effect (that was not addressed in this study) or a random

coincidence rather than a physical essence. A very interesting mechanism suggested by Troshichev and Janzhura [24] needs further considering and improving. Authors in the paper [24] noted that the solar wind dynamic pressure effect on the cloud layer would be opposite to that of the interplanetary electric field. Thus, now we can note that, probably, there is some connection between processes at the bow shock front region and meteorological processes at the lower atmosphere, because the magnetospheric plasma convection generation is associated with processes at the bow shock front (e.g., see [27]).

A further study of geomagnetospheric storm effects on tropical cyclogenesis is necessary, because the character of some influence has regional peculiarities (see Figures 8–10). In the next paper, we are going to investigate a probable effect of such magnetospheric disturbances as sawtooth events, extraordinary powerful magnetic storms, and some others on meteorological processes in the lower atmosphere.

Acknowledgments

This study was supported by the Russian Foundation for Basic Research, Project no. 12-05-31096 mol_a. The authors are grateful to creators and developers of the following websites: (1) http://www.nhc.noaa.gov/pastall.shtml; (2) http://www.csc.noaa.gov/hurricane_tracks; (3) http://www.aoml.noaa.gov/hrd/hurdat/ushurrlist18512007.txt; (4) http://russian.wunderground.com/hurricane/hurrarchive.asp?region=at; (5) ftp://ftp.cdc.noaa.gov/; (6) http://www.kosmofizika.ru/; (7) http://www.viems.ru/asnti/ntb/ntb502/oboc5.html for providing observational data.

References

[1] V. Bucha, "Geomagnetic activity and the global temperature," *Studia Geophysica et Geodaetica*, vol. 53, no. 4, pp. 571–573, 2009.

[2] V. Bucha, "Changes in geomagnetic activity and global temperature during the past 40 years," *Studia Geophysica et Geodaetica*, vol. 56, no. 4, pp. 1095–1107, 2012.

[3] D. R. Palamara, "An interhemispheric comparison of the geomagnetic activity signature in the lower atmosphere," *Earth, Planets and Space*, vol. 56, no. 9, pp. e25–e28, 2004.

[4] J. D. Haigh, M. Blackburn, and R. Day, "The response of tropospheric circulation to perturbations in lower-stratospheric temperature," *Journal of Climate*, vol. 18, no. 17, pp. 3672–3685, 2005.

[5] E. A. Sharkov, "Global tropical cyclogenesis: evolution of scientific views," *Issledovanie Zemli iz Kosmosa*, no. 6, pp. 1–9, 2005 (Russian).

[6] K. G. Ivanov, "Correlation between tropical cyclones and magnetic storms during cycle 23 of solar activity," *Geomagnetism and Aeronomy*, vol. 47, no. 3, pp. 371–374, 2007.

[7] K. G. Ivanov, "Generation of the Katrine hurricane during the geomagnetic extrastrom at crossing of the heliospheric current sheet: is it an accidental coincidence or physical essence?" *Geomagnetism and Aeronomy*, vol. 46, no. 5, pp. 609–615, 2006.

[8] M. I. Pudovkin, "Solar activity effect on the condition of the lower atmosphere and weather," *Soros's Educational Journal*, no. 10, pp. 106–113, 1996.

[9] D. Hauglustaine and J. C. Gerard, "Possible composition and climatic changes due to past intense energetic particle precipitation," *Annales Geophysicae*, vol. 8, no. 2, pp. 87–96, 1990.

[10] A. D. Richmond, G. Lu, B. A. Emery, and D. J. Knipp, "The Amie procedure: prospects for space weather specification and prediction," *Advances in Space Research*, vol. 22, no. 1, pp. 103–112, 1998.

[11] B. H. Ahn, H. W. Kroehl, Y. Kamide, and D. J. Gorney, "Estimation of ionospheric electrodynamics parameters using ionospheric conductance deduced from Bremsstrahlung X-ray image data," *Journal of Geophysical Research: Space Physics*, vol. 94, no. A3, pp. 2565–2586, 1989.

[12] R. W. Spiro, P. H. Reiff, and L. J. Maber, "Precipitating electron energy flux," *Journal of Geophysical Research: Space Physics*, vol. 87, no. A10, pp. 8215–8227, 1982.

[13] P. A. Sedykh and L. V. Minenko, "Studying CDAW9C substorms," in *Proceedings of the BSFP*, vol. 2, pp. 562–576, Irkutsk, Russia, 1999.

[14] Yu. P. Maltsev, L. L. Lazutin, V. G. Vorobyev et al., *Physics of the Near-Earth Space*, vol. 1, RAS KRC, PSP, 2000.

[15] P. A. Sedykh, "On two-stage evolution of the substorm active phase," in *Proceedings of the BSFP*, pp. 90–94, Irkutsk, Russia, 2009.

[16] Yu. P. Tsvetkov, A. N. Zaitsev, V. I. Odintsov, C. K. Hao, and N. T. K. Thoa, "Comparison of magnetic variations in the equatorial zone and the polar cap for the magnetic storm of March 13, 1989," *Geomagnetism and Aeronomy*, vol. 38, no. 2, pp. 192–200, 1998.

[17] L. A. Hajkowicz, "Global onset and propagation of large-scale travelling ionospheric disturbances as a result of the great storm of 13 March 1989," *Planetary and Space Science*, vol. 39, no. 4, pp. 583–593, 1991.

[18] M. I. Panasyuk, C. N. Kuznetsov, L. L. Lazutin et al., "Magnetic storms in October 2003: Collaboration 'Solar Extreme Events in 2003 (SEE-2003)'," *Space Research*, vol. 42, no. 5, pp. 489–534, 2004.

[19] M. I. Pudovkin and S. V. Babushkina, "Influence of solar flares and disturbances of the interplanetary medium on the atmospheric circulation," *Journal of Atmospheric and Solar-Terrestrial Physics*, vol. 54, no. 7-8, pp. 841–846, 1992.

[20] M. V. Vorotkov and V. L. Gorshkov, "Solar flare activity effect on cyclone activity of the Earth's atmosphere," in *Proceedings of the Problems of Astronomy*, pp. 231–239, Irkutsk, Russia, 2011.

[21] B. A. Tinsley, G. M. Brown, and P. H. Scherrer, "Solar variability influences on weather and climate: possible connections through cosmic ray fluxes and storm intensification," *Journal of Geophysical Research*, vol. 94, no. 12, pp. 14–792, 1989.

[22] B. A. Tinsley and G. W. Deen, "Apparent tropospheric response to MeV-GeV particle flux variations: a connection via electrofreezing of supercooled water in high-level clouds?" *Journal of Geophysical Research*, vol. 96, no. 12, pp. 22283–22296, 1991.

[23] B. A. Tinsley and R. A. Heelis, "Correlations of atmospheric dynamics with solar activity evidence for a connection via the solar wind, atmospheric electricity, and cloud microphysics," *Journal of Geophysical Research*, vol. 98, no. 6, pp. 10375–10384, 1993.

[24] O. A. Troshichev and A. Janzhura, "Temperature alterations on the Antarctic ice sheet initiated by the disturbed solar wind," *Journal of Atmospheric and Solar-Terrestrial Physics*, vol. 66, no. 13-14, pp. 1159–1172, 2004.

[25] J. W. Dungey, "Interplanetary magnetic field and the auroral zones," *Physical Review Letters*, vol. 6, no. 2, pp. 47–48, 1961.

[26] M. A. Doyle and W. J. Burke, "S3-2 measurements of the polar cap potential," *Journal of Geophysical Research*, vol. 88, no. 11, pp. 9125–9133, 1983.

[27] E. A. Ponomarev, P. A. Sedykh, and V. D. Urbanovich, "Generation of electric field in the Earth magnetosphere, caused by processes in the bow shock," *Journal of Atmospheric and Solar-Terrestrial Physics*, vol. 68, no. 6, pp. 679–684, 2006.

[28] P. A. Sedykh, "On the role of the bow shock in power of magnetospheric disturbances," *Sun & Geosphere*, vol. 6, no. 1, pp. 27–31, 2011.

[29] P. A. Sedykh and E. A. Ponomarev, "A structurally adequate model of the geomagnetosphere," *Studia Geophysica et Geodaetica*, vol. 56, no. 4, pp. 110–126, 2012.

Analysis of Convective Thunderstorm Split Cells in South-Eastern Romania

Daniel Carbunaru,[1,2] **Sabina Stefan,**[1] **Monica Sasu,**[1,2] **and Victor Stefanescu**[1,2]

[1] *Department of Atmospheric Physics, Faculty of Physics, University of Bucharest, P.O. Box MG-11, 077125 Bucharest, Romania*
[2] *National Meteorological Administration, Bucuresti-Ploiesti Avenue, No. 97, 013686 Bucharest, Romania*

Correspondence should be addressed to Daniel Carbunaru; daniel.carbunaru@meteoromania.ro

Academic Editor: Helena A. Flocas

The mesoscale configurations are analysed associated withthe splitting process of convective cells responsible for severe weather phenomena in the south-eastern part of Romania. The analysis was performed using products from the S-band Doppler weather radar located in Medgidia. The cases studied were chosen to cover various synoptic configurations when the cell splitting process occurs. To detect the presence and intensity of the tropospheric jet, the Doppler velocity field and vertical wind profiles derived from radar algorithms were used. The relative Doppler velocity field was used to study relative flow associated with convective cells. Trajectories and rotational characteristics associated with convective cells were obtained from reflectivity and relative Doppler velocity fields at various elevations. This analysis highlights the main dynamic features associated with the splitting process of convective cells: the tropospheric jet and vertical moisture flow associated with the configuration of the flow relative to the convective cells for the lower and upper tropospheric layers. These dynamic characteristics seen in the Doppler based velocity field and in the relative Doppler velocity field to the storm can indicate further evolution of convective developments, with direct implications to very short range forecast (nowcasting).

1. Introduction

In south-eastern Romania, in the convective season (from May to September), severe weather phenomena develop frequently and evolve [1–5], sometimes leading to significant damages. Analysis of severe convective events and their structure, at least for certain classes of phenomena (isolated convective cells), taking into account the mesoscale configuration can improve the nowcasting procedures. Using an S-band Doppler radar (10 cm wavelength), we see that the mesoscale phenomena (supercellular thunderstorms) have certain features during their dynamic evolution. Of great importance in understanding the evolution of supercells is the splitting process (separation of the convective cells in two other ones rotating opposite to one another, namely, cyclonic and anticyclonic), a process that takes place during the evolution of convective developments and is closely linked to the state of mesoscale dynamic configuration, especially vertical wind shear [6, 7]. This process has been studied both qualitatively [8–18] and by numeric analysis [19–21], highlighting the interaction between horizontal vorticity from the vertical profile of horizontal wind and the updraft.

Depending on the curvature of the hodograph, the development of convective cells is favoured when the propagation is either to the right, or to the left oftroposphericshear; both cells may be favoured as well, when the hodograph is linear. Along with this particular interaction, there may also exist secondary ones [22] that lead to the diminishing of the cell that propagates to the left of shear vector, in the Northern Hemisphere. Observations of the split process when the Doppler radar is used [23–32] highlight the cyclonic and anticyclonic rotation, within the field of Doppler velocity, associated with convective cells compatible with vertical shear.

The above studies also showed the asymmetry between the cells that propagate to the right and those that propagate

FIGURE 1: The spatial domain of radar analysis (geographic projection). The radar range is 230 km, the spatial resolution is 1 km, and the temporal resolution is 6 minutes, between two scans. NOAA Weather and Climate Toolkit (http://www.ncdc.noaa.gov/oa/wct/).

to the left of the tropospheric shear vector in the Northern Hemisphere, the latter generally has a shorter lifespan than the former, but is associated with severe hail.

In some cases the inflow at low levels and the outflow at upper levels [33] to and from the storm were seen with the aid of the Doppler relative velocity field; these two flows are in direct correlation with the intensity of updraft.

In this paper, several cases are analysed which featured the splitting process of strongly convective cells (supercells). The basic criteria in choosing these cases are the different mesoscale configuration (various directions of the tropospheric flow). The study focused on the analysis from fields and products of the meteorological Doppler radar that can provide relevant data on the severity of convective phenomena.

This paper is structured as follows. The first part contains the radar data that were used during the analysis of splitting processes, and the methodology thereof. At the second part, several case studies are presented that emphasize certain characteristics of the mentioned process. The paper ends with conclusions derived from these case studies, namely, mesoscale dynamic features and some particularities of convective cells trajectories. Since most of the convective cells appearing in the radar fields featured supercell traits, the common elements that were identified could be used in nowcasting.

2. Data and Methods

For the dynamic mesoscale analysis and that of the splitting process, the WSR-98D Doppler radar in S band (10 cm) was used, located in Medgidia (south-eastern Romania) (Figure 1). The range of the radar is shown in Figure 1.

The splitting process was analyzed in the fields of radar reflectivity, Doppler base velocity, and that of Doppler relative velocity at different scan elevations, in order to see the formation of cyclonic and anticyclonic vortices. The Doppler relative velocity field is obtained from the Doppler base velocity field minus the average velocity of all detected storms and represents Doppler velocity field of the reference storm [34]. The Principal Users Processor (PUP) software (METSTAR WSR-98D PUP version 10.8.R) was used. The study of Doppler velocity field at various elevations, in parallel with the Vertical Azimuth Display (VAD) Wind Velocity product resulted in information about the vertical shear of horizontal velocities and tropospheric jets. The VAD Wind Velocity is a Doppler weather radar product which gives the vertical profile of horizontal wind vectors above radar station [34–36]. Strong shear in the wind field assumes the existence of tropospheric low-level jet (LLJ) or upper-level jet (ULJ). Also the Doppler relative velocity field gave information about the low and upper relative flow associated to the storms in radar field. The presence of relative low-level flow was analysed (arrow-shaped tube, in the lower part of Figures 2(a) and 2(b)) and the relative upper-level flow (arrow-shaped tube, at top of Figures 2(a) and 2(b)). For the first three analysed cases, vertical vorticity profiles were made, further to be compared with circular movement associated with supercells (horizontal circles with cyclonic and anticyclonic vorticity, in Figure 2).

The evolution of convective cells was observed in the reflectivity field at elevations of 0.5 and 1.5 degrees, also in the Doppler radial velocities field and in the Doppler relative velocity field at elevations of 0.5, 1.5, 2.4, and 3.4 degrees.

The evolution in time and space of each convective cell that was characterised by the splitting process was also observed. Identification of splitting was done by tracking both reflectivity and cyclonic or anticyclone rotation, as seen in Figures 3(a) and 3(b). Figure 3(c) shows how the vertical vorticity of rotational movements was calculated.

Thus, for each rotational motion, maximum inbound and outbound velocities were identified in the Doppler relative velocity field at elevation where rotational movement had maximum intensity. The PUP application was used. For the actual calculation, radar algorithm is dividing the difference between the maximum inbound velocity and the maximum outbound velocity, by the distance between the two (mesocyclone diameter in Figure 3(c)).

Trajectories of convective cells in each case were obtained using polar coordinates, relative to the radar position, of the cells before and after the splitting. For the three cases studied in detail, the result was a map showing the individual characteristics of convective developments, taking into account the stage before splitting and the after-splitting stage (which is associated with the cyclonic and anticyclone rotation). Vertical variation of the horizontal velocity (vertical shear) was derived from the VAD Wind Profile algorithm and vertical profiles were then constructed from wind shear at particular moments, representative of the evolution of vertical shear during the life of the convective cell. The vertical domain associated with the z-coordinate was separated into a domain with directional shear (in the lower tropospheric layer) and a domain with nondirectional shear (in the middle tropospheric level). This division was possible because the dynamic configuration near the radar site, at least in the three cases studied, has a nondirectional shear above 1800 m associated with an ULJ, and directional shear below 1800 m.

(a)   (b)

FIGURE 2: Tiling of the horizontal vorticity due to updraft (a). Negative vorticity maxima (on the left of shear) and positive vorticity maxima (on the right of shear) and (b) the convective cell split in two other cells with opposite vorticity (from Lin 2007, adapted by Klemp 1987, and from Rotuno 1981).

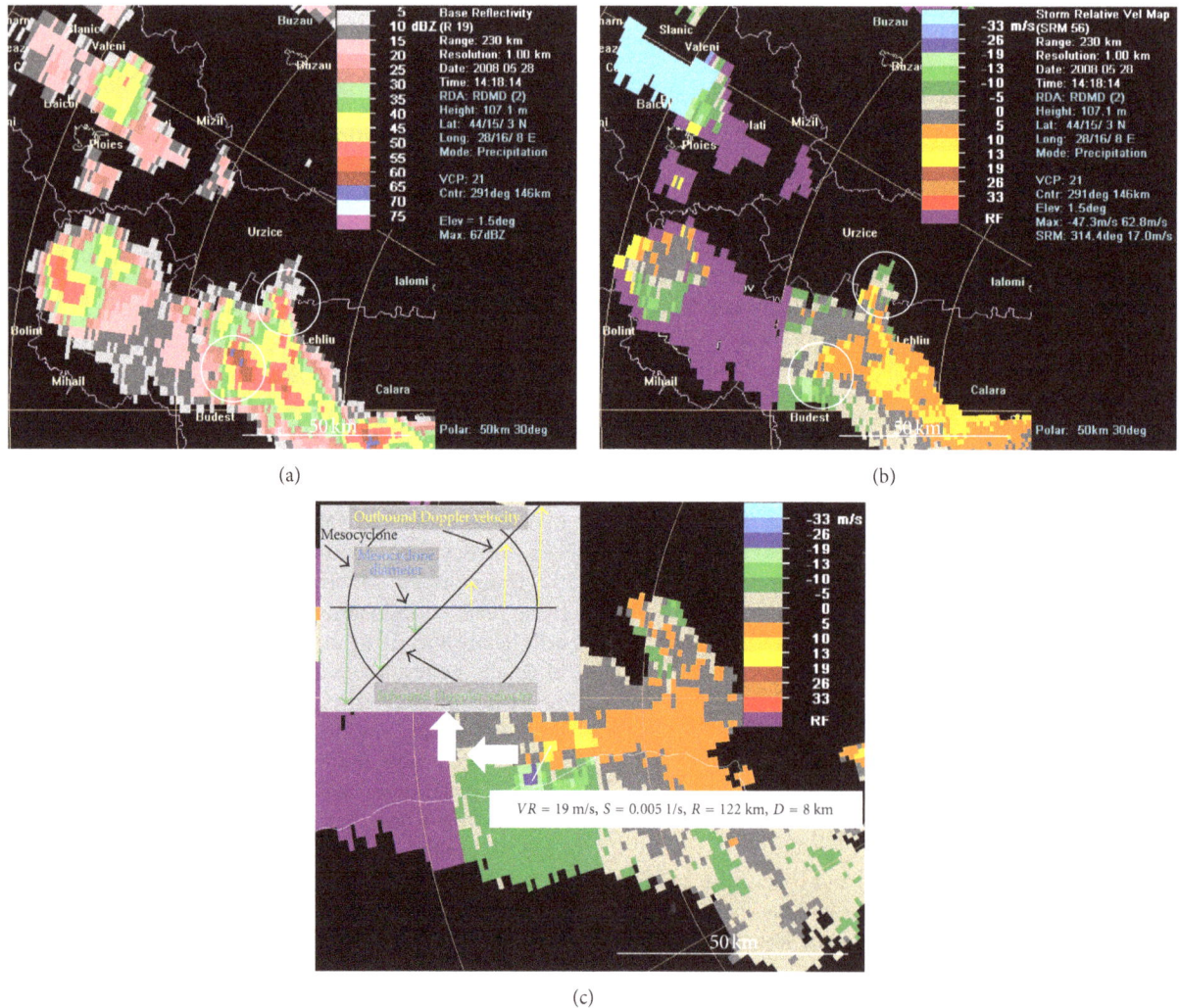

(a)   (b)

(c)

FIGURE 3: Radar reflectivity at 1.5-degree elevation (a), relative velocities at 1.5-degree elevation, (b) the white circles mark cyclonic and anticyclonic rotation associated with the two cells after the split, and shear due to the cyclonic and anticyclonic rotation (c).

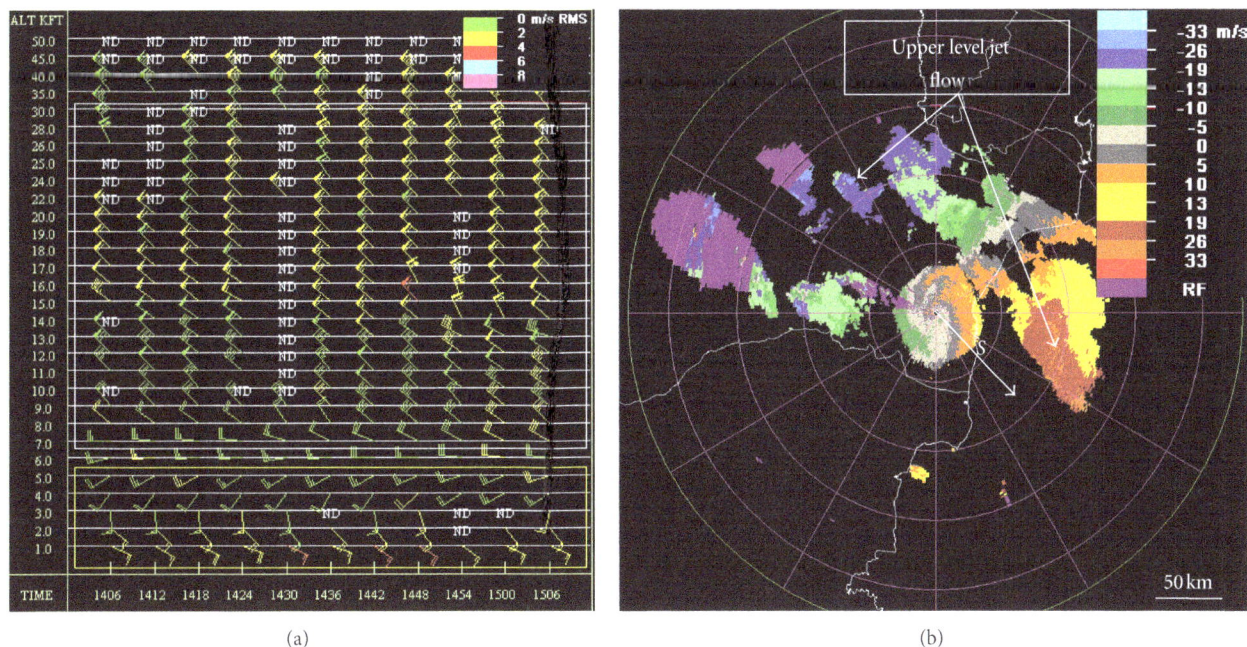

(a)   (b)

FIGURE 4: May 28, 2008, vertical profile of horizontal wind (VAD Wind Profile) for the 1406 UTC–1506 UTC (a) and (b) Doppler velocities at 3.4-degree elevation, at 1341 UTC. S vector represents the tropospheric mean shear at 1800–6000 m. The flow is from north west and the jet is present at 5000–6000 m.

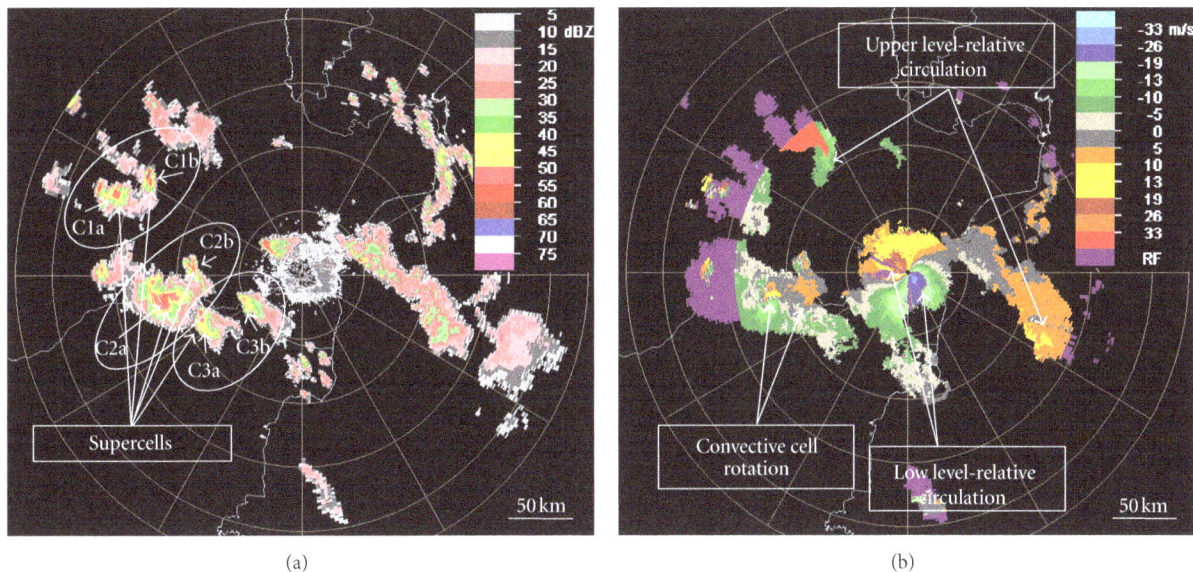

(a)   (b)

FIGURE 5: May 28, 2008, (a) radar reflectivity at 1.5-degree elevation, at 1500 UTC; (b) relative velocities at 2.4 degrees elevation, at 1500 UTC.

In the first three cases, this dynamic configuration is qualitatively compatible with Doppler velocity field at 3.4-degree elevation.

To analyse the presence of the cell splitting process of convective cells in south-eastern Romania and its role in supercell evolution, cases of severe weather were analysed where this process was shown clearly in the radar fields. For the first three cases, in the reflectivity field the convective cells were identified that split, and those that resulted from the split. Using the Doppler base velocity field at 3.4-degree elevation one can tell if there is a quality equivalence between the vertical profile of horizontal velocity near the radar station and the mesoscale flow. Trajectories of split cells were rebuilt, to identify some of their peculiarities. The trajectories were compared with the precipitation field, as obtained from WSR-98D S-band radar precipitation algorithm (Precipitation Processing System, PPS [37]).

The vertical profiles associated with horizontal vorticity were compared with vertical vorticities specific to cyclonic or anticyclonic rotation. In the last two cases, the presence of an upper-level jet was analysed, and the intensity of relative flow was examined.

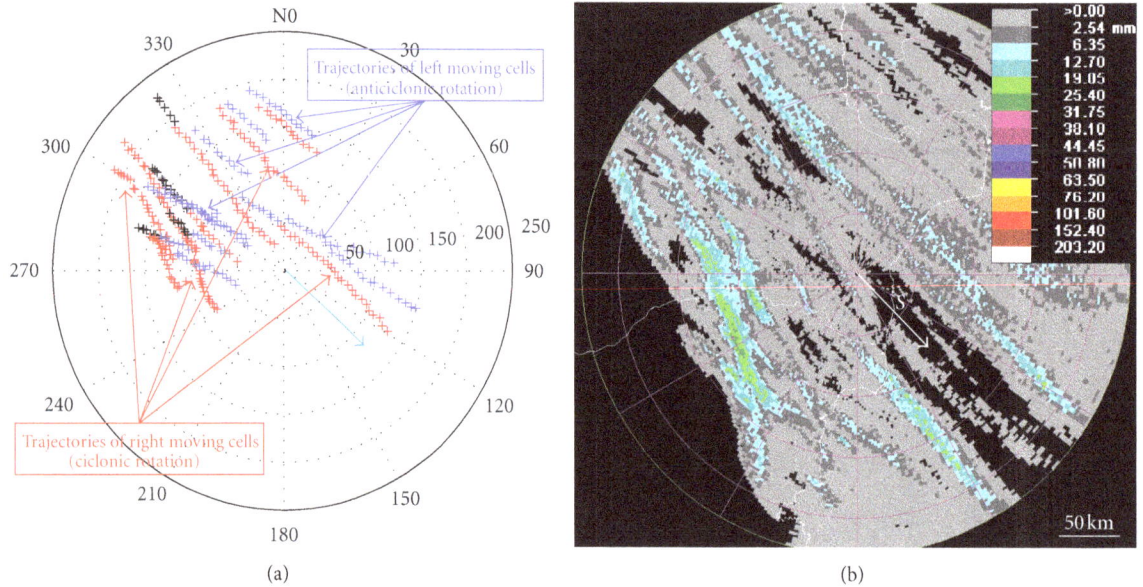

(a)   (b)

FIGURE 6: May 28, 2008, the convective cells evolution for 0900 UTC–2000 UTC using the Medgidia Doppler radar in (a). The center of images is the radar position. S vector represents the mean shear at 1800–6000 m altitude. Red stands for cyclonic rotation, blue for anticyclonic, and black signifies cells before split. The distance is measured in km. The accumulated precipitations (mm), for 0900 UTC–2000 UTC (b).

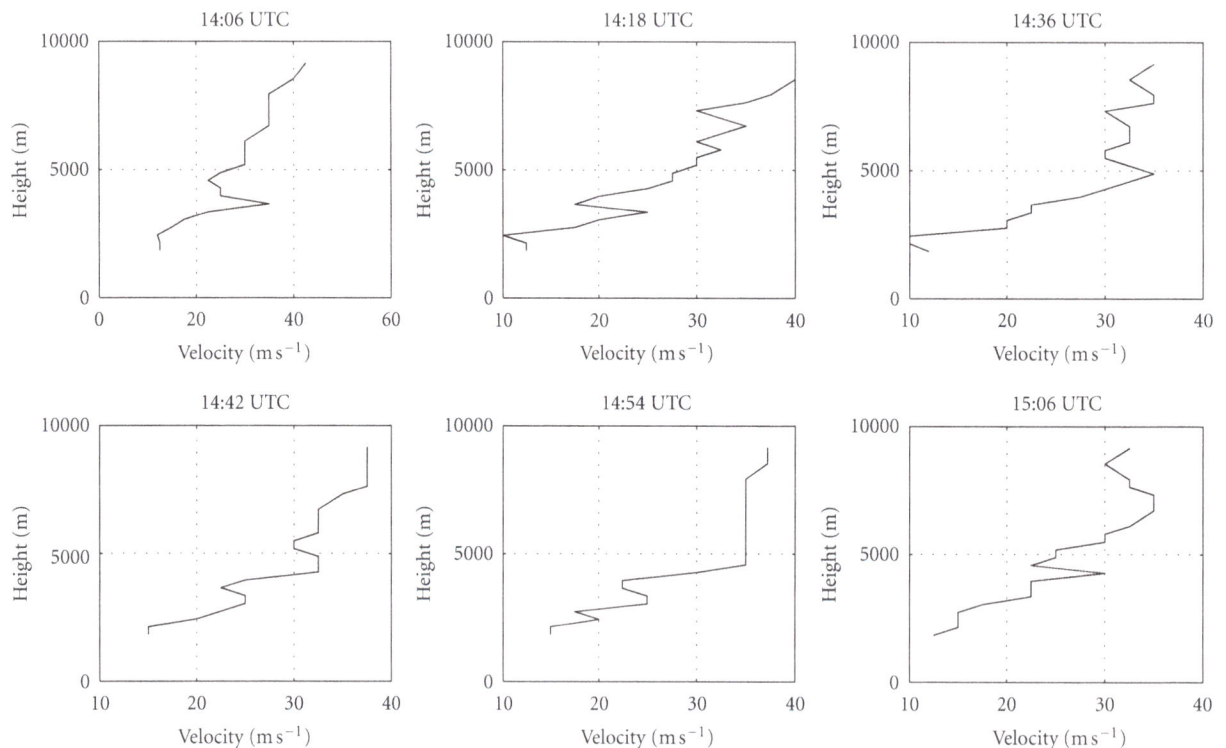

FIGURE 7: May 28, 2008, vertical velocity size distribution for 1800–9000 m altitude at 1406 UTC–1506 UTC.

3. Results and Discussions

3.1. The First Studied Case, May 28, 2008. A cyclone located in north-east Romania contributed to the instability conditions present in the Doppler radar area of the Medgidia site. Starting at 1130 UTC, on an NW-SE intense tropospheric circulation, convective cells developed in SE Romania that

were split and became supercells. The most severe supercells were the ones with cyclonic rotation, such as those having the V shape in reflectivity field. Besides these, multicells systems were present in the radar field, with a lower degree of severity.

The temporal evolution of mesoscale flow is represented in Figure 4, between 1406 UTC and 1506 UTC, as seen in the VAD Wind Profile product and in the base Doppler velocities

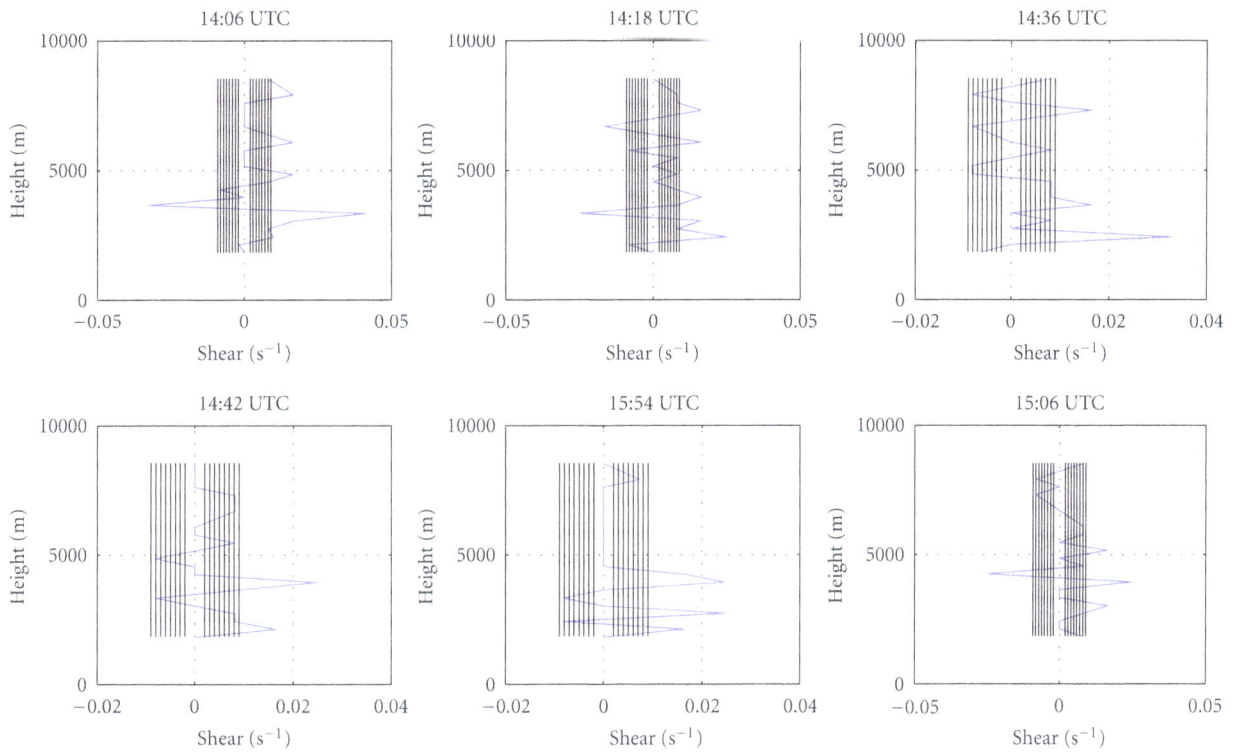

FIGURE 8: May 28, 2008, vertical shear distribution for 1800–9000 m layer, at 1406 UTC–1506 UTC (blue chart). Shear is calculated every 305 m. Horizontal shear distribution associated with convective cells (vertical lines).

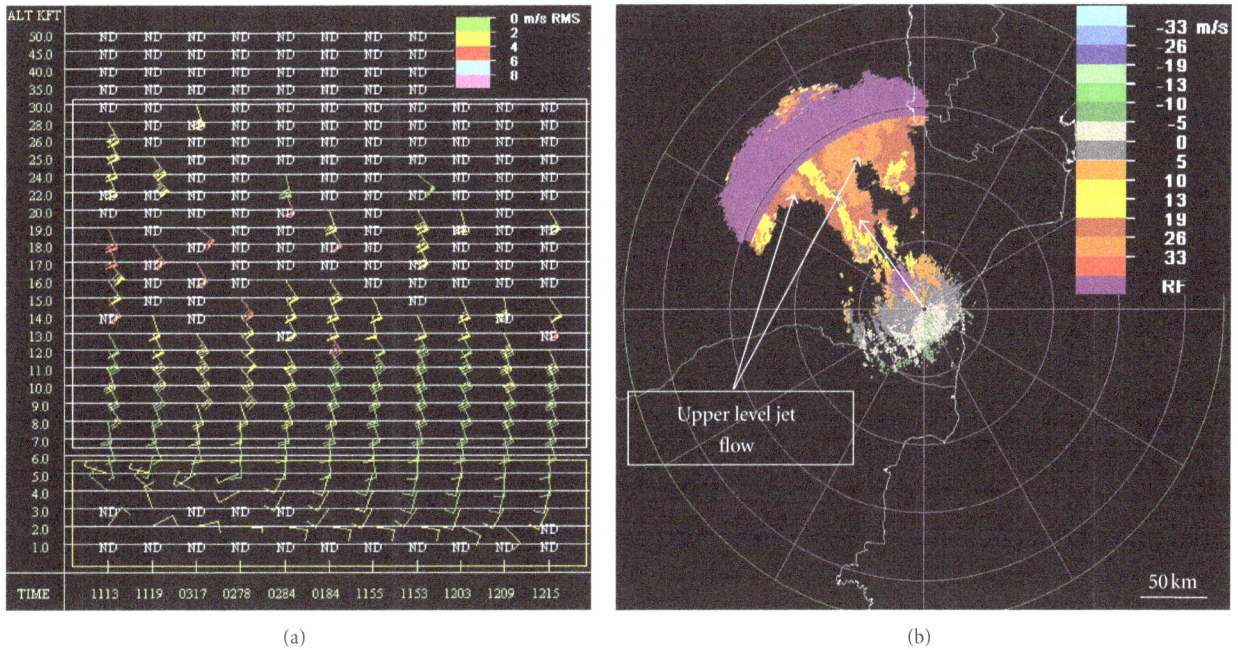

FIGURE 9: June 25, 2009, vertical profile of horizontal wind (VAD Wind Profile) for 1113 UTC–1215 UTC, using the Medgidia Doppler radar in S band in (a) and (b) Doppler velocities at 3.4 degrees elevation at 1341 UTC. The vector represents the mean shear for 1800–5000 m; the flow is from southeast and the jet is present at 4000–6000 m.

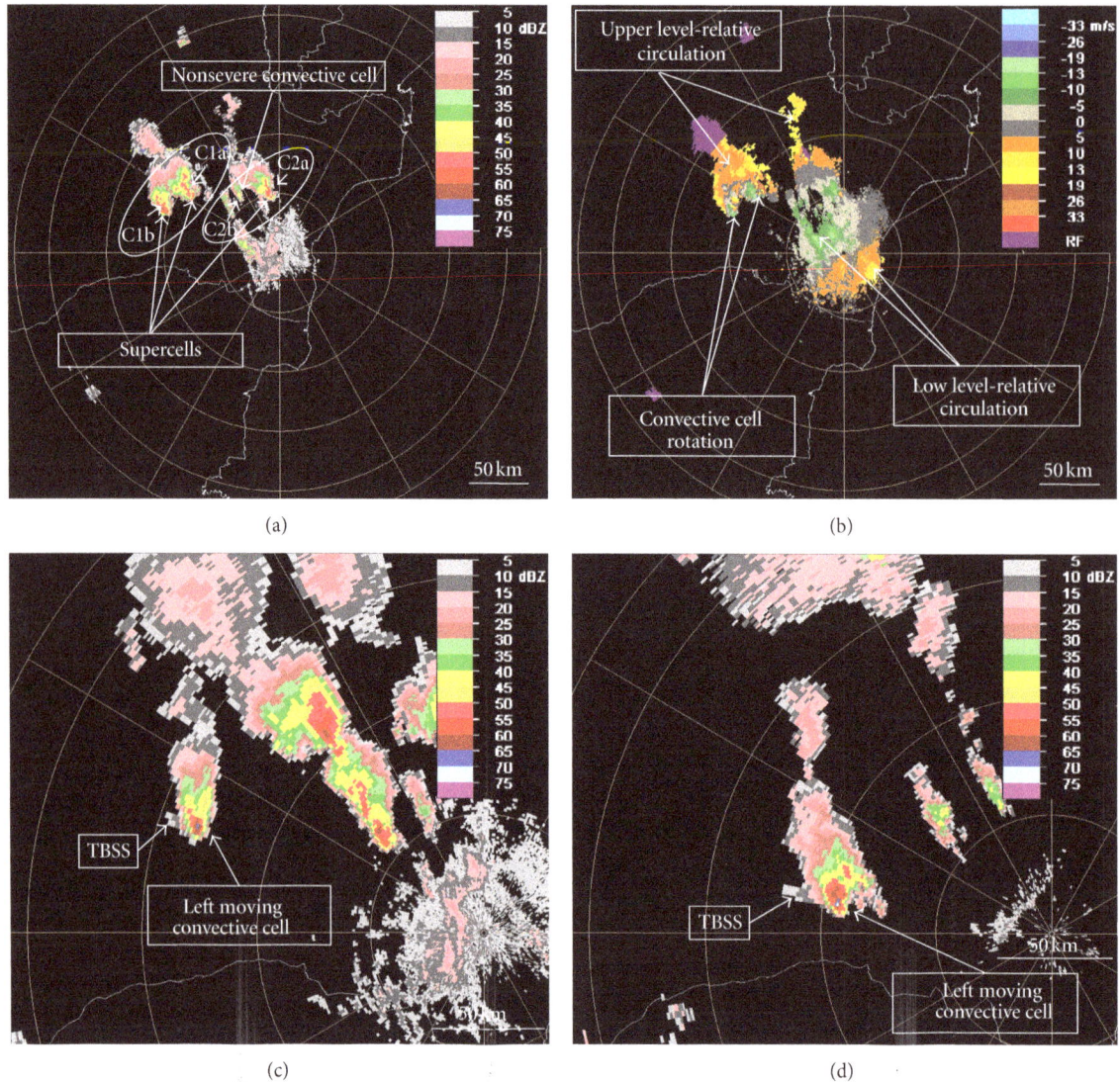

(a)

(b)

(c)

(d)

FIGURE 10: June 25, 2009, radar reflectvities at 2.4-degree elevation, at 1209 UTC (a). Doppler relative velocities at 2.4-degree elevation at 1209 UTC (b). Radar reflectvities at 1.5-degree elevation, at 1257 UTC, (c) and radar reflectvities at 2.4-degree elevation, at 1523 UTC (d).

field. The vertical wind profile was characterized by intense mesoscale flow from the north west to south east, above 1800 m (6 KFT) (Figure 4(a), in the white box). Below this level, from the VAD Wind Profile and base Doppler velocities, one can notice the anticyclonic rotation of velocity vector (Figure 4(a), in the yellow box).

At the 5000 m level, Figure 4(a), the values of velocity reach 35 m s^{-1} growing to 40–45 m s^{-1} at 35 KFT (10500 m). The colour for the wind vectors represents the level of confidence in the VAD Wind Velocity data; the level of confidence is larger in the lower troposphere (green, dispersion of 0–2 m s^{-1}) and in the upper troposphere (yellow, dispersion of 2–4 m s^{-1}). A larger dispersion of velocity is recorded in the lower layers (red, dispersion of 4–6 m s^{-1}), which can be associated with different gust fronts caused by convective developments. In the same figure, both the anticyclonic rotation of the velocity vector in the lower two km and the

upper tropospheric jet are consistent with the warm sector of the cyclone located north east of Romania. This vertical configuration of the horizontal velocity is also associated with increased thermodynamic instability in the troposphere below approximately 500 hPa; the main factor for the increase in conditional instability is moisture advection in the lower layers. However, the presence of a vertically expanded upper tropospheric jet is associated with ageostrophic velocities, which are the result of horizontal variation of the geostrophic velocities that lead to changes in the thermal gradient within the middle troposphere [18].

Vertical wind profile configurations of the mesoscale flow can be seen in the Doppler velocity field at 3.4-degree elevation (Figure 4(b)), and they are qualitatively equivalent with the vertical wind profile given by the VAD Wind Profile product. This equivalence is the result of the zero isodope analysis [34] in the Doppler velocity field. The zero isodope represents the area where the Doppler velocities are zero (gray

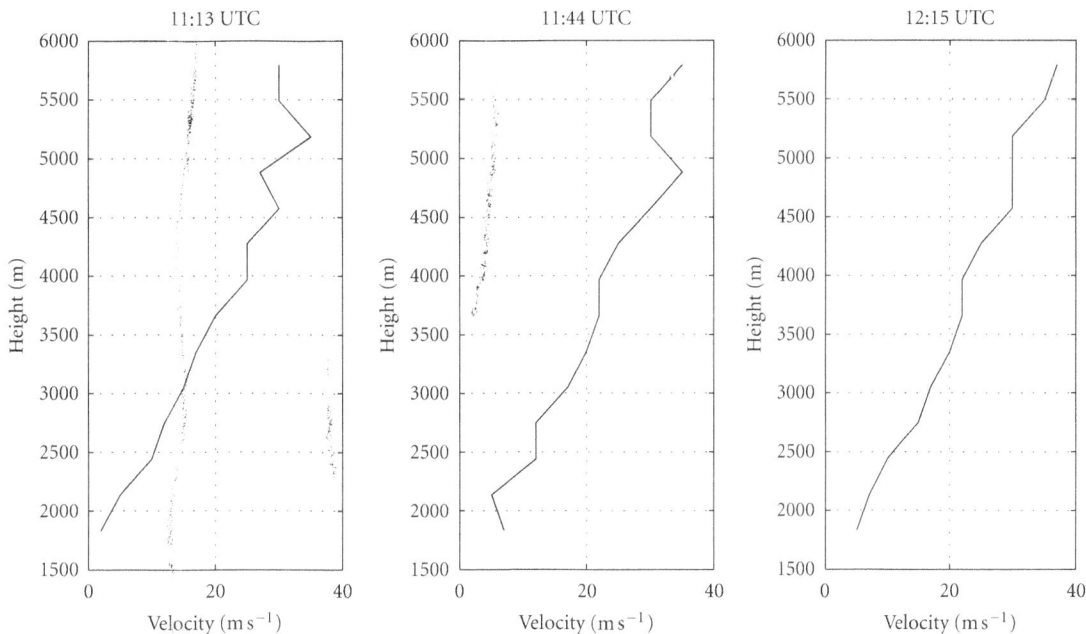

FIGURE 11: June 25, 2009, vertical velocity size distribution for 1800–6000 m, at 1113 UTC–1215 UTC.

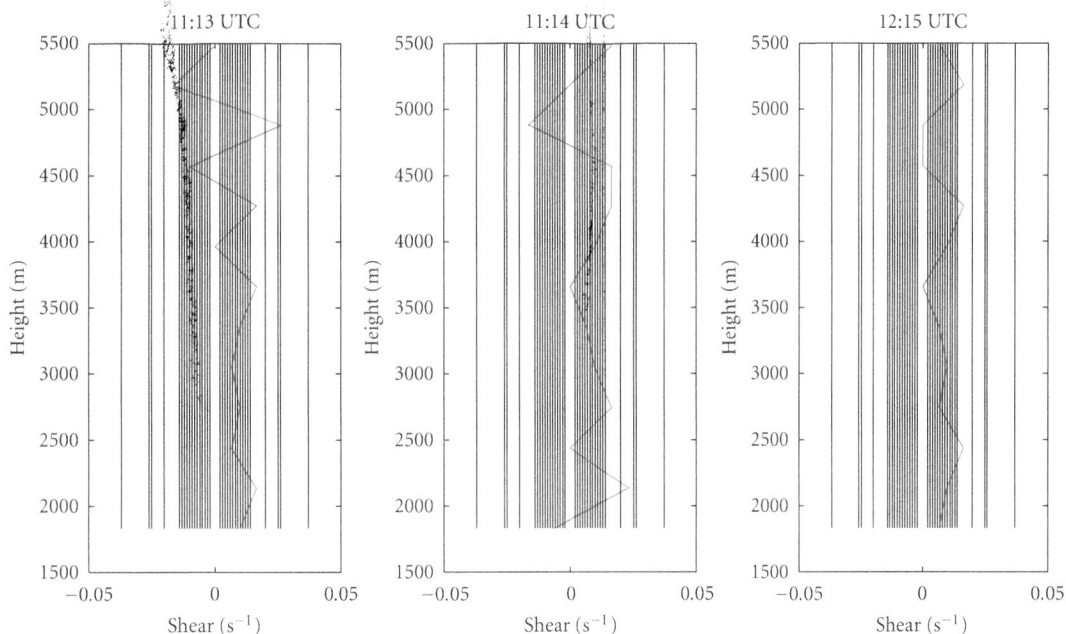

FIGURE 12: June 25, 2009, vertical shear distribution for 1800–6000 m, at 1113 UTC–1215 UTC (blue chart). Shear is calculated every 305 m. Horizontal shear distribution associated with convective cells (vertical lines).

colours, in Figure 4(b)). Near the radar site (in the boundary layer) the isodope is S shaped, compatible with the velocities profile (within the yellow box, in Figure 4(a)). Going away from the radar (increasing the height), this has a linear shape on the NE-SW direction and is compatible with the upper tropospheric jet direction (Figure 4(a)). The intensity of the jet stream is obtained by measuring the Doppler velocities at 90 degrees from the isodope, having the same height (same distance from the radar in the Doppler velocity field).

The dynamic field of the Doppler velocities was maintained over the interval with the mentioned convective phenomena (a few hours). In Figure 5(a) the reflectivity field is shown at 1500 UTC at 1.5-degree elevation, and in Figure 5(b) the Doppler relative velocity field is shown at 2.4-degree elevation at 1500 UTC. The reflectivity field at each elevation has a 1 km^2 resolution with a coverage area of 230 km radius around the radar site and the corresponding product is built every 6 minutes. The angle of the radar azimuth beam is about

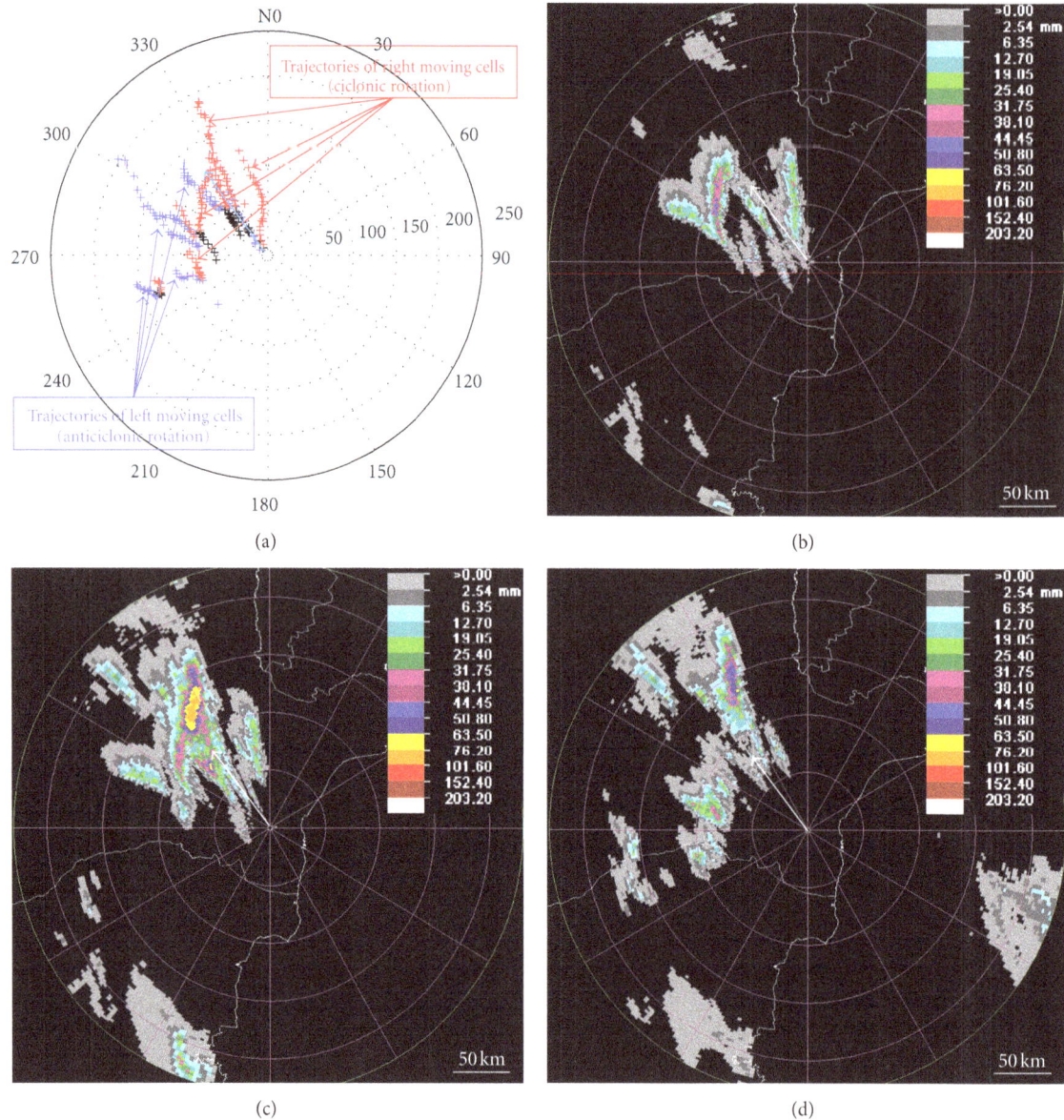

FIGURE 13: June 25, 2009, the convective cells evolution at 1030 UTC–1800 UTC using the Medgidia Doppler radar in (a). The white vector represents the mean shear for 1800–5000 m altitude. Red is the cyclonic rotations and blue is the anticyclonic rotations. The distance is in km. Accumulated precipitations are for every three hours: (b) 1040 UTC–1340 UTC, (c) 1205 UTC–1505 UTC, and (d) 1405 UTC–1705 UTC.

0.97 degrees, so that 366 azimuths are needed to cover the entire radar area. At 1.5-degree elevation (Figure 5(a)), as the height of the radar echo increases along with the distance to the site at 230 km from the site the height is 9.6 km. In Figure 5(a), the Doppler velocities field relative to the storm is at 2.4-degree elevation, and at the 230 km the height is 13.2 km.

In the reflectivity field, six convective cells can be seen, resulting from three convective cells that had split, and in the relative Doppler velocity field the rotation can be seen, associated with resulted convective cells. A fundamental feature is the relative jet at 100–1000 m with velocities of up to 29 m s^{-1}. This relative jet is the storm input flow and can only be observed in the storm's reference frame; it may be associated with the moisture inflow. Above 6000 m there is a relative flow with velocities up to 11 m s^{-1}.

Trajectories of convective cells that had split during evolution present symmetry to the mean shear in the 1800–6000 m layer (blue arrow), with cells characterised by cyclonic rotation left of the vector and anticyclone rotation right of it (Figure 6(a)). In Figure 6(a), advection of convective cells by the tropospheric circulation is on the NW-SE direction. In this case, because the flow is parallel wih the shear vector on the mean troposphere, the convective cells with cyclonic rotation (red) deviate to the right and the ones with anticyclonic rotation (blue) to the left of this direction. Following the cell evolution, the intensity of the cyclonic

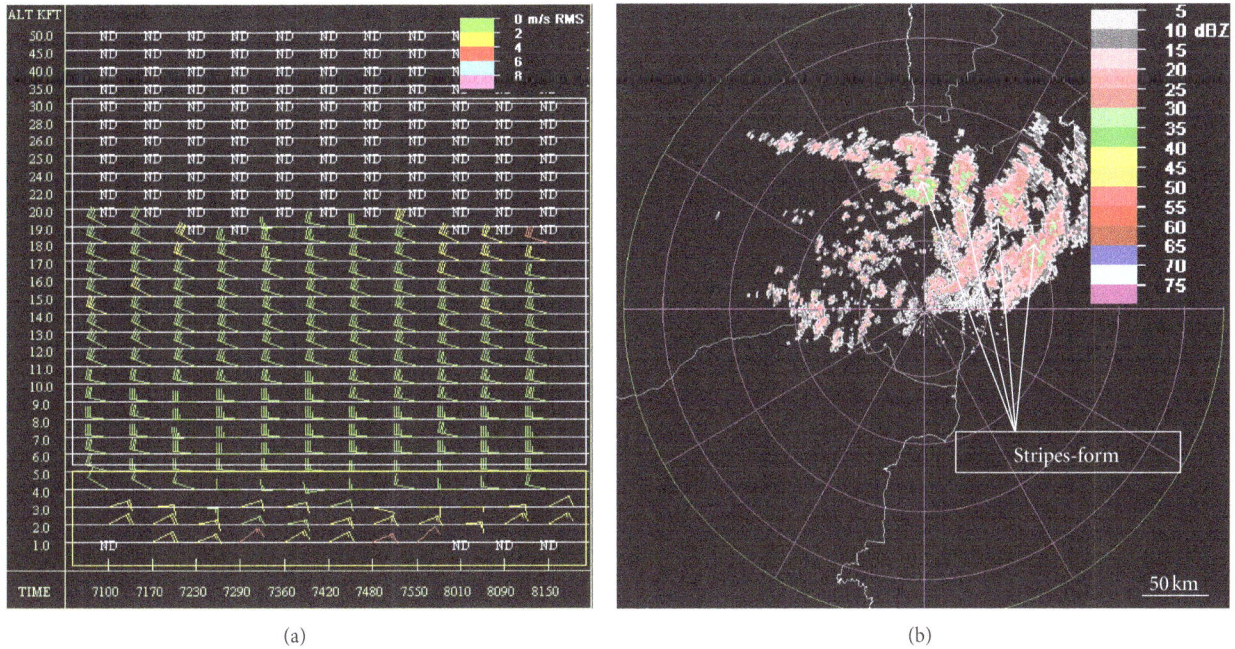

(a)   (b)

FIGURE 14: August 11, 2003, (a) vertical profile of horizontal wind (VAD Wind Profile) for 300–5800 m, at 0710 UTC–0815 UTC; (b) reflectivity field at 1.5-degree elevation.

rotation is larger than that of anticyclonic rotation. The cause for this difference can be associated with directional shear in the 300–1800 m layer and is favoured by advection of vorticity, whose vector is parallel to the wind relative to the storm [17, 18].

The splitting process may occur several times in the cells following from a previous split. For multiple splits, cells are favoured whose movement occurs on the left of the mean tropospheric shear. One possible explanation is that these cells draw more vorticity from their surroundings, vorticity which is perpendicular to the relative storm flow [18].

Cumulative rainfall in the evolution area, associated to the convective cells, is presented in Figure 6(b). The maximums are associated with convective cells, appearing as lines with symmetrical shear. Trajectories associated with more precipitation are to the right of shear in the 1800–6000 m layer. The amount of precipitation is amplified by the intersection of convective cells paths (cells with cyclonic rotation intersect those with anticyclonic rotation); indeed, in Figure 6(a), these intersections are more likely to the right of the mean shear vector.

Vertical wind variation associated with VAD Wind Profile, Figure 4(a), is represented for 6 times in Figure 7. Note the presence of the tropospheric jet in the 1800–9000 m layer, with values of about 40 m s^{-1} at 9000 m. The vertical profile of horizontal vorticity versus vertical vorticity associated with cell rotation is given in Figure 8; here, vorticity was calculated every 305 m on the z-axis. One can see several maximum values of vorticity. The highest values are found in the 2000–4000 m layer. The horizontal vorticity of the environment (Figure 8, blue chart) was superimposed over

cyclonic or anticyclonic vertical vorticity associated with mesocyclones (Figure 8, vertical lines) in the radar area. The parallel lines with vertical axes signify the vorticity of mesocyclones calculated with the radar algorithm in Figure 3(c). Vorticity was plotted both with a plus and minus sign for symmetry, to be more easily compared with the horizontal vorticity of the environment.

3.2. The Second Studied Case, June 25, 2009. The presence of a stationary cyclone in the SE of Romania modulated the mesosynoptic configuration in which severe convective cells developed. These cells had supercellular attributes and had an SE-NE direction in the S-band WSR-98D radar area (located in Medgidia). In the first stage, convection had an isolated character and formed near the radar site. Afterwards, the majority of cells split and became supercells. In the end, the cells with cyclonic rotation formed a convective system. Similarly to the previous case, the vertical velocity profile is characterized by an intense mesoscale flow above 1800 m (6 KFT) (Figure 9(a), the white box). Below this level, between 1113 UTC and 1215 UTC, in the same figure one can see the rotation of velocity vector along the z-axis (this rotation may be associated with directional shear) (the yellow box, in Figure 9(a)).

Above 1800 m, there is strong intensification of wind, associated with the presence of a ULJ from south east; this ULJ reaches 35 m s^{-1} at 5000 m. The dynamic field that can be viewed in the Doppler velocities, Figure 9(b), retained its structure between 1030 UTC and 1800 UTC.

Radar reflectivities show four secondary convective cells that are resulted from the splitting of two initial cells

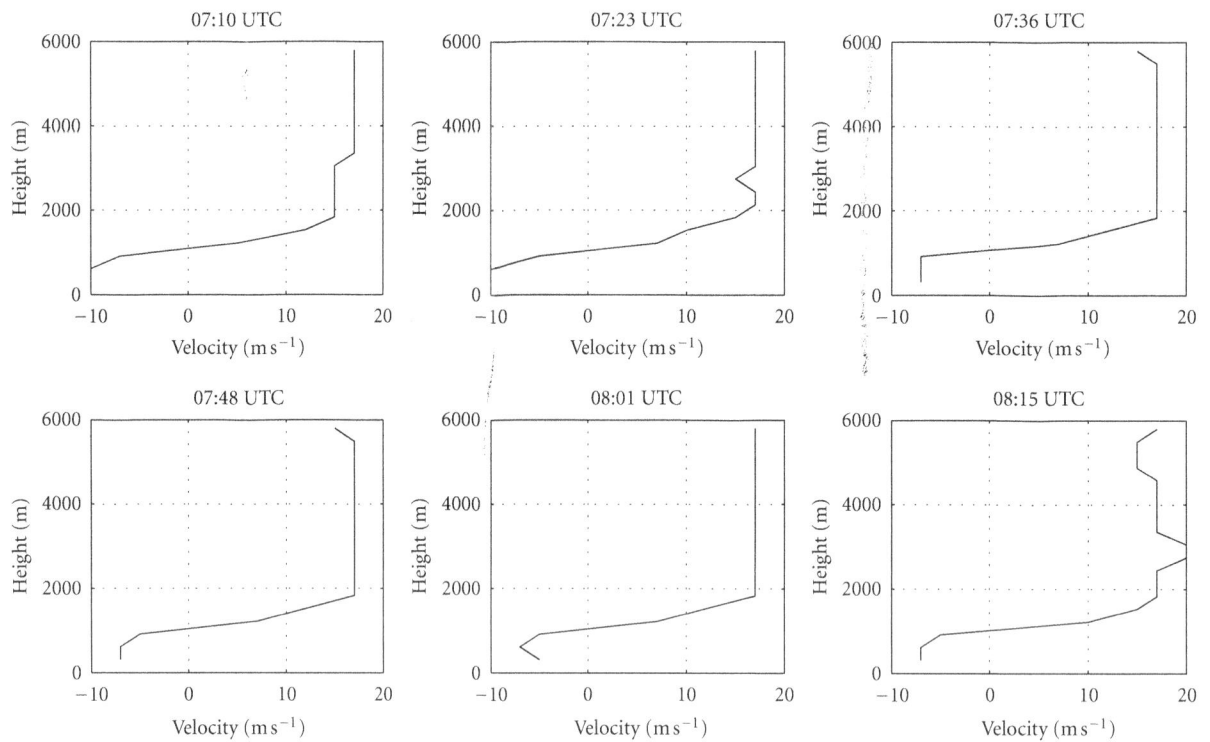

FIGURE 15: August 11, 2003, vertical velocity size distribution for 300–5800 m, at 0710 UTC–0815 UTC.

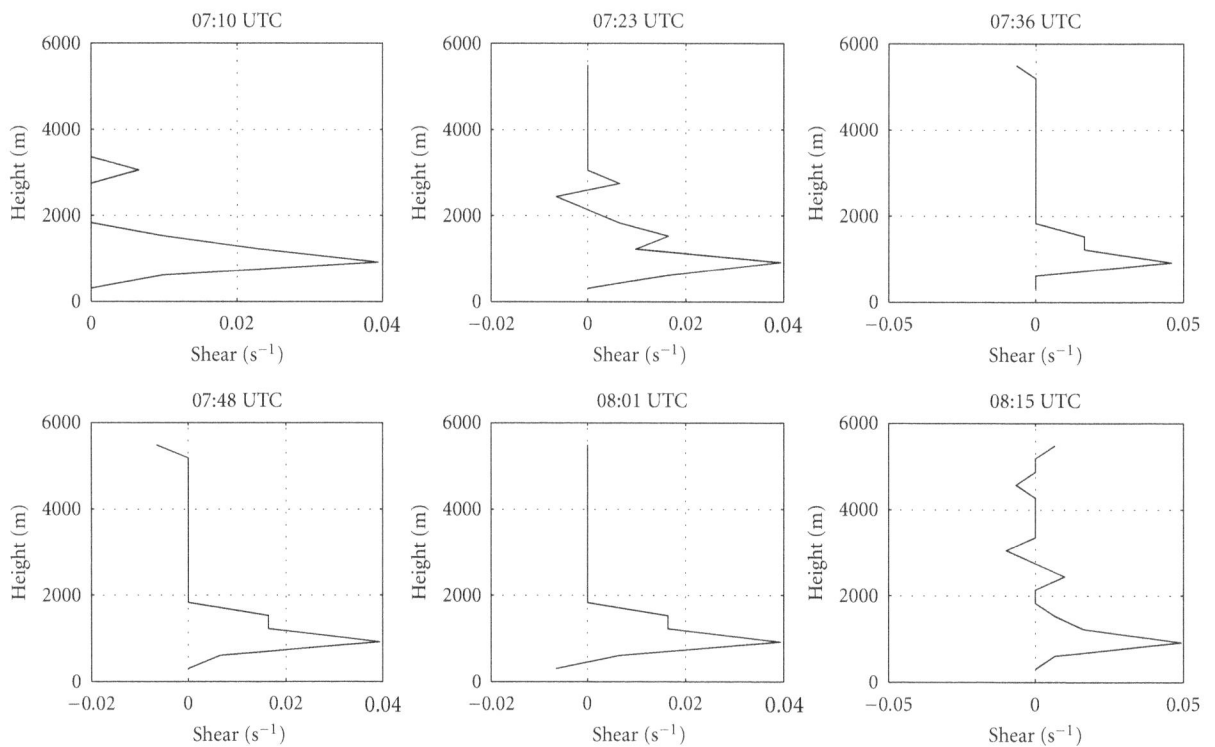

FIGURE 16: August 11, 2003, vertical shear distribution for 300–5800 m, at 0710 UTC–0815 UTC. Shear is calculated every 305 m.

(a)

(b)

(c)

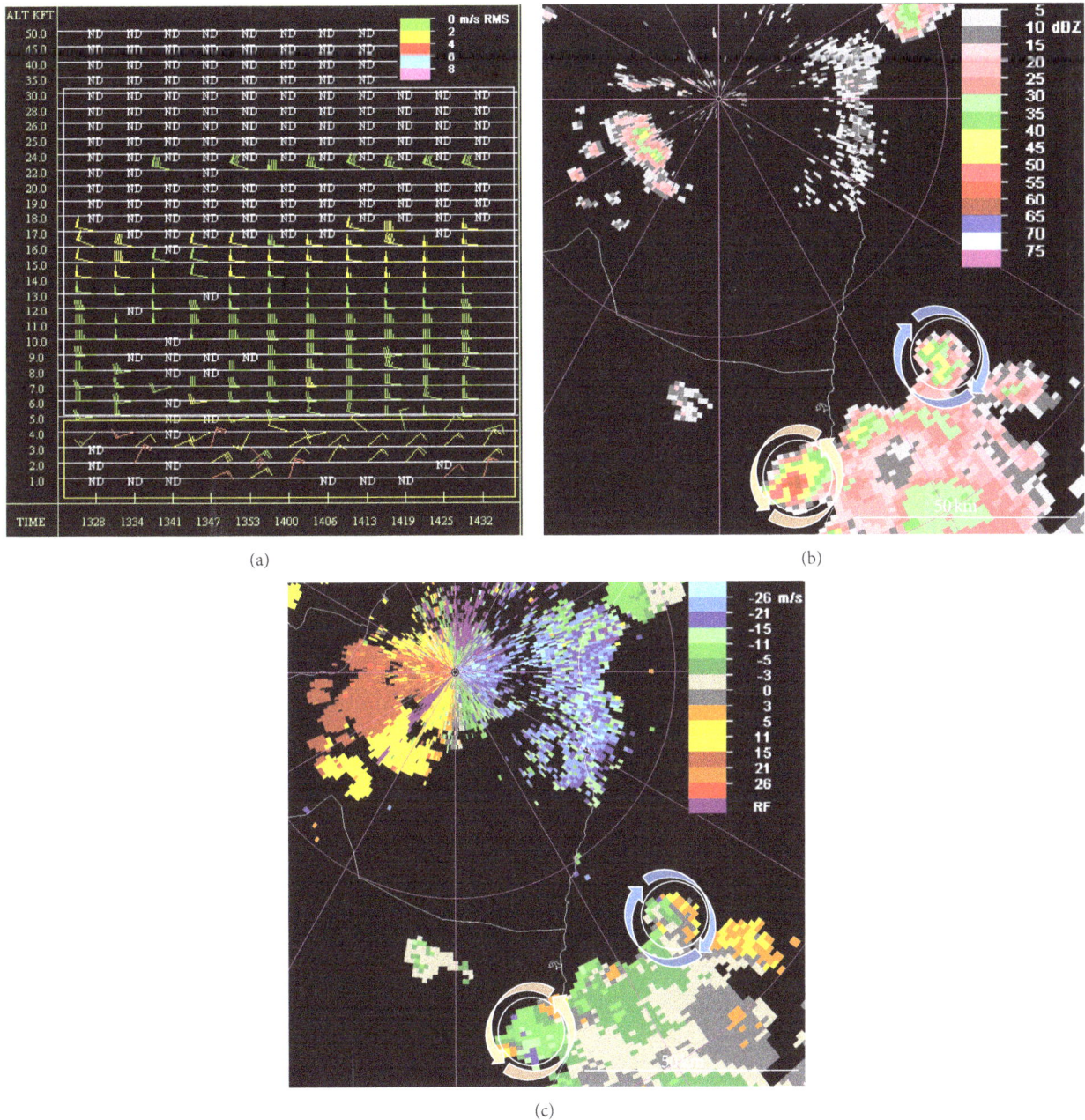

FIGURE 17: August 11, 2003, (a) vertical profile of horizontal wind (VAD Wind Profile) for 1500–6700 m, at 1328 UTC–1432 UTC; (b) radar reflectivity at 1.5-degree elevation, at 1400 UTC and (c) Doppler relative velocities at 1.5-degree elevation, at 1400 UTC.

(Figure 10(a)). Three of them are supercells, and one is nonsevere. In the field of relative Doppler velocities a relative jet is present whose speed is up to $16 \, \text{m s}^{-1}$ at 1000 m, while above 5000 m intense circulation is present, with up to $16 \, \text{m s}^{-1}$, (Figure 10(b)). The refletivity field and the relative velocity characteristics are the same as in the pevious case.

Severe storms moving towards the left of mean shear (anticyclonic rotation) were characterized by high intensity and shorter lifespan and were associated with the presence of *Three-Body Scatter Spike* (TBSS) radar artifact (Figures 10(c) and 10(d)). This artifact is a radar signature in reflectivity field, generally a 10–30 km long, aligned radially downrange

from a strong echo region in reflectivity field (>60 dBZ) associated with a severe storm. It is caused by non-Rayleigh radar microwave scattering or Mie scattering and is asociated with high probability of large hail [38–40].

In Figure 11 three vertical velocity profiles are presented, at certain times representative for this case. Here, the tropospheric ULJ has 35–$40 \, \text{m s}^{-1}$ at 6000 m. The vorticity profile is shown in Figure 12 (blue chart).

The mesocyclone vorticities associated with convective cells that splited are represented by parallel lines with the vertical axes (Figure 12). Note that most values of vorticity lie between $-0.005 \, \text{s}^{-1}$ to $-0.015 \, \text{s}^{-1}$ and $0.005 \, \text{s}^{-1}$ to $0.015 \, \text{s}^{-1}$.

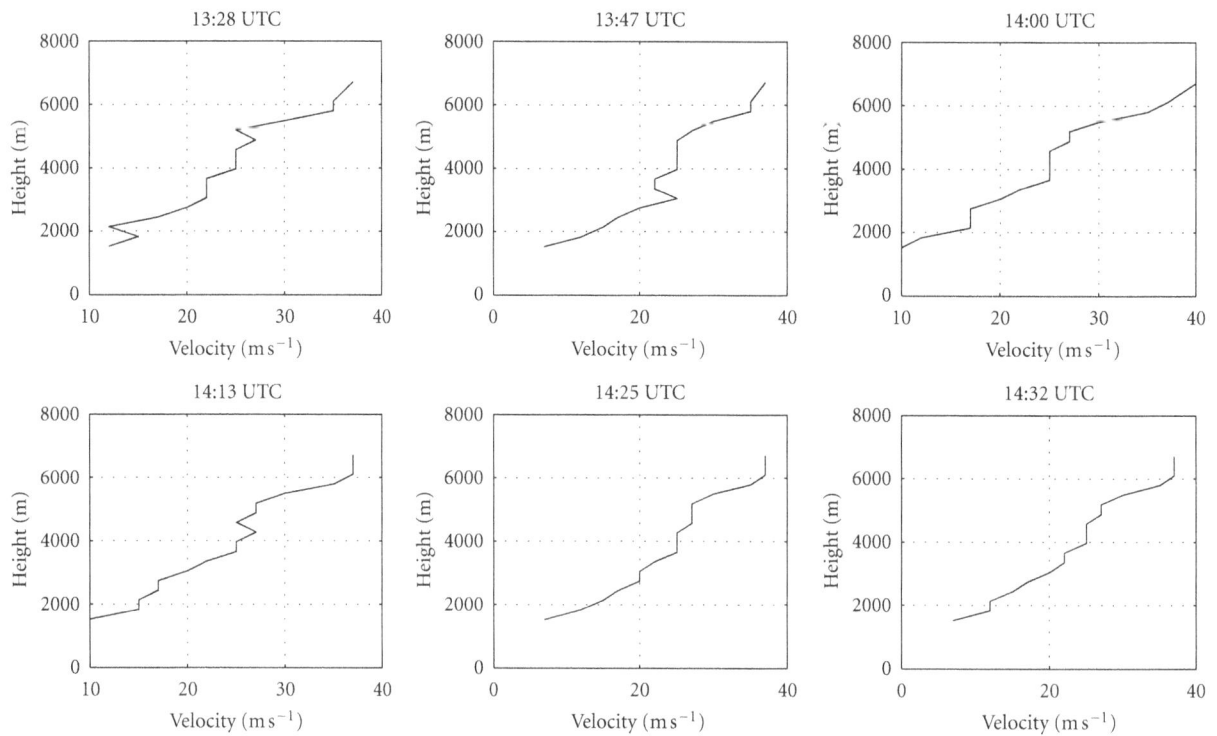

FIGURE 18: August 11, 2003, vertical velocity size distribution for 1500–6700 m, at 1328 UTC–1432 UTC.

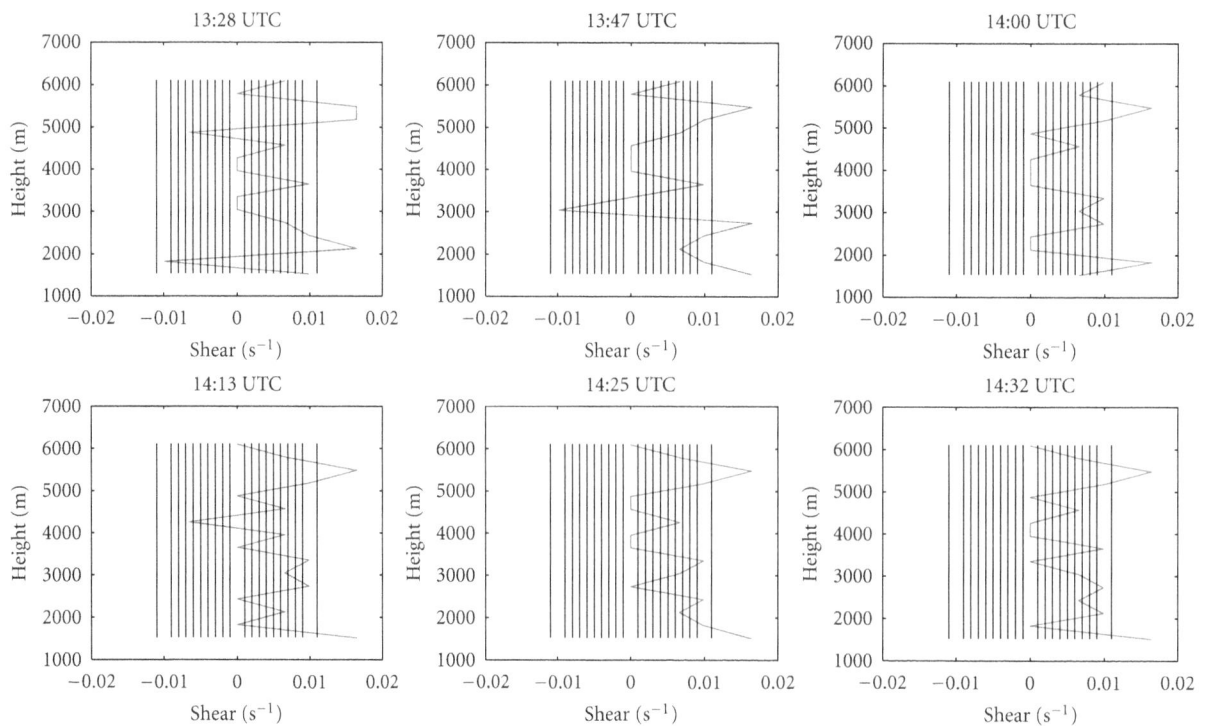

FIGURE 19: August 11, 2003, vertical shear distribution for 1500–6700 m, at 1328 UTC–1432 UTC (blue chart). Shear is calculated every 305 m. Horizontal shear distribution associated with convective cells (vertical lines).

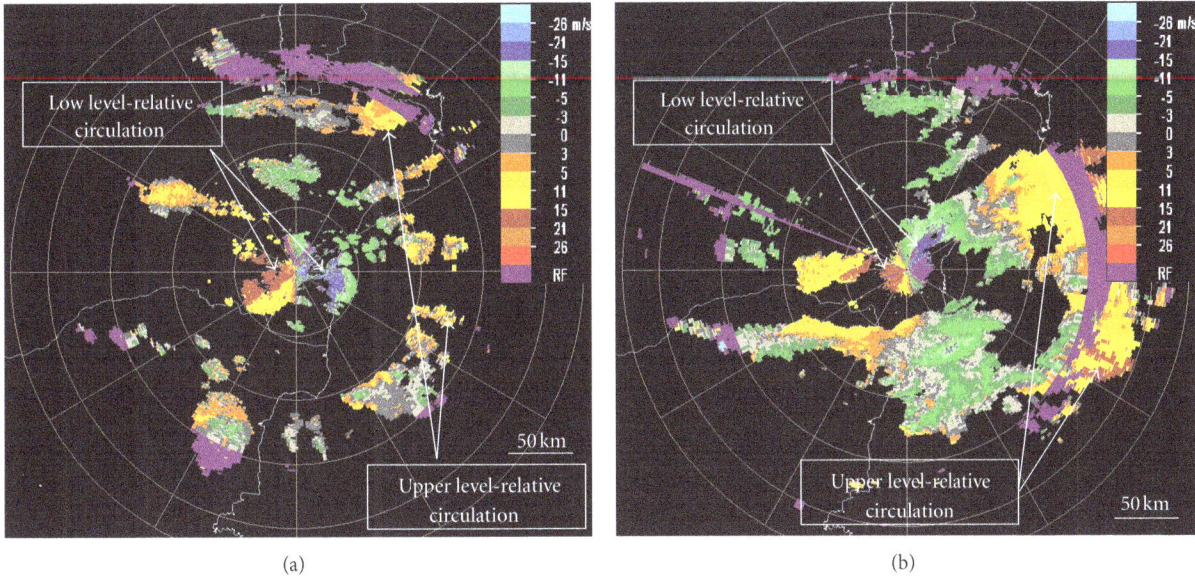

FIGURE 20: August 11, 2003, Doppler relative velocities, at 1432 UTC (a) and 1712 UTC (b) at 1.5-degree elevation.

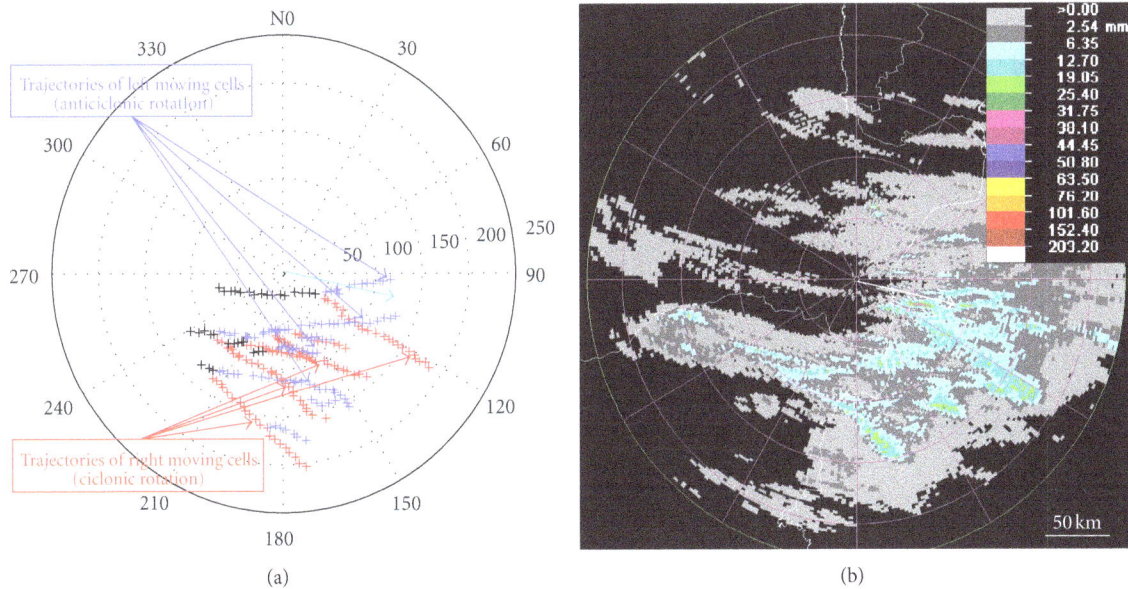

FIGURE 21: August 11, 2003, (a) the convective cells distribution, between 1126 UTC and 1925 UTC, using the Medgidia Doppler radar. The center of the image marks both the radar site and the origin of coordinates. The vector represents the mean shear for 1500–6700 m. Red stands for cyclonic rotation, blue for anticyclonic, and black signifies cells before split. The distance is measured in km. (b) Accumulated precipitation in mm, between 1547 UTC and 1847 UTC.

There is also rotation with values of shear around $0.04 \, \text{s}^{-1}$. In a first approximation, the vorticities of the vertical wind profile are comparable with those of the mesocyclone.

Trajectories of convective cells resulted from the split, present symmetry to the mean shear on the 1800–6000 m layer, with cells characterized by cyclonic rotation to the left of the vector and cells with anticyclonic rotation to the right (Figure 13(a)).

The evolution of these cells shows that the intensity of cyclonic rotation is greater than that of anticyclonic rotation.

The precipitation distribution function, by shear in the 1800–5000 m layer is represented in Figures 13(b), 13(c), and 13(d). Grouping in three-hour intervals was made in order to highlight the symmetry of precipitation distributions from shear vector. Large amounts were recorded at the intersection of several trajectories caused by several splits and also to the right of shear.

3.3. The Third Studied Case, August 11, 2003. In this case, starting at 1300 UTC, in SE Romania and NE Bulgaria, during

(a)

(b)

(c)

(d)

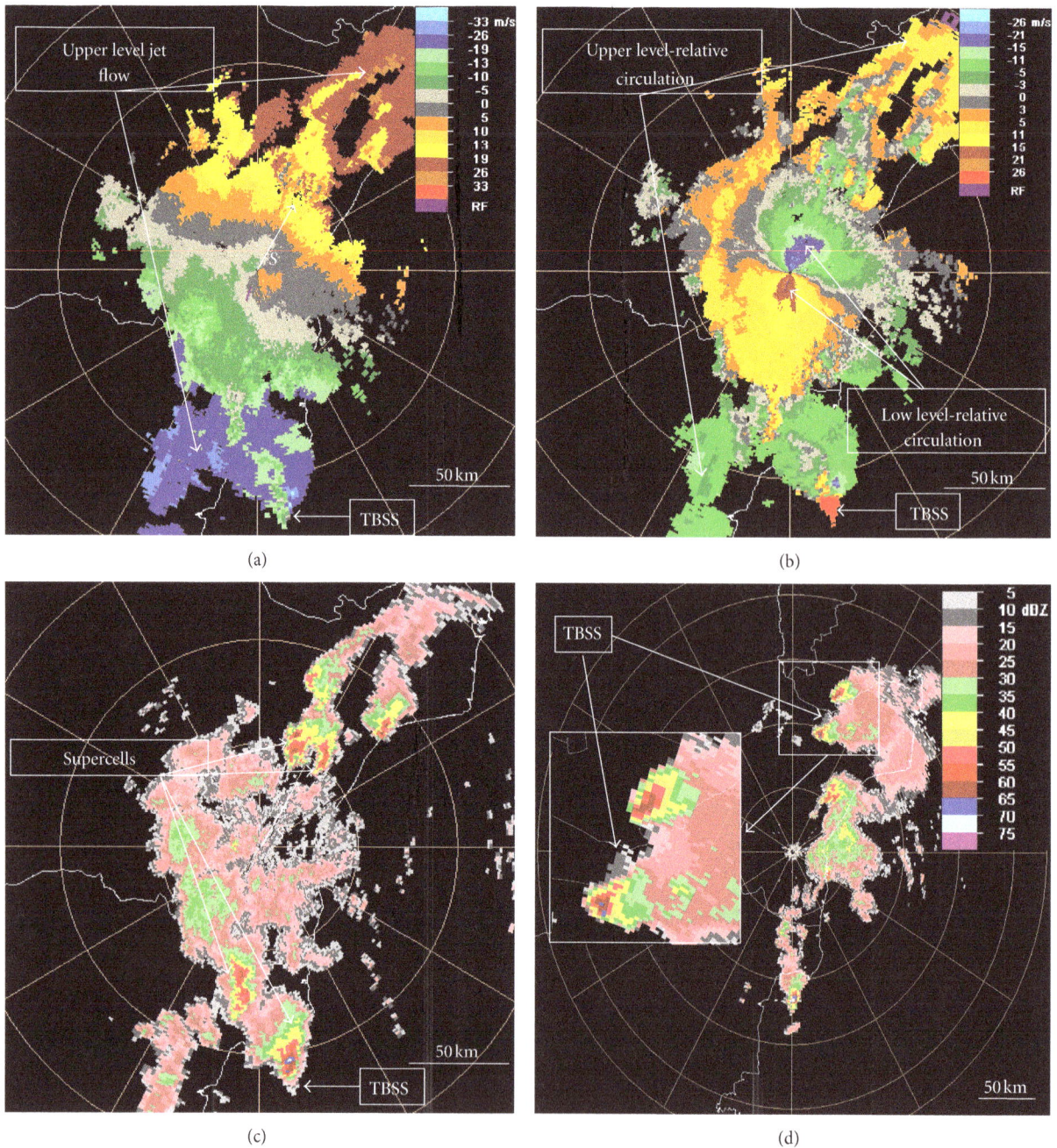

FIGURE 22: September 27, 2004: (a) Doppler velocities at 1347 UTC, at 3.4-degree elevation; (b) Doppler relative velocities at 1347 UTC, at 3.4-degree elevation; (c) reflectivity at 1347 UTC, at 1.5-degree elevation.

the fast evolution of a trough which moved from W to E, convective cells have developed with supercell characteristics, visible in the radar-swept area.

About this event it can be readily said that, depending on the wind distribution, vertical shear, and time of day, two dynamic behaviors are distinguishable. As such, in the interval 0400 UTC–1115 UTC, the vertical velocity profile is characterized by a moderate mesoscale flow above 1500 m (Figure 14(a)). The radar reflectivity at 1.5-degree elevation is in echoes with a strip form, perhaps associated with

conditional symmetric instability [41, 42] or with an internal-gravitational wave [18] (Figure 14(b)). Here, no splitting process occurred.

As it can be seen in Figure 15, wind velocities have a maximum at $20 \, \text{m s}^{-1}$ at 2000 m. The vorticity distribution, shown in Figure 16, has a pronounced maximum of $0.04 \, \text{s}^{-1}$ at 1000 m.

Between 1328 UTC and 1432 UTC, the mesoscale dynamic state changes, a ULJ being present in the high troposphere, that can be seen in the vertical profile of

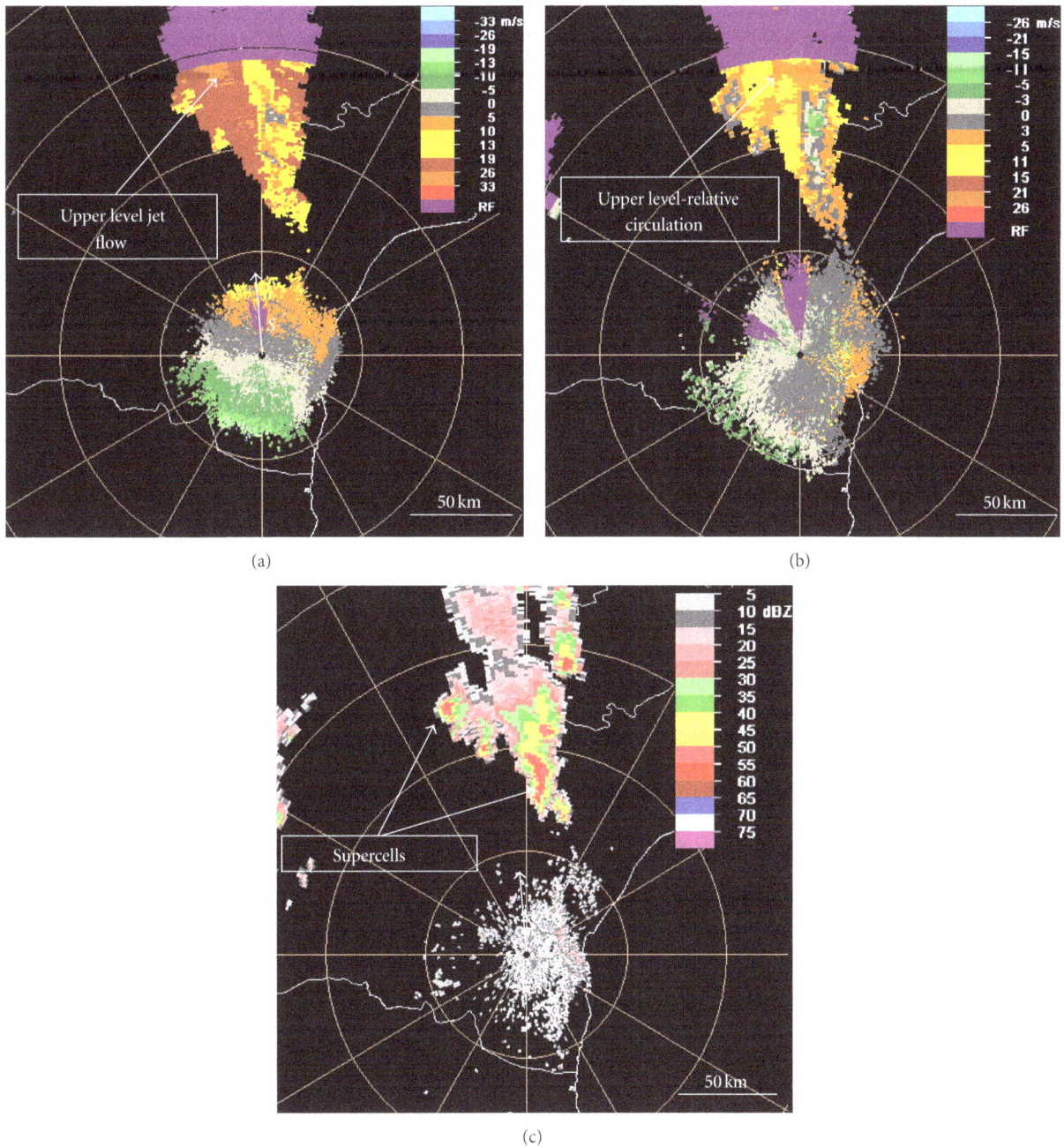

FIGURE 23: June 23, 2009: (a) Doppler velocities, at 1004 UTC at 3.4-degree elevation; (b) Doppler relative velocities, at 1004 UTC at 2.4-degree elevation; (c) reflectivity, at 1004 UTC at 1.5-degree elevation.

horizontal wind (Figure 17(a)). Between 304 and 1500 m, neither the measurements of wind nor the VAD Wind Profile are reliable. Here, the influence of convective phenomena that can generate gust fronts is present. In Figure 18 the presence of the tropospheric jet is shown by velocities of 37 to 40 m s^{-1} at 6700 m. Due to the tropospheric jet, the reflectivity field has changed, emphasizing now the splitting process both in radar reflectivities and Doppler relative velocities (Figures 17(b) and 17(c)).

Mesocyclone vorticities associated with convective cells that split are represented by parallel lines with the vertical axes (Figure 19). Note that most of vorticity values lie between -0.001 s^{-1} and -0.009 s^{-1}, and between 0.001 s^{-1} and 0.009 s^{-1}. The maximum vorticity here had a value around 0.011 s^{-1}.

The Doppler relative velocities to the elevation of 1.5 degrees, Figures 20(a) and 20(b), show a relative jet in the low levels, whose speed is up to 23 ms^{-1} at 500 m, while the relative flow at higher levels (above 6000 m) gets to 18 m s^{-1}. Spatial distribution of convective cell trajectories before and after splitting is shown in Figure 21(a). There is overlapping of the mean tropospheric shear vector on the 1500–6700 m layer. Symmetric trajectories to the mean shear are seen. In the precipitation field (Figure 21(b)), cells having higher water amounts are to the right side of the mean shear.

3.4. Other Cases (September 27, 2004; June 23, 2009). In the first case, a quasi-stationary trough of a big amplitude was present to the west of Romania. The first severe convective cells were seen by the radar at 1020 UTC in Bulgaria and were advected in Romania by the tropospheric circulation. The convective cells which later split were initiated nearby the radar site at 1220 UTC. In this case, the convective cells with anticyclonic rotation had a bigger life time and a greater severity as well.

In the second case, the mesosynoptic configuration was modulated by the cyclone west of Romania. The circulation due to this cyclone was southern over the south east of Romania in the boundary layer as well as at higher altitude. In the first part of the day, the Doppler radar caught a mesoconvective system with supercells. After the convective system dissipated, other new convective cells have been initiated and split; the most severe ones of them have cyclonic rotation.

As in the cases above, for these two other cases the main feature associated with the mesoscale configuration was the ULJ, whose speed exceeded $33 \, \text{m s}^{-1}$. In the first case (September 27, 2004), intense flow is visible at lower levels in the field of Doppler relative velocity, and in upper layers at approximately 11–$15 \, \text{m s}^{-1}$, as visible in Figure 22(b).

The convective cells in Romania that evolved to the left of tropospheric shear were associated with nonnegligible probability of hail as indicated by the presence of *Three-Body Scatter Spike* (Figure 22(d)). Cyclonic shear at low levels (Figure 22(a)) can be a direct indication of the severity of convective cells that propagate to the left of the shear vector. The cells in Bulgaria show strong propagation to the right of the tropospheric shear and also showing the *Three-Body Scatter Spike* (Figure 22(c)). For these, the splitting process was not seen. In the second case (June 23, 2009), intense flow is not present at lower levels in the field of Doppler velocity relatively, but a relatively high speed of flow of about $15 \, \text{m s}^{-1}$ is seen at upper levels (Figure 23(b)). The radar reflectivity field at 1.5-degree elevation (Figures 22(c) and 23(c)) highlights the splitting process; the secondary cells evolve to the right (cyclonic rotation) and to the left (anticyclone rotation) of the shear vector S in the 1800–6000 m layer.

4. Conclusions

This study shows the splitting process in SE Romania, linked to convective developments which can be seen using the S-band Doppler radar. This process is important as a dynamic one, because it can modulate the energy transfer between the synoptic scale and the mesoscale. The main mesosynoptic conditions associated with the generation of this dynamic process can be distinguished, and also its evolution. Information about all of these can be used as an additional tool in nowcasting.

From the analysis of convective cells resulted from the split it can be said that the splitting process of convective cells observed in radar fields indicates a corresponding severe weather event. As a fundamental condition, splitting of convective cells occurred when an upper level jet ($30 \, \text{m s}^{-1}$ $40 \, \text{m s}^{-1}$) was detected in the high troposphere (above 6000 m). In the first stages of the convective cells, the presence of the jet stream in the upper troposphere and especially its vertical extent lead them to be strongly advected by the flow. Also, the presence at lower levels of a directional variation in the horizontal wind vector leads to a stronger relative flow. In the case of a tropospheric configuration that favours the splitting process, the lower relative flow has a high intensity and is opposite to the upper relative flow; this configuration is seen in the relative Doppler velocities field at different elevations. In all the three detailed cases, below 1800 m the horizontal wind vector had a rotation associated to the asymmetry of the splitting process, and also to the strong relative flow between 100 and 1000 m, providing a powerful moisture flow through updraft and strong vorticity advection.

All the analysed cases have shown the independence of the splitting process to the atmosphere circulation. This is characteristic to baroclinic waves with a west-to-east evolution, and also to cyclones near Romania: north-west circulation for May 1–28, 2008, south-east circulation for June 2–25, 2009, west circulation for August 3–11, 2003, south-south-east circulation for September 27, 2004, and south circulation for June 23, 2009.

The largest amounts of precipitation are related to cell cross-trajectories associated with the splitting process and lie to the right of shear in 1800–6000 m layer (Figures 6, 13, and 21). The splitting process may lead to a mesoscale convective system. If this happens the secondary cells interact constructively afterwards; this will happen to the right side of the mean tropospheric shear.

The local maxima of vertical vorticity (Figures 9, 13, and 21) are most positive (to the right side of the 0,0 coordinate). Keeping the same sign of horizontal vorticity is essential in formation of cyclonic and anticyclonic vortices, confirmed by Figure 1. The magnitude of the horizontal vorticity associated with vertical wind profiles and that of vertical vorticity for cyclonic and anticyclonic rotation are comparable; the majority of values is lying in the $0.001 \, \text{s}^{-1}$–$0.009 \, \text{s}^{-1}$ interval (in a first approximation, these magnitudes verify the dynamic process associated to the formation of cyclonic and anticyclonic vortices).

Both the June 25, 2009 and in the September 27, 2004 cases highlight the TBSS associated with convective cells (anticyclonic rotation) with a motion to the left of the shear vector in the mean troposphere, which means that these cells have attributes that indicate the presence of severe hail. An explanation could be the fact that the convective cells with motion left to the mean shear are more isolated than those with motion to the right of it. The former cells are moving to the zone where the values of geopotential and temperature are falling. Here, the thermal gradient is perpendicular to the shear vector (thermal wind relation), something that can accelerate the increase in thermodynamic instability.

To have a good image of the splitting process, in the next studies observational data from meteorological stations and soundings would have to be used.

References

[1] A. Stan-Sion and C. Soci, "Supercell environment in South-eastern Romania," in *Proceedings of the 5th Annual Meeting of the European Meteorological Society (EMS '05), 7th European Conference on Applications of Meteorology (ECAM '05)*, pp. 12–16, Utrecht, The Netherland, September 2005.

[2] A. Stan-Sion and B. Antonescu, "Mesocyclones in Romania—characteristics and environments," in *Proceedings of the 23rd AMS Conference on Severe Local Storms (SLS '06)*, Louis, Mo, USA, November 2006.

[3] A. Bell, D. Carbunaru, B. Antonescu, and S. Burcea, "The Black Sea influence in convective storms initiation in South Eastern Romania," in *Proceedings of the 34th AMS Conference on Radar Meteorology*, Williamsburg, Va, USA, October 2009.

[4] A. Bell, "Mesoscale environment for tornadic supercells in SE Romania," in *Proceedings of the 6th European Conference on Radar in Meteorology and Hydrology: Advances in Radar Technology*, Sibiu, Romania, September 2010.

[5] B. Antonescu, D. Carbunaru, M. Sasu, S. Burcea, and A. Bell, "A climatology of supercells in Romania," in *Proceedings of the 6th European Conference on Radar in Meteorology and Hydrology: Advances in Radar Technology*, Sibiu, Romania, September 2010.

[6] C. A. Doswell III, *On the Use of Hodographs—Vertical Wind Profile Information in Severe Storms Forecasting*, U. S. Dept. Of Commerce/NOAA, National Severe Storms Laboratory and National Weather Service, Southern Region, Fort Worth, Tex, USA, 1988.

[7] M. J. Bunkers, "Vertical wind shear associated with left-moving supercells," *Weather and Forecasting*, vol. 17, no. 4, pp. 845–855, 2002.

[8] R. Rotunno, "On the evolution of thunderstorm rotation," *Monthly Weather Review*, vol. 109, pp. 577–586, 1981.

[9] R. Rotunno and J. B. Klemp, "The influence of shear-induced pressure gradient on thunderstorm motion," *Monthly Weather Review*, vol. 110, no. 2, pp. 136–151, 1982.

[10] E. A. Brandes, "Vertical vorticity generation and mesocyclone sustenance in tornadic thunderstorms: the observational evidence," *Monthly Weather Review*, vol. 112, no. 11, pp. 2253–2269, 1984.

[11] R. Davies-Jones, "Streamwise vorticity: the origin of updraft rotation in supercell storms," *Journal of the Atmospheric Sciences*, vol. 41, no. 20, pp. 2991–3006, 1984.

[12] J. B. Klemp, "Dynamic of tornadic thunderstorms," *Annual Review of Fluid Mechanics*, vol. 19, pp. 369–402, 1987.

[13] R. Davies-Jones, "Linear and nonlinear propagation of supercell storms," *Journal of the Atmospheric Sciences*, vol. 59, no. 22, pp. 3178–3205, 2002.

[14] R. P. Davies-Jones, "Reply to comment on Linear and nonlinear propagation of supercell storms," *Journal of the Atmospheric Sciences*, vol. 60, pp. 2420–2426, 2003.

[15] R. Rotunno and M. L. Weisman, "Comment on Linear and nonlinear propagation of supercell storms," *Journal of the Atmospheric Sciences*, vol. 60, pp. 2413–2419, 2003.

[16] J. M. L. Dahl, *Supercells—Their Dynamics and Prediction*, Free University of Berlin, Institute of Meteorology, Department of Theoretical Meteorology, 2006.

[17] Y.-L. Lin, *Mesoscale Dynamics*, Cambridge University Press, 2007.

[18] P. Markowski and Y. Richardson, *Mesoscale Meteorology in Midlatitudes*, Royal Meteorological Society, 2010.

[19] J. B. Klemp and R. B. Wilhelmson, "Simulation of right- and left-moving storms produced through storm splitting," *Journal of the Atmospheric Sciences*, vol. 35, pp. 1097–1110, 1978.

[20] M. L. Weisman and J. B. Klemp, "The dependence of numerically simulated convective storms on vertical wind shear and buoyancy," *Monthly Weather Review*, vol. 110, no. 6, pp. 504–520, 1982.

[21] M. L. Weisman and J. B. Klemp, "The structure and classification of numerically simulated convective storms in directionally varying wind shears," *Monthly Weather Review*, vol. 112, no. 12, pp. 2479–2498, 1984.

[22] L. D. Grasso, "The dissipation of a left-moving cell in a severe storm environment," *Monthly Weather Review*, vol. 128, no. 8, pp. 2797–2815, 2000.

[23] D. L. Andra Jr., "Observation of an anticyclonically rotating severe storm," in *Proceedings of the 17th Conference onn Severe Local Storms*, pp. 186–190, American Meteorological Society, St. Louis, Mo, USA, 1993.

[24] R. P. Kleya, "A radar and synoptic scale analysis of a splitting thunderstorm over north-central Texas on November 10, 1992," in *Proceedings of the 17th Conference On Severe Local Storms*, pp. 211–213, American Meteorological Society, St. Louis, Mo, USA, 1993.

[25] F. H. Glass and S. C. Truett, "Observation of a splitting severe thunderstorm exhibiting both supercellular and multicellular traits," in *Proceedings of the 17th Conference On Severe Local Storms*, pp. 224–228, American Meteorological Society, St. Louis, Mo, USA, 1993.

[26] R. A. Brown and R. J. Meitín, "Evolution and morphology of two splitting thunderstorms with dominant left-moving members," *Monthly Weather Review*, vol. 122, pp. 2052–2067, 1994.

[27] G. W. Carbin, "Analysis of a splitting severe thunderstorm using the WSR-88D," NWS Eastern Region WSR-88D Operational Note 08, National Weather Service, Eastern Region Headquarters, Bohemia, NY, USA, 1997.

[28] L. D. Grasso and E. R. Hilgendorf, "Observations of a severe left moving thunderstorm," *Weather and Forecasting*, vol. 16, no. 4, pp. 500–511, 2001.

[29] J. P. Monteverdi, W. Blier, G. Stumpf, W. Pi, and K. Anderson, "First WSR-88D documentation of an anticyclonic supercell with anticyclonic tornadoes: the Sunnyvale-Los Altos, California, tornadoes of 4 May 1998," *Monthly Weather Review*, vol. 129, no. 11, pp. 2805–2814, 2001.

[30] J. F. Dostalek, J. F. Weaver, and G. L. Phillips, "Aspects of a tornadic left-moving thunderstorm of 25 may 1999," *Weather and Forecasting*, vol. 19, no. 3, pp. 614–626, 2004.

[31] D. T. Lindsey and M. J. Bunkers, "Observations of a severe, left-moving supercell on 4 May 2003," *Weather and Forecasting*, vol. 20, no. 1, pp. 15–22, 2005.

[32] R. Edwards and S. J. Hodanish, "Photographic documentation and environmental analysis of an intense, anticyclonic supercell on the Colorado plains," *Monthly Weather Review*, vol. 134, no. 12, pp. 3753–3763, 2006.

[33] E. N. Rasmussen and J. M. Straka, "Variations in supercell morphology. Part I: observations of the role of upper-level storm-relative flow," *Monthly Weather Review*, vol. 126, no. 9, pp. 2406–2421, 1998.

[34] National Oceanic Atmospheric Administration (NOAA), http://www.wdtb.noaa.gov/courses/dloc/index.html.

[35] R. M. Lhermitte and D. Atlas, "Precipitation motion by pulse Doppler radar," in *Proceedings of the 9th Conference on Radar Meteorology*, pp. 218–223, American Meteorological Society, 1961.

[36] K. A. Browning and R. Wexler, "The determination of kinematic properties of a wind field using Doppler radar," *Journal of Applied Meteorology*, vol. 7, pp. 105–113, 1968.

[37] R. A. Fulton, P. Breidenbach, D. J. Sco, D. A. Miler, and T. O'Bannon, "The WSR-98D rainfall algorithm," *Weather Forecast*, vol. 13, pp. 337–395, 1988.

[38] D. S. Zrnic, "Three-body scattering produces precipitation signature of special diagnostic value," *Radio Science*, vol. 22, no. 1, pp. 76–86, 1987.

[39] J. B. Wilson and D. Reum, "The flare echo: reflectivity and velocity signature," *Journal of Atmospheric and Oceanic Technology*, vol. 5, pp. 197–205, 1988.

[40] L. R. Lemon, "The Radar "Three-Body Scatter Spike": an operational large-hail signature," *Weather and Forecasting*, vol. 13, pp. 327–340, 1988.

[41] D. A. Bennetts and J. C. Sharp, "The relevance of conditional symmetric instability to the prediction of mesoscale frontal rainbands," *Quarterly Journal Royal Meteorological Society*, vol. 108, no. 457, pp. 595–602, 1982.

[42] D. M. Schultz and J. A. Knox, "Banded convection caused by frontogenesis in a conditionaly, symmetrically, and inertialy unstable environment," *Monthly Weather Review*, vol. 135, pp. 2095–2110, 2007.

3

Short- and Medium-Term Induced Ionization in the Earth Atmosphere by Galactic and Solar Cosmic Rays

Alexander Mishev[1,2]

[1] *Institute for Nuclear Research and Nuclear Energy, Bulgarian Academy of Sciences, 1784 Sofia, Bulgaria*
[2] *Sodankyla Geophysical Observatory (Oulu Unit), University of Oulu, 90014 Oulu, Finland*

Correspondence should be addressed to Alexander Mishev; alex_mishev@yahoo.com

Academic Editor: Prodromos Zanis

The galactic cosmic rays are the main source of ionization in the troposphere of the Earth. Solar energetic particles of MeV energies cause an excess of ionization in the atmosphere, specifically over polar caps. The ionization effect during the major ground level enhancement 69 on January 20, 2005 is studied at various time scales. The estimation of ion rate is based on a recent numerical model for cosmic-ray-induced ionization. The ionization effect in the Earth atmosphere is obtained on the basis of solar proton energy spectra, reconstructed from GOES 11 measurements and subsequent full Monte Carlo simulation of cosmic-ray-induced atmospheric cascade. The evolution of atmospheric cascade is performed with CORSIKA 6.990 code using FLUKA 2011 and QGSJET II hadron interaction models. The atmospheric ion rate is explicitly obtained for various latitudes, namely, 40°N, 60°N and 80°N. The time evolution of obtained ion rates is presented. The short- and medium-term ionization effect is compared with the average effect due to galactic cosmic rays. It is demonstrated that ionization effect is significant only in subpolar and polar atmosphere during the major ground level enhancement of January 20, 2005. It is negative in troposphere at midlatitude, because of the accompanying Forbush effect.

1. Introduction

Cosmic rays are high, ultrahigh, and extremely high energy particles of extraterrestrial origin, mostly protons. Cosmic rays (CRs) constantly impinge the Earth's atmosphere. While the low-energy particles are absorbed in the atmosphere, those with energies greater than 1 GeV/nucleon generate new particles through interactions with atomic nuclei. They are an important source of ionization in the Earth atmosphere [1]. The ionization in the stratosphere and troposphere is governed by galactic cosmic rays [2]. They initiate a complicated nuclear-electromagnetic-muon cascade resulting in an ionization of the ambient air. In such a cascade a small fraction of the initial primary particle energy reaches the ground as high energy secondary particles. Most of the primary energy is released in the atmosphere by ionization and excitation of the air molecules, resulting in an ionization of the ambient air. The maximum in secondary particle energy release is observed at altitudes of 15–26 km depending on latitude and solar activity level. This is the Pfotzer maximum.

The galactic cosmic ray (GCR) is affected by solar activity. They follow the 11-year solar cycle and respond to long and short time scale solar-wind variations. They are modulated with the opposite phase, that is, the higher solar activity, the lower the intensity of galactic CRs is. Solar energetic particles (SEPs) are accelerated during explosive energy release on the Sun and by acceleration processes in the interplanetary space. They enter the atmosphere sporadically, with a greater probability during periods of increased solar activity. In addition the heliosphere transient phenomena lead to a strong, relatively short suppression of GCR intensity in the vicinity of Earth, followed by a slower recovery on the time scale of several days known as Forbush decrease [3]. These events are generally interpreted as a result of the influence of coronal mass ejections (CMEs) and/or high-speed streams of the solar wind from the coronal holes on the background CRs.

The abundances of CR are nearly independent of the energy. For lower energies below 1 GeV/nucleon, the relative abundance of heavier nuclei increases, particularly around solar maximum, because they are less modulated than protons. In addition for a given energy, protons produce an atmospheric cascade that develops deeper in the atmosphere than cascades from heavier nuclei.

The investigation of ionization processes in atmosphere is important for better understanding of various processes and space weather mechanisms [1]. As example galactic cosmic rays influence via ionization the electrical parameters of planetary atmospheres. They are also related to atmospheric chemistry, as example the ozone depletion in the stratosphere. In addition to CR particles, solar electromagnetic X and UV radiations can affect the ionosphere and atmosphere, specifically the upper atmosphere. The effect varies geographically following the insolation pattern. It is beyond the topic of this study.

In the work presented here the ionization rate in the atmosphere during the major solar energetic particle event on January 20, 2005 is estimated. The obtained ion rates are compared with the average GCR ion production. This event is considered for study, because it is among the largest solar energetic particles events. In addition, the event occurred during the winter period; therefore the contribution of solar UV radiation could be neglected at polar latitudes for further atmospheric studies.

2. Modeling of CR Induced Atmospheric Ionization

The estimation of cosmic-ray-induced ionization, as was recently demonstrated, could be performed on the basis of a full Monte Carlo simulation of the atmospheric cascade [4]. At present, with the development of numerical methods, evolution of the knowledge of high energy interactions and nuclear processes, an essential progress in models for cosmic-ray-induced ionization in the Earth atmosphere is carried out [5–8]. As was recently demonstrated the models agree with 10–20%, the difference is mainly due to the various hadron generators and atmospheric models [2].

The full Monte Carlo simulation of the atmospheric cascade permits to follow the longitudinal cascade evolution in the atmosphere and obtains the energy deposit by different atmospheric cascade components from ground level till the upper atmosphere. These full target models apply the formalism of ionization yield function Y

$$Y(x, E) = \Delta E(x, E) \frac{1}{\Delta x} \cdot \frac{1}{E_{\text{ion}}} \cdot \Omega, \qquad (1)$$

where ΔE is the deposited energy in layer Δx in the atmosphere and Ω is a geometry factor, integration over solid angle, $E_{\text{ion}} = 35\,\text{eV}$, which is the energy necessary for production of one ion pair [9, 10]. The ionization yield function represents the number of ion pairs produced at given altitude x in the atmosphere by given primary cosmic ray nuclei with kinetic energy E on the top of the atmosphere. In fact the ionization yield function represents the ionization capacity in air by given primary CR particle.

The estimation of atmospheric ion rate production is obtained by convolution of $Y(x, E)$ and SEP spectrum as shown in

$$q(h, \lambda_m) = \int_{E_0}^{\infty} D(E, \lambda_m) Y(h, E) \rho(h) \, dE, \qquad (2)$$

where $D(E, \lambda_m)$ is the differential primary cosmic ray spectrum at given geomagnetic latitude λ_m for a given component of primary cosmic ray, Y is the yield function (1), and $\rho(h)$ is the atmospheric density $(\text{g} \cdot \text{cm}^{-3})$. We express x, during the simulations in $\text{g} \cdot \text{cm}^{-2}$, which is a residual atmospheric depth, that is, the amount of matter (air) overburden above a given altitude in the atmosphere. This is naturally related to the development of the cascade. Subsequently the mass overburden is transformed as altitude above sea level (a.s.l.) in [km].

In this study the simulation of the development of atmospheric cascade is carried out with CORSIKA 6.990 code [11] with corresponding hadron interaction generators FLUKA 2011 [12] and Quark Gluon String with JETs QGSJET II [13]. COsmic Ray SImulations for KASKADE (CORSIKA) code is the most widely used atmospheric cascade simulation tool. The code simulates the interactions and decays of various nuclei, hadrons, muons, electrons, and photons in the atmosphere. The particles are tracked through the atmosphere until they undergo reactions with an air nucleus or in the case of unstable secondary particles, they decay. The result of the simulations is detailed information about the type, energy, momenta, location, and arrival time of the produced secondary particles at given selected altitude a.s.l. The primary particles that can be considered are protons, light, middle, and heavy nuclei up to iron.

3. Ionization Effect in the Atmosphere due to Galactic and SEP

The described above formalism is applicable for the whole atmosphere where CR could affect from ground level to altitude of 100 km a.s.l. After obtaining the energy deposition from all secondary cosmic ray particles, respectively, ionization yield function Y (1) on the basis of (2), the ion rate as ion pairs per second in cm^3 is obtained. The ion rates are presented as ion pairs per second, per cm^3, and per atmosphere.

3.1. Ion Rates at Solar Minimum and Maximum at Various Atmospheric Depths. The cosmic-ray-induced ionization by GCR at solar minimum and maximum is obtained for ground level to the upper atmosphere. In Figure 1 the obtained ion rates are presented for various atmospheric depths, namely, $100\,\text{g}\,\text{cm}^{-2}$ (16 km a.s.l.), $300\,\text{g} \cdot \text{cm}^{-2}$ (10 km a.s.l.), $500\,\text{g} \cdot \text{cm}^{-2}$ (5 km a.s.l.), and $700\,\text{g} \cdot \text{cm}^{-2}$ (3 km a.s.l.) as a function of the rigidity cut-off.

It is demonstrated that solar activity affects ion rate production significantly at high latitudes, while at the Equator (15 GV cut-off), this effect is not so strong. Such computation of GCR-induced ionization permits to estimate the ion rate production at various time scales. Therefore the average ion

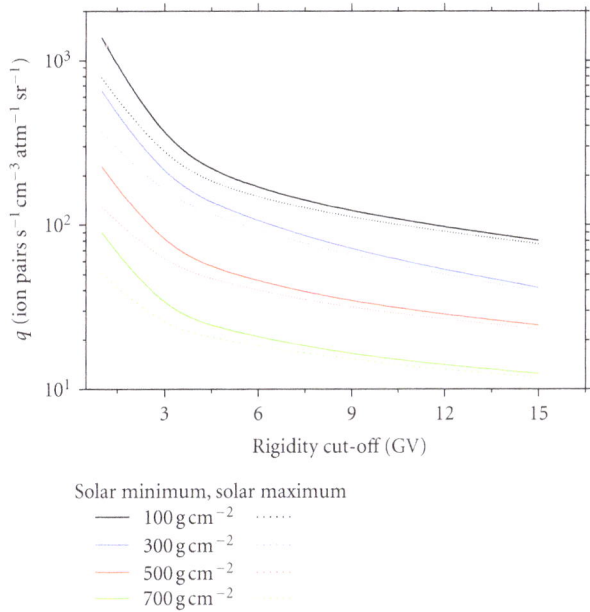

FIGURE 1: Cosmic-ray-induced ionization by GCR at solar minimum and maximum at various atmospheric depths.

rate due to GCR could be estimated for specific conditions (solar minimum and maximum). Hence, the obtained ion rates due to GCR permit to compare the SEP ionization effect with average ionization CR effect in a realistic manner.

3.2. Ionization Effect on January 20, 2005, during the Major GLE 69. In addition to continuous ionization in the Earth's atmosphere caused by galactic cosmic ray a sporadic ionization occurred during solar energetic particle events, potentially affecting the Earth's environment [14]. In general such events are low energy and are not able to initiate atmospheric cascades. Their ionization effect is limited to the upper polar atmosphere. Hence the studies of the effects caused by solar energetic particles are usually limited to the upper atmosphere above 30 km. However, it was recently demonstrated [15–18] that during some major ground level enhancements (GLEs), which are characterized by very high energy of solar particles, the ionization effect is important specifically over polar atmosphere.

On January 20, 2005 all energy channels of GOES satellite simultaneously register enhancement of proton flux. The (SEP) onset was registered at 6:50 UT, 3 minutes before maximum of X-ray flare. During the first hour the proton spectrum parameters change dramatically [19]. The changes of the spectra could be connected with some particle acceleration mechanism preceding the CME launch. The next 10 hours of acceleration produce spectra with very stable parameters [20], most likely formed at CME driven shock site.

A well-studied event [19] such as ground level enhancement 69 on January 20, 2005 was considered for analysis. Since this event is among the largest (the second largest in observation history), the ionization effect is expected to be maximal. The event on January 20, 2005 is characterized by

an anisotropic component with a very hard spectrum at the onset of the event, followed by a long isotropic emission with a softer spectrum [21]. The spectral index for SEP during GLE is typically between 4 and 6. In this work the solar proton spectrum is obtained on the basis of GOES 11 satellite measurements (high energy channels) and additional data [22, 23].

The spectrum of solar protons is expressed in two different moments: at 08:00 UT a high energy part with a slope of 2.32 and at 23:00 UT low energy part with a slope of 3.43. During the simulation, a realistic winter profile of the Earth atmosphere is considered [24, 25]. It is demonstrated that the ionization effect on event onset at 08:00 UT (Figure 2(a)) is greater than the one produced by delayed component at 23:00 UT (Figure 2(b)). Because the event on January 20, 2005 occurred during the recovery phase of the Forbush decrease and in the following days an additional suppression of the cosmic ray intensity was observed leading to a complicated time profile of cosmic ray flux, the net ionization effect is calculated as a superposition of ion rate from solar energetic particles and from reduced galactic cosmic rays.

In the case of 40°N latitude the effect at 08:00 UT is comparable to the average of GCR (Figures 1 and 2(a)). However, the ion rates quickly decrease with altitude (below 5 km a.s.l.). The ionization effect due to low energy component of the SEP spectrum, namely, at 23:00 UT is negligible (Figure 2(b)). In this case the ion rates are due mostly from reduced GCR.

The situation is quite different for latitude of 60°N (Figure 3). The ion rate due to SEP at 08:00 UT is significant. The ion rates from solar particles are greater than ion rates from GCR roughly an order of magnitude. The ionization is significant at altitudes above some 12 km a.s.l. and decreases in the troposphere. The ion rate at 23:00 UT due to a low energy component, as in a previous case, is negligible. In the case of 80°N latitude both components, hard at 08:00 UT, respectively, soft at 23:00 UT cause a significant excess on ion rates in the atmosphere. The effect at 08:00 UT is due mainly to solar protons (Figure 4(a)). It is significant at altitudes above 10–12 km a.s.l. Because the hard spectrum is with a slope of 2.32, it causes significant ionization even in a low atmosphere. At 23:00 UT the effect due to solar particles is significant at altitudes above 12 km a.s.l. and decreases in the troposphere, because of the softer spectrum. The ionization effect in a low troposphere is due to a reduced GCR (Figure 4(b)). Computed ion rates permit to estimate the net ionization effect at various time scales and to compare it to the average effect due to GCR.

3.3. Short and Medium Time Ionization Effect due to GLE 69 on January 20, 2005. The ionization effect is maximal during the first 24 h from the event onset [14, 16, 17]. The 24 h effect is computed as superposition of SEP and GCR ionization. Since the event occurred during the recovery phase of a strong Forbush decrease, a reduced GCR flux is considered. The 48 h effect is superposition of 24 h effect and reduced GCR ionization effect for the following 24 h. The weekly effect is computed taking into account the 24 h ionization effect from solar protons and complicated GCR flux, considering explicitly the transient phenomena during the week.

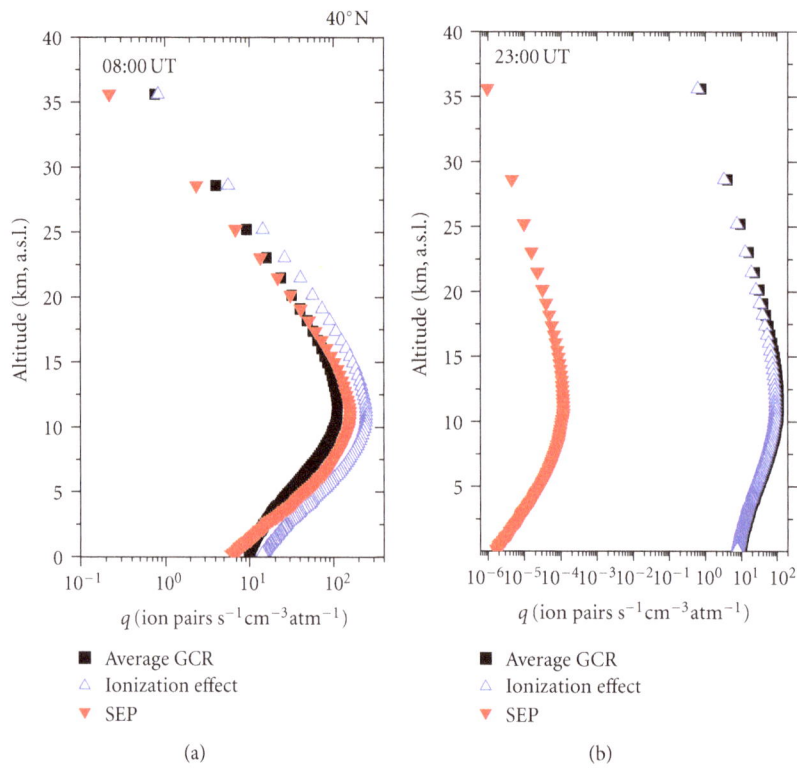

FIGURE 2: Ionization effect at 40°N due to GCR and SCR during GLE 69 on January 20, 2005.

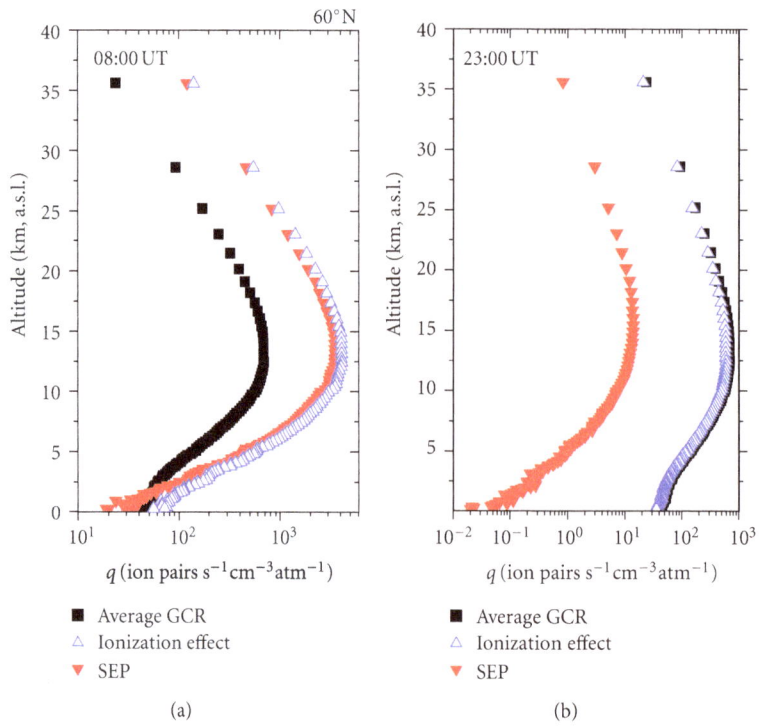

FIGURE 3: Ionization effect at 60°N due to GCR and SCR during GLE 69 on January 20, 2005.

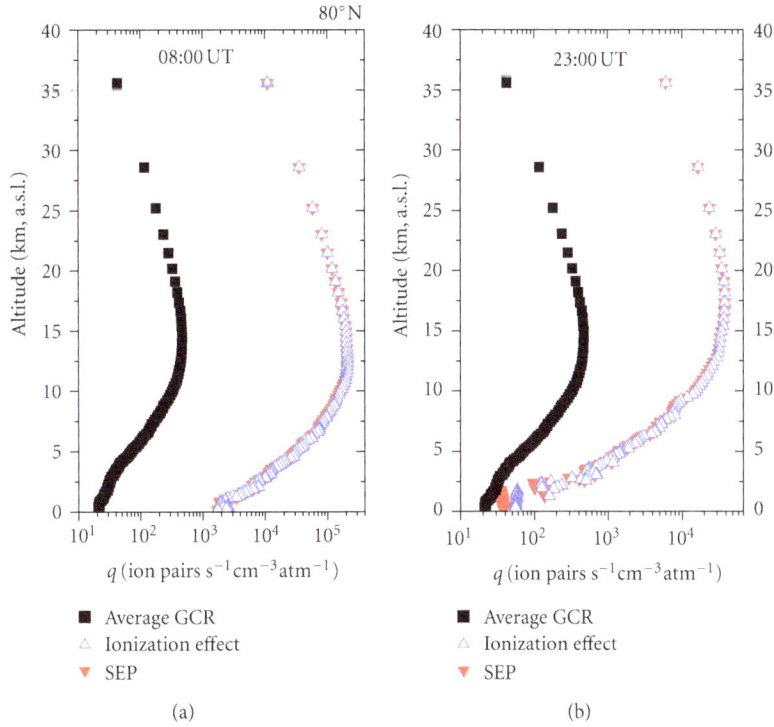

FIGURE 4: Ionization effect at 80°N due to GCR and SCR during GLE 69 on January 20, 2005.

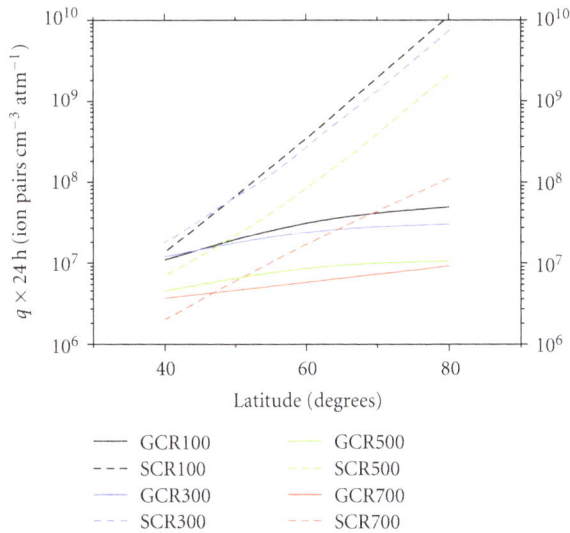

FIGURE 5: 24 h ionization effect due to solar protons during GLE69, compared with average GCR, as a function of latitude for various observation depths.

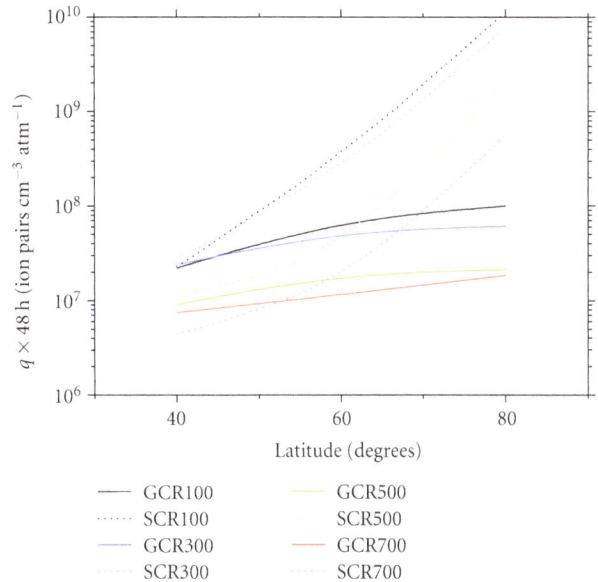

FIGURE 6: 48 h ionization effect due to solar protons during GLE69, compared with average GCR, as a function of latitude for various observation depths.

The ionization effect at various timescales is computed at various altitudes a.s.l., namely, at $100\,\mathrm{g\,cm^{-2}}$ (16 km a.s.l.), $300\,\mathrm{g\cdot cm^{-2}}$ (10 km a.s.l.), $500\,\mathrm{g\cdot cm^{-2}}$ (5 km a.s.l.), and $700\,\mathrm{g\cdot cm^{-2}}$ (3 km a.s.l.) as a function of the rigidity cut-off. This permits to estimate with good precision in which region of the atmosphere the ionization effect is significant, weak or negative. The Figures 5–7 demonstrate the net ionization effect as a function of the rigidity cut-off at various altitudes compared with average ionization effect from GCR. The average ionization effect from GCR is computed for period corresponding to the week before GLE 69 onset.

In general, as was shown in (Figure 2), the effect at 40°N is weak, with excess in the region of Pfotzer maximum.

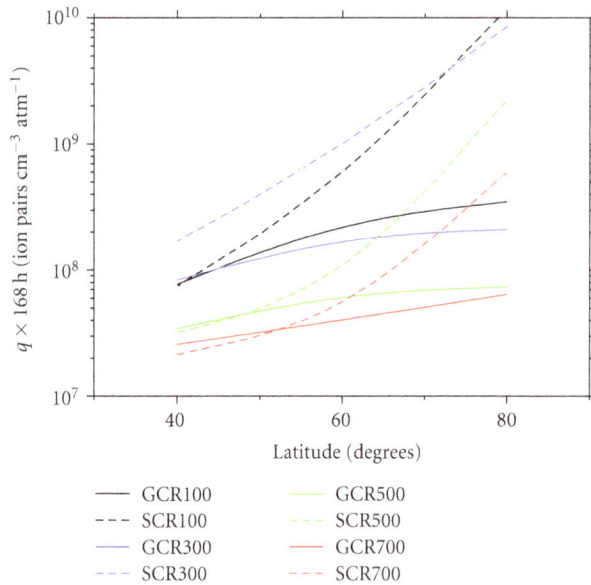

FIGURE 7: Weekly ionization effect due to solar protons during GLE69, compared with average GCR, as a function of latitude for various observation depths.

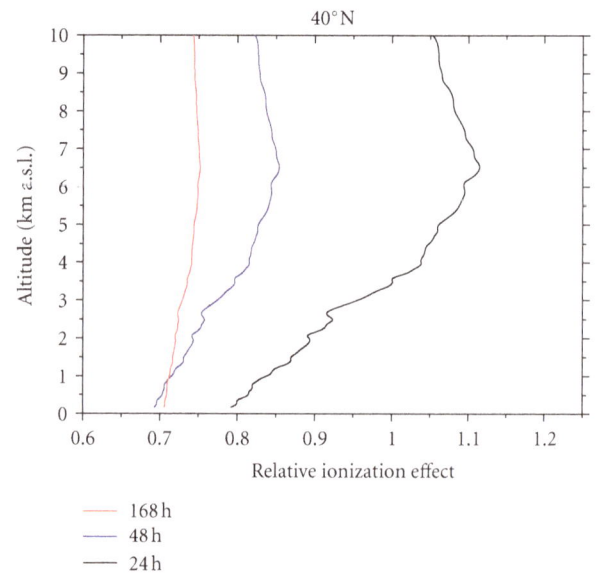

FIGURE 8: Relative to the average GCR ionization effect for 24 h, 48 h, and 168 h at 40°N in troposphere.

Therefore the 24 h is weak in a whole midlatitude atmosphere (Figure 5). We observe negative ionization effect compared to average effect from GCR at $700\,g\cdot cm^{-2}$ (red lines on Figure 5). The effect from SEP is weak at $500\,g\cdot cm^{-2}$ (green lines). Only at altitudes around some 10 km a.s.l. the ionization effect due to solar protons is greater than the average due to GCR at 40°N latitude. As was expected the effect increases for low rigidity cut-offs, that is, in subpolar and polar regions. We conclude that the 24 h ionization effect is important in the subpolar and polar atmosphere during major ground level enhancement of January 20, 2005, specifically at high altitudes (Figure 5—black and blue dashed lines).

The 48 h ion production at 40°N is comparable with the average due to GCR at altitudes above $500\,g\cdot cm^{-2}$ (Figure 6). The 48 h ion production is below the average due to GCR at altitude of 3 km for 40°N. Because the significant ion production at 60°N and 80°N; the 48 h ionization effect is still strong at these regions, specifically at high altitudes. The 48 h effect results on 50% increase compared to the average due to GCR at 60°N. The effect is even stronger at 80°N. Therefore the 48 h ionization effect is important only in the subpolar and polar atmosphere during major ground level enhancement of January 20, 2005 at high altitudes.

The weekly effect at 40°N (Figure 7) is clearly negative at low altitude (green and red lines at Figure 7). It is comparable with average GCR (black lines at Figure 7) effect at high altitude. However, the significant ion production during the event onset results on important ionization effect in the region of the Pfotzer maximum at 40°N (blue lines at Figure 7). The weekly effect is weak at the troposphere even at subpolar latitudes. However at 80°N the weekly effect is still important.

An illustration of the above conclusions is shown in Figure 8. Figure 8 demonstrates the relativity to the average GCR ionization effect for 24 h, 48 h, and 168 h at 40°N in the low troposphere. It is shown that the 24 h ionization effect is positive in a tight region of the troposphere above some 5 km a.s.l. In this region the interplay between reduced GCR flux and SEP leads to slight increase of the ionization. As was expected the 48 h is negative, as well as the weekly ionization effect (red and blue lines). In the upper troposphere the 24 h ionization effect is positive compared to the average of GCR only is a small region (Figure 9). The ionization effect is negative in the stratosphere. The 48 h ionization effect is clearly negative as was mentioned above.

Similarly it is demonstrated that 24 ionization effect at 60°N is not important in the lower troposphere (Figure 10). The 48 h ionization effect is not important as well and is slightly negative in planetary boundary layer.

Taking into account the time evolution of the obtained ion rates, the conclusion is the ionization effect is nearly negative for 40°N, especially in the troposphere and small for 60°N, because of the accompanying Forbush decrease during the event. The ionization effect is important only in the subpolar and polar atmosphere during major ground level enhancement of January 20, 2005 at high altitudes.

4. Applications and Discussion

The application of recent Monte Carlo CR ionization models is related to the study of various processes in the atmosphere, specifically related to atmospheric chemistry and physics [26, 27]. The influence of GCR should not be neglected in investigations of the tropospheric and stratospheric chemistry and dynamics. The CR ionization models show that the effects of galactic cosmic rays on the atmosphere are statistically

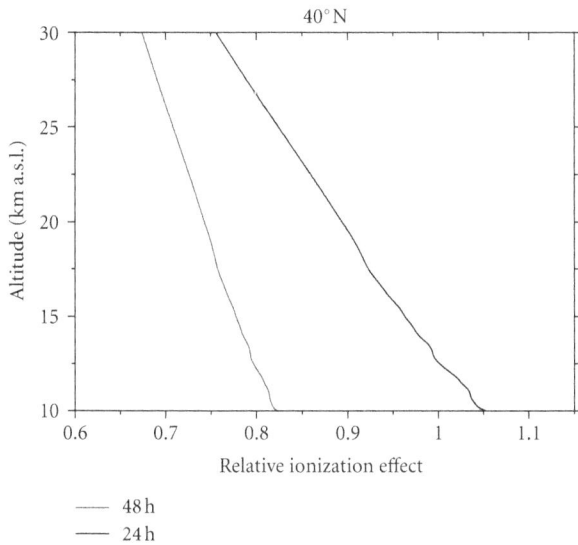

FIGURE 9: Relative to the average GCR ionization effect for 24 h and 48 h at 40°N in the upper troposphere and stratosphere.

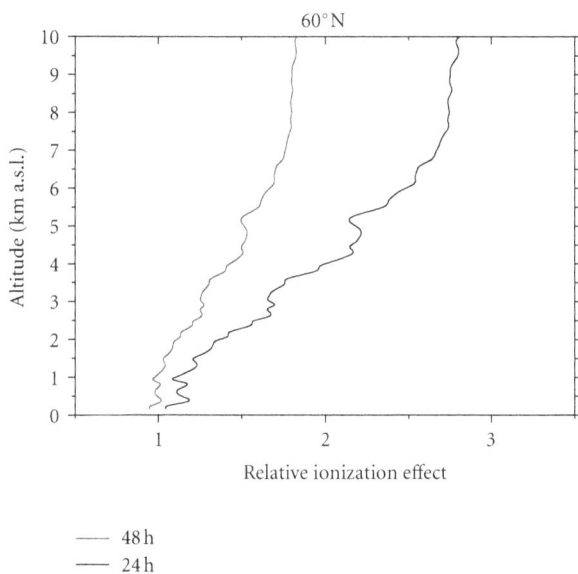

FIGURE 10: Relative to the average GCR ionization effect for 24 h and 48 h at 60°N in the troposphere.

significant in large geographic regions and for a number of relevant atmospheric species.

It was recently demonstrated for Southern hemispheric troposphere that NO_x increases more than 20% in the polar region [26]; HO_x decreases of 3% in the midlatitude upper troposphere; HNO_3 increases by more than 10% between the South Pole and subtropics and O_3 increases by up to 3% throughout the troposphere to 20 km between the South Pole to 20°N. Respectively, for Northern hemispheric troposphere HNO_3 marginally significant increases in the midlatitude upper troposphere and O_3 significant decreases in the polar upper troposphere. Undoubtedly these effects are related with the planetary distribution of GCR.

In addition it was demonstrated that major solar events leading to GLEs could also impact the Earth atmosphere [28]. According to the current knowledge, the cosmic-ray-induced ionization of the atmosphere, due to solar and galactic cosmic rays produces additional chemical sources and influences ozone molecules in the atmosphere [29] via photochemistry [30]. Recently, was demonstrated, that the ionization of the polar atmosphere by SEPs is responsible for the impact on neutral constituents. In addition it was shown that even events with limited particle flux could change the abundance of the minor constituents in the mesosphere and the upper stratosphere [31].

The obtained results presented here are related to these studies, since the atmospheric ionization is related to the mentioned above phenomena. Recent studies at the upper atmosphere on short-medium time [32, 33] demonstrate that the January 2005 SPEs caused large enhancements in the northern polar mesospheric HO_x and NO_x constituents (both observed and modeled). Observations indicated large mesospheric increases in OH (up to 4 ppbv) and HO_2 (>0.5 ppbv) as a result of the SPEs during the time period 16–21 January in the 60–85N latitude. The WACCM3 simulations showed quantitatively similar, although somewhat larger enhancements in OH and HO_2. These large HO_x enhancements led to considerable ozone decreases of greater than 40% throughout most of the northern polar mesosphere during the SPE period. All these studies demonstrate that SEP impact on the atmosphere is more important than GLE particles in the upper atmosphere. However, for better understanding of the atmospheric chemistry and physics process is the investigation if their impact is important.

While the recent model studies of cosmic-ray-induced ionization during the major GLE 69 on January 20, 2005 focus on the onset of the event and the next following hours, when the effect is maximal, in the study presented here we focus on short- to mid-term ionization effect. Previous works did not concern the mid-term ionization effect in details for various altitudes in the atmosphere. In addition several of the previous studies are limited to a high atmosphere [32] or only on the contribution of heavy ions [34]. Therefore the results presented here extend the recent studies of such events and give basis for further improvement of model studies related to atmospheric physics phenomena.

5. Conclusion

The application of Monte Carlo methods for investigation of cosmic-ray-induced ionization is very important, because it is possible to consider explicitly the hadron component. It gives the possibility to estimate middle and low atmospheric effects related to cosmic-ray-induced ionization realistically. As was recently demonstrated their application in a specific conditions [15, 17] permits detailed study of the ionization effect, specifically at middle and low altitudes [16].

The effect of sporadic solar energetic particle events is limited on a global scale but most energetic events could be strong locally, specifically in subpolar and polar regions, affecting the physical-chemical properties of the upper atmosphere. Since the large SEP events leading to GLEs are different on spectra and composition, their detailed study is connected with detailed information about heliospheric and geospace conditions. In this connection extension of the existing models to the upper atmosphere is very important, as well as their comparison with analytical models.

The ionization effects in the upper atmosphere during GLE 69 on January 20, 2005 are well studied [32, 33]. In this study it is shown that ionization effect is significant at subpolar and polar atmosphere, with fast tropospheric decrease [16, 17]. Previous studies are focused on a maximal phase of ionization effect during the first hours from GLE onset. In this study the obtained ion rates are used to estimate the mid-term ionization effect. It is demonstrated that the ionization effect is nearly negative compared to GCR at middle latitudes for short- to mid-term periods. This is very important for further studies of the influence of GLE particles on atmospheric chemistry and physics.

It has a general agreement that cosmic-ray-induced ionization of the atmosphere, due to solar and galactic cosmic rays, produces additional chemical sources and influences ozone molecules in the atmosphere. As was mentioned above the ionization effect due to SEPs is important only in a subpolar and polar atmosphere. This is very important for recent composition changes studies. The January 20, 2005 SPEs caused large enhancements in the northern polar HO_x and NO_x constituents in mesosphere, which lead to ozone decrease in the order of 40%. However on the basis of the obtained ion rate it is demonstrated that the tropospheric and stratospheric ionization effects due to SEPs are important on short to medium time scales in polar atmosphere and are nearly negative compared to the average GCR on medium time scales at middle latitudes (because of the accompanying Forbush decrease). More detailed studies of the influence on minor components of the atmosphere deserve special interest.

In addition, it was recently demonstrated [34] that the contribution of light nuclei, specifically Helium could be important for total ionization even at middle latitudes during GLE 69. The extension of such models, specifically in the upper atmosphere on the basis of a full Monte Carlo simulation of the atmospheric cascade with combination with analytical procedures [35] will contribute to deep understanding of the impact of solar and galactic cosmic rays on the atmosphere of the Earth and their connection to the space weather [36–38] and atmospheric physics [39].

Acknowledgments

The author warmly acknowledges the high energy Division of the Institute for Nuclear Research and Nuclear Energy of Bulgarian Academy of Sciences, namely, V. Guenchev, S. Piperov, E. Puncheva, and P. Konstantitov for the given computational time.

References

[1] G. A. Bazilevskaya, I. G. Usoskin, E. O. Flückiger et al., "Cosmic ray induced ion production in the atmosphere," *Space Science Reviews*, vol. 137, no. 1–4, pp. 149–173, 2008.

[2] I. G. Usoskin, L. Desorgher, P. Velinov et al., "Ionization of the earth's atmosphere by solar and galactic cosmic rays," *Acta Geophysica*, vol. 57, no. 1, pp. 88–101, 2009.

[3] S. E. Forbush, "Cosmic-ray intensity variations during two solar cycles," *Journal of Geophysical Research*, vol. 63, no. 4, pp. 651–669, 1958.

[4] I. G. Usoskin, O. G. Gladysheva, and G. A. Kovaltsov, "Cosmic ray-induced ionization in the atmosphere: spatial and temporal changes," *Journal of Atmospheric and Solar-Terrestrial Physics*, vol. 66, no. 18, pp. 1791–1796, 2004.

[5] L. Desorgher, E. O. Flückiger, M. Gurtner, M. R. Moser, and R. Bütikofer, "Atmocosmics: a geant 4 code for computing the interaction of cosmic rays with the earth's atmosphere," *International Journal of Modern Physics A*, vol. 20, no. 29, pp. 6802–6804, 2005.

[6] I. G. Usoskin and G. A. Kovaltsov, "Cosmic ray induced ionization in the atmosphere: full modeling and practical applications," *Journal of Geophysical Research D*, vol. 111, Article ID D21206, 2006.

[7] A. Mishev and P. I. Y. Velinov, "Atmosphere ionization due to cosmic ray protons estimated with Corsika code simulations," *Comptes Rendus de L'Academie Bulgare des Sciences*, vol. 60, no. 3, pp. 225–230, 2007.

[8] P. I. Y. Velinov, A. Mishev, and L. Mateev, "Model for induced ionization by galactic cosmic rays in the Earth atmosphere and ionosphere," *Advances in Space Research*, vol. 44, no. 9, pp. 1002–1007, 2009.

[9] P. I. Y. Velinov, G. Nestorov, and L. Dorman, *Cosmic Ray Influence on the Ionosphere and on the Radio-Wave Propagation*, Bulgarian Academy of Sciences Publishing House, Sofia, Bulgaria, 1974.

[10] H. S. Porter, C. H. Jackman, and A. E. S. Green, "Efficiencies for production of atomic nitrogen and oxygen by relativistic proton impact in air," *The Journal of Chemical Physics*, vol. 65, no. 1, pp. 154–167, 1976.

[11] D. Heck, J. Knapp, J. N. Capdevielle et al., "CORSIKA: a monte carlo code to simulate extensive air showers," Forschungszentrum Karlsruhe Report FZKA 6019, 1998.

[12] G. Battistoni, S. Muraro, P. R. Sala et al., "The FLUKA code: description and benchmarking," in *Proceedings of theHadronic Shower Simulation Workshop*, M. Albrow and R. Raja, Eds., vol. 896 of *AIP Conference Proceeding*, pp. 31–49, September 20062007.

[13] S. Ostapchenko, "QGSJET-II: towards reliable description of very high energy hadronic interactions," *Nuclear Physics B*, vol. 151, no. 1, pp. 143–146, 2006.

[14] R. Vainio, L. Desorgher, D. Heynderickx et al., "Dynamics of the Earth's particle radiation environment," *Space Science Reviews*, vol. 147, no. 3-4, pp. 187–231, 2009.

[15] A. Mishev, P. I. Y. Velinov, and L. Mateev, "Atmospheric ionization due to solar cosmic rays from 20 January 2005 calculated with Monte Carlo simulations," *Comptes Rendus de L'Academie Bulgare des Sciences*, vol. 63, no. 11, pp. 1635–1642, 2010.

[16] I. G. Usoskin, G. A. Kovaltsov, I. A. Mironova, A. J. Tylka, and W. F. Dietrich, "Ionization effect of solar particle GLE events in low and middle atmosphere," *Atmospheric Chemistry and Physics*, vol. 11, no. 5, pp. 1979–1988, 2011.

[17] A. L. Mishev, P. I. Y. Velinov, L. Mateev, and Y. Tassev, "Ionization effect of solar protons in the Earth atmosphere—case study of the 20 January 2005 SEP event," *Advances in Space Research*, vol. 48, no. 7, pp. 1232–1237, 2011.

[18] A. Mishev, P. I. Y. Velinov, and L. Mateev, "Ion production rate profiles in the atmosphere due to solar energetic particles on 28 october 2003 obtained with CORSIKA 6.52 simulations," *Comptes Rendus de L'Academie Bulgare des Sciences*, vol. 64, no. 6, pp. 859–866, 2011.

[19] R. Butikofer, E. O. Fluckiger, L. Desorgher, and M. R. Moser, "The extreme solar cosmic ray particle event on 20 January 2005 and its influence on the radiation dose rate at aircraft altitude," *Science of the Total Environment*, vol. 391, no. 2-3, pp. 177–183, 2008.

[20] N. K. Bostanjyan, A. A. Chilingarian, V. S. Eganov, and G. G. Karapetyan, "On the production of highest energy solar protons at 20 January 2005," *Advances in Space Research*, vol. 39, no. 9, pp. 1456–1459, 2007.

[21] C. Plainaki, A. Belov, E. Eroshenko, H. Mavromichalaki, and V. Yanke, "Modeling ground level enhancements: event of 20 January 2005," *Journal of Geophysical Research A*, vol. 112, no. 4, Article ID A04102, 2007.

[22] R. A. Mewaldt, M. D. Looper, C. M. S. Cohen et al., "Solar-particle energy spectra during the large events of October-November 2003 and January 2005," in *Proceedings of the 29th International Cosmic Ray Conference*, vol. 1, pp. 111–114, Pune, India, 2005.

[23] V. S. Makhmutov, G. A. Bazilevskaya, B. B. Grozdevsky et al., "Solar cosmic ray spectra in the 20 January GLE: comparison of simulations with ballon and neutron monitor observations," in *Proceedings of the 31th International Cosmic Ray Conference*, pp. 1–4, Lodz, Poland, 2009.

[24] A. Mishev and P. I. Y. Velinov, "Effects of atmospheric profile variations on yield ionization function Y in the atmosphere," *Comptes Rendus de L'Academie Bulgare des Sciences*, vol. 61, no. 5, pp. 639–644, 2008.

[25] A. L. Mishev and P. Velinov, "The effect of model assumptions on computations of cosmic ray induced ionization in the atmosphere," *Journal of Atmospheric and Solar-Terrestrial Physics*, vol. 72, no. 5-6, pp. 476–481, 2010.

[26] M. Calisto, I. Usoskin, E. Rozanov, and T. Peter, "Influence of Galactic Cosmic Rays on atmospheric composition and dynamics," *Atmospheric Chemistry and Physics*, vol. 11, no. 9, pp. 4547–4556, 2011.

[27] L. I. Dorman, *Cosmic Rays in the Earth's Atmosphere and Underground*, Kluwer Academic, Dordrecht, The Netherlands, 2004.

[28] B. Funke, A. Baumgaertner, M. Calisto et al., "Composition changes after the "halloween" solar proton event: the High-Energy Particle Precipitation in the Atmosphere (HEPPA) model versus MIPAS data intercomparison study," *Atmospheric Chemistry and Physics Discussions*, vol. 11, no. 3, pp. 9407–9514, 2011.

[29] A. Krivolutsky, A. Kuminov, and T. Vyushkova, "Ionization of the atmosphere caused by solar protons and its influence on ozonosphere of the Earth during 1994–2003," *Journal of Atmospheric and Solar-Terrestrial Physics*, vol. 67, no. 1-2, pp. 105–117, 2005.

[30] A. A. Krivolutsky, A. V. Klyuchnikova, G. R. Zakharov, T. Y. Vyushkova, and A. A. Kuminov, "Dynamical response of the middle atmosphere to solar proton event of July 2000: three-dimensional model simulations," *Advances in Space Research*, vol. 37, no. 8, pp. 1602–1613, 2006.

[31] A. Damiani, M. Storini, M. Laurenza, and C. Rafanelli, "Solar particle effects on minor components of the Polar atmosphere," *Annales Geophysicae*, vol. 26, no. 2, pp. 361–370, 2008.

[32] C. H. Jackman, D. R. Marsh, F. M. Vitt et al., "Short- and medium-term atmospheric constituent effects of very large solar proton events," *Atmospheric Chemistry and Physics*, vol. 8, no. 3, pp. 765–785, 2008.

[33] C. H. Jackman, D. R. Marsh, F. M. Vitt et al., "Northern Hemisphere atmospheric influence of the solar proton events and ground level enhancement in January 2005," *Atmospheric Chemistry and Physics*, vol. 11, no. 13, pp. 6153–6166, 2011.

[34] A. Mishev, P. I. Y. Velinov, and L. Mateev, "Atmospheric ionization due to solar cosmic rays from 20 January 2005 calculated with Monte Carlo simulations," *Comptes Rendus de L'Academie Bulgare des Sciences*, vol. 63, no. 11, pp. 1635–1642, 2010.

[35] P. I. Y. Velinov, S. Asenovski, and L. Mateev, "Simulation of cosmic ray ionization profiles in the middle atmosphere and lower ionosphere on account of characteristic energy intervals," *Comptes rendus de l'Academie bulgare des Sciences*, vol. 64, no. 9, pp. 1303–1310, 2011.

[36] K. Kudela, M. Storini, M. Y. Hofer, and A. Belov, "Cosmic rays in relation to space weather," *Space Science Reviews*, vol. 93, no. 1-2, pp. 153–174, 2000.

[37] A. L. Mishev and J. N. Stamenov, "Present status and further possibilities for space weather studies at BEO Moussala," *Journal of Atmospheric and Solar-Terrestrial Physics*, vol. 70, no. 2-4, pp. 680–685, 2008.

[38] L. I. Miroshnichenko, "Solar cosmic rays in the system of solar-terrestrial relations," *Journal of Atmospheric and Solar-Terrestrial Physics*, vol. 70, no. 2-4, pp. 450–466, 2008.

[39] A. L. Mishev, "A study of atmospheric processes based on neutron monitor data and Cherenkov counter measurements at high mountain altitude," *Journal of Atmospheric and Solar-Terrestrial Physics*, vol. 72, no. 16, pp. 1195–1199, 2010.

4

Heavy Rainfall Simulation over Sinai Peninsula Using the Weather Research and Forecasting Model

Gamal El Afandi,[1,2] **Mostafa Morsy,**[2] **and Fathy El Hussieny**[2]

[1] *Division of International Research and Development, College of Agriculture, Environment and Nutrition Sciences and College of Engineering, Tuskegee University, Tuskegee, AL 36088, USA*
[2] *Astronomy and Meteorology Department, Faculty of Science, Al Azhar University, Cairo 11884, Egypt*

Correspondence should be addressed to Gamal El Afandi; gamalafandy@yahoo.com

Academic Editor: Prodromos Zanis

Heavy rainfall is one of major severe weather over Sinai Peninsula and causes many flash floods over the region. The good forecasting of rainfall is very much necessary for providing early warning before the flash flood events to avoid or minimize disasters. In the present study using the Weather Research and Forecasting (WRF) Model, heavy rainfall events that occurred over Sinai Peninsula and caused flash flood have been investigated. The flash flood that occurred on January 18, 2010, over different parts of Sinai Peninsula has been predicted and analyzed using the Advanced Weather Research and Forecast (WRF-ARW) Model. The predicted rainfall in four dimensions (space and time) has been calibrated with the measurements recorded at rain gauge stations. The results show that the WRF model was able to capture the heavy rainfall events over different regions of Sinai. It is also observed that WRF model was able to predict rainfall in a significant consistency with real measurements. In this study, several synoptic characteristics of the depressions that developed during the course of study have been investigated. Also, several dynamic characteristics during the evolution of the depressions were studied: relative vorticity, thermal advection, and geopotential height.

1. Introduction

Heavy rainfall is one of the major severe weather in Egypt particularly in arid and semiarid regions especially if it is steep and mountainous regions such as Sinai Peninsula and the Eastern desert of Egypt. Short duration of heavy rainfall over a relatively small drainage area can lead to devastating flash flood, consequently causing a number of fatalities and tremendous damages.

It can destabilize soils along mountain slopes, resulting in landslides and mudslides that cause severe damage to nearby villages.

Heavy rainfall is usually resulting from individual mesoscale storms or mesoscale convective systems embedded in synoptic-scale disturbances [1]. High-resolution observations and numerical modeling technique are required to better predict heavy rainfall events, where the forecasting of heavy rainfall is very important for many decision makers who are sensitive to the occurrence of precipitation. An accurate quantitative precipitation forecast can identify the potential for heavy precipitation and possible associated flash flooding, as well as provide information for hydrological interests.

Since heavy rainfall and flash flood can lead to severe damage and losses for both life and infrastructure, the need to warn people in advance is, thus, an important goal, but heavy rainfall and flash floods forecasting is considered a difficult task, particularly in arid and semiarid regions, since they take place in very short time interval. Hence, rainfall forecasting is very much necessary for providing an early warning before the flash flood events to avoid or minimize disasters.

An early warning system (EWS) for flash floods has been developed for part of the Sinai Peninsula of Egypt, a hyperarid area confronted with limited availability of field data, limited understanding of the response of the valley (wadi) to rainfall, and a lack of correspondence between rainfall data and observed flash flood events [2].

Sinai Peninsula contains many complex terrains; so rainfall forecasting for complex terrains is challenging and requires numerical simulations at very high resolution.

In arid areas such as Sinai in Egypt, rainfall is mainly caused by squall line and convective cloud mechanisms and by low-intensity frontal rain, causing storm floods [3]. The complexity of terrain influences the weather in a variety of ways. Under stable atmospheric conditions, the terrain generates internal gravity waves that distribute momentum over wider areas. These processes may be related to strong winds and turbulence that influence the air traffic. Under unstable conditions, convective clouds and precipitation are generated over complex terrain which can grow into severe thunderstorms. All the mentioned processes occur on spatial length scales smaller than 100 km, usually even smaller than 10 km.

Because of the above-mentioned challenges, the mesoscale meteorological model named the Advanced Research Weather Research and Forecasting (WRF-ARW) is selected for rainfall forecasting in this research. WRF is developed mainly by the US National Centre for Atmospheric Research (NCAR) in collaboration with many other research centres and universities. WRF allows forecasting weather in complex terrains such as the one in Sinai and in the same time is considering orographic features. It is suitable for a broad spectrum of applications across scales ranging from meters to thousands of kilometres.

The main objectives of this study can be summarized as follows:

(a) the understanding of the processes leading to the flood event at Sinai using WRF?

(b) the investigation of the capability of WRF to simulate successfully the rainfall amount during a flood event?

2. Study Area

The Sinai Peninsula (Figure 1), study area, is located in the far northeast of Egypt. It represents about 6% of Egypt's area which is about $61,000\,km^2$.

The Sinai Peninsula comprises a wedge-shaped block of territory with its base along the Mediterranean Sea coast to the north and its apex bounded by the Gulfs of Suez to the west and Aqaba to the east.

Its southern portion consists of rugged, sharply serrated mountains. These reach elevations of more than 2,400 meters; among them is Mount Catherine, Egypt's highest mountain, which has an elevation of 2,642 meters. The central area of Sinai consists of two plateaus, Al-Tih and Al-Ajmah, both deeply indented and dipping northward toward Wadi (Valley) El Arish.

Toward the Mediterranean Sea, the northward plateau slope is broken by dome-shaped hills; between them and the coast are long, parallel lines of dunes, some of which are more than 100 meters high.

It is dissected by the largest wadi in Sinai, Wadi El Arish, which emerges from elevated gravelly plains and terraces in the south to a distance of about 20 km till the coast Mediterranean Sea to the north. The rainfall and floods

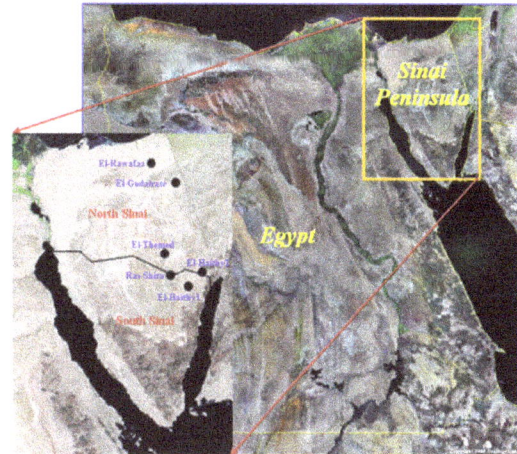

FIGURE 1: Study area and rain gauges location.

are the only sources of renewable water resources in Sinai Peninsula.

2.1. Climate of Sinai Peninsula. It is characterized by the Mediterranean climate in the northernmost of Sinai to be close to the desert and semidesert climate to the south.

Much of the Sinai is hot, or very hot with a higher temperature inland, but there is a more temperate region near the north coast and over the mountains. During the period May/June to September/October, the mean daily maximum is 28°C to 37°C in the North, 31°C to 42°C near the south coast, and 35°C to 41°C inland. Minimum temperatures average between 20°C and 25°C in the summer.

The winter season is a little less harsh with day high's in the mid-teens and possible 20's, and evenings often falling to around 6°C–10°C, and may drop below 0°C.

The amount of rainfall in Sinai decreases from the northeast towards the southwest. The greatest amount of the annual rainfall was found at Rafah station (304 mm) in the northeast. The annual rainfall average is about 120 mm along the Mediterranean coast. It decreases in the uplands to the south to about 32 mm. The annual average of rainfall all over Sinai is about 40 millimeters, 27 millimeters from it is estimated to come from one storm that may provide 10 millimeters at a time.

Along the Mediterranean Coast, 60% of the rain occurs in the winter, while 40% falls during the transitional seasons.

The annual rainfall in the southern region is significantly less than the northern one, where it reaches 20 mm in the coastal areas over the Gulfs of Aqaba and Suez. Its amount increases to 70 millimeters over the mountain regions.

The rainfall in autumn and winter seasons is ranging from medium to heavy especially at some high terrain areas. The rainfall is completely nonsexist in summer season.

3. Data and Methodology

WRF model is used in this research with its nesting capability to simulate the heavy rainfall with fine grid spacing.

During the storm mesoscale, convective systems are highly interacted with synoptic-scale environment.

During this study we conducted multinested experiment for four domains with different horizontal resolutions of 81 km (DM1), 27 km (DM2), 9 km (DM3), and 3 km (DM4), respectively, as shown in Figure 2. The second domain (DM2) will be used to simulate the synoptic situation over Egypt on 18 January, 2010, while the last domain (DM4) will be used to investigate the ability of the WRF model to simulate and predict rainfall to compare it with rain gauges at Sinai.

The model output is adopted to produce its output every hour to be compatible with the rain gauges observations and two-way nesting domain starting from 17 to 19 January, 2010. The 27 vertical layers with the model top of 50 hPa are used in this study.

The initial and lateral boundary meteorological data which used to run the model has been downloaded from the National Centers for Environmental Prediction (NCEP), global final analyses on $1° \times 1°$ degree from Global Forecasting System (GFS), and it is updated every six hours.

The hourly rainfall data, measured in millimeter (mm), was obtained from the rain gauges stations installed by the Water Resources Research Institute (WRRI); WRRI is the Egyptian governmental research institute with the mandate for flash flood management.

WRF offers multiple physics options that can be combined in any way. The options typically range from simple and efficient to sophisticated and more computationally costly and from newly developed schemes to well-tried schemes such as those in current operational models. Table 1 represents the selected physics of WRF model.

4. Results and Discussions

Egypt is subjected to thunderstorms and heavy rains started at the north coast, the Red Sea, and the Sinai Peninsula during 18 January, 2010. This extreme weather and intensive rainfall led to flash flood events over Sinai Peninsula.

The development of intensive weather events that invade Egypt during January 18, 2010, were characterized by "exceptional and extremely heavy rainfall," which affected a wide part of Egypt, including Sinai Peninsula and fatal to some Bedouin tribes located in its path. This heavy rainfall is accompanied by strong winds and thunderstorms.

The Mediterranean region is considered to be one of the most cyclogenetic areas in the world, usually favoring the development of weak low-pressure systems. Occasionally, these systems develop into deep cyclogenesis that cause a series of severe weather events as they cross the Mediterranean.

Generally, the winter season in Middle East is known to be associated with the development of low-pressure systems (Cyprus lows) over the eastern Mediterranean Sea [4]. Such systems have been observed at upper levels, particularly at 500 hPa. Rainfall has been observed to be heavy over Sinai and extended to southern parts of Egypt during our case study in winter season, January 18, 2010.

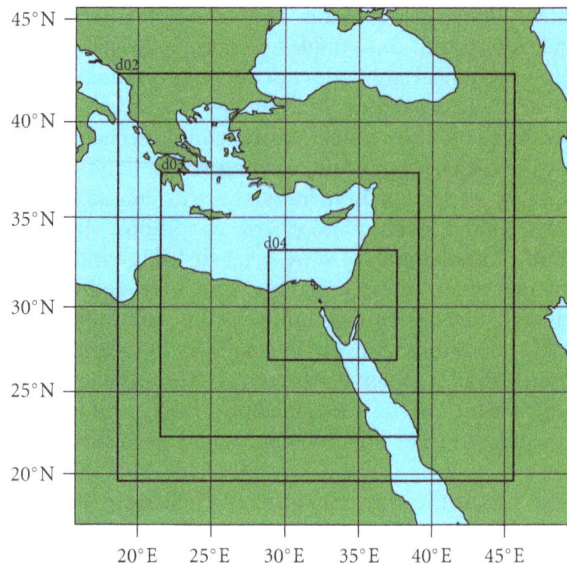

FIGURE 2: Model domains of 81 km, 27 km, 9 km, and 3 km horizontal resolution.

TABLE 1: Selected physics of the WRF model options.

Physics scheme	Type of the scheme that used in WRF run
Microphysics	Single-moment 3-class WSM3 scheme
Longwave radiation	Rapid Radiative Transfer Model (RRTM) scheme
Shortwave radiation	Dudhia scheme
Land surface	Noah Land Surface Model scheme
Surface and boundary layers	Yonsei University scheme
Cumulus parameterization	Kain-Fritsch scheme

Kahana et al. [5] found that Mediterranean Sea cyclones deepen in general, in association with upper level troughs. Upper level troughs are regarded as the key factors in activity of midlatitudes, such as the Mediterranean Sea system, particularly in their front sides, where positive vorticity advection and enhanced convection take place [6]. Ferraris et al. [7], however, found that the polar continental air from central Asia moving toward the eastern Mediterranean, and interacting with the relatively warm sea surface temperature (SST), produces enhanced lower level instability.

Because of the importance of the Mediterranean cyclogenesis, many studies on weather systems in the region have been conducted. The synoptic situation started by the development of the subtropical jet stream in the upper troposphere. The maximum wind speed ranged between 120 and 135 knots as shown in Figure 3, with the southern extension of the polar jet to the south of the Mediterranean.

Figure 3 shows the wind speed of the subtropical branch of the jet stream at 200 hPa for 4 consecutive times per day for our case study, with the southern extension of the polar jet to the east of the Mediterranean Sea.

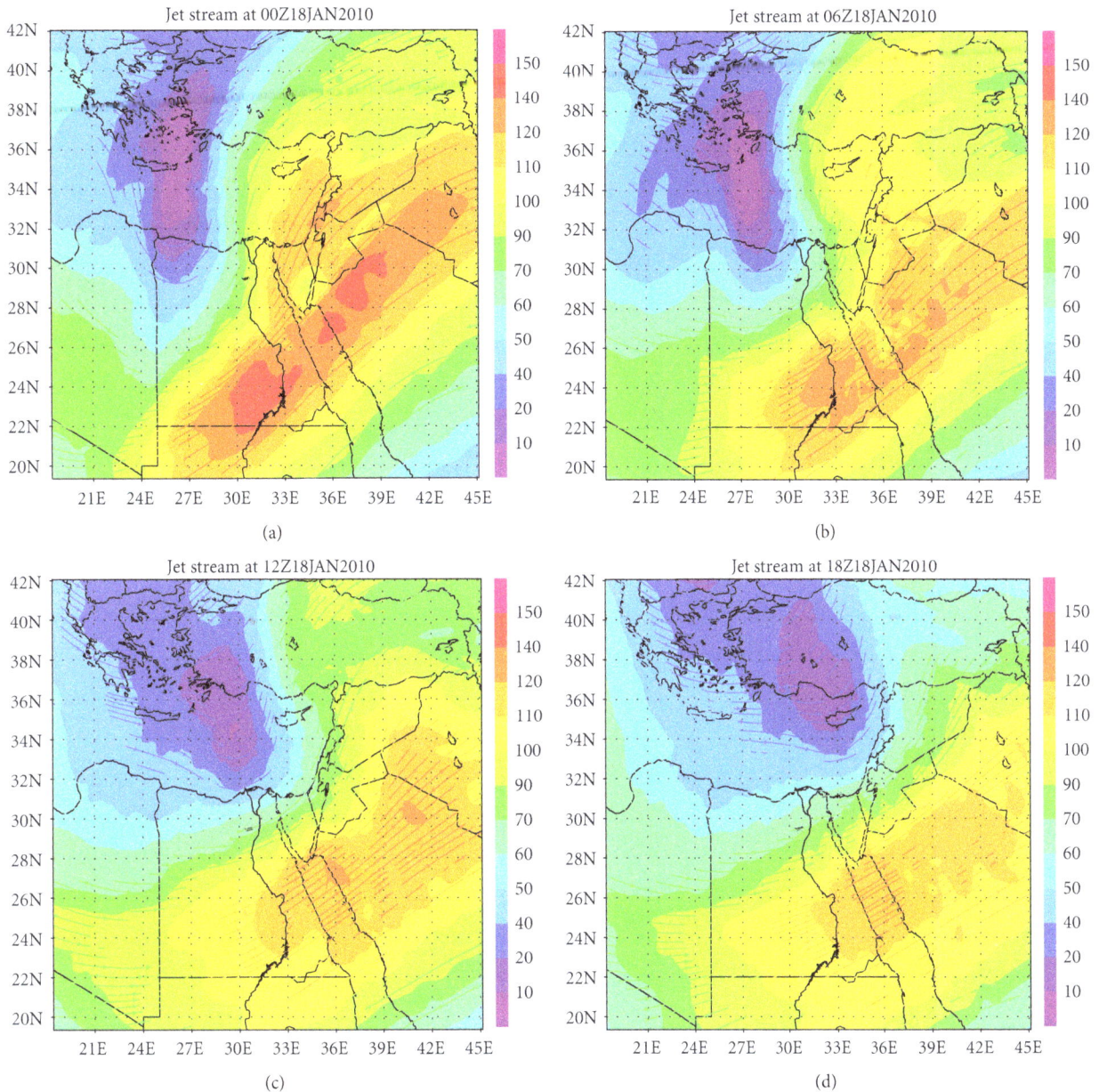

FIGURE 3: 200 hPa subtropical Jet stream (knots) at over northeastern Africa with the southern extension of the polar jet to the east of the Mediterranean at January 18, 2010.

The synoptic situation can also be analyzed by referring to the midtroposphere. Figure 4 shows the geopotential height at 500 hPa level for the same day; the synoptic system at 500 hPa level also follows the subtropical jet development.

The axis of the trough which is located over east Europe shifts southeast. This trough line orientation indicates that the affected area has extended and engulfed the whole eastern Mediterranean area. This development agrees with Krichak et al. [8, 9] and Krichak and Alpert [10] who showed that the meridional orientation of the upper trough line represents the major synoptic process of the development of the East Mediterranean cyclone. As the subtropical jet stream shifts south and splits, the system with high geopotential heights

moves eastward and allows the development of a low system in the eastern Mediterranean.

The relative vorticity advection is considered as one of the most important parameter that leads to occurrence of heavy rainfall and flash flood events in Sinai Peninsula. As shown in Figure 5, the positive relative vorticity advections leads to move the upper atmospheric trough from west to east (steering of trough) and reached to Sinai Peninsula.

From Figure 6, we can also follow the temperature advection that calculated using centered finite difference scheme of second order. In the stage of rapid development, there is a cooperative interaction between the upper level and surface flows; strong cold advection is seen to occur west of

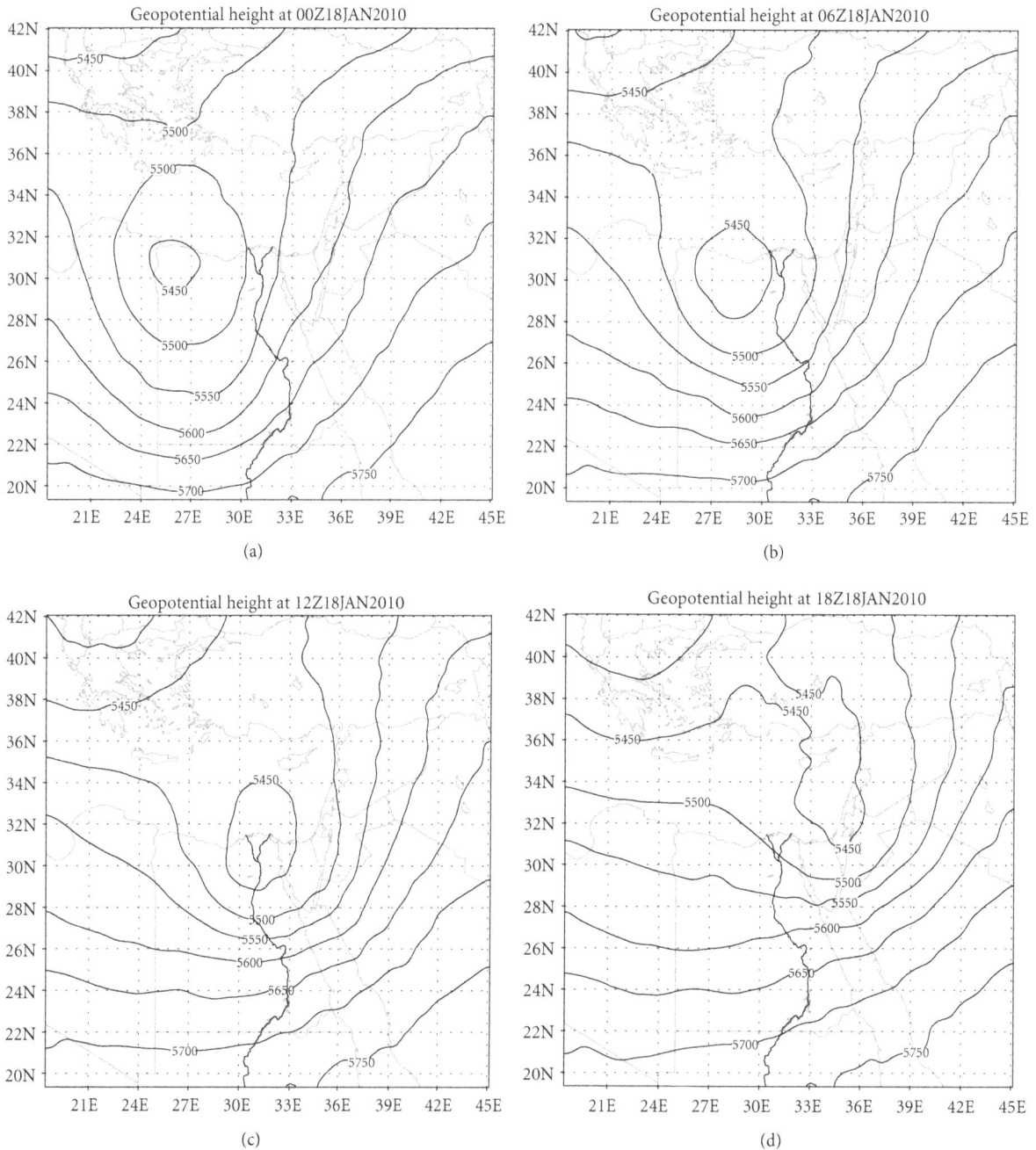

FIGURE 4: Geopotential height (m), at 500 hPa level every six hours at January 18, 2010.

the trough at the surface, with warm advection to the east. This pattern of thermal advection is a direct consequence of the fact that the trough at 500 hPa lags (lies to the west of) the surface trough. These are consistent with the results of Charney [11], Holton [6, 12], and Bluestein [13], where the growth of midlatitude cyclone is associated with a meridional gradient of temperature. We can expect that the cold advection on the west of the trough is responsible for the decreasing of temperature over Egypt in that period.

The mean sea level pressure (MSLP) of the case study is shown in Figure 7 and follows the system development at

the 850 hPa level as shown by the heat advection (Figure 6). The intensification of Azores high, to the west of the region, leads the low system to be developed in the eastern of the Mediterranean region. In addition, this produces a larger air mass temperature difference between the east and west of the low trough line and allows the Red Sea trough to extend northward. The Red Sea trough extension and the meridional orientation of the upper trough permit favorable conditions for the formation and development of the eastern Mediterranean cyclone.

Figure 8 shows the total precipitation associated with the east Mediterranean storm. The rainfall amount increases as

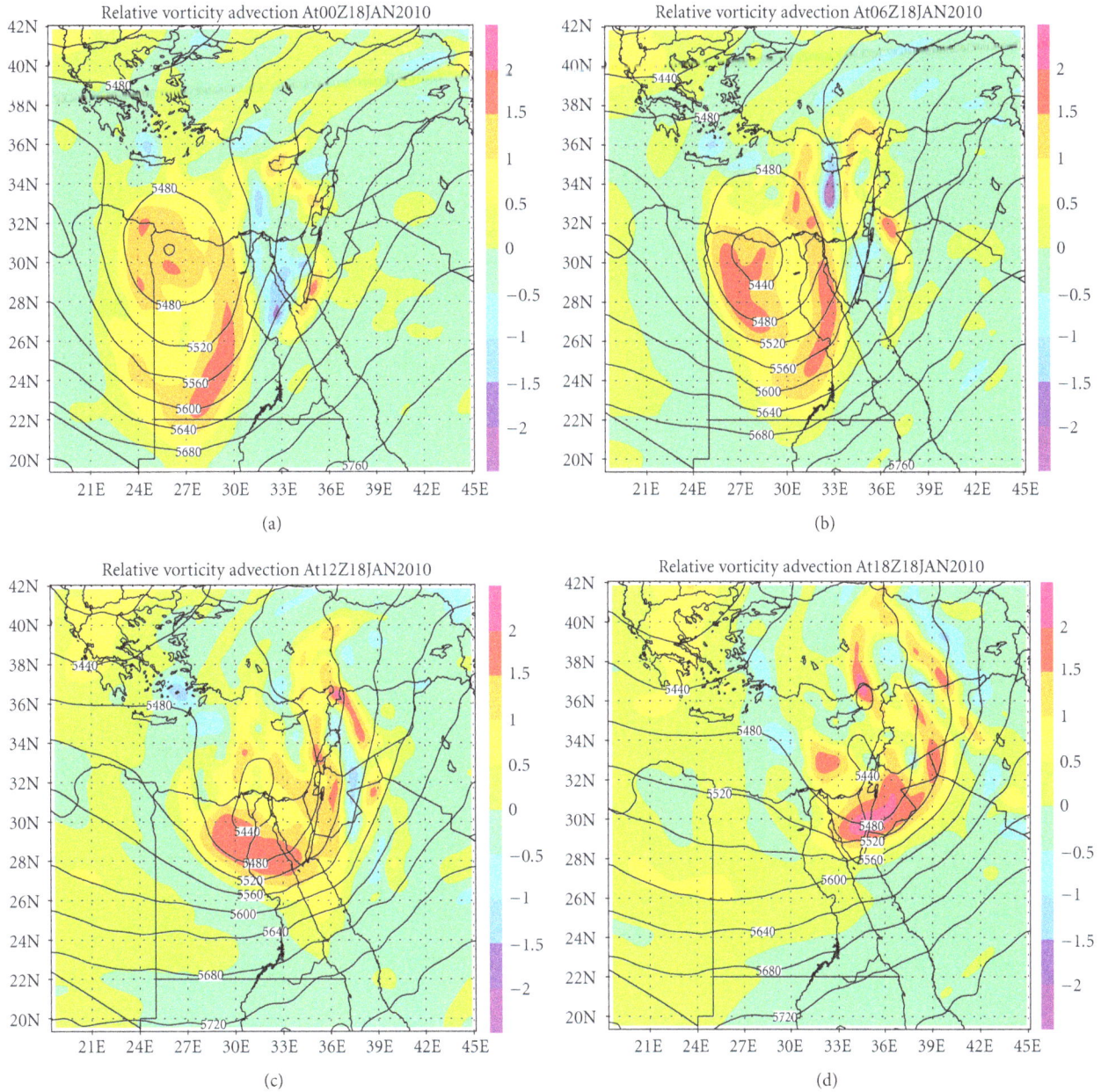

Figure 5: Relative vorticity advections ($10^{-5}\,S^{-1}$) with geopotential height contour at 500 hPa level every six hours at January 18, 2010.

the low-pressure system deepens and the precipitation area extends eastward following the surface low.

Evaluation of the performance of WRF model for prediction of heavy rainfall events over Sinai Peninsula will be done by comparison of the rainfall predicted by the WRF model and the rain gauges measurements; the complete information of the rain gauges is shown in Table 2.

One of the very important statistical tools is the root mean square error (RMSE) which gives a good overall measure of model performance. The weighting of (prediction-observation) by its square tends to inflate RMSE, particularly when extreme values are present. With respect to a perfect model, the root mean square error should approach zero.

Table 2: Rain gauges name and location which used in the comparison.

No.	Station name	Lon.	Lat.
1	El-Rawafaa	34° 08′ 51″	30° 49′ 37″
2	El-Gudairate	34° 24′ 35″	30° 38′ 28″
3	El-Themed	34° 18′ 19″	29° 40′ 37″
4	El-Haithy1	34° 36′ 36″	29° 29′ 11″
5	El-Haithy2	34° 42′ 24″	29° 28′ 04″
6	Ras-Shira	34° 27′ 57″	29° 31′ 02″

On the other hand, the mean bias (MB) is considered also to evaluate the model performance which represents the

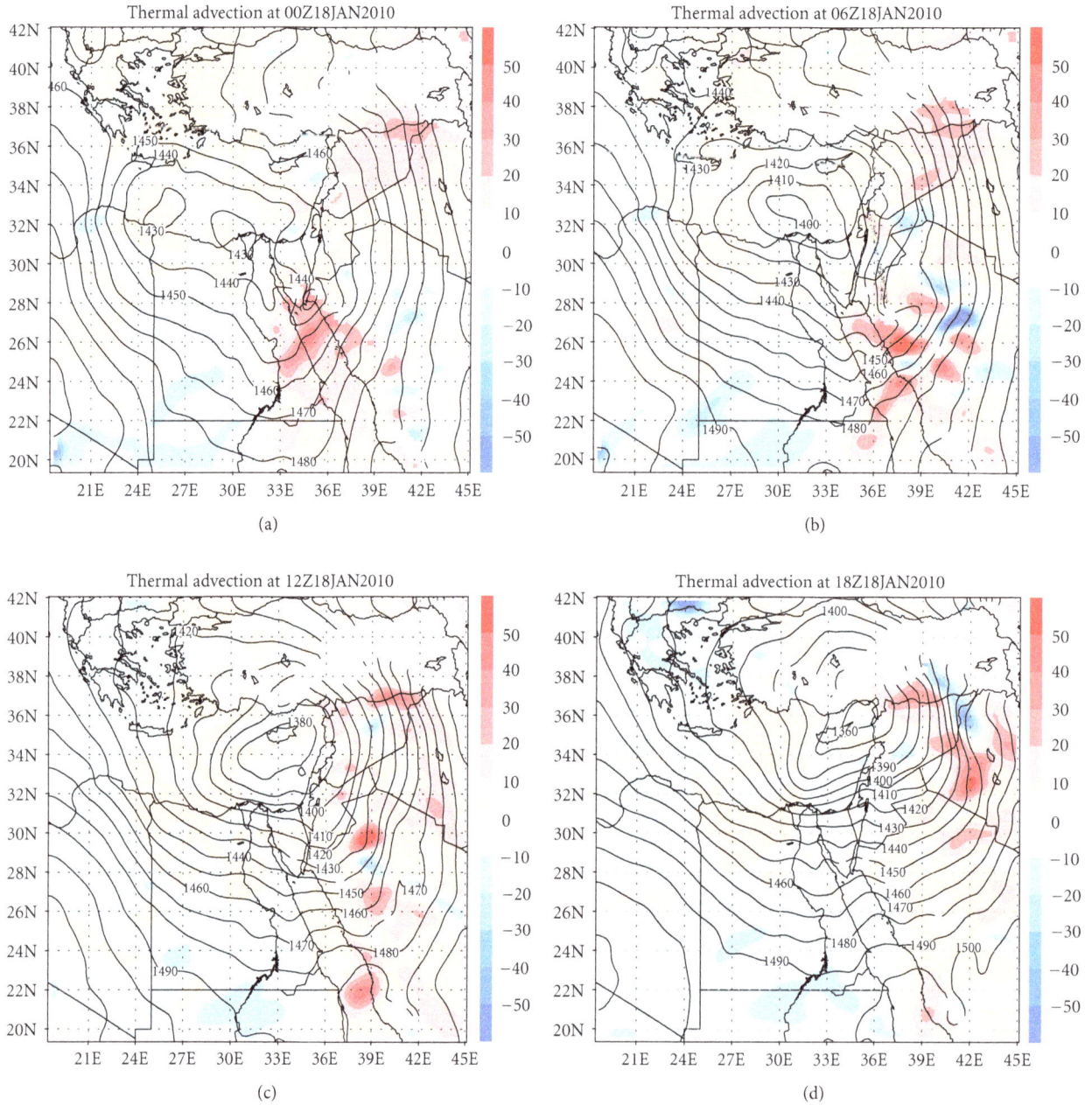

FIGURE 6: Thermal advection (°C) with geopotential height contour at 850 hPa level every six hours at January 18, 2010.

degree of correspondence between the mean prediction and the mean observation. Lower numbers are best and values less than 0 indicate underprediction.

The equations for RMSE, MB and its percentages are given as follows:

$$RMSE = \left(\sum_{i=1}^{n} \frac{\left(X_p - X_o \right)^2}{n} \right)^{0.5},$$

$$RMSE \% = \left(\frac{RMSE}{\overline{X_o}} \right) * 100,$$

$$MB = \sum_{i=1}^{n} X_p - \frac{X_o}{n},$$

$$MB \% = \left(\frac{MB}{\overline{X_o}} \right) * 100,$$

$$(1)$$

where n is number of observations, X_p and X_o are the predicted and observed values, respectively, and $\overline{X_o}$ is the average of observed values.

These statistical measures are used extensively for evaluation of model forecasts of precipitation (e.g., [14–16]). The

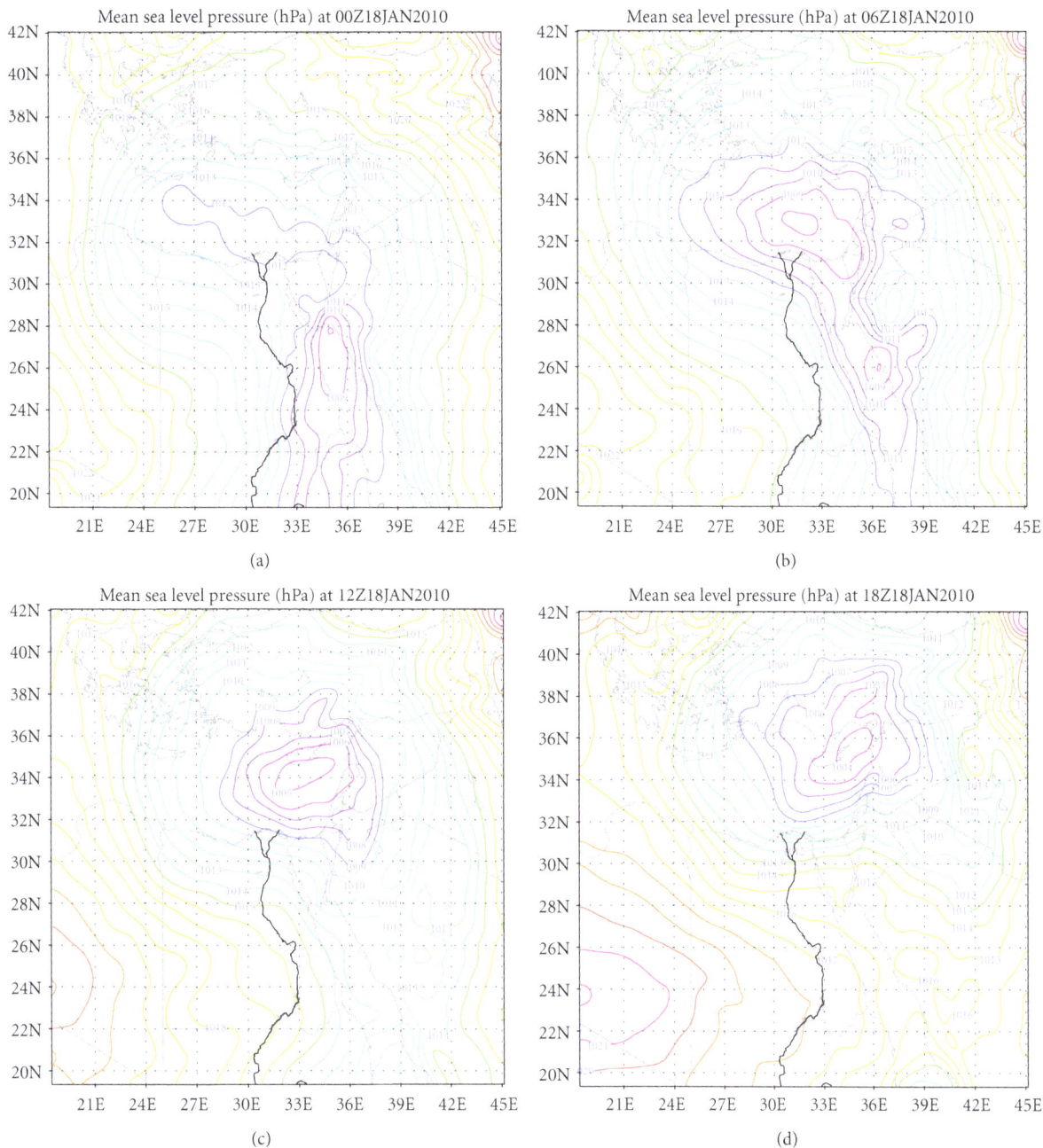

FIGURE 7: Mean sea level pressure (hPa) every six hours at January 18, 2010.

quantitative validation of the simulated precipitation is done by calculating their differences with the corresponding rain gauges.

Figure 9 shows the comparisons between the predicted hourly rainfall and the measured ones in different regions over Sinai Peninsula during January 18, 2010.

From Figure 9, it can be noticed that the predicted rainfall by WRF model is in a good consistency and harmony with observed rainfall for all rain gauge stations.

In El-Gudairate, El-Rawafaa, El-Haithy2, and El-Themed Stations rainfall prediction by WRF is very close to the measured ones. But in the other stations such as El-Haithy1

and Ras-Shira, it gave small differences with the measured rainfall during January 18, 2010.

Table 3 shows the percentages of root mean square error and mean bias for predicted rainfall and the measured ones at different stations.

It can be deduced from this table that the maximum values of root mean square error and mean bias are for El-Haithy1 station, while the minimum values of root mean square error and mean bias are for El-Gudairate, El-Rawafaa, El-Haithy2, El-Themed, and Ras-Shira Stations.

This means the WRF model has high-performance rainfall prediction for all stations except El-Haithy1 station.

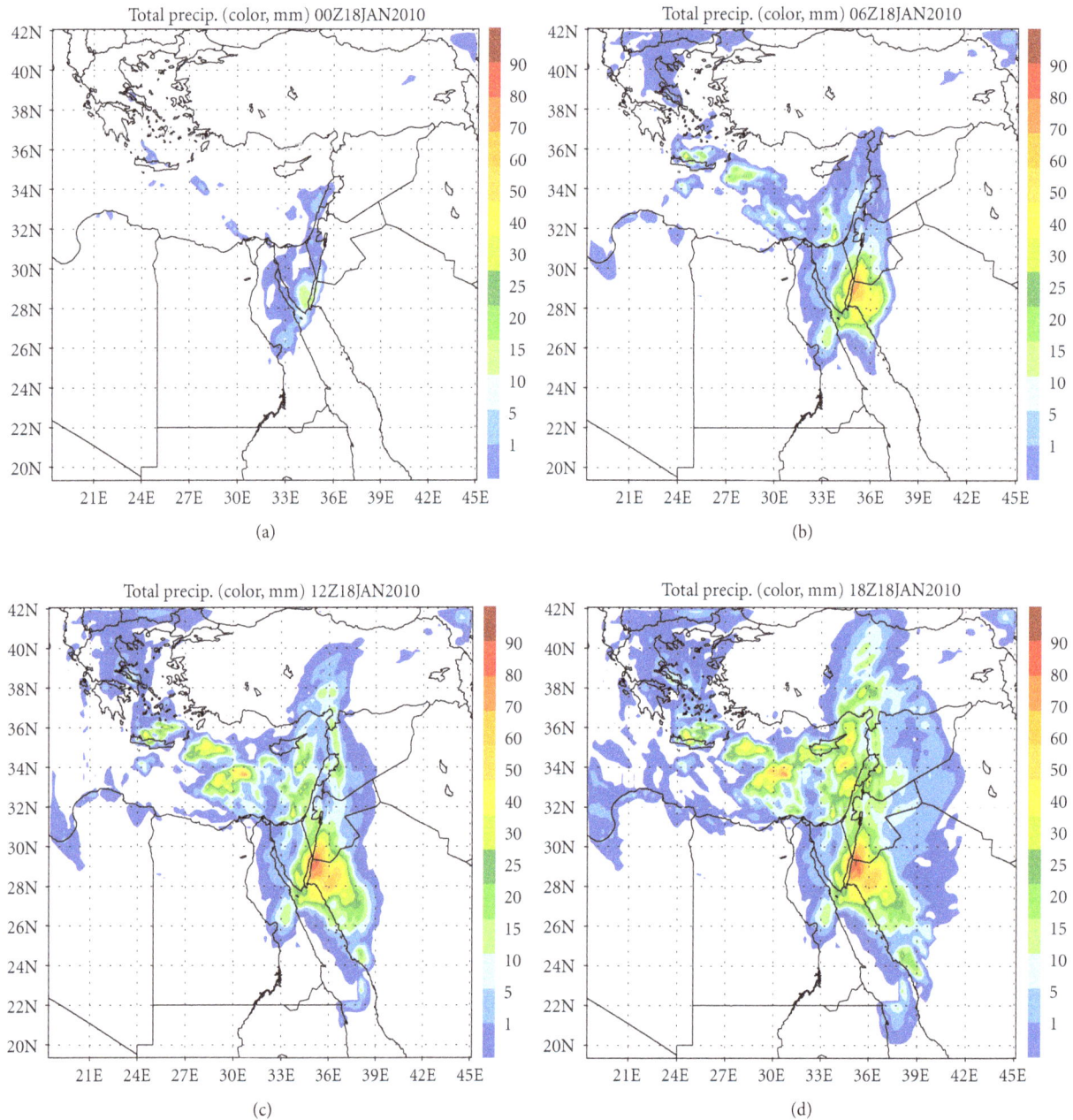

FIGURE 8: Accumulated rainfall (mm) during January 18, 2010 plotted every six hours with the maximum volume of rainfall over Sinai Peninsula.

TABLE 3: The percentages of root mean square error and mean bias for rainfall from WRF and the measurements at different stations.

Station	RMSE %	MB %
El-Gudairate	4.5	−0.3
El-Rawafaa	8.6	0.7
El-Themed	11.8	1.2
El-Haithy1	30.7	20.4
El-Haithy2	4.7	1.3
Ras-Shira	15.8	−4.2

The WRF model was able to simulate the synoptic situation which invaded Egypt and accompanied by heavy rains started at the north coast, the Red Sea, and the Sinai Peninsula during 18 January, 2010.

In the same time, the model simulation of the synoptic situation was in a good compatibility with the previous studies.

It may conclude that the performance of WRF model approximately gave a good and reasonable rainfall prediction compared to measurements at all stations over Sinai Peninsula.

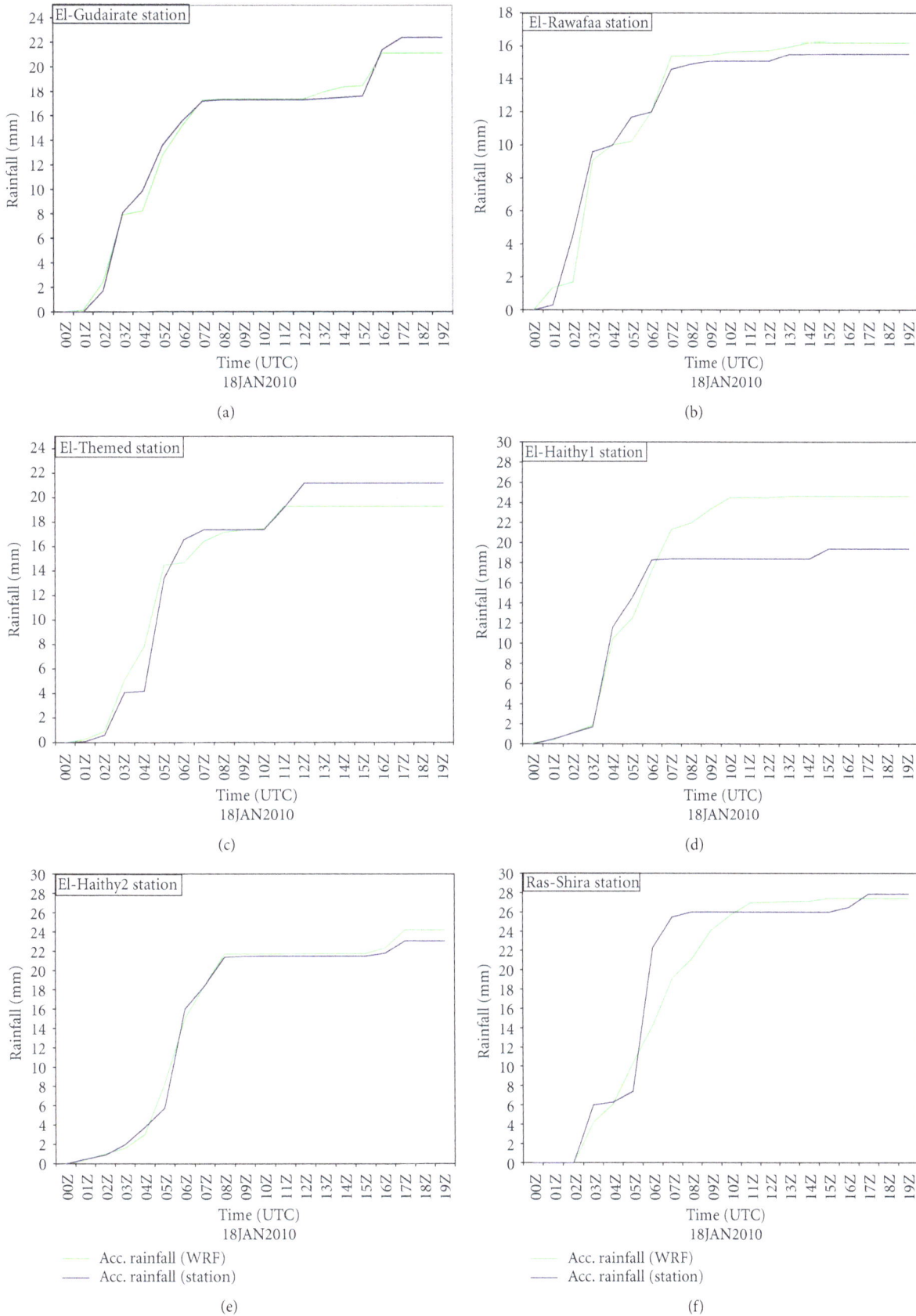

FIGURE 9: Comparison between accumulated observed and predicted precipitations (mm) for El-Gudairate, El-Rawafaa, El-Themed, El-Haithy1, El-Haithy2, and Ras-Shira stations through January 18, 2010.

5. Conclusions

The WRF model was able to simulate the synoptic situation which invaded Egypt and accompanied by heavy rains started at the north coast, the Red Sea, and the Sinai Peninsula during January 18, 2010.

In the same time, the model simulation of the synoptic situation was in a good compatibility and agreement with the previous studies over the Mediterranean region.

It may be concluded that the performance of WRF model approximately gave a good and reasonable rainfall prediction compared to measurements at all stations over Sinai Peninsula.

References

[1] D.-K. Lee, H.-R. Kim, and S.-Y. Hong, "Heavy rainfall over Korea during 1980–1990," *Korean Journal of the Atmospheric Sciences*, vol. 1, no. 1, pp. 32–50, 1998.

[2] J. Cools, P. Vanderkimpen, G. El Afandi et al., "An early warning system for flash floods in hyper-arid Egypt," *Natural Hazards and Earth System Sciences*, vol. 12, no. 2, pp. 443–457, 2012.

[3] G. El Afandi, "Developing flash-flood guidance in Egypt's Sinai Peninsula with the Weather Research forecast (WRF) model," *International Journal of Meteorology*, vol. 35, no. 347, pp. 75–82, 2010.

[4] C. A. Babu, A. A. Samah, and H. Varikoden, "Rainfall climatology over Middle East region and its variability," *International Journal of Water Resources & Arid Environments*, vol. 1, no. 3, pp. 180–192, 2011.

[5] R. Kahana, B. Ziv, U. Dayan, and Y. Enzel, "Atmospheric predictors for major floods in the Negev Desert, Israel," *International Journal of Climatology*, vol. 24, no. 9, pp. 1137–1147, 2004.

[6] J. Holton, *An Introduction to Dynamic Meteorology*, Academic Press, New York, NY, USA, 2nd edition, 1992.

[7] L. Ferraris, R. Rudari, and F. Siccardi, "The uncertainty in the prediction of flash floods in the northern Mediterranean environment," *Journal of Hydrometeorology*, vol. 3, pp. 714–727, 2002.

[8] S. O. Krichak, P. Alpert, and T. N. Krishnamurti, "Interaction of topography and tropospheric flow—a possible generator for the Red Sea trough?" *Meteorology and Atmospheric Physics*, vol. 63, no. 3-4, pp. 149–158, 1997.

[9] S. O. Krichak, P. Alpert, and T. N. Krishnamurti, "Red Sea Trough/cyclone development–numerical investigation," *Meteorology and Atmospheric Physics*, vol. 63, no. 3-4, pp. 159–169, 1997.

[10] S. O. Krichak and P. Alpert, "Role of large scale moist dynamics in November 1-5, 1994, hazardous Mediterranean weather," *Journal of Geophysical Research D*, vol. 103, no. 16, pp. 19453–19468, 1998.

[11] J.G. Charney, "The dynamics of long waves in a baroclinic westerly current," *Journal of Atmospheric Sciences*, vol. 4, no. 5, pp. 136–162, 1947.

[12] J. Holton, *An Introduction to Dynamic Meteorology*, Academic Press, New York, NY, USA, 3rd edition, 1992.

[13] H. B. Bluestein, *Synoptic-Dynamic Meteorology in Midlatitudes, Vol. II: Observations and Theory of Weather Systems*, Oxford University Press, Oxford, UK, 1992.

[14] F. Mesinger, T. L. Black, D. W. Plummer, and J. H. Ward, "Eta model precipitation forecasts for a period including tropical storm Allison, Wea," *Forecasting*, vol. 5, pp. 483–493, 1990.

[15] F. Mesinger, "Improvements in quantitative precipitation forecasts with the eta regional model at the national centers for environmental prediction: the 48-km upgrade," *Bulletin of the American Meteorological Society*, vol. 77, no. 11, pp. 2637–2649, 1996.

[16] K. Lagouvardos, V. Kotroni, A. Koussis, C. Feidas, A. Buzzi, and P. Malguzzi, "The meteorological model BOLAM at the National Observatory of Athens: assessment of two-year operational use," *Journal of Applied Meteorology*, vol. 42, pp. 1667–1678, 2003.

5

Temporal Patterns of Energy Balance for a Brazilian Tropical Savanna under Contrasting Seasonal Conditions

Thiago R. Rodrigues,[1] **Sérgio R. de Paulo,**[1] **Jonathan W. Z. Novais,**[1] **Leone F. A. Curado,**[1] **José S. Nogueira,**[1] **Renan G. de Oliveira,**[1] **Francisco de A. Lobo,**[2] **and George L. Vourlitis**[3]

[1] *Instituto de Física, Universidade Federal de Mato Grosso, 78060-900 Cuiabá-MT, Brazil*
[2] *Departamento de Agronomia e Medicina Veterinária, Universidade Federal de Mato Grosso, 78060-900 Cuiabá-MT, Brazil*
[3] *Department of Biological Sciences, California State University, San Marcos, San Diego, CA 92096, USA*

Correspondence should be addressed to Thiago R. Rodrigues; thiagorangel@pgfa.ufmt.br

Academic Editor: Dimitris G. Kaskaoutis

The savanna of Central Brazil (locally known as cerrado) has a long history of land cover change due to human activity. These changes have led to the degradation of cerrado forests and woodlands, leading to the expansion of grass-dominated cerrados and pastures. Thus, the aim of this study was to evaluate the temporal variation in energy flux in areas of degraded, grass-dominated cerrado (locally known as *campo sujo*) in Central Brazil. The amount of R_n partitioned into H declined as monthly rainfall increased and reached a level of approximately 30% during the wet season, while the amount of R_n partitioned into L_e increased as monthly rainfall increased and reached a level of approximately 60% during the wet season. As a result, H was significantly higher than L_e during the dry season, resulting in a Bowen ratio ($\beta = H/L_e$) of 3-5, while Le was higher than H during the wet season, resulting in a $\beta \approx 1$. These data indicate that the energy partitioning of grass-dominated cerrado is relatively more sensitive to water availability than cerrado woodlands and forests, and have important implications for local and regional energy balance.

1. Introduction

Tropical savannas cover about 12% of the global land surface [1] and are characterized by high plant species diversity [2]. In Brazil, savanna (locally known as cerrado) covers about 24% of the territory, mostly in the central portion, and is the dominant vegetation in areas where the dry season causes prolonged periods of plant water stress [3–5]. The biodiversity of cerrado is extremely high and is estimated to be 160,000 species, including known plants, animals, and fungi [6].

Over the last few decades, cerrado has been converted to cattle pasture, and more recently, soybean and sugar cane agriculture [2, 7], and has experienced deforestation rates much higher than in the Amazon rainforest [8]. Another major threat to the remaining areas of cerrado is the decrease of the woody component due to the increase of anthropogenic fire frequency [9], converting the vegetation to a more open and shallow-rooted ecosystem. These changes have

the potential to cause multiple changes in the structure and function of cerrado [7, 10, 11]; however, this biome has received relatively little attention from researchers in comparison with tropical rainforests [12, 13].

The change in land cover has the potential to change energy partitioning, by affecting the seasonal pattern and magnitude of radiation balance and albedo [12], and energy partitioning in the form of latent [14–16], sensible [17, 18], and soil heat flux [19], which, in turn, will feedback on local, and perhaps regional, climate [20, 21]. For example, tropical forest conversion to pasture can cause a 1.5–2.0 kPa increase in vapor pressure deficit and a 5–10°C increase in soil surface temperature relative to intact forest [22]. Land cover change can also lead to an increase in the duration of the dry season, cause more rainfall to be partitioned into runoff, affect the development of the nocturnal and convective boundary layer, and destabilize regional rainfall regimes and surface water availability [12, 20].

Given the potential for land cover change to alter surface energy balance in cerrado, we evaluated the seasonal and interannual variations of energy partitioning in a degraded grass-dominated cerrado (locally known as *campo sujo*) of Central Brazil. We used Bowen ratio energy balance (BREB) methods over two consecutive years to characterize the seasonal and interannual variations in energy flux dynamics. We hypothesized that *campo sujo* cerrado would exhibit higher rates of sensible and ground heat flux than latent heat flux, especially during the dry season when surface water availability required for sustaining latent heat flux would be minimal. Little is known about the seasonal and interannual variations in energy balance for *campo sujo* cerrado, and testing these hypotheses is important for understanding how land degradation will affect energy balance.

2. Materials and Methods

2.1. Site Description. The experimental site was located in Santo Antonio de Leverger, MT, Brazil, which is 15 km south of Cuiabá (15°43′S : 56°04′W). The study site is within a grass-dominated cerrado that was degraded approximately 35 years ago after the partial clearing of cerrado woodland vegetation. According to Koppen, the climate of region is characterized as Aw, tropical semihumid, with dry winters and wet summers. Mean annual rainfall and temperature are 1420 mm and 26.5°C, respectively, and rainfall is seasonal with a dry season extending from May to September [11]. The research area is on flat terrain at an elevation of 157 m above sea level. The regional soil type is a rocky, dystrophic red-yellow latosol locally known as a Solo Concrecionário Distrófico [23].

2.2. Data Collection. A micrometeorological tower enabled the collection of data on air temperature (T_a), relative humidity (RH), wind speed (u), precipitation (P), soil temperature (T_s), soil heat flux (G), net radiation (R_n), global solar radiation (R_g), and soil moisture (q). T_a and RH were measured 5 m and 18 m above the ground level using thermohygrometers (HMP45AC, Vaisala Inc., Woburn, MA, USA). G was measured using two heat flux plates (HFP01-L20, Hukseflux Thermal Sensors BV, Delft, The Netherlands) installed 1.0 cm below the soil surface, with one placed in a sandy soil type and the other placed in a laterite soil type, which were typical of the local soil. R_n and R_g were measured 5 m aboveground using a net radiometer (NR-LITE-L25, Kipp & Zonen, Delft, The Netherlands) and a pyranometer (LI200X, LI-COR Biosciences, Inc., Lincoln, NE, USA), respectively. Precipitation was measured using a tipping-bucket rainfall gauge (TR-525M; Texas Electronics, Inc., Dallas, TX, USA). The sensors are connected to a datalogger (CR1000, Campbell Scientific, Inc., Logan, UT, USA) that scanned each sensor every 30 seconds and stored average and, in the case of P, total quantities every 30 minutes.

2.3. Data Processing. Data were collected between May 2009 and April 2011. Fluxes of latent (L_e) and sensible (H) heat were calculated over 30-minute intervals using Bowen ratio

and energy balance (BREB) techniques [24] following the guidelines and modifications described by Perez et al. [25]. Bowen ratio methods have been used for decades, and while other methods, such as eddy covariance, may be more direct and amenable to analysis of measurement error, there are objective methods that are available for minimizing errors associated with resolving small gradients in vapor pressure or temperature caused by poor instrument performance and/or atmospheric conditions [25, 26].

The balance of energy was calculated as

$$R_n = G + H + L_e, \tag{1}$$

where R_n (J m^{-2} s^{-1}) was measured by net radiometer and G (J m^{-2} s^{-1}) was the mean heat flux in the soil measured by the soil heat flux plates installed in the sandy and laterite soils. H and L_e were calculated as a function of the Bowen ratio (β)

$$\beta = \frac{H}{L_e} \tag{2}$$

which, in turn, can be calculated as a function of the air temperature (ΔT) and vapor pressure (Δe) gradients and the psychrometric constant [24]

$$\beta = \gamma \frac{\Delta T}{\Delta e}. \tag{3}$$

For (3), actual vapor pressure (e) was calculated as a function of saturation vapor pressure (e_s) and RH using (4) and (5), respectively:

$$e_s = 2.172 \times 10^7 \times e^{-4157/((T-273)-33.91)}, \tag{4}$$

$$e = \frac{RH \times e_s}{100} \tag{5}$$

while psychrometric constant (γ) was calculated as a function of the specific heat at constant pressure ($C_p = 1010$ J kg^{-1} C^{-1} according to [27]), the local atmospheric pressure ($p = 103$ kPa at the research site),

$$\gamma = \frac{C_p \times p}{0.622 \times L}, \tag{6}$$

and the latent heat of vaporization (L), which varies as a function of temperature [28],

$$L = 1.919 \times 10^6 \times \left(\frac{T + 273}{(T + 273) - 33.91} \right)^2. \tag{7}$$

With estimates of β (3), L_e can be calculated as

$$L_e = \frac{(R_n - G)}{(\beta + 1)}, \tag{8}$$

and H can be calculated as the difference between R_n, G, and L_e using (1) [25].

The criteria for accepting data collected from the Bowen ratio method were based on those described by Perez et al. [25]. Briefly, the Bowen ratio method fails when (1) sensor resolution is inadequate to resolve gradients in e and T_a, (2) stable atmospheric conditions, such as during the dawn and dusk, cause $\beta \approx -1$, and (3) conditions change abruptly leading to errors in measurement [25, 26]. Using this filtering method, physically realistic values of β can be obtained in an objective, quantitative manner which limits the potential for bias and error in estimating energy balance terms [25, 29]. Gaps in estimates of H and L_e were filled by using linear relationships between retained values of H and/or L_e and measured values of R_n-G.

The percentage of available energy (R_n) partitioned into L_e and H was determined using linear regression, where diel (24 h) average H or L_e (dependent variables) was regressed against R_n over monthly intervals. The slope of these regressions indicates the relative partitioning of R_n into H or L_e. Seasonal and annual differences between energy balance terms were statistically analyzed using bootstrap randomization techniques, where the mean and the 95% confidence interval were calculated by randomly resampling each energy flux variable time series over 1000 iterations [30].

3. Results and Discussion

3.1. System Performance. Approximately 63% of all possible β values were retained after filtering for inadequate resolution, stable atmospheric conditions, and abrupt changes in measurement conditions [25]. Independent measurements of L_e were obtained from eddy covariance in April 2011 to assess the performance in the Bowen ratio energy balance estimates. Using linear regression with the Bowen ratio estimates of L_e as the dependent variable, the mean (±95% confidence interval) intercept and slope were -13.54 ± 4.89 W/m^2 and 0.96 ± 0.05, respectively ($R^2 = 0.80$; $n = 1326$ observations). These data indicate that estimates of L_e derived from two independent measurement systems were comparable and provide confidence in the Bowen ratio time series reported here.

3.2. Seasonal and Interannual Variations in Micrometeorology. Seasonal and interannual variations in micrometeorology were large over the study period (Figure 1). For example, precipitation varied from 0 mm in June 2010 to 380 mm in March 2011, in general, the months of June–August were the driest, and the months of January–March were the wettest (Figure 1(a)). Both years had a pronounced dry season (defined as the number of months when precipitation was <100 mm/month); however, the dry season in 2010 (April–December, 9 months) was substantially longer than the dry season in 2009 (April–August, 5 months) (Figure 1(a)). The dry season in 2009 was about 1 month shorter than the long-term (30 year) average, and 2010 dry season was about 4 months longer than the long-term average [11, 31]. Normally, the rainy season commences in October–November in the Cuiaba Basin [31], indicating an early transition to the rainy season in 2009 and a late transition in 2010 (Figure 1(a)). Total rainfall was 1415 m in 2009-2010 (May 1–31 April) and 1353 mm in 2010-2011, even with large interannual differences

in dry season length. These rainfall totals are similar to the long-term average of 1420 mm for the region [11, 31].

Temporal trends in relative humidity were positively correlated ($r = 0.67$; $P < 0.05$) with trends in precipitation, with the highest values (70–80%) during the peak of the wet season and the lowest values (ca. 60% in 2009 and 45% in 2010) in August-September of the dry season (Figure 1(a)). Air temperature trends were positively correlated with RH ($r = 0.65$; $P < 0.05$); however, cross-correlation analysis indicated that RH lagged behind air temperature by approximately two months (Figure 1(a)). Air temperature typically reached a minimum in June and increased consistently at the end of the dry season (August-September) to a peak in the wet season. The increase in temperature toward the end of the dry season, coupled with corresponding increases in convection and humidity, serves as an important trigger in the transition to the wet season [31].

Seasonal patterns in solar radiation (R_g) were negatively correlated with rainfall ($r = -47$; $P < 0.05$; Figure 1(b)), due to frequent cloud cover during the wet season [31]. However, R_n was negatively correlated with R_g ($r = -0.67$), presumably because low leaf area index (LAI) during the dry season [32], when R_g was at a seasonal maximum, causes a higher proportion of R_g to be reflected [33].

3.3. Average Diel Patterns in Energy Partitioning. Average diel (24 h) patterns of R_g and R_n were consistent from month to month with maximum values observed during the midday (1200 h) local time (Figure 2(a)). Diel peaks in R_g ranged from a maximum of 823 J m^{-2} s^{-1} in April 2009 to a minimum of 568 J m^{-2} s^{-1} in June 2009, while peaks in R_n ranged from a maximum of 619 J m^{-2} s^{-1} in April 2009 to a minimum of 407 J m^{-2} s^{-1} in June 2009 (Figure 2(a)). However, peaks in average diel R_g were typically lower during the wet season, while peaks in average diel R_n were higher in the wet season, which is consistent with patterns observed in average monthly R_n and R_g described above (Figure 1(b)). These seasonal dynamics were likely due to variations in cloud cover, and LAI was described above [31–33].

Average diel trends in H and L_e followed average diel trends in radiation closely (Table 1); however, seasonal variations in energy fluxes were large (Figure 2(b)). For example, coefficients of determination (r^2) for linear regressions between R_n and H were typically >0.94, while r^2 values for linear regressions between R_n and L_e were typically >0.87, expect during the peak of the dry season (August and/or September) when water limitation caused reductions in L_e (Table 1; Figure 2(b)). While diel variations in R_n controlled the diel variations in L_e and H (Table 1), the proportion of energy partitioned into L_e or H varied depending on rainfall. Substantially more R_n was partitioned into L_e during the wet season, while substantially more R_n was dissipated by H during the dry season (Table 1). Midday peaks in H exceeded peaks in L_e, at times by as much as 2-fold, during the dry season, but, during the wet season, the midday peak in L_e exceeded the peak in H by the same amount (Figure 2(b)). The relative difference in the amount of R_n partitioned into L_e and H was found to be in part controlled

(a)

(b)

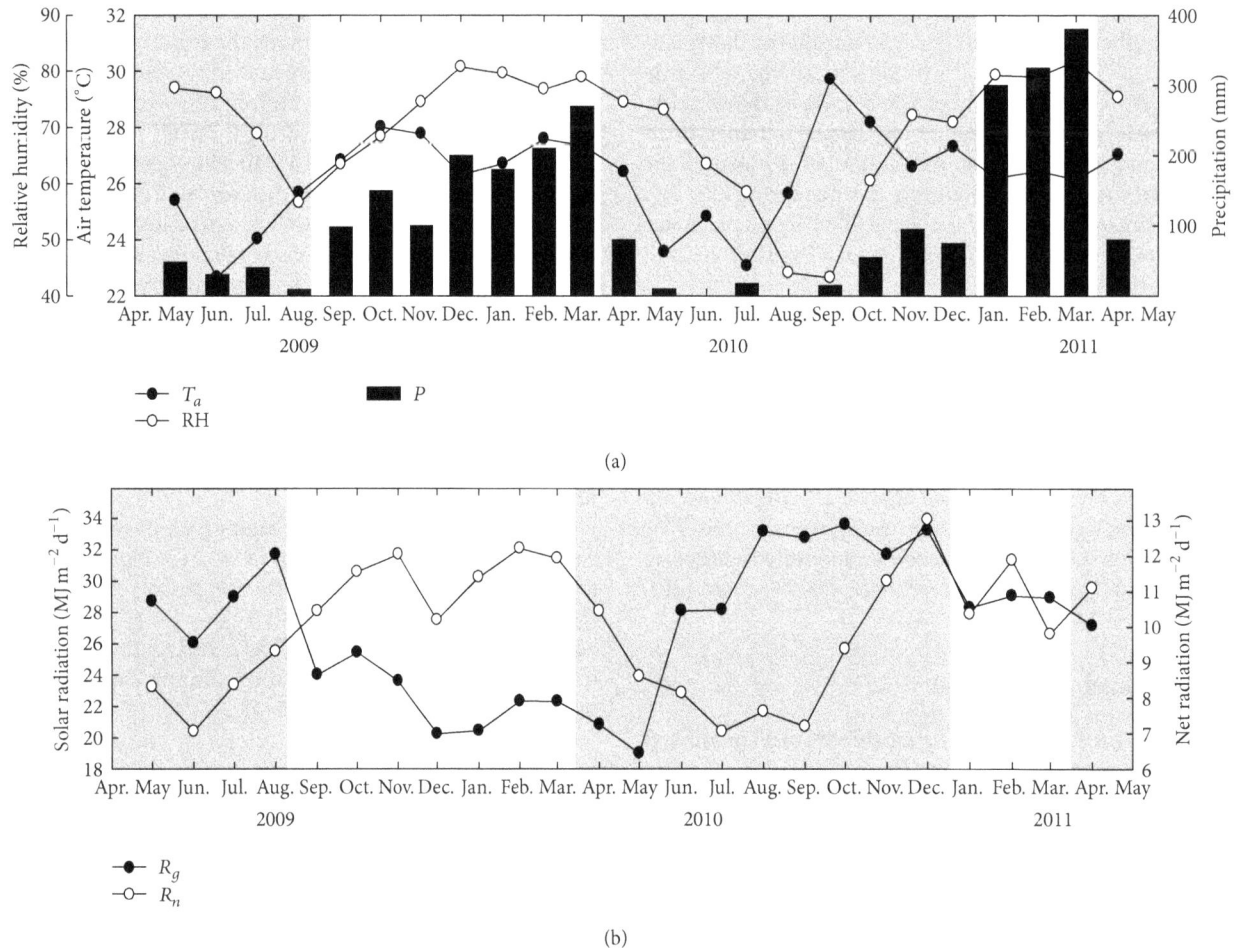

FIGURE 1: (a) Average monthly air temperature (T_a; closed symbols), relative humidity (RH; open symbols), and total monthly precipitation (P; closed bars); (b) average monthly solar radiation (R_g; closed symbols) and net radiation (R_n; open symbols) for the *campo sujo* cerrado at Fazenda Miranda during the study period. The shaded areas define the climatological dry season defined as the consecutive months when precipitation < 100 mm/month.

by monthly rainfall (Figure 3). For example, the proportion of R_n partitioned into L_e increased as rainfall increased up to 100 mm/month and then leveled off at approximately 0.50 at higher monthly rainfall rates (Figure 3(a)). Similarly, the proportion of R_n partitioned into H declined as rainfall increased to approximately 100 mm/month and then stabilized at 0.30 at higher monthly levels of rainfall (Figure 3(b)). These seasonal dynamics in energy partitioning are consistent with data reported from topical pastures, grass-dominated savanna, and semiarid temperate ecosystems [33–36] but are much more variable compared to cerrado woodlands and tropical forests [11, 14, 37–40].

Diel variations in G also followed average diel trends in radiation closely (Figure 2(b)), but, as with H and L_e, there were large seasonal variations. G exhibited wider diel variations during the dry season, ranging from $-60\,\mathrm{J\,m^{-2}\,s^{-1}}$ at night to as much as $200\,\mathrm{J\,m^{-2}\,s^{-1}}$ during the day, while during the wet season G ranged from approximately $-30\,\mathrm{J\,m^{-2}\,s^{-1}}$ at night to on average $100\,\mathrm{J\,m^{-2}\,s^{-1}}$ during the day (Figure 2(b)). Such large seasonal fluctuations in G reflect the seasonal

variations in soil thermal conductivity, which are affected by variations in rainfall and soil moisture, and seasonal variation in vegetation coverage, which influences exposure of soil to R_n [19]. In general, soil thermal conductivity increases with soil moisture [41], which should result in higher G during the wet season; however, increased plant growth and cover during the wet season cause shading of the soil surface, which counteracts the increase in soil thermal conductivity.

3.4. Seasonal Patterns in Energy Balance. Temporal variations in energy fluxes were large over daily, seasonal, and annual time scales (Figure 4; Table 2). During both years, H was significantly larger than L_e during the dry season, reflecting the larger surface-to-air temperature gradient [31] and the lower water availability that are typical of the dry season in south-central Mato Grosso [14]. Mean (±95% confidence interval (CI)) values of H were 4.81 ± 0.37 and $6.00 \pm 0.28\,\mathrm{MJ\,m^{-2}\,d^{-1}}$ during the 2009 and 2010 dry seasons, respectively, resulting in a mean (±95% CI) β of 3.04 ± 0.81 in 2009 and 5.17 ± 1.00 in 2010 (Table 2). These large β values are comparable to those

(a)

(b)

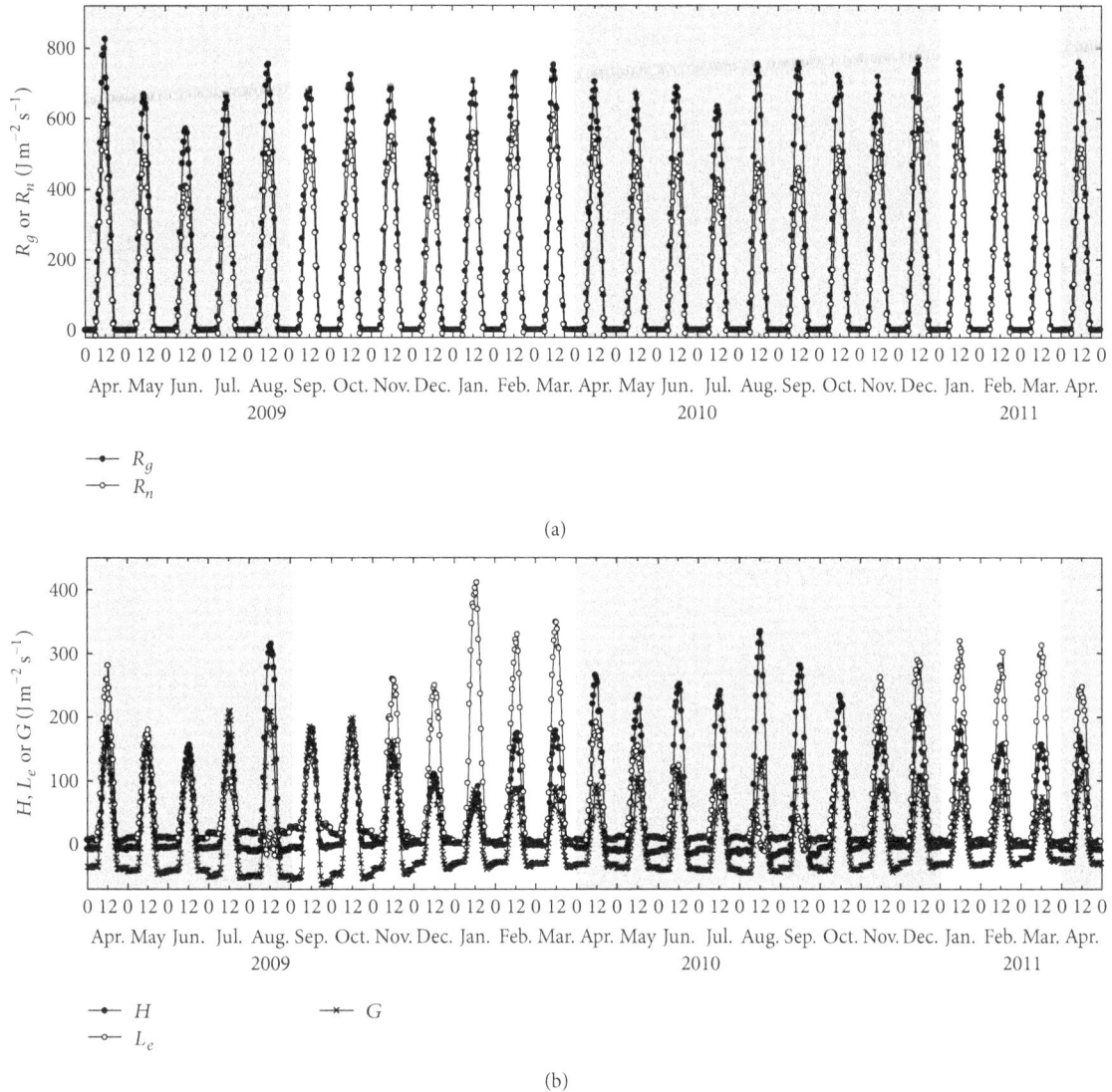

FIGURE 2: Mean monthly diel (24 hour) variation in (a) solar radiation (R_g; closed symbols) and net radiation (R_n; open symbols); (b) latent (L_e; open symbols), sensible (H; closed symbols), and ground (G; crosses) heat fluxes for the *campo sujo* cerrado at Fazenda Miranda during the study period. The shaded areas define the climatological dry season defined as the consecutive months when precipitation< 100 mm/month.

observed for grass-dominated cerrado [2, 4, 42] and semiarid temperate ecosystems [34] but substantially higher than those observed in cerrado woodlands and forests [11, 14, 43]. These variations reflect a decline in the H as the density of woody vegetation declines [2].

The significantly higher H in 2010 presumably reflected the 4-month longer dry season experienced that year (Figure 4). However, it is interesting to note that H was statistically similar in the wet and dry seasons of 2009, but, in 2010, H was significantly lower during the wet season (Table 2). The lower wet season H in 2010 appeared to be due to heavy rainfall that occurred during the 2010-2011 wet season. For example, after a long (9 months) dry season, approximately 1005 mm of rain was recorded for January–March 2011 (Figure 1), accounting for nearly 75% of all of the rainfall for the 2010-2011 measurement year.

In reality, H began to decline relative to L_e as early as November 2010, when rainfall increased but was still below the 100 mm/month threshold for the dry season (Figure 1); however, such a high amount of rainfall during the 2011 dry season would act to increase surface water availability and hence energy partitioning to L_e and decrease surface-air temperature gradients that drive H [31].

In contrast, L_e exhibited the largest and most consistent seasonal and interannual variations that were coincident with seasonal and interannual variations in rainfall (Table 2; Figure 4). The decline in dry season L_e was presumably due to declines in both transpiration and evaporation during the long dry season. For example, declines in surface water availability lead directly to a decline in surface evaporation and transpiration in shallow-rooted grasses [42]. Similarly, while cerrado trees are thought to be deeply rooted [9],

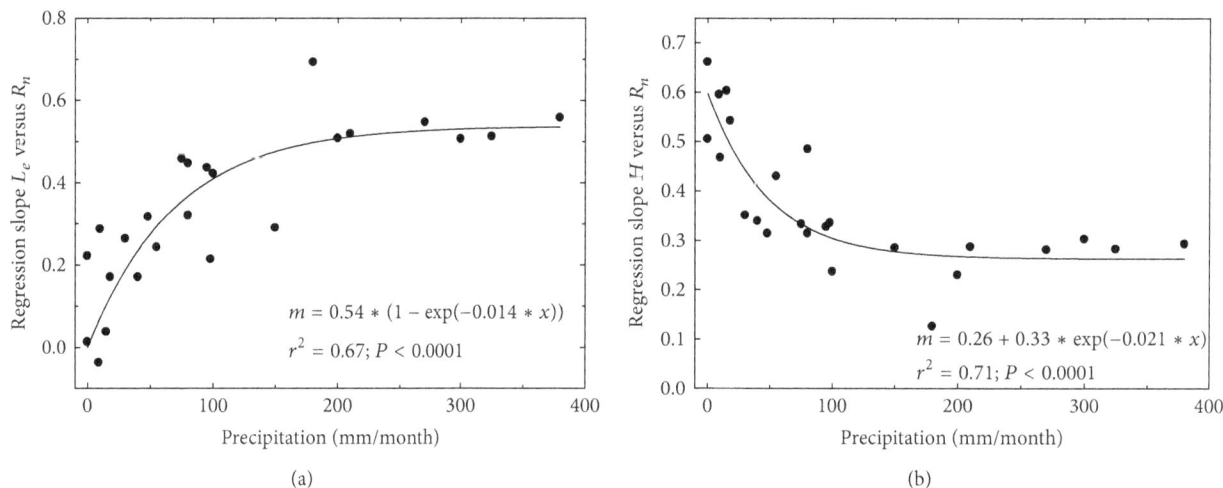

$$m = 0.54 * (1 - \exp(-0.014 * x))$$
$$r^2 = 0.67; \, P < 0.0001$$

(a)

$$m = 0.26 + 0.33 * \exp(-0.021 * x)$$
$$r^2 = 0.71; \, P < 0.0001$$

(b)

FIGURE 3: The proportion of net radiation (R_n) portioned into (a) latent heat flux (L_e) and (b) sensible heat flux (H), calculated as the slope of the slope of the linear regression between R_n (independent variable) and H or L_e (dependent variables), as a function of total monthly rainfall. Linear regression statistics are from Table 1.

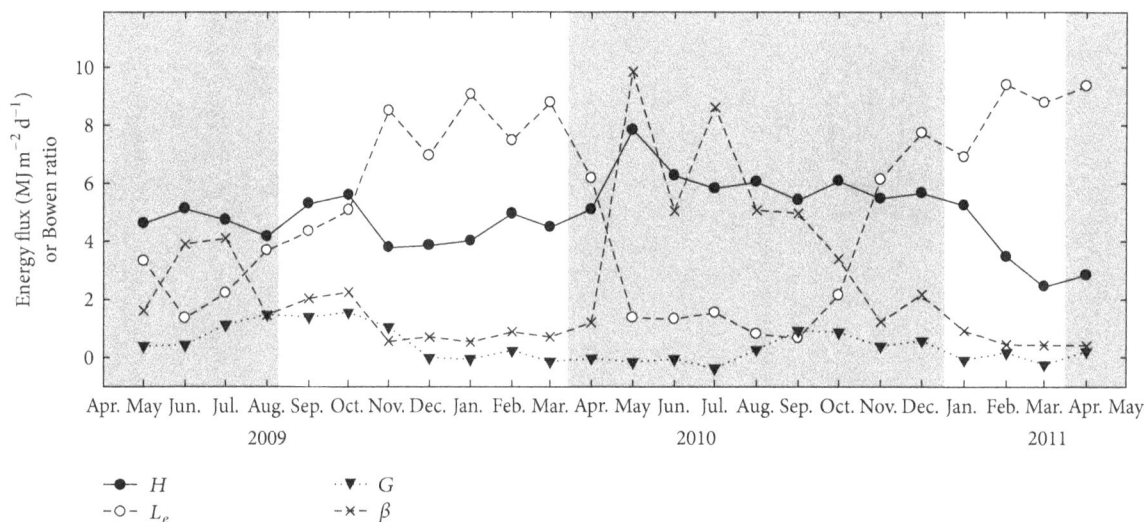

FIGURE 4: Average monthly values of latent (L_e; open symbols, dashed lines), sensible (H; closed symbols, solid lines), and ground (G; inverted triangles; dotted lines) heat fluxes, and the Bowen ratio (β; cross symbols, long-dashed lines) for the *campo sujo* cerrado at Fazenda Miranda during the study period. The shaded areas define the climatological dry season defined as the consecutive months when precipitation < 100 mm/month.

declines in root and stem hydraulic conductance with soil drying often lead to a concomitant decline in stomatal conductance, transpiration, and leaf area during the dry season [44, 45].

4. Conclusions

Energy fluxes were measured over a two-year period in a Brazilian savanna (*campo sujo* cerrado) using Bowen ratio energy balance techniques. Our data indicate that rainfall was the primary control on the partitioning of available energy (R_n) into sensible (H), latent (L_e), and ground (G) heat fluxes.

The amount of R_n partitioned into H declined as monthly rainfall increased and reached a level of approximately 30% during the wet season, while the amount of R_n partitioned into L_e increased as monthly rainfall increased and reached a level of approximately 60% during the wet season. As a result, H was significantly higher than L_e during the dry season, resulting in a Bowen ratio (β) > 1, while L_e was higher than H during the wet season, resulting in a $\beta \approx 1$. Our data are comparable to data collected from other grass-dominated cerrado ecosystems, but seasonal variations in energy fluxes are much higher in our system compared to tree-dominated cerrado and tropical forest because of the importance of water limitation to grasses and evaporation. Given that land cover

TABLE 1: Linear regression statistics for mean diel latent (L_e) and sensible (H) fluxes (dependent variables) versus net radiation for each month during the 2009–2011 study period at Fazenda Miranda.

Year	Month	L_e			H		
		Slope	Intercept ($J\,m^{-2}\,s^{-1}$)	r^2	Slope	Intercept ($J\,m^{-2}\,s^{-1}$)	r^2
2009	April	0.40	21.06	1.00	0.29	2.41	0.99
	May	0.32	23.58	0.99	0.31	5.19	0.99
	June	0.26	19.19	0.99	0.35	6.32	0.98
	July	0.17	25.20	0.98	0.34	6.46	0.99
	August	−0.04	16.74	0.52	0.59	14.01	0.99
	September	0.21	33.21	0.99	0.33	5.68	0.99
	October	0.29	26.98	0.99	0.28	6.11	0.99
	November	0.42	23.91	0.99	0.24	6.33	1.00
	December	0.51	18.39	0.99	0.23	6.16	1.00
2010	January	0.69	19.56	1.00	0.13	5.27	0.94
	February	0.52	18.56	1.00	0.29	5.41	1.00
	March	0.55	19.85	1.00	0.28	6.08	1.00
	April	0.32	18.57	0.99	0.48	5.85	1.00
	May	0.29	22.50	0.97	0.47	3.90	1.00
	June	0.22	18.96	0.93	0.50	7.48	1.00
	July	0.17	17.73	0.87	0.54	9.48	1.00
	August	0.01	11.14	0.02	0.66	13.34	0.99
	September	0.04	10.17	0.19	0.60	9.38	0.99
	October	0.24	18.54	0.98	0.43	8.68	1.00
	November	0.44	22.27	1.00	0.33	6.06	1.00
	December	0.46	20.45	1.00	0.33	4.51	1.00
2011	January	0.51	20.72	1.00	0.30	5.32	1.00
	February	0.51	20.40	1.00	0.28	5.20	1.00
	March	0.56	16.07	1.00	0.29	3.89	1.00
	April	0.45	19.99	1.00	0.31	1.74	1.00

TABLE 2: Mean (+95% confidence interval) net radiation (R_n), sensible heat flux (H), latent heat flux (L_e), ground heat flux (G), and the Bowen ratio for the dry and wet seasons and annual cycle in 2009-2010 and 2010-2011. The number of days for each season and year is shown in parentheses. Confidence intervals were calculated using bootstrap randomization techniques ($n = 1000$ iterations).

Variable	2009-2010			2010-2011		
	Dry (107)	Wet (231)	Annual (338)	Dry (229)	Wet (126)	Annual (355)
R_n ($MJ\,m^{-2}\,d^{-1}$)	8.19 ± 0.55	11.31 ± 0.56	10.32 ± 0.44	8.82 ± 0.46	11.01 ± 0.60	9.60 ± 0.39
H ($MJ\,m^{-2}\,d^{-1}$)	4.81 ± 0.37	4.63 ± 0.27	4.68 ± 0.21	6.00 ± 0.28	3.90 ± 0.50	5.25 ± 0.27
L_e ($MJ\,m^{-2}\,d^{-1}$)	2.67 ± 0.22	6.17 ± 0.32	5.08 ± 0.29	2.74 ± 0.30	7.12 ± 0.48	4.28 ± 0.32
G ($MJ\,m^{-2}\,d^{-1}$)	0.79 ± 0.26	0.56 ± 0.18	0.63 ± 0.15	0.28 ± 0.16	0.06 ± 0.13	0.20 ± 0.11
β	3.04 ± 0.81	1.04 ± 0.17	1.67 ± 0.30	5.17 ± 1.00	0.96 ± 0.27	3.68 ± 0.71

and climate changes are expected to lead to an increase in the dry season duration and a decrease in rainfall, the high sensitivity energy partitioning to water availability in grass-dominated cerrado has important implications to local and regional energy balance.

Acknowledgments

The research was supported by Universidade Federal de Mato Grosso (UFMT) Programa de Pós Graduação em Física Ambiental (PPGFA), UFMT-Grupo de Ecofisiologia vegetal (GPEV), and Coordenação de Aperfeiçoamento de Pessoal do Ensino Superior (CAPES). Special thanks are due to Dr. Clovis Miranda and his family for allowing this work to be conducted at Fazenda Miranda.

References

[1] R. J. Scholes and S. R. Archer, "Tree-grass interactions in Savannas," *Annual Review of Ecology and Systematics*, vol. 28, pp. 517–544, 1997.

[2] T. W. Giambelluca, F. G. Scholz, S. J. Bucci et al., "Evapotranspiration and energy balance of Brazilian savannas with contrasting tree density," *Agricultural and Forest Meteorology*, vol. 149, no. 8, pp. 1365–1376, 2009.

[3] R. J. Lascano, *Review of Models for Predicting Soil Water Balance. Soil Water Balance in the Sudano-Shaelian Zone*, IAHS Press, 1991.

[4] J. J. San José, N. Nikonova, and R. Bracho, "Comparison of factors affecting water transfer in a cultivated paleotropical grass (Brachiaria decumbens Stapf) field and a neotropical savanna during the dry season of the Orinoco lowlands," *Journal of Applied Meteorology*, vol. 37, no. 5, pp. 509–522, 1998.

[5] T. R. Rodrigues, L. F. A. Curado, J. W. Z. Novais et al., "Distribuição dos componentes do balanço de energia do Pantanal Mato-grossense," *Revista De Ciências Agro-Ambientais*, vol. 9, no. 2, pp. 165–175, 2011.

[6] P. S. Oliveira and R. J. Marquis, *The Cerrados of Brazil*, Columbia University Press, New York, NY, USA, 2002.

[7] C. A. Klink and A. G. Moreira, *Past and Current Human Occupation, and Land-Use. The Cerrados of Brazil: Ecology and Natural History of a Neotropical Savanna*, Columbia University Press, New York, NY, USA, 2002.

[8] C. Mueller, "Expansion and modernization of agriculture in the cerrado—the case of soybeans in Brazil's center-west," Working Paper 306, Department of Economics, University of Brasilia, Brasilia, Brazil, 2003.

[9] R. S. Oliveira, L. Bezerra, E. A. Davidson et al., "Deep root function in soil water dynamics in cerrado savannas of central Brazil," *Functional Ecology*, vol. 19, no. 4, pp. 574–581, 2005.

[10] J. A. Ratter, J. F. Ribeiro, and S. Bridgewater, "The Brazilian cerrado vegetation and threats to its biodiversity," *Annals of Botany*, vol. 80, no. 3, pp. 223–230, 1997.

[11] G. L. Vourlitis and H. R. da Rocha, "Flux dynamics in the cerrado and cerrado-forest transition of Brazil," in *Ecosystem Function in Global Savannas: Measurement and Modeling at Landscape To Global Scales*, M. J. Hill and N. P. Hanan, Eds., pp. 97–116, CRC, Boca Raton, Fla, USA, 2011.

[12] C. Von Randow, A. O. Manzi, B. Kruijt et al., "Comparative measurements and seasonal variations in energy and carbon exchange over forest and pasture in South West Amazonia," *Theoretical and Applied Climatology*, vol. 78, no. 1–3, pp. 5–26, 2004.

[13] M. Zeri and L. D. A. Sá, "The impact of data gaps and quality control filtering on the balances of energy and carbon for a Southwest Amazon forest," *Agricultural and Forest Meteorology*, vol. 150, no. 12, pp. 1543–1552, 2010.

[14] G. L. Vourlitis, J. de Souza Nogueira, F. de Almeida Lobo et al., "Energy balance and canopy conductance of a tropical semi-deciduous forest of the southern Amazon Basin," *Water Resources Research*, vol. 44, no. 3, Article ID W03412, 2008.

[15] M. A. Minor, "Surface energy balance and 24-h evapotranspiration on an agricultural landscape with SRF willow in central New York," *Biomass and Bioenergy*, vol. 33, no. 12, pp. 1710–1718, 2009.

[16] S. Chen, J. Chen, G. Lin et al., "Energy balance and partition in Inner Mongolia steppe ecosystems with different land use types," *Agricultural and Forest Meteorology*, vol. 149, no. 11, pp. 1800–1809, 2009.

[17] R. A. Memon, D. Y. C. Leung, and L. Chunho, "A review on the generation, determination and mitigation of Urban Heat Island," *Journal of Environmental Science*, vol. 20, pp. 120–128, 2008.

[18] J. L. Schedlbauer, S. F. Oberbauer, G. Starr, and K. L. Jimenez, "Controls on sensible heat and latent energy fluxes from a short-hydroperiod Florida Everglades marsh," *Journal of Hydrology*, vol. 411, no. 3-4, pp. 331–341, 2011.

[19] B. G. Heusinkveld, A. F. G. Jacobs, A. A. M. Holtslag, and S. M. Berkowicz, "Surface energy balance closure in an arid region: role of soil heat flux," *Agricultural and Forest Meteorology*, vol. 122, no. 1-2, pp. 21–37, 2004.

[20] M. H. Costa and G. F. Pires, "Effects of Amazon and Central Brazil deforestation scenarios on the duration of the dry season in the arc of deforestation," *International Journal of Climatology*, vol. 30, no. 13, pp. 1970–1979, 2010.

[21] S. K. Kharol, D. G. Kaskaoutis, K. V. S. Badarinath, A. R. Sharma, and R. P. Singh, "Influence of land use/land cover (LULC) changes on atmospheric dynamics over the arid region of Rajasthan state, India," *Journal of Arid Environments*, vol. 88, pp. 90–101, 2013.

[22] A. D. Culf, J. L. Esteves, A. O. Marques Filho, and H. R. da Rocha, "Radiation, temperature and humidity over forest and pasture in Amazonia," in *Amazonian Climate and Deforestation*, J. H. C. Gash, C. A. Nobre, J. M. Roberts, and R. L. Victoria, Eds., pp. 175–192, J. M. Wiley and Sons, New York, NY, USA, 1996.

[23] Radambrasil, Levantamentos dos Recursos Naturais Ministério das Minas de Energia. Secretaria Geral. Projeto RADAMBRASIL. Folha SD 21 Cuiabá, Rio de Janeiro, Brazil, 1982.

[24] I. S. Bowen, "The ratio of heat losses by conduction and by evaporation from any water surface," *Physical Review*, vol. 27, no. 6, pp. 779–787, 1926.

[25] P. J. Perez, F. Castellvi, M. Ibañez, and J. I. Rosell, "Assessment of reliability of Bowen ratio method for partitioning fluxes," *Agricultural and Forest Meteorology*, vol. 97, no. 3, pp. 141–150, 1999.

[26] J. Z. Drexler, R. L. Snyder, D. Spano, and U. Kyaw Tha Paw, "A review of models and micrometeorological methods used to estimate wetland evapotranspiration," *Hydrological Processes*, vol. 18, no. 11, pp. 2071–2101, 2004.

[27] J. L. Monteith and M. Unsworth, *Principles of Environmental Physics*, Arnold, London, UK, 1990.

[28] R. G. Allen, L. S. Pereira, D. Raes, and M. Smith, "Evapotranspiración del Cultivo," in *Guías Para la Determinación de los Requerimientos de Agua de los Cultivos*, p. 298, Organización de las Naciones Unidas para la Agricultura y La Alimentación (FAO), 2006.

[29] P. J. Perez, F. Castellvi, and A. Martínez-Cob, "A simple model for estimating the Bowen ratio from climatic factors for determining latent and sensible heat flux," *Agricultural and Forest Meteorology*, vol. 148, no. 1, pp. 25–37, 2008.

[30] B. Efron and R. Tibshirani, *An Introduction to the Bootstrap*, Chapman & Hall, New York, NY, USA, 1993.

[31] L. A. T. Machado, H. Laurent, N. Dessay, and I. Miranda, "Seasonal and diurnal variability of convection over the Amazonia: a comparison of different vegetation types and large scale forcing," *Theoretical and Applied Climatology*, vol. 78, no. 1–3, pp. 61–77, 2004.

[32] P. Ratana, A. R. Huete, and L. Ferreira, "Analysis of cerrado physiognomies and conversion in the MODIS seasonal-temporal domain," *Earth Interactions*, vol. 9, no. 3, 2005.

[33] M. S. Biudes, "Balanço de energia em área de vegetação monodominante de Cambará e pastagem no norte do Pantanal," Tese (doutorado)—Universidade Federal de Mato Grosso, Faculdade de Agronomia e Medicina Veterinária, Pós-graduação em Agricultura Tropical, 2008.

[34] W. Eugster, W. R. Rouse, R. A. Pielke et al., "Land-atmosphere energy exchange in Arctic tundra and boreal forest: available

data and feedbacks to climate," *Global Change Biology*, vol. 6, no. 1, pp. 84–115, 2000.

[35] N. Priante-Filho, G. L. Vourlitis, M. M. S. Hayashi et al., "Comparison of the mass and energy exchange of a pasture and a mature transitional tropical forest of the southern Amazon Basin during a seasonal transition," *Global Change Biology*, vol. 10, no. 5, pp. 863–876, 2004.

[36] M. H. Costa, A. Botta, and J. A. Cardille, "Effects of large-scale changes in land cover on the discharge of the Tocantins River, Southeastern Amazonia," *Journal of Hydrology*, vol. 283, pp. 206–217, 2003.

[37] A. C. Miranda, H. S. Miranda, J. Lloyd et al., "Fluxes of carbon, water and energy over Brazilian cerrado: an analysis using eddy covariance and stable isotopes," *Plant, Cell and Environment*, vol. 20, no. 3, pp. 315–328, 1997.

[38] Y. Malhi, E. Pegoraro, A. D. Nobre et al., "Energy and water dynamics of a central Amazonian rain forest," *Journal of Geophysical Research D*, vol. 107, no. 20, p. 8061, 2002.

[39] H. R. da Rocha, M. L. Goulden, S. D. Miller et al., "Seasonality of water and heat fluxes over a tropical forest in eastern Amazonia," *Ecological Applications*, vol. 14, no. 4, pp. S22–S32, 2004.

[40] H. R. Rocha, A. O. Manzi, O. M. Cabral et al., "Patterns of water and heat flux across a biome gradient from tropical forest to savanna in Brazil," *Journal of Geophysical Research-Biogeosciences*, vol. 114, no. 1, 2009.

[41] D. Hillel, "Thermal properties and processes," in *Encyclopedia of Soils in the Environment*, D. Hillel, C. Rosenzweig, D. Powlson, K. Scow, M. Singer, and D. Sparks, Eds., pp. 156–163, Academic Press, San Diego, Calif, USA, 2005.

[42] A. J. B. Santos, G. T. D. A. Silva, H. S. Miranda, A. C. Miranda, and J. Lloyd, "Effects of fire on surface carbon, energy and water vapour fluxes over campo sujo savanna in central Brazil," *Functional Ecology*, vol. 17, no. 6, pp. 711–719, 2003.

[43] H. R. Rocha, H. C. Freitas, R. Rosolem et al., "Measurements of CO_2 exchange over a woodland savanna (Cerrado Sensu stricto) in southeast Brazil," *Biota Neotropica*, vol. 2, pp. 1–11, 2002.

[44] S. J. Bucci, F. G. Scholz, G. Goldstein et al., "Controls on stand transpiration and soil water utilization along a tree density gradient in a Neotropical savanna," *Agricultural and Forest Meteorology*, vol. 148, no. 6-7, pp. 839–849, 2008.

[45] H. J. Dalmagro, F. A. Lobo, G. L. Vourlitis et al., "Photosynthetic parameters for two invasive tree species of the Brazilian Pantanal in response to seasonal flooding," *Photosynthetica*, vol. 51, pp. 281–294, 2013.

6

Warm Season Temperature-Mortality Relationships in Chisinau (Moldova)

Roman Corobov,[1] Scott Sheridan,[2] Kristie Ebi,[3] and Nicolae Opopol[4]

[1] *Eco-TIRAS International Environmental Association, 9/1 Independentii Street, Apartment. 133, 2060 Chisinau, Moldova*
[2] *Department of Geography, Kent State University, Kent, OH 44242, USA*
[3] *Department of Global Ecology, Carnegie Institution for Science, Stanford, CA 94305, USA*
[4] *Hygiene and Epidemiology Department, State Medical and Pharmaceutical University, 67a Gh. Asachi Street, 2028 Chisinau, Moldova*

Correspondence should be addressed to Roman Corobov; rcorobov@gmail.com

Academic Editor: Sunling Gong

Results of the epidemiological study of relationships between air temperature and daily mortality in Chisinau (Moldova) are presented. The research's main task included description of mortality dependence on different temperature variables and identification of thermal optimum (minimal mortality temperature, MMT). Total daily deaths were used to characterize the mortality of urban and rural populations in April–September of 2000–2008, excluding the extremely warm season of 2007. The simple moving average procedure and 2nd-order polynomials were used for daily mean (T_{mean}), maximum (T_{max}), and minimum (T_{min}) temperatures and mortality approximation. Thermal optimum for mortality in Chisinau (15.2 deaths) was observed at T_{mean}, T_{max}, and T_{min} about 22°C, 27-28°C, and 17-18°C, respectively. Considering these values as certain cut-points, the correlations between temperature and mortality were estimated below and above MMTs. With air temperatures below its optimal value, each additional 1°C increase of T_{mean} (T_{max}, T_{min}) was accompanied by 1.40% (1.35%, 1.52%) decrease in daily mortality. The increase of T_{mean} and T_{max} above optimal values was associated with ~2.8% and 3.5% increase of mortality; results for T_{min} were not statistically significant. The dependency of mortality on apparent temperature was somewhat weaker below MMT; a significant relationship above MMT was not identified.

1. Introduction

Mortality rates are ambient temperature dependent and have long been associated with the effects of both heat and cold. Research by epidemiologists and climatologists has grown rapidly following the European heat waves in 2003 [1–11]. That summer many western European countries experienced dramatic death tolls, and temperatures were considered as "a shape of things to come" [12].

However, while an analysis of isolated heat waves provides a useful insight into the short-term response of populations to these events, the time-series epidemiological analysis of temperature-mortality association over a long time period enables the investigation and quantification of not only general temperature-related mortality dependencies, but also additional meteorological, environmental and social confounding risk factors (e.g., [1, 13–16]). In such studies, a J- or U-shaped relationship between temperature and mortality is often identified [4, 6].

Air temperature is usually expressed in terms of its mean (T_{mean}), maximum (T_{max}), or minimum (T_{min}) values as well as the composite indices such as apparent temperature (AT) that takes into account humidity; in particular, it was shown that heat-related mortality is generally better identified when effects of high humidity are taken into account [17]. A greater response of mortality to daily T_{max} was shown by Kyselý and Kříž [18] and Michelozzi et al. [19]; Gosling et al. [4] noted that very little attention is paid to the explicit role of a diurnal temperature range.

TABLE 1: Mean air temperature and aggregated daily deaths in April–September in Chisinau.

	Years								
	2000	2001	2002	2003	2004	2005	2006	2007	2008
Air temperature	18.5	18.1	18.5	18.4	17.3	18.2	18.0	19.9	18.2
Mortality counts	2802	2748	2966	2939	2880	3115	2978	3108	3117

This paper presents a part of the comprehensive analysis of impacts on human health carried out in the framework of climatological and epidemiological justification of the development of Heat Health Warning System (HHWS) for Moldova. The research was motivated by observations of a general warming of Moldova's climate, including the record heat waves of 2007 [20], and a complete lack of national biometeorological research in the country in recent times, especially concerning the impact of elevated summer temperatures. Therefore, the main goal of this paper was to present the first modern study of the regional relationships between population mortality and a warm period temperature regime in the capital city of Chisinau.

2. Materials and Methods

2.1. Initial Data. The air temperature exposure was examined during the warm seasons (April 1 to September 30) over the nine-year study period (2000–2008).

Daily mortality data comprised total daily counts of deaths from all causes in the resident population of Chisinau. This information was retrieved from the death certificates archived at the National Center of Management in Health. The data represented both urban and rural population. As of 1 January 2009, of the 785,400 residents of Chisinau, 716,920 (91.3%) resided in the city itself, with the remainder in the suburban area. On the whole, the study encompassed about a quarter of Moldova's population at that time, including about half of the urban one.

Daily meteorological data were provided by the State Hydrometeorological Service. The daily values were calculated as the average of eight 3h measurements. The chosen period is long enough for statistical processing and does not include a significant long-term trend in air temperature (Table 1). Since the goal of research was to find the shape of relationships between ambient temperature and mortality for "typical" years, the year 2007, when the extremely intensive heat waves were recorded, was excluded.

In addition, apparent temperature (AT), as a measure of perceived exposure, was derived according to Steadman [21]:

$$AT = -2.653 + \left(0.994 * T_a\right) + \left(0.0153 * T_d^2\right), \quad (1)$$

where T_a is air temperature and T_d is dew point temperature. Because apparent temperature was one of the main exposure measures, humidity was not modeled separately.

2.2. Temperature-Mortality Relationships Identification. This research involves explaining mortality as a health outcome based upon ambient air temperature, considered as a predictor, and potentially confounding variables, for example, month (season). To identify temperature-mortality relationships, both dependent (death) and independent (daily temperature) variables were averaged over the entire sample period: 2000–2008 years without the year 2007 (hereafter, for the sake of simplicity, in all descriptions of this period it is implied that 2007 is omitted). Such averaging smoothed the possible long-term trends and year-to-year variability in data and improved "signal-to-noise" ratio.

Because mortality has its inherent seasonal cycle that is not directly related to immediate atmospheric conditions, some forms of smoothing are usually used to account for seasonal patterns [2, 13, 22, 23]. In our study, a simple moving average procedure was used to choose the optimal degree of smoothing. To identify the statistically significant differences among monthly averaged deaths, or the presence of seasonality in daily mortality, the one-way analysis of variance (ANOVA) statistical tool [24] was applied.

The different forms of regression analysis, which utilized temperature variables as independent variables and mortality as a dependent variable, serve as a reliable research tool for the adequate description of temperature-mortality relationships. Given that the temperature-mortality relationship tends to be U-shaped, and the left and right slopes of the U-like curve represent respectively the cold- and heat-related impacts of temperature, these two segments of impacts need to be accounted for separately [13, 25, 26].

To calculate these slopes separately, it is important to find the breakpoint where, while temperature increases, mortality no longer decreases, thus reaching its *minimum value*, and increases thereafter. This *thermal optimum* [8] corresponds to the average temperature with the lowest mean mortality. Some research, for example, Vigotti et al. [26], uses for this point the term "minimum mortality temperature" (MMT). As far back as in the 1990s, the Europe-wide Euro summer project [27] revealed the existence of a relatively narrow temperature band in which mortality is the lowest. This band varies substantially within Europe, the USA and other countries (e.g., [28]). Different methods are used to identify MMT. For instance, Donaldson et al. [28] calculated the temperature at which daily mortality was the lowest by computing the mean daily mortality over a range of 3°C at successive 0.1°C intervals. The upper margin of this band was taken as the temperature of heat-related mortality onset. They also found that using narrower bands gives data with excessive random variability. The same approach was used by Laaidi et al. [8] to study temperature-related mortality in France. To smooth the high variability in daily mortality, sometimes evident at higher temperatures, the 2°C class interval was used by Gosling et al. [6].

In our research, the narrow-band approach has proved itself to be a good identifier of thermal optima as well as excess death thresholds in the heat-event study [29]. We grouped daily deaths into 2°C temperature class intervals with 0.01°C increments; such increments allow preserving all

TABLE 2: Descriptive statistic of monthly averages of total daily mortality in Chisinau, 2000–2008.

Month	T_{mean} °C	Death counts								
		Sum	Average	Sd	CV, %	Min	Max	Range	Ssk	Sku
April	10.8	536	17.9 ± 0.28	1.69	9.4	14.5	20.5	6.0	−0.08	−1.11
May	16.8	517	16.7 ± 0.27	1.76	10.5	13.1	20.5	7.4	−0.09	−0.43
June	19.8	480	16.0 ± 0.28	1.64	10.2	12.8	18.8	6.0	−0.95	−0.60
July	22.6	462	14.9 ± 0.27	1.15	7.7	11.6	16.5	4.9	−1.71	0.63
August	22.4	475	15.3 ± 0.27	1.40	9.2	12.2	18.2	6.0	−0.66	−0.71
September	16.2	475	15.8 ± 0.28	1.44	9.1	13.1	19.4	6.3	1.10	0.23
Period	*16.1*	*2946*	*16.1 ± 0.26*	*1.79*	*11.1*	*11.6*	*20.5*	*8.9*	*1.53*	*−0.35*

Sd: standard deviation; CV: coefficient of variation; Ssk: standardized skewness; Sku: standardized kurtosis.

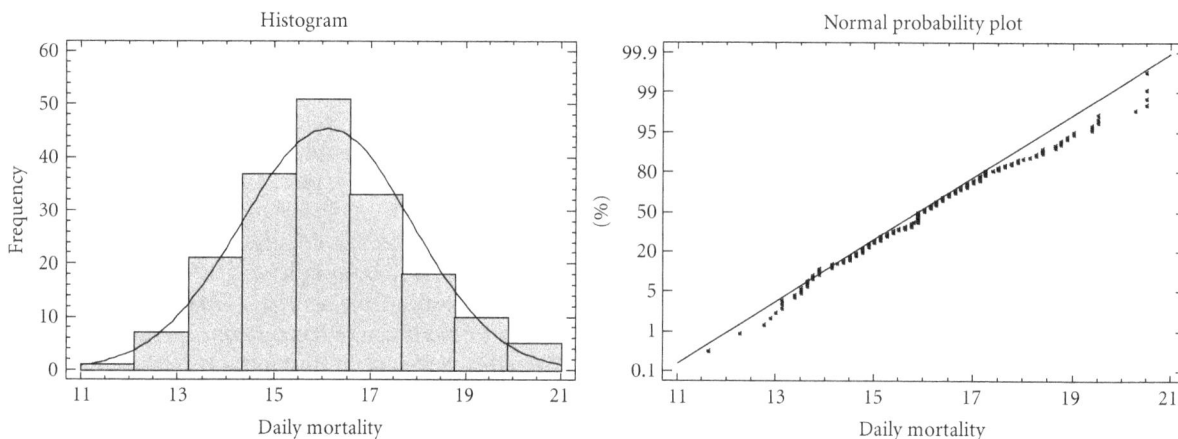

FIGURE 1: Normal distribution fitting of daily mortality in Chisinau in warm period, 2000–2008.

initial information given that air temperature is registered with 0.1°C resolution.

Statistical computations were performed, using the *StatGraphics Centurion Data Analysis and Statistical Software* [24].

3. Results and Discussion

3.1. Summary Statistics of Mortality. The average minimum mortality (about 15 deaths a day) was observed in July, the warmest month in Chisinau, and the maximum (about 18 deaths a day) in April (Table 2). Thus, the range of monthly averages is only 3 deaths; the range of their year-to-year variation is nearly double, at 6-7 deaths. Coefficient of correlation (CV), derived as the percentage ratio of standard deviation (Sd) to the average value, is 9-10%. Standardized skewness and kurtosis values within the range of −2 to +2 suggest that monthly deaths are close to being normally distributed. Across the warm season as a whole, daily death totals also follow a normal distribution (Figure 1).

Figure 2 demonstrates two outputs of ANOVA. In our case, mean mortality in April is statistically different from all other months; May is different from all months except June, and July is different from all months except August. All other combinations of monthly deaths show no significant differences between them. Both plots demonstrate clearly a seasonal course in mortality.

3.2. Comparison of the Information Content of Different Temperature Parameters. Undoubtedly, the quantification of temperature-mortality relationships requires the selection of an optimal metric. We found [30] that all daily temperature variables (T_{mean}, T_{max}, and T_{min}) and corresponding apparent temperatures (AT_{mean}, AT_{max}, AT_{min}) are highly ($r > .95$) and statistically significantly ($P < .001$) correlated. Slightly weaker correlations, though still generally statistically significant, are observed between daily ranges of air and apparent temperature. However, the correlations between observed temperature values and daily ranges are weak ($r < 0.3$-0.4).

Principal component analysis [24] yields similar results (Table 3). From eight components, only the first two with eigenvalues >1.0 can be used; together they account for 98.2% of the variability in the original data set. Weights of components show that the first one, accounting for 76.6% of general variability, is formed by direct characteristics of air temperature and apparent temperature, the second component with weight 21.6% by their diurnal ranges. The practically equal information adequacy of direct temperature variables, demonstrated by component weights, assumes *a priori* their adequacy in mortality description. This assumption is further supported by our research and is in full accord with the conclusion of Barnett et al. [31].

3.3. Identification of Mean Mortality Temperature. Figure 3(a) demonstrates the scatterplot of daily mortality

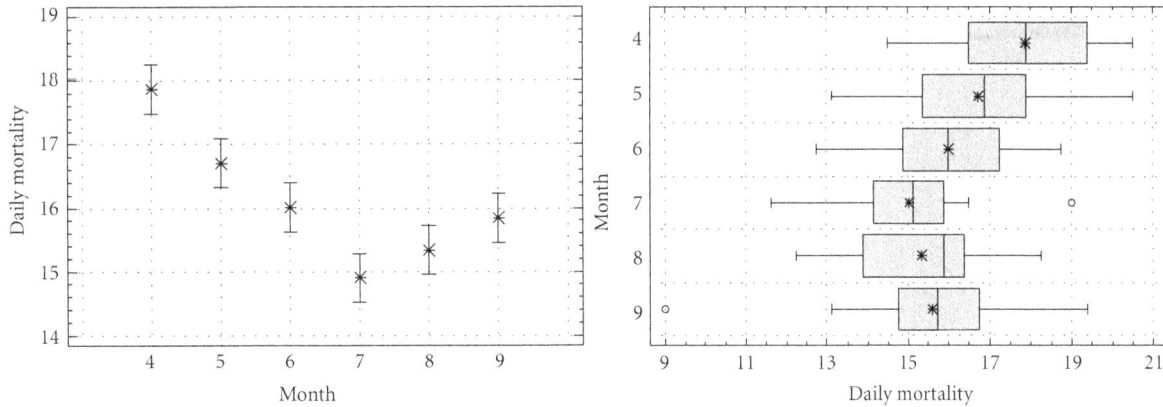

FIGURE 2: Mean (left) and Box-and-Whisker (right) Plots of daily deaths in warm period in Chisinau. (*Means plot* shows monthly death averages and Fisher's Least Significance Differences intervals. The overlapping of intervals signifies that two means are the same with 95% confidence.) (*The Box-and-Whisker Plot* divides data into four equal areas of frequency. The central boxes cover the middle 50% of the mortality, the box's sides are lower and upper 25% quartiles, the vertical line—the median, and the whiskers—the range. The means and outliers are marked as a single point (∗ and ∘, resp.)).

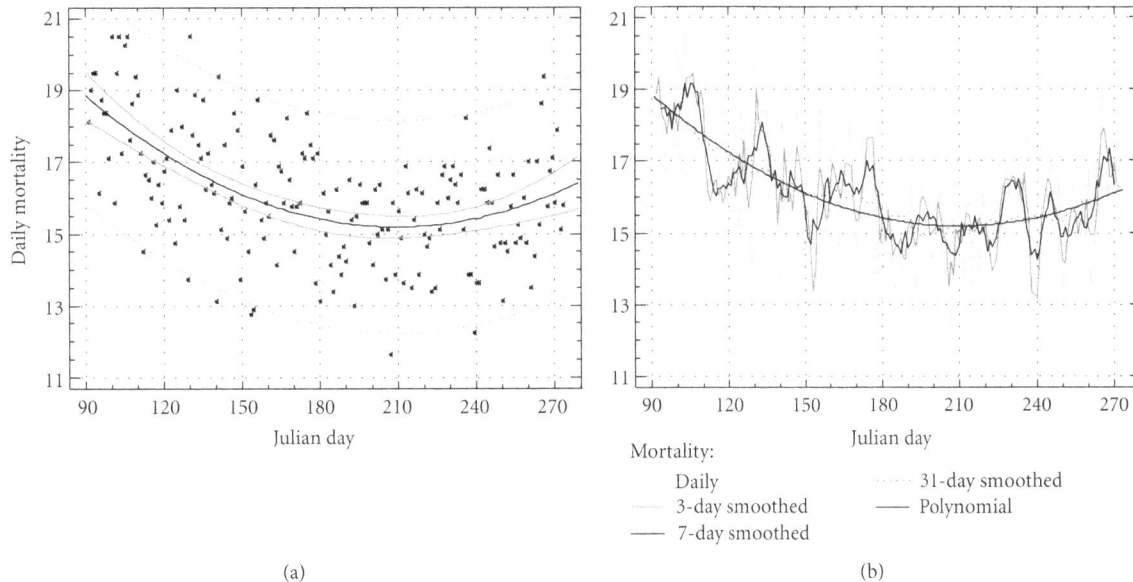

(a)   (b)

FIGURE 3: Scatter plot of warm period daily mortality in Chisinau (2000–2008), approximated by 2nd-order polynomial (a) and simple moving averages of different length (b).

TABLE 3: Principal components analysis of the set of temperature characteristics: description of components (left) and components weights (right).

Component number	Eigenvalue	Percentage		Parameter	Component weight	
		Of variance	Cumulative		Component 1	Component 2
1	6.130	76.62	76.62	T_{mean}	0.402	−0.060
2	1.725	21.56	**98.18**	T_{max}	0.403	0.020
3	0.130	1.632	98.81	T_{min}	0.392	−0.172
4	0.012	0.149	99.96	AT_{mean}	0.401	−0.091
5	0.002	0.029	99.99	AT_{max}	0.403	−0.026
6	0.001	0.009	100.0	AT_{min}	0.388	−0.205
7	0.000	0.000	100.0	T_{range}	0.165	0.668
8	0.000	0.000	100.0	AT_{range}	0.146	0.685

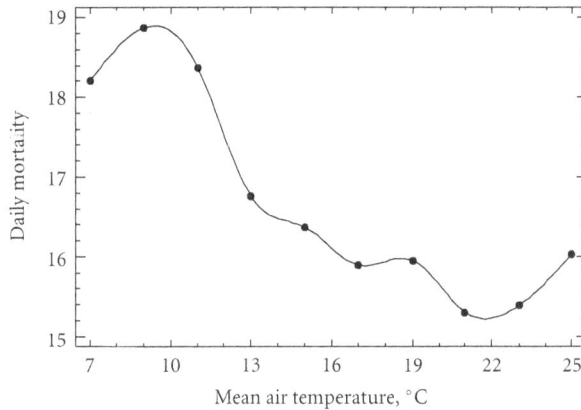

FIGURE 4: The third-order spline of daily mortality grouped by 2°C class intervals of mean daily temperature.

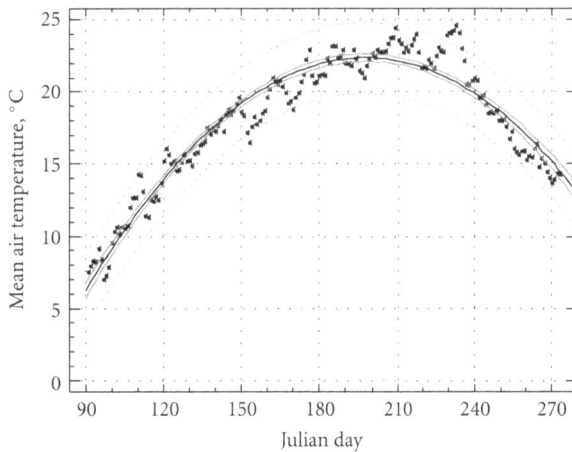

FIGURE 5: Plot of the dependence of mean air temperature on Julian day in Chisinau.

averaged for the whole period. The seasonality in data is evident and well approximated by the 2nd-order polynomial described by the equation:

$$\text{Md} = 26.47 - 0.108 * \text{Day} + 0.00026 * \text{Day}^2,$$

$$R^2 = .3072, \qquad P \leq .001, \tag{2}$$

where Md is daily total deaths and Day is day of the beginning of year (Julian day).

The polynomial curve can be considered as a hypothetical seasonal course of daily mortality with an unlimited extension of the period of observations as well as with a reasonable extension of the period of smoothing. This assumption is well demonstrated in Figure 3(b) where daily mortality is smoothed by simple moving averages of different length. As the window in which the data are averaged increases from one to 31 days, the corresponding plots approach the polynomial approximation.

We can also note that in the warm period, starting from April 1, each subsequent day the total mortality is decreasing up to a certain moment (around 210 Julian Day, Figure 3),

which could be treated as a thermal optimum or MMT, that is, the day (period) with temperature at which mortality is at a minimum. Thus, in Chisinau, using the polynomial approximation (Figure 3), the mean daily mortality (about 15.2 deaths) is observed in the late July–early August period when mean temperature reaches about 22°C. An alternative approach—the 3rd-order spline of death counts in 2°C temperature intervals (Figure 4)—shows a MMT of 21.8°C, with a mean mortality of 15.2 deaths as well.

3.4. Dependence of Daily Mortality on Mean Temperature. Over the summer, the mean air temperature is also very well approximated by a parabolic curve (Figure 5) that can be described by the equation:

$$T_{\text{mean}} = 32.06 + 0.55 * \text{Day} - 0.0014 * \text{Day}^2,$$

$$R^2 = .9204, \qquad P \leq .001, \tag{3}$$

where T_{mean} is daily mean temperature and Day is Julian Day.

The comparison of two curves (Figures 3(a) and 5) presupposes the complex, mainly inverse relationship between mortality and temperature. We can also presuppose unacceptability of estimating this relationship as uniform for the entire temperature range because such an approach ignores the differences in mortality responses to temperature increase at two slopes of the dependency curve: below and above the thermal optimum (Figure 4). A linear regression analysis of daily mortality on T_{mean} (Figure 6) shows that *cold* (descending) and *heat* (ascending) parts of the mortality curve need to be analyzed separately.

Table 4 demonstrates results of the regression analysis of relationships between daily mortality and T_{mean} below the thermal optimum. All regression models, regardless of the level of smoothing, demonstrate very similar regression parameters, showing that prognostic power of the models is low sensitive to the length of moving average. At the same time, the 7-day smoothing was selected as optimal in comparison with three other alternatives as it is associated with high correlation and small errors yet still includes the entire weekly cycle of mortality. The 31-day averaging should be rejected as an evident "oversmoothing".

Thus, in Chisinau the dependency of daily mortality (Md) on warm period's temperatures below MMT ($\text{Md}_{<\text{MMT}}$) is well described by the following equation (notations as in Table 4):

$$\text{Md}_{<\text{MMT}} = 20.17 - 0.226 * T_{\text{mean}},$$

$$r = .719, \qquad P \leq 0.001, \qquad \text{SE} = 0.80, \tag{4}$$

$$\text{MA} = 0.66.$$

The regression models estimate *per se* the mean response of mortality to ambient temperature change. In particular, the regression coefficients express human sensitivity to temperature exposure. So, proceeding from the regression coefficient (−0.226) in (4), we can state at the 99% confidence level that in spring–early summer period the increase of ambient temperature up to its optimal value, for example by 4°C,

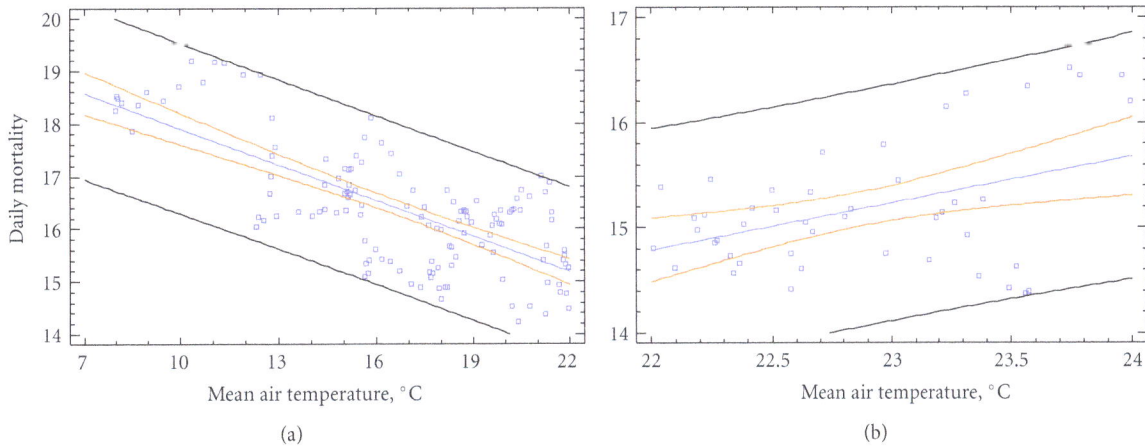

(a)   (b)

FIGURE 6: Dependencies of daily mortality (*death cases*) on air temperatures below (a) and above (b) thermal optimum. Both variables are smoothed by 7-day moving averages.

TABLE 4: Summary of simple regression analyses of daily mortality on mean air temperature below the thermal optimum for different lengths of smoothing.

Period of smoothing, days	Parameters of regressions						
	Constant	Regression coefficients	r	r^2, %	P	Standard error, SE	Mean abs error, MA
0	20.29	−0.237	−0.742	55.00	>0.001	0.80	0.65
3	20.30	−0.237	−0.740	54.71	>0.001	0.80	0.65
7	20.17	−0.226	−0.719	51.63	>0.001	0.80	0.66
31	19.70	−0.200	−0.600	36.05	>0.001	0.79	0.67

TABLE 5: Simple linear regression models of 7-day moving averages of daily mortality on analogously smoothed daily air temperatures.

Air temperature range	Parameters of regressions						
	Constant	Regression coefficient	r	r^2, %	P	Standard error	Mean abs error
	Mean temperature						
<MMT	20.17	−0.226	−0.719	51.63	>0.001	0.80	0.66
≥MMT	4.987	0.446	0.419	17.54	0.004	0.55	0.43
	Maximum temperature						
<MMT	21.18	−0.218	−0.725	52.7	>0.001	0.80	0.65
≥MMT	−1.05	0.568	0.466	21.7	0.028	0.57	0.46
	Minimum temperature						
<MMT	19.34	−0.244	−0.717	51.5	>0.001	0.81	0.67
≥MMT	11.80	0.189	0.174	3.04	0.216	0.57	0.45

TABLE 6: Simple linear regression models of 7-day moving averages of daily mortality on analogously smoothed daily apparent temperatures.

Air temperature range	Parameters of regressions						
	Constant	Regression coefficient	r	r^2, %	P	Standard error	Mean abs error
	Mean apparent temperature						
<MMT	19.36	−0.195	−0.699	48.91	>0.001	0.83	0.68
≥MMT	14.96	0.010	0.014	0.020	0.920	0.58	0.44
	Maximum apparent temperature						
<MMT	20.25	−0.191	−0.713	50.96	>0.001	0.82	0.67
≥MMT	10.99	0.150	0.198	3.92	0.187	0.59	0.46
	Minimum apparent temperature						
<MMT	18.50	−0.206	−0.690	47.58	>0.001	0.84	0.70
≥MMT	15.66	−0.027	−0.041	0.17	0.768	0.56	0.44

causes a decrease in daily mortality by one death. This also means that for average daily mortality in the observed period (16.1 deaths) a 1°C change in mean daily temperature causes 1.4% decrease in total death cases.

A somewhat different picture takes place on the ascending slope of temperature–mortality relationship after air temperatures have crossed MMT. Here, the analogous regression of daily mortality on mean temperature ($Md_{\geq MMT}$) is well described by the equation:

$$Md_{\geq MMT} = 4.99 + 0.446 * T_{mean},$$

$$r = .419, \qquad P \leq .01, \qquad SE = 0.55, \qquad (5)$$

$$MA = 0.43.$$

In other words, a mean temperature increase above thermal optimum leads to the corresponding increase of daily mortality (about 0.5 deaths per 1°C) that is practically twice more than its initial decrease caused by seasonal warming. Transformation of this value in relative daily mortality shows that high mean temperatures (those above MMT) cause on the average about 2.8% increase of daily mortality. This figure is in the range of European estimations—between 0.7% and 3.6% [32]. In Chisinau's "normal warm season" the range of these temperatures, smoothed in weekly windows (Figure 6), is only two degrees (from 22°C to 24°C). Because of the lower range of temperatures and smaller sample size, there is less certainty in the slope of the ascending limb of the regression line in comparison with that for the descending one ((5) versus (4)).

The effects of higher temperatures, describable as heat days, are considered in a special paper [29].

3.5. Maximum and Minimum Temperature as Predictors of Mortality. The splines of daily mortality, grouped by 2°C class intervals of maximum and minimum temperatures, showed their optimal values around 28°C and 18°C, respectively. These values, received as corresponding temperatures on the day of minimum mortality, were similar, 27.6°C and 17.4°C.

The parameters of regression models of daily mortality on increasing T_{max} and T_{min} are shown in Table 5 where, for easy comparison, the regressions' parameters for T_{mean} are repeated. Obviously, both variables are good predictors of a temperature-conditioned change in daily mortality for temperatures below MMT. Again, based on the warm season average daily deaths in Chisinau and regression coefficients, it can be stated that each 1°C increase of T_{max} and T_{min} up to the thermal optima causes respectively 1.3-1.4% and ~1.5% decreases in daily mortality. Thus, in the observed temperature range, mortality is slightly more sensitive to minimal night temperatures, while the sensitivity to mean and maximum daily temperatures is practically the same.

The increase of maximum temperatures above their optimum value causes an expected heightened mortality. Under such weather conditions, each additional 1°C of T_{max} yields a 2.9% increase of daily deaths—again about twice more that the positive effect of tolerable heat. The effect of minimum temperatures is not statistically significant.

3.6. Apparent Temperature versus Air Temperature in an Analysis of Mortality Sensitivity. The values of AT_{mean}, AT_{max}, and AT_{min} corresponding to minimal mortality, identified concurrently through a narrow-bands and long-term apparent temperature-mortality relationships, were respectively 21.5°C (21.9°C), 27.8°C (26.8°C), and 15.6°C (16.6°C). Given the strong correlation between air temperature and apparent temperature, their relationships with mortality below MMTs are rather similar, with slightly weaker correlation for the latter (Table 6). For all AT above MMT, no statistically significant relationships between their increase and daily mortality were found.

A somewhat weaker relationship between mortality and apparent temperature, in part explained by the relatively low humidity of the Moldavian warm season, emphasizes the necessity of direct accounting for humidity in assessing the weather-health relationships.

4. Conclusions

The statistical analysis of temperature-mortality relationships in Chisinau (Moldova) in the warm period suggests the following principal conclusions.

(1) An initial increase of ambient temperature from spring to summer months occurs with a decrease in human mortality, with its minimal mean values observed in July (14.9 ± 0.27 deaths per day). A transition from daily mortality decrease to its increase as the season progresses is observed in late July-early August. The second-order polynomial is a good describer both of the seasonal course of mortality and of its minimal value.

(2) The linear regression analysis is a good "estimator" of daily mortality dependence on air temperature in the warm season, but these dependencies must be estimated independently for descending and ascending parts of the mortality-temperature curve. A narrow-band approach, based on the distribution of daily mortality in 2°C temperature intervals is a good identifier of the minimum mortality temperature.

(3) In a warm period, due to high multicolinearity, the prognostic power of mean, maximum, and minimum temperatures is adequate. Inclusion of air humidity in the analysis (through an apparent temperature) has not resulted in the strengthening of regression models and in this case performed worse than temperature variables alone.

(4) The analysis of historical dependence of daily mortality on air temperature is a reliable tool for epidemiological studies and develops a good baseline for heat impacts early warning. At the same time, such analysis is insufficient to solve this task in full because it "hides" individual heat events, and an additional heat-episode analysis is needed.

(5) Also, the identified relationships between ambient temperatures and human mortality may not be stationary in time, being only relevant to the time

period studied (2000–2008). There are good reasons to expect that the equations derived in this work may change over time, both as a function of societal/technological changes as well as of climate change. Undoubtedly, with further accumulation of reliable daily meteorological and medical information, such research should be continued.

Acknowledgment

The research described in this paper was made possible in part by Award no. MOB-2928-CS-08 of the U.S. Civilian Research and Development Foundation for the Independent States of the Former Soviet Union (CRDF).Any opinions, finding and conclusions, or recommendations expressed in this material are those of the authors and do not necessary reflect those of the CRDF.

References

[1] R. Basu, W. Y. Feng, and B. D. Ostro, "Characterizing temperature and mortality in nine California counties," *Epidemiology*, vol. 19, no. 1, pp. 138–145, 2008.

[2] C. Carson, S. Hajat, B. Armstrong, and P. Wilkinson, "Declining vulnerability to temperature-related mortality in London over the 20th century," *American Journal of Epidemiology*, vol. 164, no. 1, pp. 77–84, 2006.

[3] U. Confalonieri, B. Menne, R. Akhtar et al., "Human health," in *Climate Change 2007: Impacts, Adaptation and Vulnerability. Contribution of Working Group II To the Fourth Assessment Report of the Intergovernmental Panel on Climate Change*, M. L. Parry, O. F. Canziani, P. J. Palutikof, P. van der Linden, and C. E. Hanson, Eds., pp. 391–431, Cambridge University Press, Cambridge, UK, 2007.

[4] S. N. Gosling, J. A. Lowe, G. R. McGregor, M. Pelling, and B. D. Malamud, "Associations between elevated atmospheric temperature and human mortality: a critical review of the literature," *Climatic Change*, vol. 92, no. 3-4, pp. 299–341, 2009.

[5] S. N. Gosling, G. R. McGregor, and J. A. Lowe, "Climate change and heat-related mortality in six cities Part 2: climate model evaluation and projected impacts from changes in the mean and variability of temperature with climate change," *International Journal of Biometeorology*, vol. 53, no. 1, pp. 31–51, 2009.

[6] S. N. Gosling, G. R. McGregor, and A. Páldy, "Climate change and heat-related mortality in six cities. Part 1: model construction and validation," *International Journal of Biometeorology*, vol. 51, pp. 525–540, 2007.

[7] G. Jendritzky and R. de Dear, "Adaptation to thermal environment," in *Biometeorology for Adaptation to Climate Variability and Change*, K. L. Ebi, J. Burton, and G. R. McGregor, Eds., pp. 9–32, Springer, 2009.

[8] M. Laaidi, K. Laaidi, and J. P. Besancenot, "Temperature-related mortality in France, a comparison between regions with different climates from the perspective of global warming," *International Journal of Biometeorology*, vol. 51, no. 2, pp. 145–153, 2006.

[9] F. Matthies, G. Bickler, N. C. Marin, and S. Hales, *Heat-Health Action Plans: Guidance*, WHO Regional Office for Europe, Copenhagen, Denmark, 2008.

[10] B. Menne, F. Apfel, S. Kovats, and F. Racioppi, Eds., *Protecting Health in Europe From Climate Change*, WHO Regional Office for Europe, 2008.

[11] C. Schär, P. L. Vidale, D. Lüthi et al., "The role of increasing temperature variability in European summer heatwaves," *Nature*, vol. 427, no. 22, pp. 332–336, 2004.

[12] M. Beniston, "The 2003 heat wave in Europe: a shape of things to come? An analysis based on Swiss climatological data and model simulations," *Geophysical Research Letters*, vol. 31, Article ID L02202, 4 pages, 2004.

[13] B. Armstrong, "Models for the relationship between ambient temperature and daily mortality," *Epidemiology*, vol. 17, no. 6, pp. 624–631, 2006.

[14] R. Basu, "High ambient temperature and mortality: a review of epidemiologic studies from 2001 to 2008," *Environmental Health*, vol. 8, article 40, 2009.

[15] R. Basu, F. Dominici, and J. M. Samet, "Temperature and mortality among the elderly in the United States: a comparison of epidemiologic methods," *Epidemiology*, vol. 16, no. 1, pp. 58–66, 2005.

[16] S. Hajat, R. S. Kovats, R. W. Atkinson, and A. Haines, "Impact of hot temperatures on death in London: a time series approach," *Journal of Epidemiology and Community Health*, vol. 56, no. 5, pp. 367–372, 2002.

[17] J. Kyselý and J. Kim, "Mortality during heat waves in South Korea, 1991 to 2005: how exceptional was the 1994 heat wave?" *Climate Research*, vol. 38, no. 2, pp. 105–116, 2009.

[18] J. Kyselý and B. Kříž, "Decreased impacts of the 2003 heat waves on mortality in theCzech Republic: an improved response?" *International Journal of Biometeorology*, vol. 52, pp. 733–745, 2008.

[19] P. Michelozzi, F. de Donato, L. Bisanti et al., "The impact of the summer 2003 heat waves on mortality in four Italian cities," *Euro Surveillance*, vol. 10, no. 7, pp. 161–165, 2005.

[20] R. Corobov, S. Sheridan, A. Overcenco, and N. Terinte, "Air temperature trends and extremes in Chisinau (Moldova) as evidence of climate change," *Climate Research*, vol. 42, no. 3, pp. 247–256, 2010.

[21] R. G. Steadman, "A universal scale of apparent temperature," *Journal of Climate & Applied Meteorology*, vol. 23, no. 12, pp. 1674–1687, 1984.

[22] S. Hajat, B. Armstrong, M. Baccini et al., "Impact of high temperatures on mortality: is there an added heat wave effect?" *Epidemiology*, vol. 17, no. 6, pp. 632–638, 2006.

[23] M. Pascal, K. Laaidi, M. Ledrans et al., "France's heat health watch warning system," *International Journal of Biometeorology*, vol. 50, no. 3, pp. 144–153, 2006.

[24] Statgraphics Centurion XVI User Manual, StatPoint Technologies, 2010, http://www.statgraphics.com.

[25] V. M. Muggeo, "A note on temperature effect estimate in mortality time series analysis," *International Journal of Epidemiology*, vol. 33, pp. 1151–1153, 2004.

[26] M. A. Vigotti, V. M. R. Muggeo, and R. Cusimano, "The effect of birthplace on heat tolerance and mortality in Milan, Italy, 1980–1989," *International Journal of Biometeorology*, vol. 50, no. 6, pp. 335–341, 2006.

[27] W. R. Keatinge, G. C. Donaldson, E. Cordioli et al., "Heat related mortality in warm and cold regions of Europe: observational study," *British Medical Journal*, vol. 321, pp. 670–673, 2000.

[28] G. C. Donaldson, W. R. Keatinge, and S. Näyhä, "Changes in summer temperature and heat-related mortality since 1971

in North Carolina, South Finland, and Southeast England,"
Environmental Research, vol. 91, no. 1, pp. 1–7, 2003.

[29] R. Corobov, S. Sheridan, K. Ebi, and N. Opopol, "Heat-related
mortality in Moldova: the summer of 2007," *International
Journal of Climatology*, 2012.

[30] R. Corobov and N. Opopol, "Some temperature-mortality
relationships in the warm season in Chisinau," *Curier Medical*,
vol. 2, pp. 35–43, 2010.

[31] A. G. Barnett, S. Tong, and A. C. A. Clements, "What measure of
temperature is the best predictor of mortality?" *Environmental
Research*, vol. 110, no. 6, pp. 604–611, 2010.

[32] WHO, *Improving Public Health Responses to Extreme
Weather/Heat-Waves—EuroHEAT Technical Summary*, WHO
Regional Office for Europe, Copenhagen, Denmark, 2009.

7

Self-Organized Criticality: Emergent Complex Behavior in PM$_{10}$ Pollution

Shi Kai,[1,2] Liu Chun-Qiong,[1,2] and Li Si-Chuan[2]

[1] *Key Laboratory of Hunan Ecotourism, Jishou University, Jishou, Hunan 416000, China*
[2] *College of Biology and Environmental Sciences, Jishou University, Jishou, Hunan 416000, China*

Correspondence should be addressed to Shi Kai; einboplure@163.com

Academic Editor: Prodromos Zanis

We analyze long-term time series of daily average PM$_{10}$ concentrations in Chengdu city. Detrended fluctuation analysis of the time series shows long range correlation at one-year temporal scale. Spectral analysis of the time series indicates $1/f$ noise behavior. The probability distribution functions of PM$_{10}$ concentrations fluctuation have a scale-invariant structure. Why do the complex structures of PM$_{10}$ concentrations evolution exhibit scale-invariant? We consider that these complex dynamical characteristics can be recognized as the footprint of self-organized criticality (SOC). Based on the theory of self-organized criticality, a simplified sandpile model for PM$_{10}$ pollution with a nondimensional formalism is put forward. Our model can give a good prediction of scale-invariant in PM$_{10}$ evolution. A qualitative explanation of the complex dynamics observed in PM$_{10}$ evolution is suggested. The work supports the proposal that PM$_{10}$ evolution acts as a SOC process on calm weather. New theory suggests one way to understand the origin of complex dynamical characteristics in PM$_{10}$ pollution.

1. Introduction

The adverse effects of PM$_{10}$ have been recognized in environmental sciences. Besides the reduction of visibility, the direct impact on human health via inhalation is an important issue [1]. It will be very useful to develop accurate PM$_{10}$ concentrations forecasting methods, which can help to put forward effective warning strategies to reduce impacts on public health during episodes or poor air quality [2]. In recent years, some PM$_{10}$ concentrations forecasting methods have been developed [3]. These methods mainly come from two approaches. One is to establish accurate atmospheric model based on meteorologic, physical, and chemical process. The other is to find inherent correlations based on the statistical analysis of the collected data. However, there are still some pending problems with predicting PM$_{10}$ concentrations. PM$_{10}$ evolution is highly complex events involving human factors as well as meteorologic, topographic, physical, and chemical conditions. Interrelationships between these processes and PM$_{10}$ concentrations are complex and nonlinear. The circumstances that determine high PM$_{10}$ concentrations

are uncertain sometimes [4]. So the microscopic physical and chemical mechanisms that drive PM$_{10}$ temporal evolutions are not well understood. However, even if such microcosmic dynamical mechanisms have been illuminated, it is likely that the system would be highly nonlinear without any simple way to predict emergent behavior [5].

As asked by Nagel [6], "is there anything we can say about these systems from first principles without knowing about the 'microscopic' details of the problem?" It is obvious that the complexity theory and related methods are needed when confronted with these complex and nonlinear problem. So many researchers have investigated the nonlinear dynamics of air pollutants evolution with different aspects [7–9]. The relational studies have observed that air pollutants concentrations exhibited long range correlation, scale-invariant, and multifractal behavior. These progresses enhance our fundamental knowledge on the complex structure of PM$_{10}$ concentrations. Thus, one has to be very careful to employ the standard statistical methods in air pollution prediction (as, e.g., ARIMA and ARIMAX models). If one wants to make robust forecasting, the first important thing to do is to identify the structure of the process. If it is approximately linear, then a linear method (e.g., an autoregressive one) can be helpful. On the contrary, if it is nonlinear, completely different types should be used. In this case, predicting models can be validated by using the information on the persistence characteristics of air pollution time series.

However, why do the complex structures of PM_{10} concentrations evolution exhibit scale-invariant?

As an introduction to the concept, self-organized criticality (SOC) has been proposed by Bak et al. [10] to provide a framework of modeling such phenomena as persistent behavior, $1/f$ noise, and scale-invariant, which are widespread in nature. The Bak-Tang-Wiesenfeld (BTW) sandpile model is a classical numerical model in SOC theory. This concept is well illustrated with a model of sandpile. One considers a grain is dropped into a sandpile randomly and slowly. As we add new grains, the pile grows more and more until the pile reaches a critical slope in a statistically stationary state. At some critical point, the addition of other grains may cause either small avalanches or trigger a very large avalanche. Some phenomena, such as power-law temporal correlations and scale-invariant, emerged from SOC state in a dissipative system. The issue of determining whether evolution of PM_{10} is governed by SOC is a difficult one. There are no unequivocal determining criteria. One approach is to compare characteristic measures of air pollution process to those obtained from a known SOC system [11, 12]. We consider that PM_{10} pollution is a complex dynamical system, which will automatically adjust itself to a critical state characterized by power-law correlations in both space and time. This state is "critical" in the sense of an equilibrium critical point where there is no characteristic length or time scale that controls the behaviour of the system. So PM_{10} pollution can be analogous to a variety of nonequilibrium relaxation processes in nature such as earthquakes and avalanches.

In this work, at first, long range correlation, $1/f$ noise, and scale-invariant of PM_{10} concentrations measured at Chengdu are examined by detrended fluctuation analysis, power spectrum, and cumulative magnitude-frequency distribution, respectively. Then, in the SOC framework, we put forward the self-organized evolution theory of PM_{10} pollution under calm condition. The SOC mechanism of PM_{10} evolution is discussed. At last, a modified sandpile model of SOC, which describes the essential mechanisms of self-organized evolution of PM_{10}, is put forwarded to simulate the scaling characteristic of PM_{10} concentrations and to illuminate the origin of scale-invariant in PM_{10} pollution. This work provides new insight and approaches to research the nonlinear dynamics and emergent complex behavior in PM_{10} evolution.

2. Materials and Methods

2.1. Data. Chengdu city is located in western Sichuan Basin of China. Sichuan Basin covers $260,000\,km^2$, generally at low altitudes of about $500\,m$. These lower lying areas are surrounded by mountains and a plateau higher than $4\,km$. The unique geographical environment directly affects meteorological condition of pollutant diffusion and increases the

frequency of calm wind at Chengdu. At Chengdu city, the average annual frequency of calm wind, namely, wind speed being $0\sim0.2\,m/s$, is 46%. So the local pollution sources play a more significant role in air quality of Chengdu [13].

There are eight automatic monitoring stations at Chengdu city. The daily average concentrations of pollutants are made at each station. These concentrations are further averaged over the stations to provide the daily average values of pollutant to represent the daily average air quality of Chengdu city. In this work, the examined data set is daily average PM_{10} concentrations data from 2001 to 2011 in Chengdu city, as shown in Figure 1. The length of data set is 4018. These data are provided by the Sichuan Environment Monitoring Center. These data are characterized by many large fluctuations with no obvious correlation that is difficult to interpret (see Figure 1).

2.2. Detrended Fluctuation Analysis. Detrended fluctuation analysis (DFA), which was proposed by Peng et al. [14], can be used to demonstrate the scaling behavior associated with SOC systems. It can avoid spurious detection of correlations that are artifacts of nonstationary, which often affects the time series data. It is that, for sample size n, the root mean square fluctuation of this integrated and detrended time series $F(n)$ behaves as a power-law function of n within the scaling region, data present scaling

$$F(n) \propto n^H. \tag{1}$$

A DFA exponent $H = 0.5$ indicates a wholly stochastic process lacking correlation; $0.5 < H \leq 1$ indicates persistent long range power-law correlations; for $0 < H \leq 0.5$, power-law anticorrelations are present such that large values are more likely to be followed by small values and vice versa. Although a DFA exponent between 0.5 and 1.0 does not absolutely prove the presence of SOC, it is a strong (in a statistical sense) indication of the presence of SOC.

2.3. Spectral Analysis. The power spectrums have been used in investigating the $1/f$ noise behavior of some time series. If a time series spectrum obeys a power-law form

$$S(f) \propto f^{-\beta}, \tag{2}$$

where f is the frequency, it indicates the absence of a characteristic time scale, that is, a scaling behavior. Thus, fluctuations at all scales are related to each other and a fractal behavior may be assumed. Malamud and Turcotte [15] have shown theoretically that one should have $\beta = 2H - 1$. However, in practice this relation is only weakly satisfied.

2.4. Cumulative Magnitude-Frequency Distribution. In some natural phenomena, cumulative magnitude-frequency distributions exhibit power-law scaling. It is regarded as the typical "critical" dynamical behavior of SOC systems [16]. A power-law applied to a cumulative distribution has the relation

$$N(c > c_0) \propto \int_{C_0}^{\infty} c^{-\tau}dc \propto c^{-(\tau+1)}, \tag{3}$$

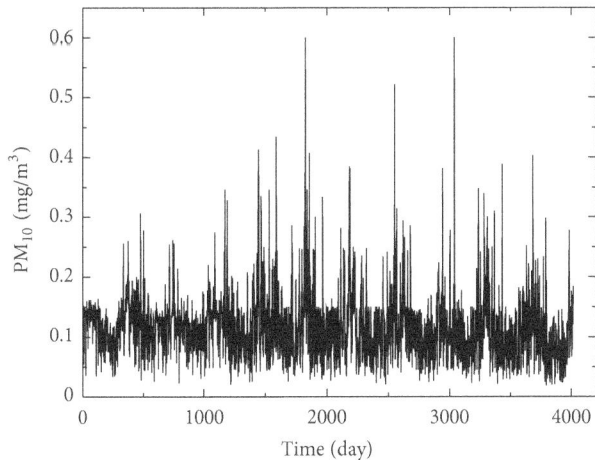

FIGURE 1: The daily average PM_{10} concentrations data from 2001 to 2011 in Chengdu city. The data lengths all are 4018 (days).

where N is the cumulative number of events with size greater than the magnitude (c) and τ is the scaling exponent.

2.5. Simplified Sandpile Model for PM_{10} Pollution.

In the study, a general sandpile model for PM_{10} pollution has been established with a nondimensional formalism.

The model is defined on a square lattice of size $L \times L$ with open boundaries in 2D. The boundaries of the lattice are open. So sands are allowed to leave the system through the boundaries. A given amount of PM_{10} pollutants $h(i, j)$ is associated with each site (i, j).

Driving mechanism: at a given time, as a consequence of pollution source emission at some site (i, j), amount of PM_{10} pollutants changes as follows:

$$h(i, j) = h(i, j) + \Delta h. \qquad (4)$$

Redistribution and relaxation mechanism: when the amounts of PM_{10} pollutants at some site (i, j) reach some threshold magnitude h_c, the site becomes unstable or critical and it relaxes by a toppling. The redistribution and relaxation rule is

$$h(i, j) \longrightarrow \frac{\Delta h}{5},$$
$$h(i \pm 1, j) \longrightarrow h(i \pm 1, j) + \left[h(i, j) + \frac{4}{5}\Delta h\right] \times 0.24, \quad (5)$$
$$h(i, j \pm 1) \longrightarrow h(i, j \pm 1) + \left[h(i, j) + \frac{4}{5}\Delta h\right] \times 0.24.$$

This rule will be circulating running in accordance with the above method until a new stable configuration is reached, namely, all $h(i, j) < h_c$. Avalanche size (s) is measured as the total number of toppling during an avalanche.

These rules represent the movements and transformation process of PM_{10} under calm condition. When PM_{10} pollutants diffuse, the site of pollution source will reserve partial pollutants, which are set to one-fifth of its original value. At the same time, owing to precipitation, adsorption, and chemical action, some PM_{10} pollutants will be lost during transportation and diffusion. In our model, we presume that

4% of PM_{10} pollutants will be lost when they topple to the four adjacent neighbor sites. So the model is local and nonconservative.

Temporal degradation mechanism: PM_{10} pollutants will decay with time owing to self-purification of atmospheric environment. We simplify this process and presume that degradation of PM_{10} follows the first level of decaying kinetics. So when a new stable configuration is reached after each relaxation rule, PM_{10} pollutants at all sites will decay to e^{-k} of the original level as follows:

$$h(i, j) \longrightarrow h(i, j) \times e^{-k}. \qquad (6)$$

After all lattice sites are stable, another grain of sand is added.

If PM_{10} evolution is an example of a SOC process, the avalanche size distribution will follow power-law distribution. In a nondimensional formalism, we select $\Delta h = 1$ and $h_c = 4$ referring to the classical BTW sandpile model.

3. Results and Discussion

3.1. Long Range Correlation as Detected by DFA Method.

Figure 2 shows the DFA for daily average PM_{10} concentrations data of Chengdu from 2001 to 2011. The observed $F(n) \propto n^H$ relationship shows obviously two different period regimes, with a critical time scale (n_c) of about one year. Fitting by the least square method the $F(n) \sim n$ plot, for $n < n_c$, $H_1 = 0.829$, while $n > n_c$, $H_2 = 0.493$. For one-year periods $(n < n_c)$, it indicates high persistence. For example, there is a tendency for increase in PM_{10} concentration to be followed by another increased tendency in PM_{10} concentration at a different time in a power-law fashion. This suggests that the correlations between the fluctuations in PM_{10} concentrations do not obey the classical Markov-type stochastic behavior (exponential decrease with time) but display more slowly decaying correlations. Over longer time periods, $n > n_c$, H_2 is close to 0.5 and indicates that the fluctuations in PM_{10} concentrations behave like the stochastic process at a large temporal scale. This phenomenon perhaps reflects an influence of the annual climate cycle.

3.2. 1/ f Noise as Detected by Spectral Analysis.

On the power spectrum plot shown in Figure 3, we can see that the power spectral density obeys two different power laws in the high-frequency and low-frequency regimes. We have power-law fits $\beta_1 = 1.021$ for 1 year$^{-1} < f <$ 1 day^{-1} and $\beta_2 = 0.583$ for $f <$ 1 year^{-1}. In shorter period, spectral analysis shows that the fluctuations in PM_{10} concentrations are characterized by $1/f$ noise and self-affine type fractal behaviors, which are similar to the results of DFA methods. For β_1 and β_2 values, we note that they are higher than the ones estimated by the relation $\beta = 2H - 1$. It shows that this relation is only weakly satisfied.

3.3. The Magnitude-Frequency Distribution of PM_{10} Concentrations.

Figure 4 shows the number (N) of PM_{10} events from 2001 to 2011, with size greater than some PM_{10} concentration (c) on a double logarithmic plot. Fitting by the least

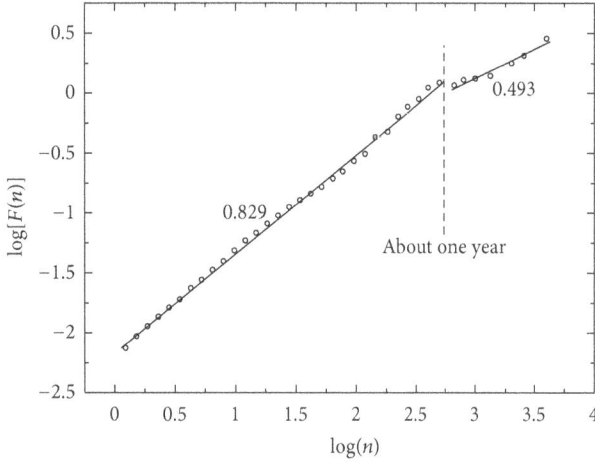

FIGURE 2: DFA of the study data. The dots are values of $\log[F(n)]$ against the corresponding $\log[n]$. The solid lines are power law $F(n) \propto n^H$, with $H_1 = 0.829$ and $H_2 = 0.493$, respectively. The vertical line indicates that n_c is about one year.

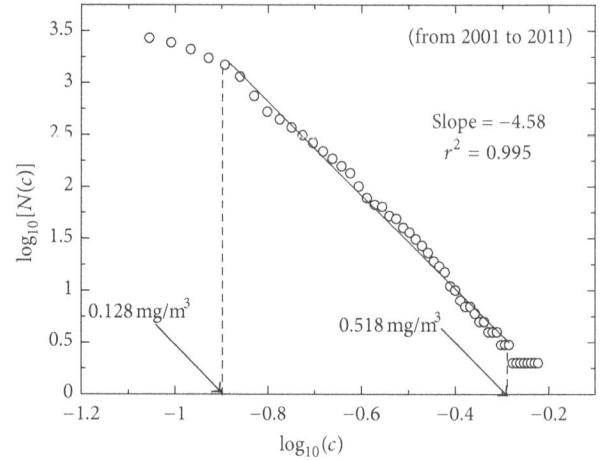

FIGURE 4: The number (N) of PM_{10} events from 2001 to 2011, with size greater than some PM_{10} concentration (c) on a double logarithmic plot.

statistical distributions all exhibit power laws and the scaling exponents are almost the same. The robustness to the time intervals is the critical behavior of SOC system.

3.4. Simulation Results of Sandpile Model. The SOC state is stationary in the sense that over long timescales, the average height $\langle h \rangle$ neither grows nor decays. The average height can be calculated according to the relation

$$\langle h \rangle = \frac{1}{L^2} \sum_{i=1}^{L} \sum_{j=1}^{L} h(i, j). \qquad (7)$$

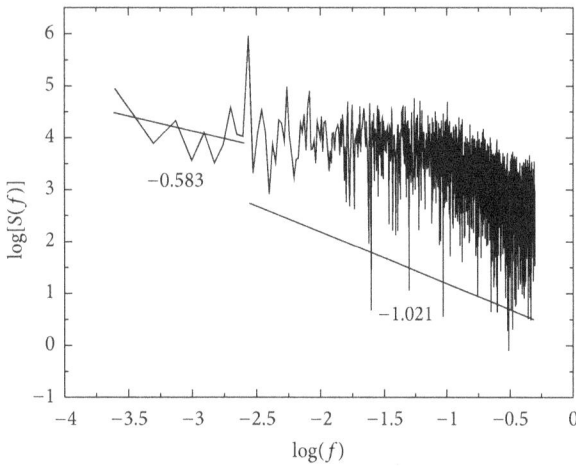

FIGURE 3: The power spectrum plot for the study data. The black lines are power law $S(f) \propto f^{-\beta}$, with $\beta_1 = 1.021$ and $\beta_2 = 0.583$ corresponding to high-frequency and low-frequency regimes, respectively.

The simulations are performed for 50×50 lattice sizes. The average height $\langle h \rangle$ is plotted against the number of avalanches up to 50000 in Figure 7 for $k = 3.5 \times 10^{-4}$. It can be seen that a constant average height is achieved and it remains constant over a large number of avalanches. The inset figure is a closeup on the average height curve, which looks like very low amplitude "ripples" propagating along the mean field. This phenomenon indicates that the SOC state is reached.

When reaching the nonequilibrium steady state, extensive data collection has been made for $L = 50$ in runs of 10^7 avalanches. The simulated result of avalanche size distribution is shown in Figure 8. We have found the value $\alpha = 3.57$ when $k = 3.5 \times 10^{-4}$ for the exponent of power-law relation $P(s > s_0) \propto s^{-(\alpha+1)}$, quite close to the exact value $\tau = 3.58$ in Figure 4. We emphasize that the parameter k is closely related to characteristics of PM_{10} pollutants in atmospheric environment. The emerged scale-invariant in the avalanches size distribution demands no external parameter tuning and should be seen as an evidence of SOC.

3.5. Possible Explanation of PM_{10} Evolution SOC. There are no unequivocal determining criteria to ascertain whether evolution of some natural phenomenon is governed by SOC. One accustomed approach is to compare characteristic

square method, the scaling exponent (τ) is 3.58 according to (1). The scale invariance region starts from $0.128 \, \text{mg/m}^3$ and ends at $0.518 \, \text{mg/m}^3$. In the PM_{10} concentrations regions, a typical scale of pollution events does not exist. There is inherent dynamical connection among the fluctuations in PM_{10} concentrations. In smaller PM_{10} concentrations regions, the power-law breaks down obviously. We consider that low monitoring frequency of PM_{10} concentrations results in the low-size tail of the frequency distribution. A similar phenomenon in rainfall has been reported by Peters and Christensen [17].

In order to investigate the robustness of the scale invariance in PM_{10} concentrations, the same analysis is performed at the different time intervals, shown in Figure 5 (from 2006 to 2011) and Figure 6 (from 2010 to 2011). These observed

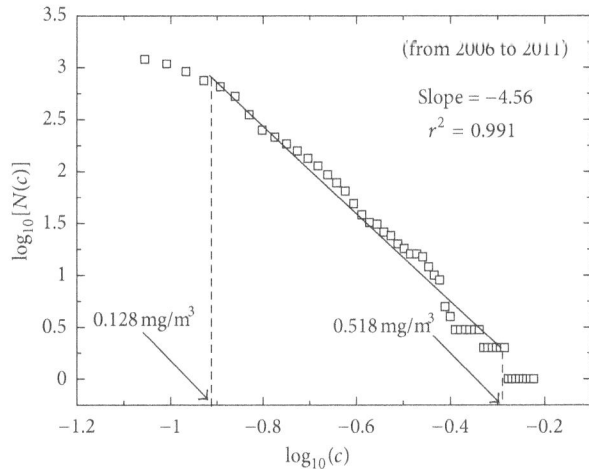

FIGURE 5: The number (N) of PM$_{10}$ events from 2006 to 2011, with size greater than some PM$_{10}$ concentration (c) on a double logarithmic plot.

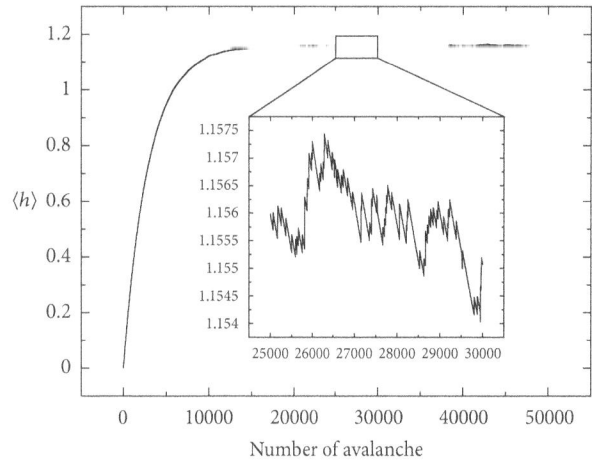

FIGURE 7: Plot of average height $\langle h \rangle$ against the number of avalanches. A closeup on the average height curve is shown in the inset.

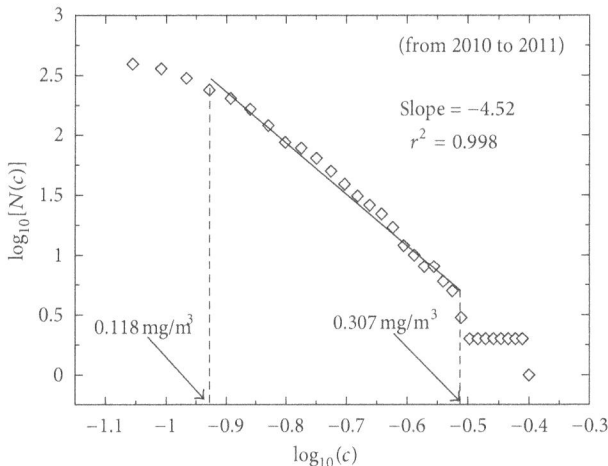

FIGURE 6: The number (N) of PM$_{10}$ events from 2010 to 2011, with size greater than some PM$_{10}$ concentration (We have found the value c) on a double logarithmic plot.

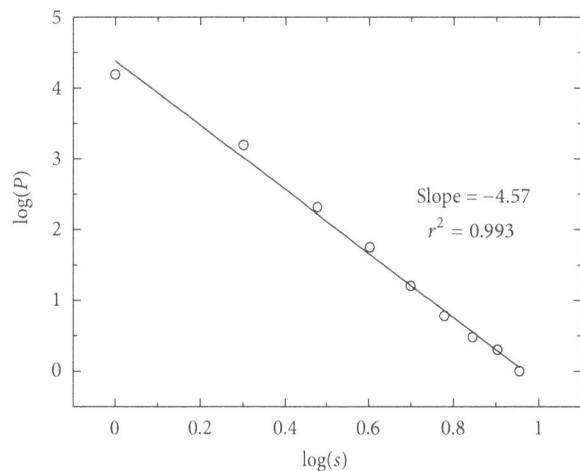

FIGURE 8: Avalanche size distribution for the PM$_{10}$ pollution sand model when $k = 3.5 \times 10^{-4}$. It follows the power-law relation $P(s > s_0) \propto s^{-(\alpha+1)}$, with an exponent of $\alpha = 3.57$ (solid line), which is consistent with that of the size distribution of PM$_{10}$ pollution events in Figure 4.

measures of some natural phenomenon to those obtained from a known SOC system.

To motivate comparisons between PM$_{10}$ pollution and SOC sandpile, we firstly take a qualitative description of the complexity in PM$_{10}$ pollution system which could give rise to SOC dynamics. PM$_{10}$ pollution system contains many components such as pollutant sources, atmospheric pollutants components, solar radiation, wind speed, temperature, atmospheric self-purification, topographical feature, and other meteorologic factors. Each component has a certain influence on the average PM$_{10}$ concentration each day. When all the components are considered together, they interact and correlate with each other on vastly different timescales. One group of comparatively fast radical chemical reactions relaxes on timescales of fractions of seconds up to hours, while another group the rather slow processes (e.g., the movement of PM$_{10}$ pollutants) relaxes on timescales of

days up to years or even longer. Thus, PM$_{10}$ pollution system is a complex system composed of a series of interconnected components. These components have some complex pattern of influence on the daily average PM$_{10}$ concentration, which result in the fluctuations of PM$_{10}$ concentrations in long term.

We make an analogy between the sandpile and PM$_{10}$ evolution. PM$_{10}$ pollutants form mainly as a result of first and (or) secondary pollutants produced from the emission of air pollution sources. The driving force is the continuously PM$_{10}$ pollutants emission in atmospheric environment, which serve as the grains continuously dropped on a pile. We can define the superposition of local PM$_{10}$ pollutants concentration which represents the chain of forces in sandpile. When the amounts of microscopic condensed PM$_{10}$ pollutants reach some threshold magnitude, the pollutants masses can be transported on microscopic scales by diffusion or convection. They reach a new location, where the local PM_{10} pollutants concentration is lower and can be diluted. If the local PM_{10} pollutants concentration in the neighborhood is also high, the amount of condensed pollutants will increase. Once the system reaches some critical point, any small perturbation, in principle, can trigger a chain reaction like the avalanches in atmospheric system. The normal atmospheric environmental capability serves as the critical state. If the local PM_{10} pollutants concentration is higher than the same critical value, pollutants are assembled and precipitated in the atmosphere. Thus, the system will adapt itself by removing these dissidents to maintain the critical state just as the sandpile adapts itself by avalanching to reach its constant angle of repose. Therefore, we can define the fluctuations of PM_{10} concentrations as avalanches events in a SOC sandpile. At the critical state, long range correlation, $1/f$ noise, and scale-invariant of PM_{10} concentrations will emerge from the dissipative system. It is important to note that PM_{10} pollution system is "tuned" to a critical state solely by its own internal dynamics rather than external dynamics. The high correspondence of the simulated results to observations supports that PM_{10} evolution acts as a SOC process on calm weather. And SOC is a useful framework to explain the nonlinear evolution of PM_{10} concentrations.

The micromechanisms of PM_{10} concentrations evolution are very complex. Some microscopic physical and chemical mechanisms are still uncertain. For example, how do photochemical reaction rate change with first and (or) secondary pollutants? How do the components of pollutants affect mass transport and chemical reaction at gas and solid two phases? Based on this traditional "reductionism" science perspective, the origin of robust scale-invariant in PM_{10} concentrations evolution can be quite hard to understand. However, when we turn sight to "holism" science perspective, the satisfactory understanding is achieved to this problem. Considering the similarities between sandpile system and PM_{10} evolution, a simplified sandpile model for PM_{10} pollution with a nondimensional formalism is put forward. This model mechanism only includes the emission of PM_{10} pollutants, the movements and transformation of PM_{10}, and temporal degradation process of PM_{10}. The high correspondence of the results to observations indicates that the model provides an effective parameterization of the key physical process that governs PM_{10} concentrations evolution.

It is important to note that the power system organizes itself to an operating point near to, but not at, a critical value. This could make the system quite robust in different time intervals. One consequence is that the measured frequency of occurrence of small events can be used to estimate the frequency of occurrence of large events. For example, the recurrence interval for serious air pollution can be estimated from the frequency of smaller air pollution.

4. Conclusion

Based on DFA, power spectrum, and statistical analysis, we have identified long range correlation, $1/f$ noise, and scale-invariant of PM_{10} concentrations measured at Chengdu. These statistics seem consistent with avalanche sizes in a running sandpile known to be SOC. In order to explain the origin of scale-invariant in PM_{10} evolution, according to the characteristics of PM_{10} evolution on calm weather, a simplified sandpile model for PM_{10} pollution with a nondimensional formalism is put forward. The simulated result is consistent with the actual monitored data very well. The work supports the proposal that PM_{10} evolution acts as a SOC process on calm weather. Far from being equilibrium forms, PM_{10} pollutants will evolve into a nonequilibrium and critical state. At the critical state, long range correlation, $1/f$ noise, and scale-invariant of PM_{10} concentrations will emerge from the dissipative system. This insight will inspire new research into the macroeffect of air pollution processes and improvement of modeling of PM_{10} pollution.

Acknowledgments

The work is supported by the National Natural Science Foundation of China (no. 41105118) and Hunan Provincial Natural Science Foundation of China (13JJB012).

References

[1] A. Nel, "Air pollution-related illness: effects of particles," *Science*, vol. 308, no. 5723, pp. 804–806, 2005.

[2] A. B. Chelani, "Change detection using CUSUM and modified CUSUM method in air pollutant concentrations at traffic site in Delhi," *Stochastic Environmental Research and Risk Assessment*, vol. 25, no. 6, pp. 827–834, 2011.

[3] A. Zolghadri and F. Cazaurang, "Adaptive nonlinear state-space modelling for the prediction of daily mean PM_{10} concentrations," *Environmental Modelling and Software*, vol. 21, no. 6, pp. 885–894, 2006.

[4] B. Sivakumar and R. Berndtsson, "Modeling and prediction of complex environmental systems," *Stochastic Environmental Research and Risk Assessment*, vol. 23, no. 7, pp. 861–862, 2009.

[5] C. Lee and S. Lin, "Chaos in air pollutant concentration (APC) time series," *Aerosol and Air Quality Research*, vol. 8, no. 4, pp. 381–391, 2008.

[6] S. R. Nagel, "Instabilities in a sandpile," *Reviews of Modern Physics*, vol. 64, no. 1, pp. 321–325, 1992.

[7] C.-K. Lee, D.-S. Ho, C.-C. Yu, and C.-C. Wang, "Fractal analysis of temporal variation of air pollutant concentration by box counting," *Environmental Modelling and Software*, vol. 18, no. 3, pp. 243–251, 2003.

[8] C. Varotsos, J. Ondov, and M. Efstathiou, "Scaling properties of air pollution in Athens, Greece and Baltimore, Maryland," *Atmospheric Environment*, vol. 39, no. 22, pp. 4041–4047, 2005.

[9] C. Lee, L. Juang, C. Wang et al., "Scaling characteristics in ozone concentration time series (OCTS)," *Chemosphere*, vol. 62, no. 6, pp. 934–946, 2006.

[10] P. Bak, C. Tang, and K. Wiesenfeld, "Self-organized criticality: an explanation of the 1/f noise," *Physical Review Letters*, vol. 59, no. 4, pp. 381–384, 1987.

[11] K. Shi, C. Q. Liu, N. S. Ai, and X. H. Zhang, "Using three methods to investigate time-scaling properties in air pollution indexes time series," *Nonlinear Analysis: Real World Applications*, vol. 9, no. 2, pp. 693–707, 2008.

[12] K. Shi and C. Liu, "Self-organized criticality of air pollution," *Atmospheric Environment*, vol. 43, no. 21, pp. 3301–3304, 2009.

[13] D. Chang, Y. Song, and B. Liu, "Visibility trends in six megacities in China 1973–2007," *Atmospheric Research*, vol. 94, no. 2, pp. 161–167, 2009.

[14] C.-K. Peng, S. V. Buldyrev, S. Havlin, M. Simons, H. E. Stanley, and A. L. Goldberger, "Mosaic organization of DNA nucleotides," *Physical Review E*, vol. 49, no. 2, pp. 1685–1689, 1994.

[15] B. D. Malamud and D. L. Turcotte, "Self-affine time series: measures of weak and strong persistence," *Journal of Statistical Planning and Inference*, vol. 80, no. 1, pp. 173–196, 1999.

[16] D. L. Turcotte and B. D. Malamud, "Landslides, forest fires, and earthquakes: examples of self-organized critical behavior," *Physica A*, vol. 340, no. 4, pp. 580–589, 2004.

[17] O. Peters and K. Christensen, "Rain viewed as relaxational events," *Journal of Hydrology*, vol. 328, no. 1-2, pp. 46–55, 2006.

8

Numerical Simulations and Analysis of June 16, 2010 Heavy Rainfall Event over Singapore Using the WRFV3 Model

B. H. Vaid[1,2]

[1] TMSI, National University of Singapore, Singapore
[2] Department of Applied Sciences, Raj Kumar Goel Institute of Technology for Women, Near Jain Tube,
  Delhi Meerut Road, Ghaziabad 200203, India

Correspondence should be addressed to B. H. Vaid; bakshi32@gmail.com

Academic Editor: Dimitris G. Kaskaoutis

The Numerical Simulations of the June 16, 2010, Heavy Rainfall Event over Singapore are highlighted by an unprecedented precipitation which produced widespread, massive flooding in and around Singapore. The objective of this study is to check the ability of Weather Research Forecasting version 3 (WRFV3) model to predict the heavy rain event over Singapore. Results suggest that simulated precipitation amounts are sensitive to the choice of cumulus parameterization. Various model configurations with initial and boundary conditions from the NCEP Final Global Analysis (FNL), convective and microphysical process parameterizations, and nested-grid interactions have been tested with 48-hour (June 15–17, 2010) integrations of the WRFV3. The spatial distributions of large-scale circulation and dynamical and thermodynamical fields have been simulated reasonably well in the model. The model produced maximum precipitation of ~5 cm over Changi airport which is very near to observation (6.4 cm recorded at Changi airport). The model simulated dynamic and thermodynamic features at 00UTC of June 16, 2010, lead to understand the structure of the mesoscale convective system (MCS) that caused the extreme precipitation over Singapore. It is observed that Singapore heavy rain was the result of an interaction of synoptic-scale weather systems with the mesoscale features.

1. Introduction

On 16th June 2010, a heavy rainfall event occurred in Singapore producing devastating flash flood and tremendous amount of property damage (Singapore's national water agency (PUB) report, Annual Weather Review, 2010, NEA, Singapore). Heavy rainfall is usually resulted from individual mesoscale storms or mesoscale convective systems (MCSs) embedded in synoptic-scale disturbances [1]. High-resolution observations and numerical modeling technique are needed to better predict heavy rainfall events and understand the evolution and development mechanisms of mesoscale convection and storms responsible for heavy rainfall. In this study, a high-resolution version of the WRFV3 (Weather Research and Forecasting Version 3) model is used to investigate the predictability of heavy rainfall over Singapore and try to exploit the mesoscale convective systems which are highly interacting with synoptic-scale environment. WRFV3 has been used successfully for predicting heavy rainfall which occurred in many different countries and for understanding the associated convective systems [2–7]. The objective of the present study is to identify the best possible microphysics, cumulus, and PBL scheme for simulation of heavy rainfall events over Singapore, to assess the predictability of the intensity of the event, and to understand the dynamic and thermodynamic characteristics of Mesoscale Convective System (MCS) that lead to heavy precipitation over Singapore.

2. Data

In the present study, 1.0×1.0 degree gridded NCEP FNL (Final) Operational Global Analysis and Global Forecast System (GFS) data has been used. FNL product is from the Global Data Assimilation System (GDAS), which continuously collects observational data from the Global Telecommunications System (GTS), and other sources, for many analyses. The FNLs are made with the same model which NCEP uses in the Global Forecast System (GFS), but the FNLs are prepared about an hour or so after the GFS is initialized. The FNLs are delayed, so that more observational

TABLE 1: Microphysics, Cu-physics, and PBL schemes.

Microphysics options	Cumulus parameterization	Boundary-layer option
WSM 6-class graupel WDM6 Lin et al. scheme	Kain-Fritsch (new Eta) scheme Grell-Devenyi ensemble scheme (GD) New Grell scheme (G3) Betts-Miller-J	YSU scheme Mellor-Yamada-Janjic (Eta) TKE scheme

data can be used. Moreover, we also used the Tropical Rainfall Measuring Mission (TRMM) rainfall remote sensing data. Rainfall is very variable in space and time. Accurate rainfall measurement in the tropics has long been and remains a difficult task. Before the existence of satellite remote sensing, there was very little rainfall measurement data over the open oceans and undeveloped countries. TRMM's TMI and PR instruments have greatly improved this situation by scanning the entire globe between 35N and 35S and supplying excellent coverage daily in the tropics. TRMM is a joint US-Japan satellite mission to monitor tropical and subtropical precipitation and to estimate its associated latent heating. TRMM is used as it gives the data of unprecedented accuracy. TRMM observations used in the study are of .25 × .25 degree resolutions. So, only the pixel which covers Singapore's domain was used and compared with model output.

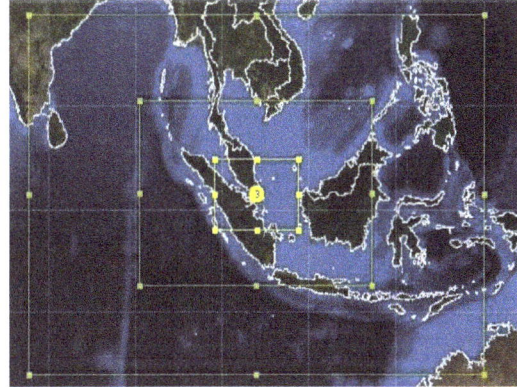

FIGURE 1: Domain used (three nested domains having resolution 27 km : 9 km : 3 km).

3. Model Configuration and Experimental Setup

Many attempts have been made to assess and evaluate the performance of numerical weather predictions from model configurations for the region, because on the basis of flood forecasting and even rainfall forecasting as well as safety evaluation model for Mumbai, the early warning could be avoiding disaster. During the last two decades, weather forecasting all over the world has greatly benefitted from the guidance provided by the Numerical Weather Prediction (NWP). Significant improvement in accuracy and reliability of NWP products has been driven by advances in numerical techniques, explosive growth in computer power, and the phenomenal increase in satellite-based soundings. The prediction of these systems is subject to the limitations of synoptic forecasting methods, which only indicate probable occurrence of heavy precipitation but not the quantity. Although numerical models provide the quantitative prediction of precipitation, they are subject to the limitations of initial data, model dynamics, and physics which can lead to uncertainties model output. Uncertainties are "data uncertainties," "modeling uncertainties," "completeness uncertainties." Data uncertainties arise from the quality or appropriateness of the data used as inputs to models. Modeling uncertainties arise from an incomplete understanding of the modeled phenomena, or from approximations that are used in the formal representation of the processes. Completeness uncertainties refer to all omissions due to the lack of knowledge. They are, in principle, nonquantifiable and irreducible. The prediction of the mesoscale systems requires the use of high-resolution atmospheric mesoscale models and observations with a mesoscale network. Some studies of the numerical

prediction of heavy rainfall event over India using high resolution mesoscale models show the predictability of events with precipitation less than 20 cm/day [2, 3, 8].

The model used in this study is the Advanced Research Weather Research and Forecast model (WRF) version 3.3. The WRF is the next generation forecast model and data assimilation system that has advanced both the understanding and prediction of weather. It has been designed to support operational forecasting and atmospheric research needs. The WRF model is a fully compressible, nonhydrostatic model [9]. The grid staggering is the Arakawa C grid. The model uses higher order numerics. These numerics include the Runge Kutta 2nd and 3rd order time integration schemes and 2nd to 6th order advection schemes in both horizontal and vertical directions [10].

The model supports both idealized and real-data applications with various lateral boundary condition options. Due to meteorological complexities involved in replicating the rainfall occurrences over Singapore, the WRFV3 modeling system is tested for different physics schemes. In the present study, we tested out the possible combination of Microphysics, Cu-physics, and PBL schemes provided in the Table 1.

It is to be noted that MYJ PBL scheme can only be used with the MYJ SFC layer scheme. Other physics options include longwave radiation from RRTM Scheme, shortwave radiation from Dudhia scheme, surface physics from unified Noah land-surface model, and Surface Clay Physics from Surface scheme Monin-Obukhov similarity theory. As grid spacings decrease, convective parameterizations become more inappropriate (and scientifically questionable given the underlying assumptions), whereas the explicit representation of microphysical processes can be computed for increasingly small clouds, cloud particles, water droplets, and so forth. So therefore in the present case, we conducted multinested experiments of 3 model domains of 27, 9, and 3 km horizontal resolution (Figure 1) to simulate a heavy rainfall case over Singapore on June 16, 2010.

In the present study, the model is forced with initial and boundary condition form NCEP Final global analysis (FNL) and model integrated 48 hr with initial and boundary condition which started from June 15, 2010. Forcing variables are air temperature, cloud, amount/frequency, cloud liquid, water/ice, convection, geopotential height, humidity, hydro-static pressure, ice extent, land cover, maximum/minimum temperature, planetary, boundary layer height, potential temperature, precipitable water, sea surface temperature, skin temperature, snow water, equivalent soil, moisture/water content, surface air temperature, surface pressure surface winds, tropospheric ozone, upper level winds, and wind shear.

4. Results and Discussions

Rainfall is an important parameter in many operational and research activities, ranging from weather forecasting to climate research. The WRFV3 model with the WSM6 microphysics is observed to provide useful information on high-resolution weather phenomena over Singapore. The study demonstrated that the WSM6 schemes are competitive options in WRF by reproducing precipitating convection and associated meteorological phenomena over Singapore. Apart from this, it has been observed that performance with GD scheme is found to be the best in both spatial and temporal bases. The spatial distributions of large-scale circulation and thermodynamics features have been simulated reasonably well in this model. The model produced maximum precipitation of ~5 cm at Changi station (the observed rainfall over Changi is 6.4 cm during the day). Figure 2 shows (a) TRMM rainfall (mm) and (b) WRFV3 total accumulated precipitation (mm) for 16th June 2010. The resolution for the model precipitation shown is 3 km. It is encouraging to note that over Singapore, the WRF-simulated accumulated rainfall on 16th June 2010 agrees well with that of observation. Note that the model simulation (shown in figure) has a much higher resolution than what the TRMM rainfall has shown. Therefore, a more detailed structure of rainfall is visible in the model simulation. Moreover, the model could also predict heavy rainfall on the central part of Singapore which seems to be in agreement with the news on floods over Singapore by Singapore national water agency (PUB).

An attempt has been carried out with available sources of data to analyse mesoscale characteristics at Singapore favorable for the formation of the event during 16th June 2010. Here, we used Finite Global Analysis (FNL) available data for the analysis. Figure 3 shows the mean sea level pressure for 00UTC of June 16, 2010 (left panel), 06UTC 16 of June, 2010 (centre panel), and 12UTC 16 June 2010 (right panel). The significant low value of mean sea level pressure is clearly evident during 00UTC of 16th June 2010 which indicates the tendency of the sudden formation of the system

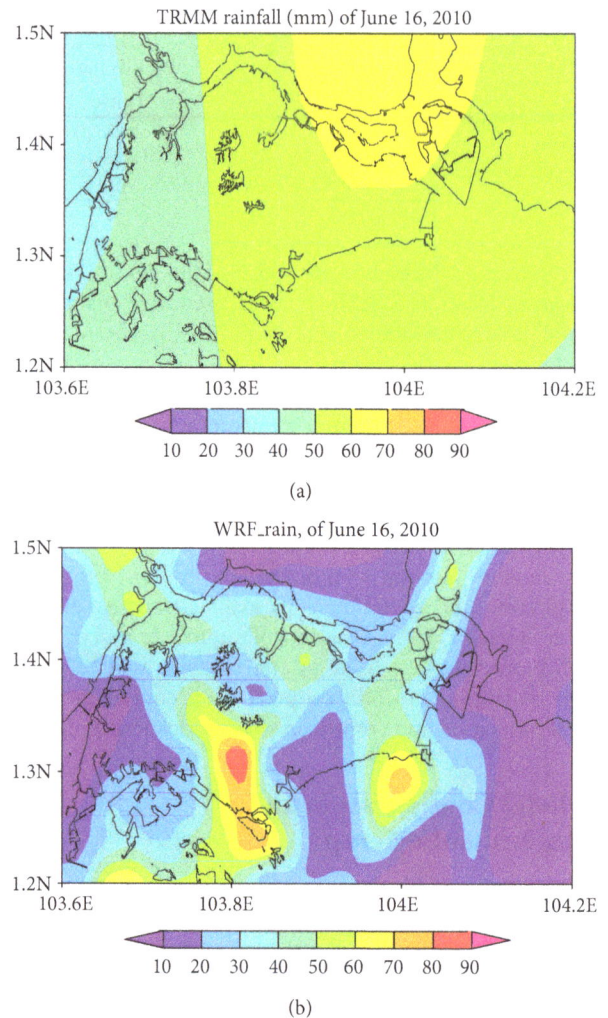

FIGURE 2: (a) TRMM rainfall (mm) and (b) WRFV3 total accumulated precipitation (mm) for 16th June 2010. Resolution for the model precipitation shown is 3 km.

responsible for the heavy rainfall. Figure 4 shows surface convective inhibition (CIN) for 00UTC 16 June 2010 (left panel), 06UTC of 16 June 2010 (centre panel), and 12UTC of 16 June 2010 (right panel). It clearly shows instability of the atmosphere during 00UTC of 16th June 2010. The model derived dynamical and thermodynamic fields were analyzed to understand the characteristics of the convective system which was responsible for the heavy rainfall event. Upper level atmosphere behavior is studied carefully as it is crucial in understanding its system which leads to heavy rainfall. The upper level flow and dynamics and thermodynamics can give key insights development of the system and how long it might have persisted. Since examining all heights of the atmosphere at any given time is not feasible, it is logical to choose a particular height that best represents the atmosphere at any given time. A higher than a normal height pattern at the 500 mb level typically represents regions at the surface in which higher pressure and warmer temperatures tend to occur. Likewise, a lower than normal height pattern at 500 mb level typically represents regions at the surface in

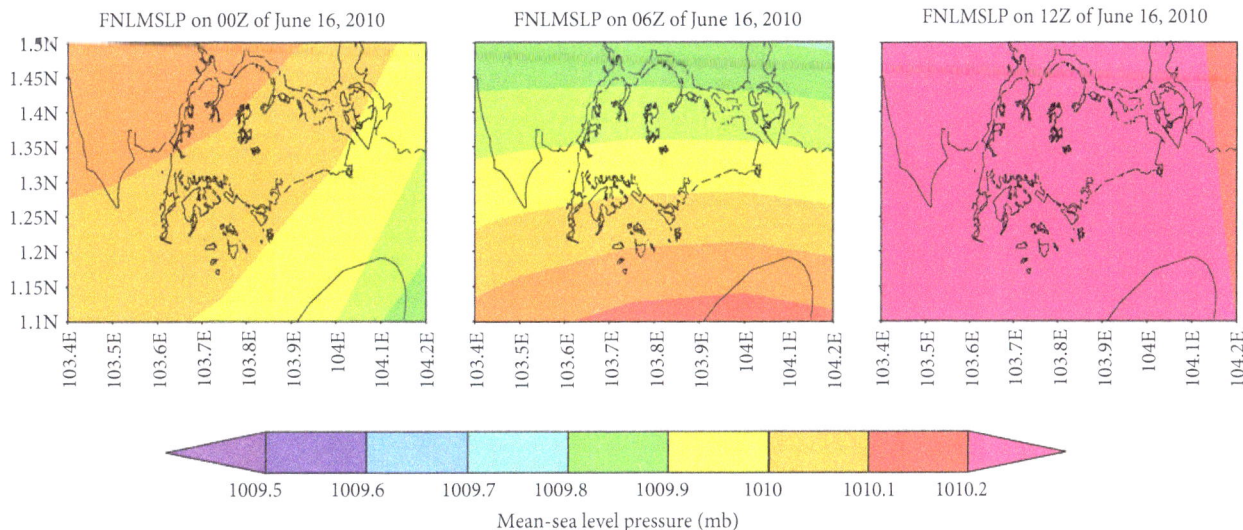

FIGURE 3: Mean sea level pressure for 00UTC of 16 June 2010 (left panel), 06UTC of 16 June 2010 (centre panel), and 12UTC of 16 June 2010 (right panel).

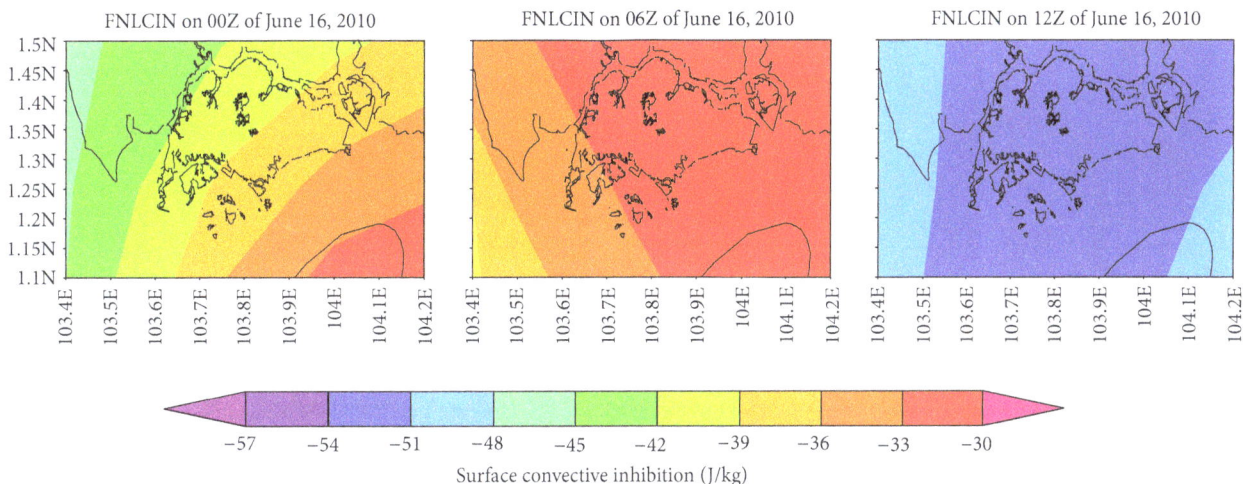

FIGURE 4: Surface convective inhibition (CIN) for 00UTC of 16 June 2010 (left panel), 06UTC of 16 June 2010 (centre panel), and 12UTC of 16 June 2010 (right panel).

which lower pressure and cooler temperatures tend to occur. The strong values of RH are observed over the 500 mb level. It is encouraging that the model could capture the dynamic and thermodynamic features which lead to form mesoscale convective system which infact produced the heavy rainfall in Singapore. Figures 5 and 6 show wind magnitude (shaded) and wind vector (arrow) at 850 mb (500 mb) level on the 16th of June, 2010 (a) GFS, (b) FNL, and (c) model. WRF simulated winds are observed to be consistent and show fairly good agreement with observation. Further, a detailed study of model derived fields was carried out to explore the mesoscale characters features during the event.

The model derived dynamical and thermodynamic fields were analyzed to understand the characteristics of the convective system which was responsible for the 16th of June, 2010 heavy rainfall over Singapore. Figure 7 shows model derived relative humidity (RH) at 500 mb level over Singapore

(left panel) and on 00UTC of the 16th of June, 2010 over Singapore with respect to vertical level (right panel). Over Singapore model predicts more than 95% RH above the vertical level of 500 mb (typically at 18,000 feet or ~5-6 kilometers). The time series of RH over Singapore at 500 mb level clearly shows the time at which high RH was prominent. Interestingly, RH values were suddenly risen during 00UTC of 16 June 2010 to 06UTC June 16, 2010 which shows the model predictability of the event in agreement with the formation of system. Model derived wind direction in degrees with respect to vertical levels over Singapore on 00UTC 16 June 2010 is shown in Figure 8. The clear indication of circulation in wind is seen in between the levels 700–550 mb.

Differential model fields which can be represented by Z-component of wind at 250 mb Minus Z-component of wind at 850 mb over Singapore latitude on 00UTC of the 16th of June, 2010 have also been analyzed (Figure 9). The Z-component of

GFS wind vectors 850 mb on June 16, 2010

FNL wind vectors 850 mb on June 16, 2010

(a)

(b)

Model wind vectors 850 mb on June 16, 2010

(c)

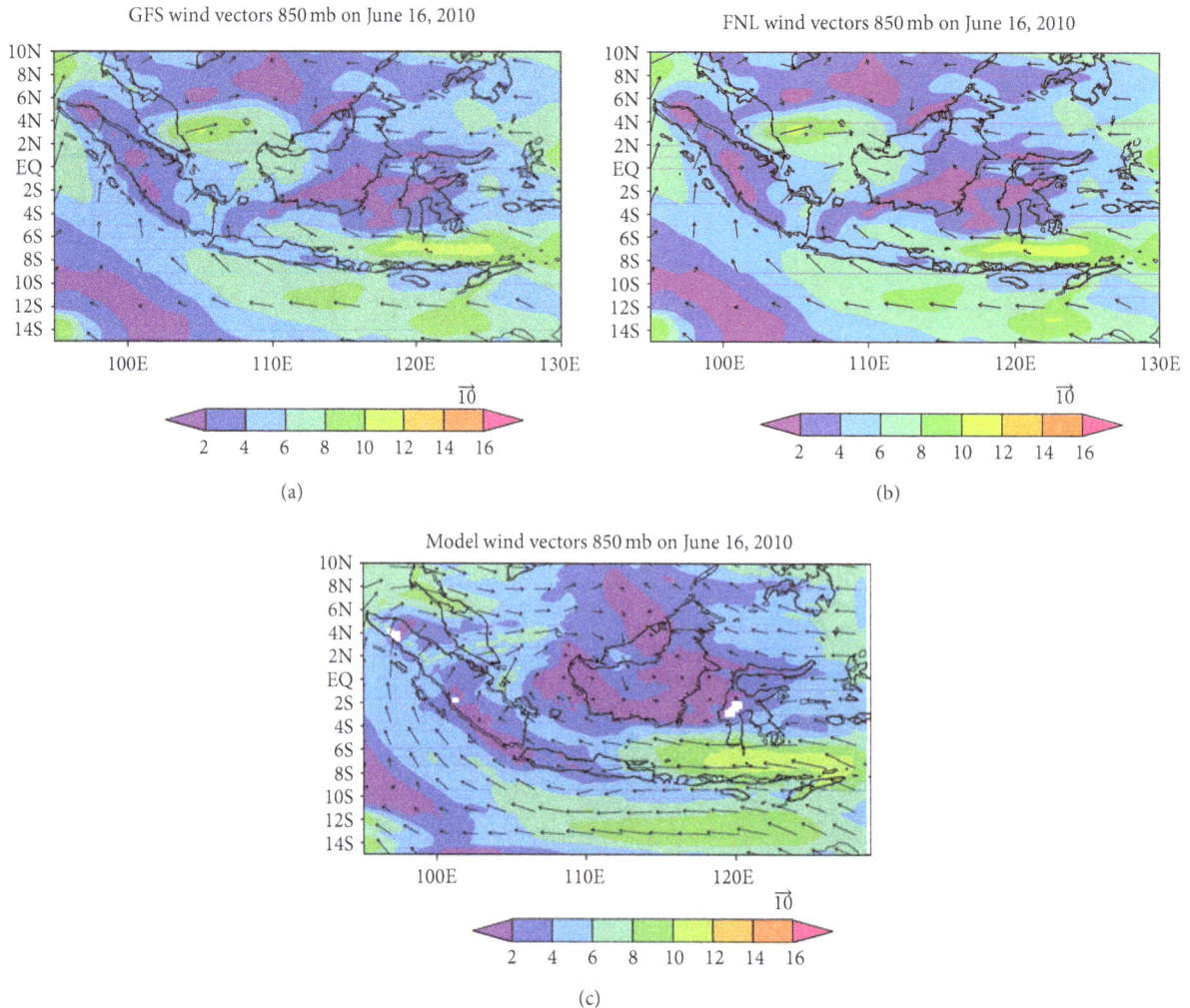

FIGURE 5: Wind magnitude (shaded) and wind vector (arrow) at 850 mb level on 16th June 2010 (a) GFS, (b) FNL, and (c) model.

wind at 250 mb level was higher than that 850 mb level, which speaks of shear presented during the time for the formation of heavy rainfall and we speculate that this strong vertical Z-component of wind causes the cloud burst during the event, which leads to heavy rainfall. Figure 10 shows the model derived high cloud fraction in percentage over Singapore. Clearly high clouds are seen during the event. If there have been cyclone detection radar in and around Singapore, then it could have reported the clouds with heights of 5-6 km around Singapore at 00UTC of 16 June 2010.

Figure 11 shows the potential temperature (in degrees) over Singapore on 00UTC of 16 June 2010 at 400 mb (upper panel) and at 700 mb (lower panel). The model predicted a decrease of potential temperature from surface to 700 hPa level and increase up to 400 hPa during 00UTC of 16 June 2010, indicating that potential instability increased at lower levels due to dry air capping at middle levels that caused the suppression of convection during this period and all this gave rise to sudden explosive deep convection with super cell structure producing a heavy precipitation spell. From model derived results, we can conclude that the Singapore heavy

rain was the result of an interaction of synoptic-scale weather systems with the mesoscale features. Moreover, the model produced dynamical structure shows the veering of wind with height, with westerlies at lower levels and easterly at levels indicating warm air advection triggering the convection (Figure 12).

5. Summary

The WRFV3 was observed to make good estimate for the MCS and its timing which leads to heavy rainfall event over Singapore. The model produced maximum precipitation of ~5 cm over Changi airport which is very near to observation (6.4 cm recorded at Changi airport). The model simulated circulation features shows the mesoscale characteristics of the convective system. The model simulated dynamic and thermodynamic features at 00UTC of 16th June 2010 which lead to understand the structure of the MCS that caused the heavy precipitation over Singapore. In meteorology, the vertical wind shear perspective is superior to other meteorological parameters, because it establishes the kind of physical cause and effect link between storm structure and

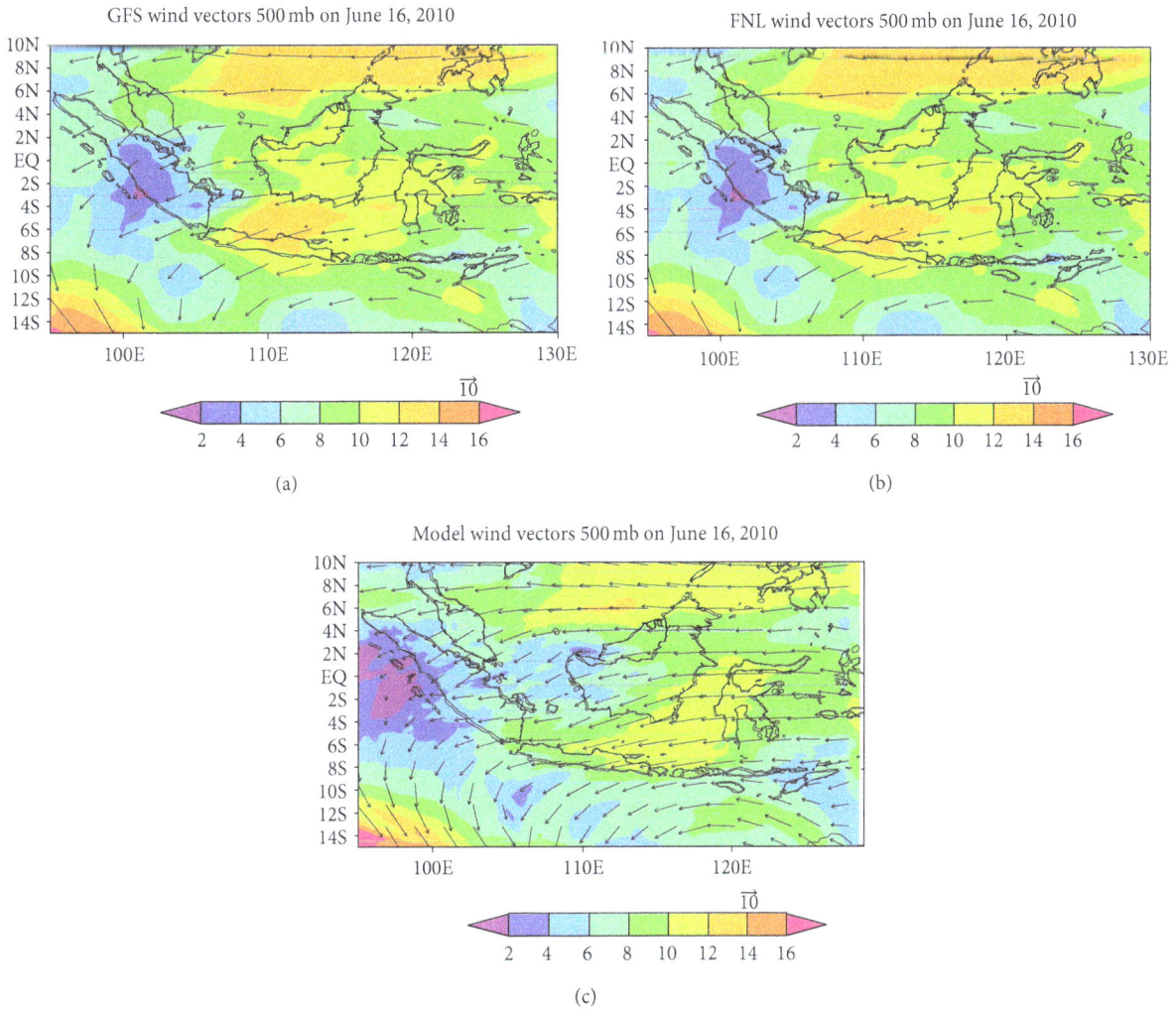

FIGURE 6: Wind magnitude (shaded) and wind vector (arrow) at 500 mb level on 16th June 2010 (a) GFS, (b) FNL, and (c) model.

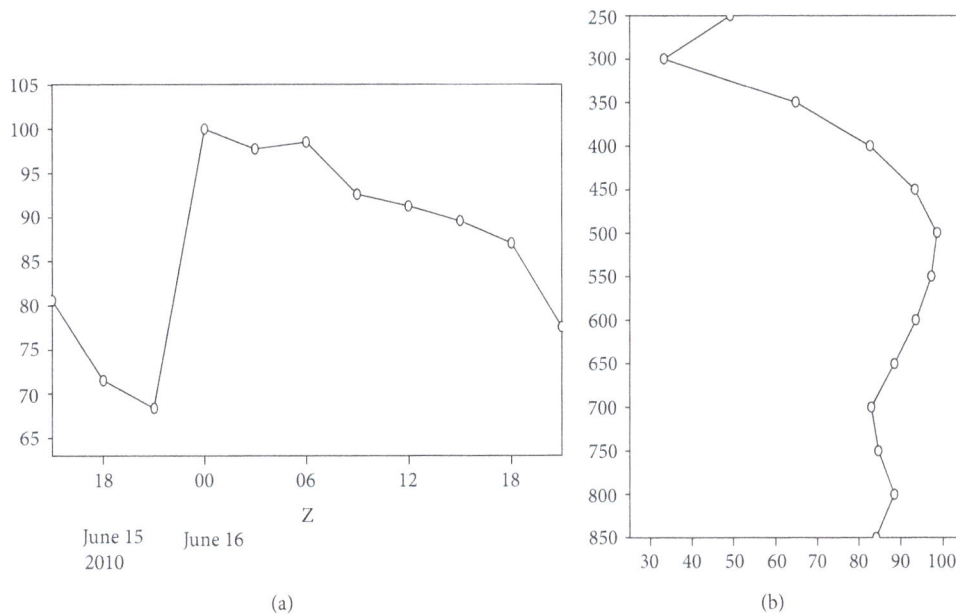

FIGURE 7: Model derived relative humidity (RH) at 500 mb level over Singapore (a) and on 00UTC of 16th June 2010 over Singapore with respect to vertical level (b).

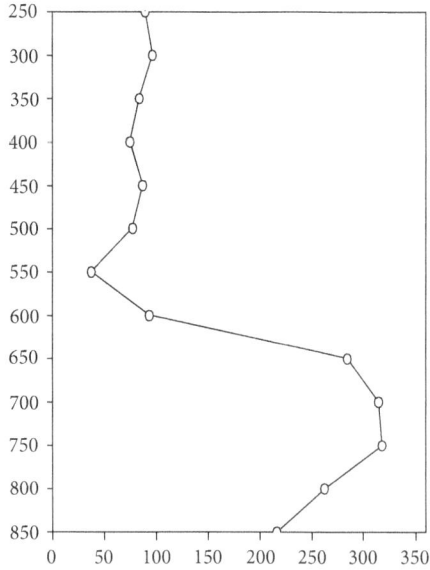

FIGURE 8: Wind direction (WD) in degrees with respect to vertical levels at Singapore on 00UTC of 16 June of 2010.

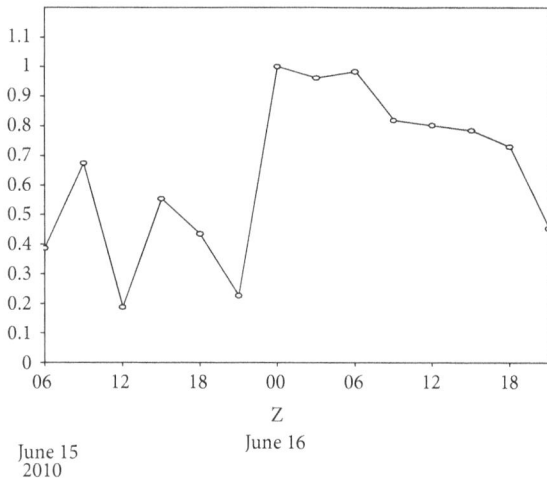

FIGURE 9: Z-component of wind at 250 mb minus Z-component of wind at 850 mb at Singapore.

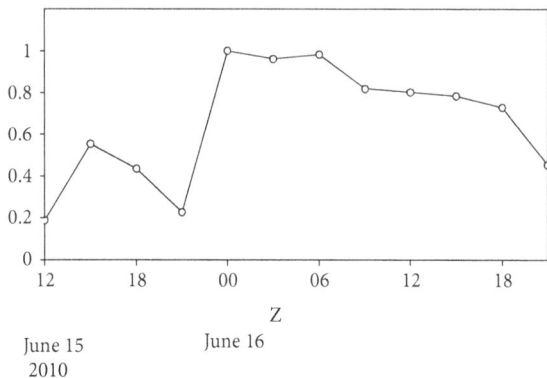

FIGURE 10: Model derived high cloud fraction in percentage over Singapore.

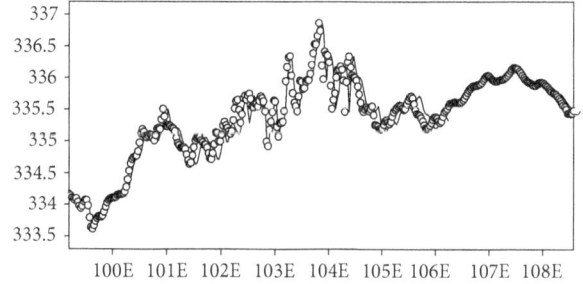

(a)

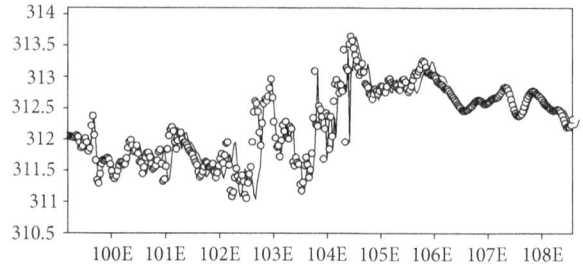

(b)

FIGURE 11: Potential temperature (in degrees) over Singapore on 00UTC of 16 June 2010 at 400 mb (a) and at 700 mb (b).

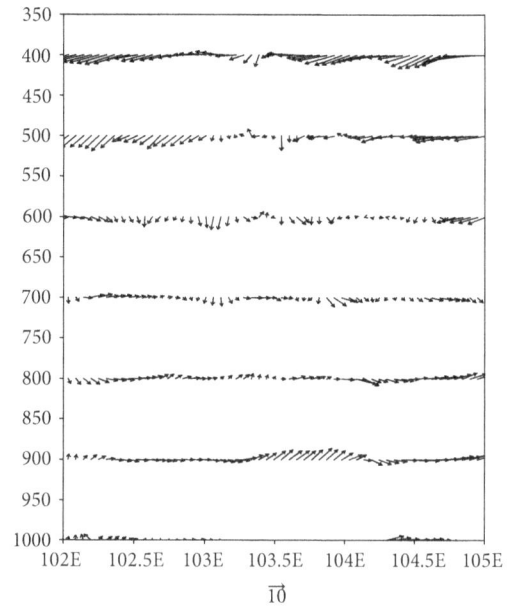

FIGURE 12: Wind vector on 00UTC of the 16th of June, 2010 with respect to vertical levels. x-axis represents longitudes.

the prestorm environment that forecasters can readily apply when attempting to assess storm potential on any given day. To the degree that the prestorm environment is known and that convection is going to occur, a forecaster can simply look at the hodograph and determine to a good approximation whether there is sufficient shear over a sufficient depth to promote supercell development (e.g., 20 ms^{-1} of wind variation over the lowest 4–6 km above ground level). Significant vertical shear presented during the time, and it is speculated that the strong vertical Z component of wind causes the cloud burst during the event which contributed to heavy rainfall. It is to be noted that model simulations are always sensitive to the initial conditions, domain size and location, model dynamics, and physics, and hence research is still continuing to improve the prediction. From model derived results, we can conclude that the highly localized, heavy rain was the result of an interaction of synoptic-scale weather systems with the mesoscale features.

Acknowledgments

Dr. B. H. Vaid would like to express his gratitude to the WRF group, GrADs, and Ferret for online support and help. Dr. S. Y. Liong is acknowledged for the kind help and support. The author very much thankful to the reviewers and editor for valuable suggestions and comments, which really helped the author to improve the paper in the present stage.

References

[1] D. K. Lee, H. R. Kim, and S. Y. Hong, "Heavy rainfall over Korea during 1980–1990," *Korean Journal of the Atmospheric Sciences*, vol. 1, pp. 32–50, 1998.

[2] A. Routray, U. C. Mohanty, A.K. Das, and N. V. Sam, "Study of HPE over west coast of India using analysis nudging inMM5 during ARMEX-I," *Mausam*, vol. 56, pp. 107–120, 2005.

[3] H. R. Hatwar, Y. V. Rama Rao, S. K. Roy Bhowmik, and D. Joardar, "An impact of ARMEX data on limited area model analysis and forecast system of India Meteorological Department—a preliminary study," *Mausam*, vol. 56, pp. 131–138, 2005.

[4] A. K. Bohra, S. Basu, E. N. Rajagopal et al., "Heavy rainfall episode over Mumbai on 26 July 2005: assessment of NWP guidance," *Current Science*, vol. 90, no. 9, pp. 1188–1194, 2006.

[5] D. R. Sikka and P. Sanjeeva Rao, "The use and performance of mesoscale models over the Indian region for two high-impact events," *Natural Hazards*, vol. 44, no. 3, pp. 353–372, 2008.

[6] A. Kumar, J. Dudhia, R. Rotunno, D. Niyogi, and U. C. Mohanty, "Analysis of the 26 July 2005 heavy rain event over Mumbai, India using the Weather Research and Forecasting (WRF) model," *Quarterly Journal of the Royal Meteorological Society*, vol. 134, no. 636, pp. 1897–1910, 2008.

[7] A. Staniforth, "Regional modeling: a theoretical discussion," *Meteorology and Atmospheric Physics*, vol. 63, no. 1-2, pp. 15–29, 1997.

[8] D. B. V. Rao and D. H. Prasad, "Impact of special observations on the numerical simulations of a HPE during ARMEX-Phase I," *Mausam*, vol. 56, pp. 121–130, 2005.

[9] Z. I. Janjic, "A nonhydrostatic model based on a new approach," *Meteorology and Atmospheric Physics*, vol. 82, pp. 271–285, 2003.

[10] A. J. Litta, S. M Ididcula, U. C. Mohanty, and S. K. Prasad, "Comparison of thunderstorm simulations from WRF-NMM and WRF-ARW models over East Indian region," *The Scientific World Journal*, vol. 2012, Article ID 951870, 20 pages, 2012.

9

Characterization of Dispersive Fluxes in Mesoscale Models Using LES of Flow over an Array of Cubes

Adil Rasheed[1] and Darren Robinson[2]

[1] *SINTEF ICT, Applied Mathematics, Strindveien 4, 7050 Trondheim, Norway*
[2] *LESO-PB, EPFL, CH-1015 Lausanne, Switzerland*

Correspondence should be addressed to Adil Rasheed; adil.rasheed@sintef.no

Academic Editor: Helena A. Flocas

Field studies have shown that local climate is strongly influenced by urban structures. This influences both energy consumption and the pedestrian comfort. It is thus useful to be able to simulate the urban environment to take these effects into account in building and urban design. But for computational reasons, conventional computational fluid dynamics (CFD) codes cannot be used directly on a grid fine enough to resolve all scales found in a city. For this, we use mesoscale models, variants of CFD codes in which the 3D conservation equations are solved on grids having a resolution of a few kilometers. At this resolution, the effects of subgrid scales need implicit representations. In other words, phenomena such as momentum and energy exchanges averaged over the mesoscale grid contribute necessary sources/sinks to the corresponding equations. Such spatial averaging results in additional terms called dispersive fluxes. Until now these fluxes have been ignored. To better understand these fluxes, we have conducted large eddy simulations (LESs) over an array of cubes for different inter-cube spacings. The study shows that these fluxes are as important as the turbulent fluxes and exhibit trends which are related to the eddy formations inside the canopies.

1. Introduction

The large and continuous variety of scales present in the atmospheric flow over a city generates an intrinsic difficulty in the numerical treatment of the atmospheric conservation equations. From scaling considerations, the ratio between the smallest flow scale and the characteristic length scale is approximately proportional to $Re^{3/4}$ [1], where Re is the Reynolds number. This means that a 3D representation of the planetary boundary layer resolving all scales will require about 10^{15} grid cells. This number is far from being handled conveniently nowadays or in the near future by any computing device. Therefore, the transport phenomena over larger distances (in our case covering a city and its bounding context) must be handled by mesoscale atmospheric codes with spatial resolutions of a few kilometers either by subgrid urban parameterizations [2, 3] or by a coupling with a higher resoltion microscale model [4, 5]. In the former approach, the mesoscale codes cannot "see" buildings explicitly. Yet buildings and urban land use significantly impact the micro-

and mesoscale flow, altering the wind, temperature, and turbulence fields and radiation exchanges [6, 7]. Since mesoscale numerical models do not have the spatial resolution to directly simulate the fluid dynamics and thermodynamics in and around urban structures, urban parameterizations are needed to approximate the drag, heating, and enhanced turbulent kinetic energy (tke) produced and dissipated by the subgrid-scale urban elements. The drag forces offered by the buildings as well as the heat transfer characteristics are a function of the local velocity field. Local turbulent fluxes, dispersive fluxes (generally ignored in mesoscale models), and drag coefficients can significantly impact the exchange of mass, momentum, and energy. However, a mesoscale code as described earlier, does not have the spatial resolution to generate the profiles of these quantities. Several attempts have been made in the recent past to estimate the velocity profiles in the urban canopy. These so-called urban canopy models are based on either single-layer [8] or multilayer considerations of the canopy [9]. In almost all these models, the dispersive fluxes which result from the averaging of the governing equation in the horizontal plane are either ignored or assumed to behave in the same way as the turbulent fluxes. However, recent work [10, 11] tends to confirm that these stresses can be significant and sometime comparable to the turbulent stresses themselves and may behave differently. However, most of these findings were based on simulations conducted using Reynolds Averaged Navier Stokes (RANS) codes whose validity for this kind of application is somewhat questionable [12]. Moreover, in the study we wanted to compare the magnitude of the dispersive fluxes and the turbulent fluxes. A RANS model only gives the modelled turbulent fluxes against which the dispersive fluxes cannot be compared. In an attempt to resolve this issue, we therefore employ an LES code which is capable of resolving the turbulent fluxes.

An LES "resolves" the large-scale fluid motions and "models" the subgrid-scale motions by filtering the Navier Stokes equations. When unsteady RANS methods are used, it is implicitly assumed that there is a fair degree of scale separation between the large timescale of the unsteady flow features and the time scale of the genuine turbulence. However, in reality it is hard to find an evident time scale gap for many turbulent flows. Furthermore, RANS generally does not intend to capture most of the genuinely turbulent fluctuation information. A RANS approach thus has obvious weaknesses and poses serious uncertainties in flows for which large-scale organized features dominate, such as flows around building like obstacles. Against this background, it may be argued that the use of a sound although computationally expensive LES approach is fully justified. Although a city might not be well represented by a regular array of buildings, this is nevertheless a sound pragmatic starting point because these shapes are the basic building blocks of any city and also because there is good availability of data for validation purposes. Thus, the study of the flow over a matrix of cubes (resembling an array of buildings) can provide some fundamental understandings of the various physical phenomena involved in the flow through an urban area. Various characteristics like the vortex shedding, flow separation and velocity profiles have been experimentally investigated for such problems in the past [13, 14].

In this study, all the simulations have been conducted with the Smagorinsky model because of its numerical simplicity and stability. It has also provided excellent results for the case we are interested in, when compared against the experimental data [13]. Since, we are interested in general behaviors rather than results for a particular point inside the domain, we present the spatially averaged profiles of the velocity, turbulent fluxes, and dispersive fluxes and explain their nature especially that of the dispersive fluxes.

2. Governing Equations for LES

2.1. Transport Equations. In LES, only the large-scales are explicitly resolved by the numerical grid while the smaller ones are represented by a subgrid-scale model. The motivation for this approach is that the large-scale vortices are dominated by geometrical constraints and boundary

conditions. Due to turbulent transport phenomena, these vortices pass their kinetic energy onto smaller scales while the orientation of the initial vortices gets lost during this energy cascade. Therefore, the small-scale turbulence is expected to be isotropic without any preferred orientation and should consequently be much easier to model than the whole spectrum of turbulence. Starting with the governing equations for an incompressible three-dimensional (3D) unsteady flow field, we apply a top-hat filter function to separate large- and small-scale motions leading to the filtered equation set

$$\frac{\partial \overline{u}_i}{\partial x_i} = 0,$$

$$\frac{\partial \overline{u}_i}{\partial t} + \frac{\partial \left(\overline{u_i u_j} \right)}{\partial x_j} = \frac{-1}{\rho} \frac{\partial \overline{p}}{\partial x_i} + \frac{\partial \left(2 S_{ij} \right)}{\partial x_j}, \qquad (1)$$

where $\overline{u}_i$ are the filtered velocity components, $\overline{p}$ is the filtered pressure, $S_{ij} = (1/2)[\nu((\partial \overline{u}_i/\partial x_j) + (\partial \overline{u}_j/\partial x_i))]$ denotes the filtered strain-rate tensor, and ν denotes the molecular viscosity. The correlation within the convective term $(\overline{u_i u_j})$ is a priori unknown and has to be modeled.

The most common way is to rewrite this term into $\tau_{ij} = \overline{u_i u_j} - \overline{u}_i \overline{u}_j$, where τ_{ij} is the unresolved stress resulting from the subgrid-scale contribution and needs to be modelled by an appropriate subgrid-scale (SGS) model. The additional stresses are split into an anisotropic part $\tau_{ij}^a = \tau_{ij} - (1/3)\tau_{kk}\delta_{ij} = -2\nu_t S_{ij}$ (where ν_t is the eddy viscosity) and an isotropic part which is added to the pressure $p^* = \overline{p} + (1/3)\tau_{kk}$, leading to the LES equation set which forms the basis of this investigation:

$$\frac{\partial \overline{u}_i}{\partial x_i} = 0,$$

$$\frac{\partial \overline{u}_i}{\partial t} + \frac{\partial \left(\overline{u}_i \overline{u}_j \right)}{\partial x_j} = \frac{-1}{\rho} \frac{\partial p^*}{\partial x_i} + \frac{\partial}{\partial x_j} \left[\nu \left(\frac{\partial \overline{u}_i}{\partial x_j} + \frac{\partial \overline{u}_j}{\partial x_i} \right) \right] - \frac{\partial \tau_{ij}^a}{\partial x_j}. \qquad (2)$$

2.2. Numerics. For conducting large eddy simulations, we used the Transat code [15] which is based on a finite volume discretization. It solves for mass, momentum, and heat transport in both single- and two-phase flow conditions and provides the option of using the reynolds averaged or unsteady turbulence modelling (LES or direct numerical simulation). For the present simulation, we used LES because of the accuracy we needed for a better understanding of the flow. A Runge Kutta 3rd-order scheme is used for time integration while the convective schemes used for density, velocity and temperature were HYBRID, CDS (central differencing scheme), and HLPA (hybrid linear parabolic approximation), respectively. A preconditioned (multigrid) GMRES (generalized minimal residual method) pressure solver is used for solving the acoustic equation. The Standard Smagorinsky model is used to simulate the effects of the subgrid scales on the flow. Although there are more accurate models available, the established accuracy of prediction of this particular model has proved to be sufficient for our purpose.

2.3. Subgrid-Scale Modelling. An eddy-viscosity-based model has been employed in the computations presented in this paper, where the anisotropic part of the SGS stress is modelled using

$$\tau_{ij} - \frac{1}{3}\delta_{ij}\tau_{kk} = -2\nu_t S_{ij}. \tag{3}$$

The eddy viscosity ν_t is determined using Smagorinsky's expression [16] $\nu_t = C_s \Delta^2 S$ with $|S| = (2S_{ij}S_{ij})^{1/2}$ and $\Delta = (\Delta x \Delta y \Delta z)^{1/3}$, determined using an explicit box filter of width twice the mesh size in wall-parallel planes, together with averaging in the spanwise direction and relaxation in time with a factor of 10^{-3}. The model coefficient C_s is taken to be equal to 0.12. The dynamic Smagorinsky model (DSM) of [17], with the modification of [18], could also be applied here, but the required averaging of the model coefficient C_s (which will now depend on the resolved invariant $|S|$) is made difficult by the absence of a clear homogeneous averaging direction. The near-wall behavior of the model is such that it yields an eddy viscosity that reduces as the wall is approached, using an explicit damping following the van Driest relationship [19]:

$$f_\mu = 1 - \exp\left(-\frac{y^+}{26}\right), \tag{4}$$

where $y^+ = yu_\tau/\nu$ is the distance from the wall in viscous wall units for which u_τ is the friction velocity.

3. Governing Equations for Mesoscale Model

Since we intend to interpret the results from the LES simulations in a mesoscale modelling context, it is worth mentioning the standard Reynolds Averaged Navier Stokes equations that form the basis of a large variant of existing Mesoscale Codes. The mass and momentum equations can be mathematically represented as:

$$\frac{\partial U_i}{\partial x_i} = 0, \tag{5}$$

$$\frac{\partial U_i}{\partial t} + \frac{\partial (U_i U_j)}{\partial x_j} + \frac{\partial \left(\overline{u_i' u_j'}\right)}{\partial x_j}$$

$$= -\frac{1}{\rho}\frac{\partial P}{\partial x_i} + \nu \frac{\partial^2 U_i}{\partial x_j^2} + Q, \tag{6}$$

where U_i is the mean part of the velocity, u_i' are the fluctuation of velocities in time, P the mean pressure, $\overline{u_i' u_j'}$ the Reynolds stresses and Q the source term. Equation (6) when averaged over space takes the following form:

$$\left\langle \frac{\partial U_i}{\partial t} \right\rangle + \left\langle \frac{\partial (U_i U_j)}{\partial x_j} \right\rangle + \left\langle \frac{\partial \left(\overline{u_i' u_j'}\right)}{\partial x_j} \right\rangle$$

$$= -\left\langle \frac{1}{\rho}\frac{\partial P}{\partial x_i} \right\rangle + \left\langle \nu \frac{\partial^2 U_i}{\partial x_j^2} \right\rangle + \langle Q \rangle, \tag{7}$$

where the angular bracket is the space averaging operator. Such space averaging as we will show later in the paper results in an additional term called the dispersive flux.

4. Geometric Description and Test Cases

For the purpose of determining the validity of the LES model for our study, we used the experimental results of [13, 14] who carried out detailed measurements of the mean flow and turbulence characteristics using Doppler anemometry throughout an array of cubes. The experimental setup consisted of 250 cubes, each of height H placed at a distance of $3H$ from their neighbouring cubes in an aligned configuration (Figure 1(a)) consisting of 25 rows of 10 columns. The depth of the plane channel was $3.4H$. Due to the high computation cost associated with LES, our numerical simulations were based on a domain of $4H \times 4H \times 3.4H$ (*streamwise × spanwise × height*) with the cube located at the center (consistent with the experimental setup) and with periodic boundary conditions in the streamwise direction. Since a symmetric boundary condition is inappropriate for the instantaneous velocity field, a periodic boundary condition was also applied across the pair of vertical boundary planes of the flow domain in the spanwise direction. For the top and bottom walls of the channel, as well as for the surfaces of the cube, no-slip and impermeability conditions were specified (for the tangential and wall normal velocity components, resp.). In accordance with the experiment, the Reynolds number for the simulation was 3800, based on the mean bulk velocity U_b and the height H of the cube. The domain was discretized into 66 nodes in each direction, with the grids being preferentially fine near the wall surfaces. Although no grid independence test was attempted, it should be pointed out that the resolution we used in this simulation is finer than that used in a similar simulation reported in another study [12]. We refer to this particular case ($4H \times 4H \times 3.4H$) which has been validated against the experimental data, as Case A.

Four more simulations were conducted for the domain sizes of $2H \times 2H \times 5H$, $2.5H \times 2.5H \times 5H$, $3H \times 3H \times 5H$, and $4H \times 4H \times 5H$ corresponding to B/H (B is the inter-cube spacing and H is the cube height) ratios of 1, 1.5, 2, and 3. These we refer to as Cases I, II, III, and IV. The number of nodes used in all these cases was $66 \times 66 \times 82$ out of which $24 \times 24 \times 24$ nodes have been used to represent the cube. It should be noted that Case IV corresponds to $B/H = 3$, which is the same as that for Case A. Indeed the only difference between these two cases is the type and placement of the upper boundary: in Case IV a free slip boundary condition is applied at a height of $5H$. Although, we have not been able to directly validate Cases I–IV because of the unavailability of experimental data, we found that the profiles of velocities and stresses inside the canopy predicted in Case IV are almost identical to those obtained in Case A, so that we may have a reasonably good degree of confidence in our own results. The volume between $3.4H$ to $5H$ in Case IV was meshed with a uniform grid of dimensions similar to those of the topmost level in the mesh of Case A. For Cases I, II, and III, the domain size was reduced but the same

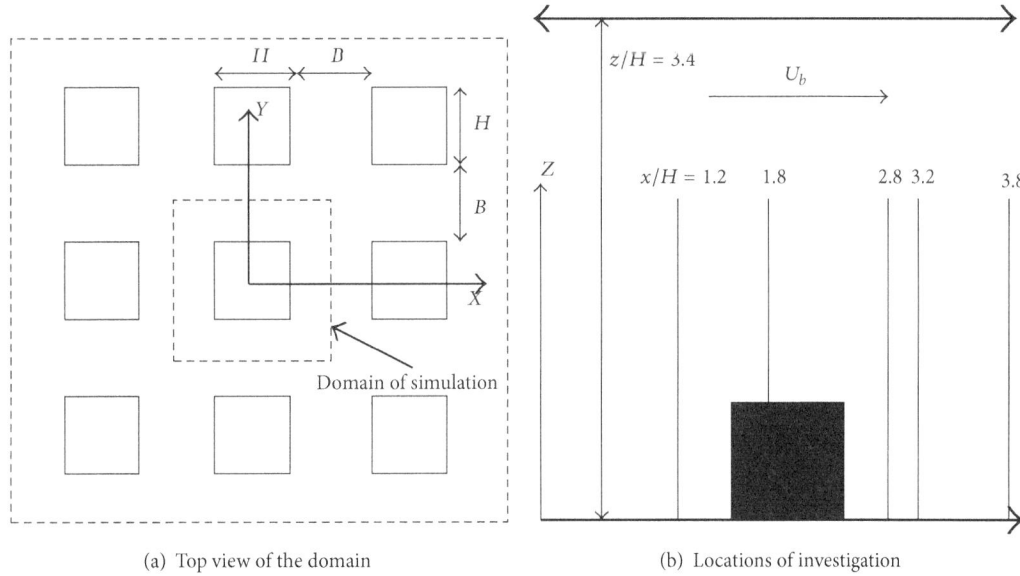

(a) Top view of the domain

(b) Locations of investigation

FIGURE 1: Domain and locations where comparisons have been made.

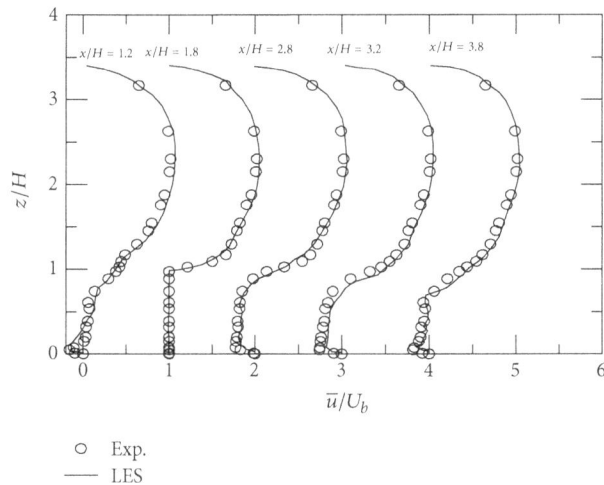

○ Exp.
——— LES

FIGURE 2: Vertical profiles of the time-mean streamwise velocity $\overline{u}/U_b$ on the vertical plane (x-z) through the center of the cube (i.e., at $y/H = 0$). Each profile has been offset by one unit.

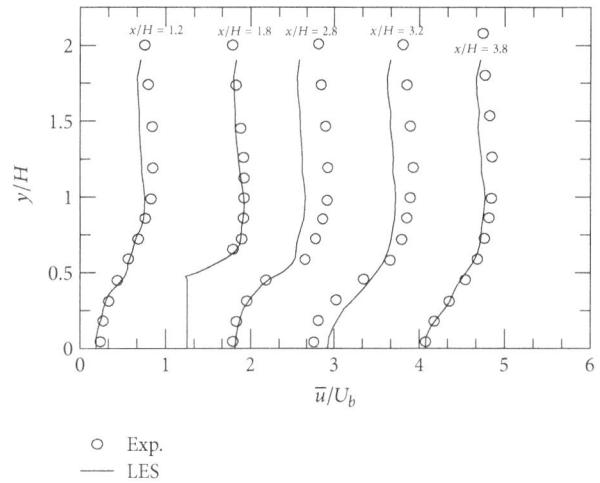

○ Exp.
——— LES

FIGURE 3: Horizontal profiles of the time-mean streamwise velocity $\overline{u}/U_b$ on the horizontal (x-y) plane at half cube height ($z/H = 0.5$). Each profile has been offset by one unit.

number of nodes was used, so resulting in finer resolutions in Cases I, II, and III as compared to Case IV. The profiles of mean velocity and turbulence statistics were obtained using a time-averaging procedure. After carrying out the simulation for several large-eddy turnover times to ensure that the final time-averaged results were independent of the initial conditions, we averaged the instantaneous quantities over 20,000 time steps. The corresponding averaged quantities were compared with those that were similarly obtained for 40,000 time steps. Very little difference was observed between the two cases implying statistical convergence. However, for the statistical convergence of the dispersive fluxes, 200,000 time steps were required.

5. Results

5.1. Validation of LES Model

5.1.1. Velocity Profiles. For the validation of the simulation results we used the experimental data of [13]. We present results, on the mid-plane at different positions (x/H), Figure 1(b). In Figure 2, one can see that the normalized velocity component remains positive over the cube, however, inside the spanwise canyons (between two cubes) the streamwise velocity component is negative implying a reverse flow or a strong vortex formation. Figure 3 presents the horizontal profiles of the mean streamwise velocity on the x-y plane at $z/H = 0.5$ at the same locations. Again one can see

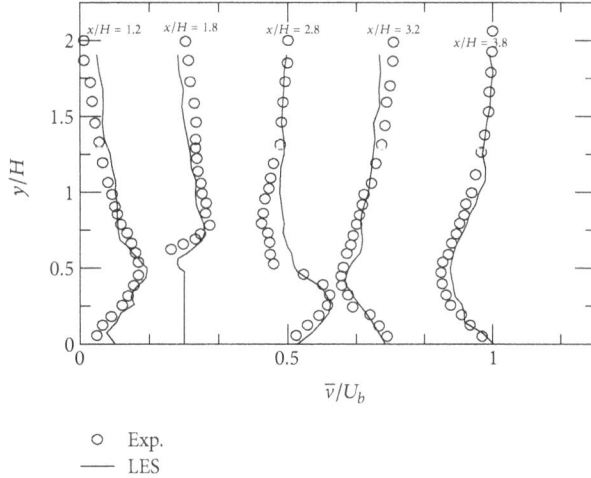

FIGURE 4: Horizontal profiles of the time-mean spanwise velocity $\bar{v}/U_b$ on the horizontal $(x\text{-}y)$ plane at half cube height $(z/H = 0.5)$. Each profile has been offset by 0.25 unit.

that the normalized streamwise velocity is positive in the streamwise canopy. However, the profiles inside the spanwise canopy is negative again due to the vortex formed between the cubes. Figure 4 displays predictions of the horizontal profiles of the mean spanwise velocity on the $x\text{-}y$ plane at $y/H = 0.5$ at the same x/H locations. A nonzero value of spanwise velocity represents recirculation, and as the figure shows this is more dominant in the vicinity of the cube. The agreement between LES results and the experimentation is generally very good. Indeed the prediction of the velocity profile (u, v, w) in general is very much acceptable.

5.1.2. Stress Profiles. Figures 5(a) and 5(b) present the Reynolds normal stress $\overline{u'u'}$, in the $x\text{-}z$ plane at $y/H = 0$ and on the $x\text{-}y$ plane at $z/H = 0.5$ at five selected x/H locations of 1.2, 1.8, 2.8, 3.2, and 3.8. From these results, it is clear that the normal stress is always at a maximum near the walls (top or side). These peaks correspond to the generation and development of thin intense vertical and horizontal shear layers along the roof and side walls of the cube, respectively. As is very clear from Figures 5(a) and 5(b), the values of the streamwise Reynolds normal stress predicted by LES are in very good agreement with the experimental data. However, the results for the spanwise Reynolds stresses (Figures 6(a) and 6(b)) are not very encouraging. While the predicted values compare well for $z/H > 1$, inside the canopy the magnitude is underpredicted. At the moment, this is an unresolved issue. Furthermore, profiles of Reynolds shear stress on the horizontal plane at half cube height (Figure 7) are in good agreement at most locations but for $x/H = 3.2, 3.8$ the agreement is not good specially inside the canopy. A more detailed physical explanation of the results can be found in the paper by [12]. In general, one can say that the LES results are in good agreement with the experimental data, and its use for further investigation of similar cases is reliable.

5.2. Effects of the Change in Street Width to Building Height Ratio on the Spatially Averaged Quantities. Since our numerical simulations (LESs) were conducted at a very high-resolution detailed profiles of the velocity field, pressure, stresses, and turbulent kinetic energy have been obtained. However, we are interested in spatially averaged quantities that can be representative of a mesoscale grid. Assuming that the values computed by the LES at every grid point are also representative of the volume average of the corresponding grid volume, we apply (7) to the domain on which LES was conducted. The equation for the streamwise velocity component can be expanded to the following form:

$$
\begin{aligned}
&\frac{\partial \langle U \rangle}{\partial t} + \frac{\partial \langle UU \rangle}{\partial x} + \frac{\partial \langle UV \rangle}{\partial y} + \frac{\partial \langle UW \rangle}{\partial z} \\
&= \frac{\partial \langle \overline{u'u'} \rangle}{\partial x} + \frac{\partial \langle \overline{u'v'} \rangle}{\partial y} + \frac{\partial \langle \overline{u'w'} \rangle}{\partial z} - \frac{1}{\rho}\frac{\partial \langle P \rangle}{\partial x} \\
&+ \nu \left\langle \frac{\partial^2 \langle \overline{U} \rangle}{\partial x^2} \right\rangle + \nu \left\langle \frac{\partial^2 \langle \overline{V} \rangle}{\partial y^2} \right\rangle \\
&+ \nu \left\langle \frac{\partial^2 \langle \overline{W} \rangle}{\partial z^2} \right\rangle + \langle Q \rangle .
\end{aligned}
\tag{8}
$$

Since we are looking for the steady-state equation, we can neglect the first term on the left-hand side. The second term on the LHS, using the flux divergence theorem, can be written as:

$$
\left\langle \frac{\partial UU}{\partial x} \right\rangle = \frac{1}{V_{\text{air}}} \int_{V_{\text{air}}} \frac{\partial UU}{\partial x} dv = \frac{1}{V_{\text{air}}} \int_{S} UU n_x ds.
\tag{9}
$$

Here, V_{air} is the volume of air over which the average is performed. S is the surface delimiting the volume over which the average is performed. n_x is the x component of the normal entering in the surface (x in this case because the derivative is respect to x). For the horizontal surfaces, the value of n_x is zero. There are two types of vertical surfaces: those at the boundaries of the domain and those delimiting the obstacle. For the surfaces at the boundary of the domain, since we have periodic boundary conditions the contribution is zero. Over the surfaces of the obstacle, the velocity is zero. So, the second term on the right-hand side of (8) is zero. Similarly, the third terms is also zero. Similarly, the first and second term on the RHS can be neglected. Because the flow is turbulent, the viscous terms (fifth, sixth, and seventh) can be neglected as well. Thus, we are left with the simplified equation:

$$
\frac{\partial \langle \overline{u'w'} \rangle}{\partial z} + \frac{\partial \langle UW \rangle}{\partial z} + \frac{1}{\rho}\frac{\partial \langle P \rangle}{\partial x} = \langle Q \rangle .
\tag{10}
$$

Split $U = \langle U \rangle + \tilde{u}$ and $W = \langle W \rangle + \tilde{w}$, where $\langle U \rangle$, $\langle W \rangle$ are spatially averaged velocity components in the direction of flow and vertical directions and $\tilde{u}$ and $\tilde{w}$ are fluctuation in space. Using these expressions, one gets

$$
\langle UW \rangle = \langle U \rangle \langle W \rangle + \langle \tilde{u}\tilde{w} \rangle .
\tag{11}
$$

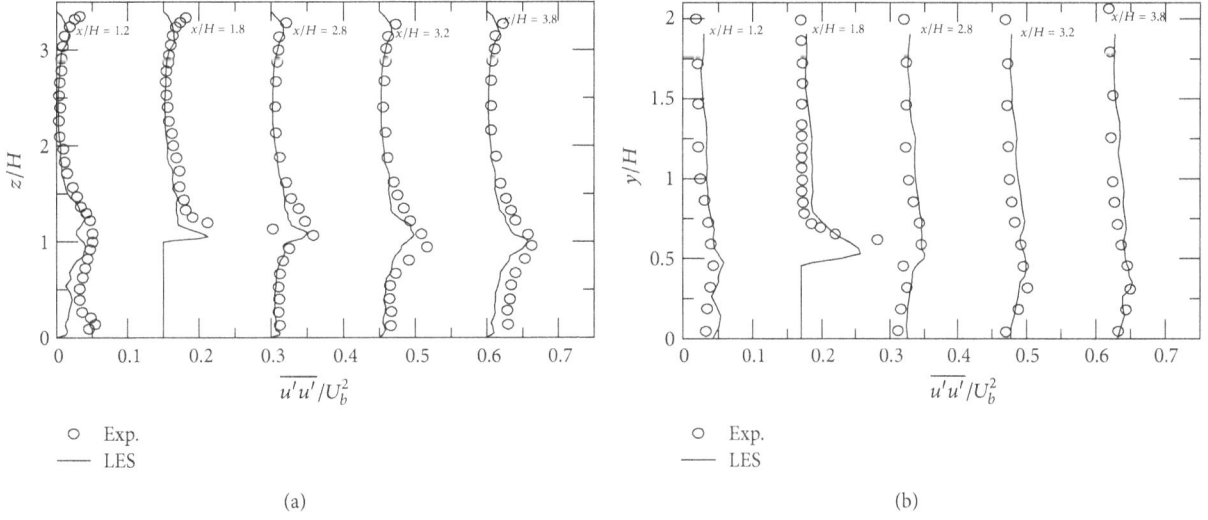

FIGURE 5: (a) Profiles of streamwise Reynolds normal stress on the vertical plane through the center of the cube (i.e. $y/H = 0$). Each profile has been offset from the previous one by 0.15 unit. (b) Profiles of streamwise Reynolds normal stress on the horizontal plane at half cube height (i.e., $z/H = 0.5$). Each profile has been offset from the previous one by 0.15 unit.

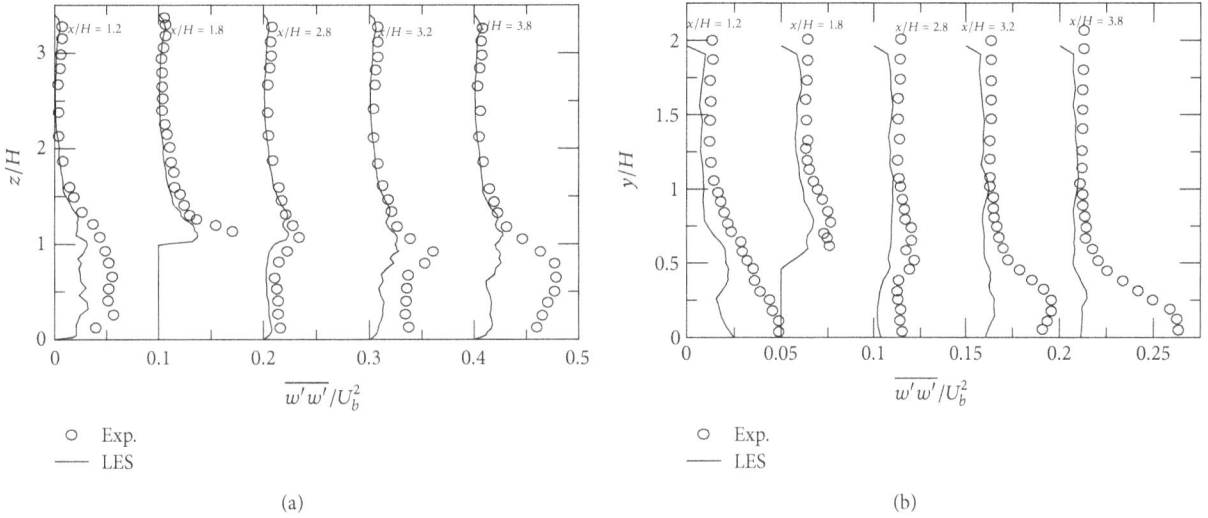

FIGURE 6: (a) Profile of spanwise Reynolds normal stress on the vertical (x-z) plane through the center of the cube ($y/H = 0$). The successive profiles in the figure have been offset by 0.1 unit. (b) Profile of spanwise Reynolds normal stress on the the horizontal (x-y plane at half cube height ($z/H = 0.5$)). The successive profiles in the figure have been offset by 0.05 unit.

Since $\langle W \rangle = 0$ (11) reduces to $\langle UW \rangle = \langle \tilde{u}\tilde{w} \rangle$. Introducing this in (10), one gets the following equation:

$$\frac{\partial \langle \overline{u'w'} \rangle}{\partial z} + \frac{\partial \langle \tilde{u}\tilde{w} \rangle}{\partial z} + \frac{1}{\rho}\frac{\partial \langle P \rangle}{\partial x} = \langle Q \rangle, \qquad (12)$$

where the first and second terms on LHS are the gradient of turbulent and dispersive fluxes, respectively, in the vertical direction and the third term is the gradient of pressure in the flow direction. To study the vertical profiles of the streamwise velocity, turbulent stress, and dispersive stress, we evaluate

these quantities from the result obtained form the simulation using (13) through (16):

$$\langle U \rangle_k = \frac{\sum_i \sum_j (U)_{i,j} V_{i,j}}{\sum_i \sum_j V_{i,j}}, \qquad (13)$$

$$\langle \tilde{u}\tilde{w} \rangle_k = \frac{\sum_i \sum_j (\tilde{u}\tilde{w})_{i,j} V_{i,j}}{\sum_i \sum_j V_{i,j}}, \qquad (14)$$

$$\langle \overline{u'w'} \rangle_k = \frac{\sum_i \sum_j (\overline{u'w'})_{i,j} V_{i,j}}{\sum_i \sum_j V_{i,j}}, \qquad (15)$$

$$\langle TKE \rangle_k = \frac{\sum_i \sum_j (TKE)_{i,j} V_{i,j}}{\sum_i \sum_j V_{i,j}}, \qquad (16)$$

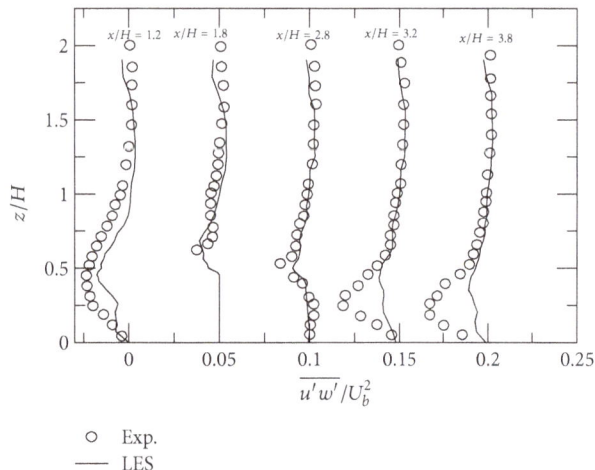

FIGURE 7: Horizontal profile of Reynolds shear stress $u'w'/U_b^2$ on the horizontal (x-y) plane at half cube height ($z/H = 0.5$). Each successive profiles in the figure has been offset by 0.05 unit from the previous one.

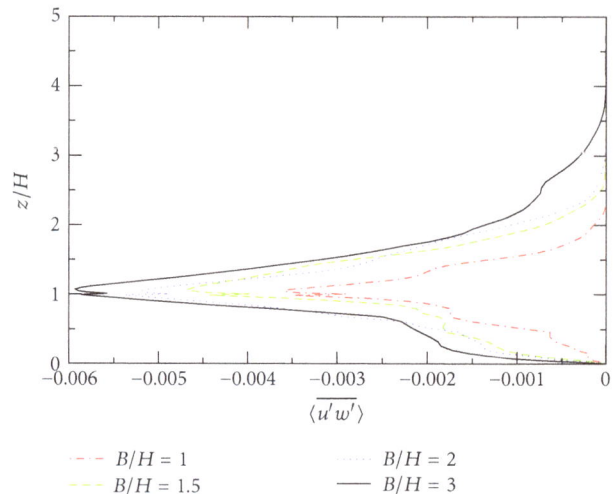

FIGURE 9: Space averaged turbulent flux.

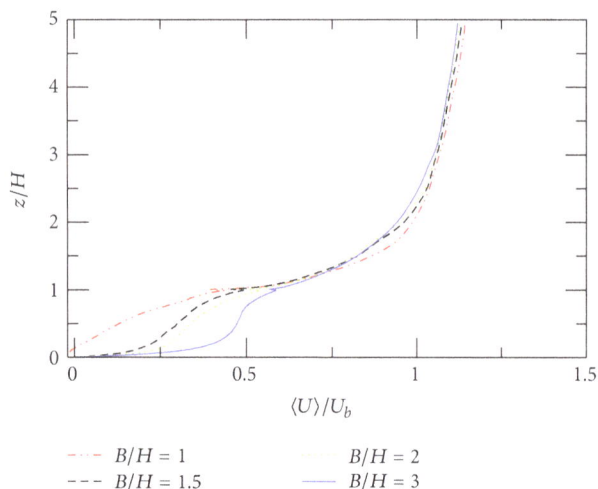

FIGURE 8: Space averaged velocity.

where $V_{i,j,k} = 0$ for the blocked regions and i, j, k are the indexes in the streamwise, spanwise, and vertical directions. As is clear from (13) through (16) the averaging is performed over horizontal planes at different heights.

5.2.1. Mean Velocity.
Figure 8 gives the spatially averaged streamwise velocity profile. This appears to be logarithmic above the canopy but inside the canopy it can be strongly affected by adjacent cubes, as can be seen comparing the cases corresponding to different B/H ratios. In the case of $B/H = 3$, the profile inside the canopy also takes a logarithmic profile because the flow has sufficient time and space to redevelop within the wide canopy. However, as B/H ratio decreases the profile starts to deviate from the normal logarithmic profile, becoming linear for $B/H = 1$.

5.2.2. Turbulent Stresses.
The mesoscale unresolved fluxes can be split into two components: the turbulent part and the dispersive part. The vertical profile as shown in Figure 9 of the turbulent stresses is negative throughout the profile (there is a downward transfer of momentum) implying that the flux is a downgradient. Within the canopy, the turbulent fluxes decrease with height until the top of the cube where a minima is seen. Above the canopy the magnitude increases in height to a height of $3.5H$. These fluxes are then absent above this height, so that there is very little sign of turbulence within this region. Also noteworthy are the linear profiles of the turbulent stress both inside and above the canopy with negative and positive slopes, respectively. These turbulent stresses, which are actually an indication of the transport property of turbulence, decrease with B/H ratio, indicating that an area of widely spaced cubes can experience more turbulence because of greater penetration of eddies within these streets. Conversely, very narrow streets will experience little turbulence because of low eddy penetration. However, there is a need for more exhaustive data analysis to support such a generalization.

5.2.3. Dispersive Stresses.
In most mesoscale models, the dispersive stresses are neglected. In order to study and understand the behavior, these stresses are plotted in Figure 10. From this it is clear that these stresses can be significant and in some cases comparable to turbulent stresses (Figure 9). These dispersive stresses are absent near the bottom wall but increase or decrease (depending upon the B/H) to attain a maxima or minima at half the cube height. Above the canopy these stresses vanish. As with turbulent stresses they reduce to zero above $3H$. Of particular interest is that these dispersive stresses do not exhibit a regular trend when expressed as a function of B/H ratio. When the cubes are wide apart ($B/H = 2, 3$), the dispersive fluxes inside the canopy are negative but upon decreasing the inter-cube spacing beyond a certain point a switch to a positive flux is experienced.

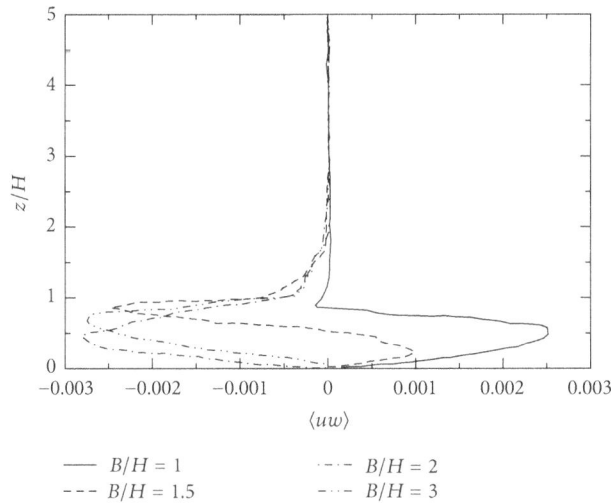

FIGURE 10: Space averaged dispersive flux.

The physical explanation of the dispersive stresses lies in the coherent vortex formed in the canyon.

In order to better understand the behavior of these dispersive fluxes contours of vertical velocity (at the top of the canopy), horizontal velocity and spatially averaged dispersive flux for different B/H are plotted side by side, as shown in Figure 12. Since dispersive flux is defined as $\tilde{u}\tilde{w} = (\langle U \rangle - U)(\langle W \rangle - W)$, its sign will also depend on $\tilde{u}$ and $\tilde{w}$. Here, $\langle U \rangle$ is always positive (Figure 8) and inside the canopies U is mostly negative implying that $\tilde{u}$ will mostly be positive. Thus, the sign of the dispersive flux will depend on the sign of $\tilde{w}$. Also, since $\langle W \rangle$ is negligible one can conclude that the sign of $\tilde{w}$ will be opposite to that of W. Now let us consider each of our cases in turn.

Case 1 ($B/H = 1$). In Figure 11(a), one can see that a jet of fluid impinges the top of the windward side of the cube and is deflected downward, resulting in two clockwise rotating vortices one above the other with the stronger one at the top. These vortices are eccentric with their "eyes" shifted toward the leeward side of the cube. The formation of such eccentric vortices results in negative U and W in most of the regions inside the canopy (Figure 12(a)). However, there are regions near the bottom and top of the leeward side of the cube where there is a positive vertical velocity. At the same time, there are regions near the mid-plane inside the canopy where the vertical velocity is mostly negative. The existence of a mixture of regions of positive and negative vertical velocity fields results in the cancelation of the dispersive fluxes during the averaging over horizontal planes near the top and bottom planes. Near the mid-plane there is no such canceling, and hence a maxima of dispersive flux is obtained.

Case 2 ($B/H = 1.5$). An increase of the B/H ratio from 1 to 1.5 results in the shift of the eye of the primary vortex towards the windward side of the cube. The primary vortex in this case is concentric (Figure 11(b)). The formation of such a nearly concentric vortex results in positive vertical velocities near the leeward side of the cube and negative velocities near the windward side as show in Figure 12(b). On any horizontal plane below $0.5H$ there is greater flux injection (negative vertical velocity) and less ejection (positive vertical velocity). However, above $0.5H$ the situation is exactly the opposite. At the mid-plane itself both the ejection and injection balance each other resulting in a cancelation of dispersive fluxes. As explained earlier, the profile of the dispersive flux depends on the sign of the vertical velocity, so that in this particular case the dispersive fluxes are positive below $0.5H$ and negative above $0.5H$.

Cases 3 and 4 ($B/H = 2, 3$). For both $B/H = 1$ and 1.5, there is a clear demarcation between the zones of positive and negative vertical velocities. Injection from the top takes place near the windward side and ejection near the leeward side of the cube. This is the result of a strong large vortex formed inside the canopy as a consequence of the strong shear forces at the top. However, this behavior changes significantly when B/H is increased to 2 and 3. Contrary to the observations in Cases 1 and 2, in these cases the injection occurs in the middle of the top plane just above the canopy (Figures 11(c) and 11(d)) and the jet impinges not on the top of the cube but on the lower half of the cube. This situation leads to the formation of several tilted vortices resulting in strong ejection at both the leeward and the windward sides of the cube. Also, the tilt in the vortices, which is a result of the injection in the middle, results in positive vertical velocities in most regions and hence negative dispersive fluxes (Figures 12(c) and 12(d)).

In all the cases (I–IV), the vertical velocity above the cube and the canopy is positive and hence it always resulted in a negative dispersive flux. Another thing to be observed in Figures 8 and 9 is that the effect of the cube can be felt up to a height of $3H$ to $4H$, which is consistent with the observations from field experiments [20].

6. Conclusion

LES with standard Smagorinsky' model was used to compute a fully developed turbulent flow over a matrix of cubes. Detailed comparsions between the numerical predictions obtained with the LES and the corresponding experimental data of [13] were conducted. Later on the numerical data generated was used to study the spatially averaged profiles of the velocity, turbulent flux, and dispersive flux. The results of this investigation allow the following conclusions to be drawn.

  (i) Validation of the numerical results: qualitatively, the profiles of mean velocities and Reynolds stresses, the latter including $\overline{u'^2}$, $\overline{w'^2}$, and $\overline{u'w'}$, are generally well represented by the LES model. The greatest discrepancy between the predictions and observations was for $\overline{w'^2}$ within the street canyon of the obstacle array. The underestimation of $\overline{w'^2}$ will lead to an underestimation of the turbulent kinetic energy and hence must be kept in mind. Overall the numerical predictions were very good for the job we are interested in.

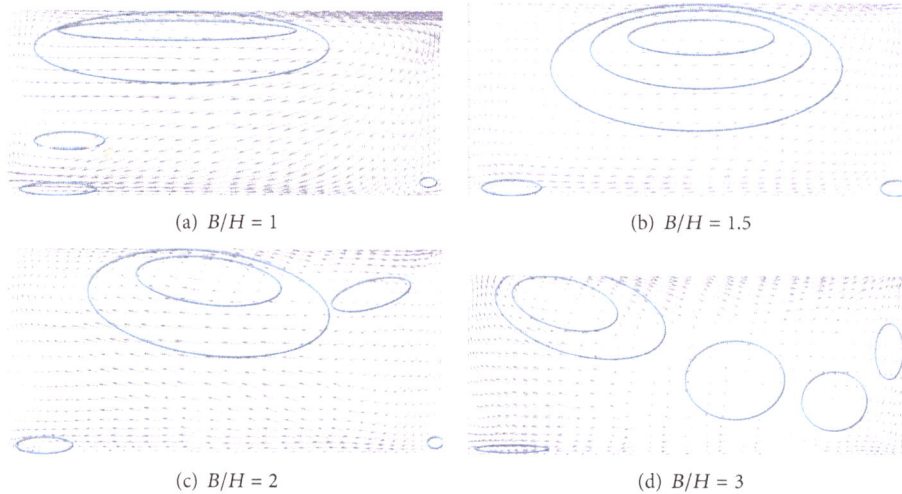

(a) $B/H = 1$

(b) $B/H = 1.5$

(c) $B/H = 2$

(d) $B/H = 3$

FIGURE 11: Time-averaged velocity field and vortices inside the canopy.

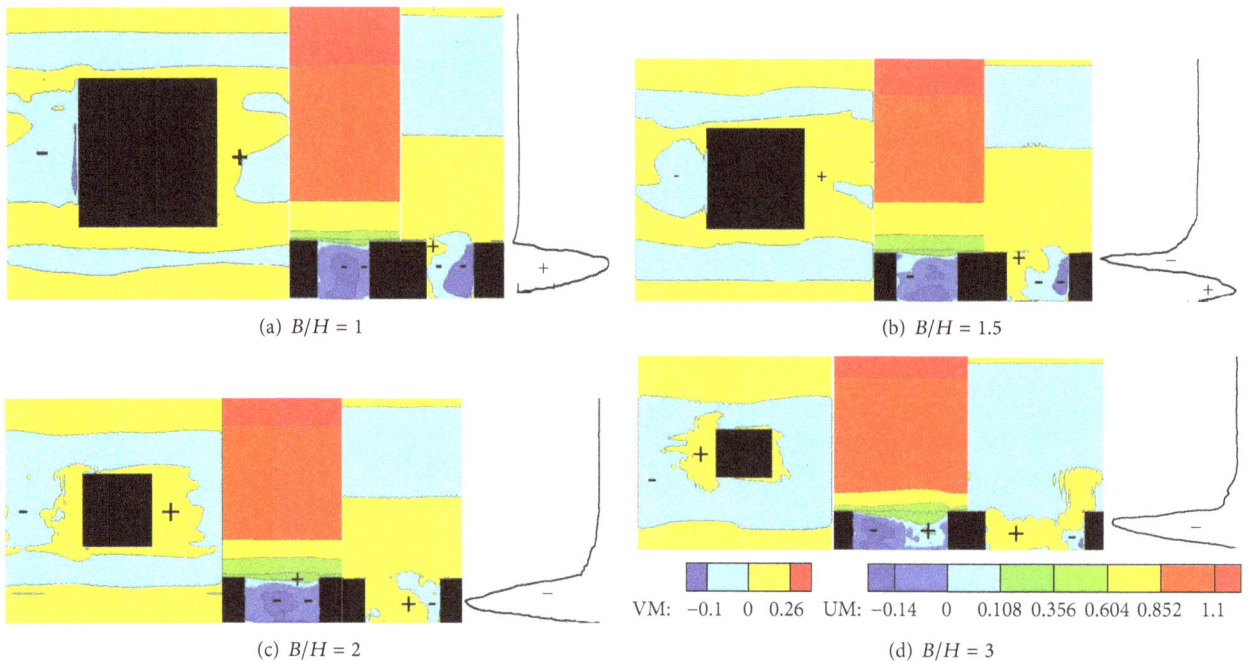

(a) $B/H = 1$

(b) $B/H = 1.5$

(c) $B/H = 2$

(d) $B/H = 3$

VM: -0.1 0 0.26 UM: -0.14 0 0.108 0.356 0.604 0.852 1.1

FIGURE 12: From extreme left to extreme right. Contours of the vertical velocity at the top of the canopy, streamwise velocity at the mid-plane, vertical velocity at the mid-plane, and profile of space averaged dispersive flux (extreme right).

(ii) Spatially averaged quantities: we then carried out the tests for an array of cubes with different inter-cube spacings. The results that were obtained from LES were spatially averaged to derive information useful for urban mesoscale simulations. It was evident that the profiles of turbulent flux and dispersive flux go to zero at nearly three times the cube height, a fact which has been observed in various field experiments [20]. Of particular significance has been the use of the results to explain the behavior of dispersive fluxes. It was observed that for widely spaced array of cubes these fluxes were negative (same sign as the turbulent fluxes). This implies that these fluxes can be modeled in the same way as the turbulent fluxes. However, for high packing density, these profiles started assuming a negative profile which may result in the canceling of other sources in the equation and therefore need to be modelled differently.

However, it must be stressed here that the conclusions drawn from this study are only valid for a regular array of cubes in a neutral atmosphere. In order to generalize such conclusion, more numerical and wind tunnel experiments are required. On a related note, this work leads to pose some interesting scientific questions, such as how much complexity must be added to produce a configuration that gives spatially averaged values similar to those of a real city? Or in other words, which is the simplest configuration that represents a real city? Which combination of parameters (building heights, building shapes, building width, street widths, etc.) is sufficient to characterize city morphology?

References

[1] H. Tennekes and J. L. Lumley, *A First Course in Turbulence*, MIT Press, 1972.

[2] M. Tombrou, A. Dandou, C. Helmis et al., "Model evaluation of the atmospheric boundary layer and mixed-layer evolution," *Boundary-Layer Meteorology*, vol. 124, no. 1, pp. 61–79, 2007.

[3] A. Dandou, M. Tombrou, E. Akylas, N. Soulakellis, and E. Bossioli, "Development and evaluation of an urban parameterization scheme in the Penn State/NCAR Mesoscale Model (MM5)," *Journal of Geophysical Research D*, vol. 110, no. 10, pp. 1–14, 2005.

[4] J. Ehrhard, I. A. Khatib, C. Winkler, R. Kunz, N. Moussiopoulos, and G. Ernst, "The microscale model MIMO: development and assessment," *Journal of Wind Engineering and Industrial Aerodynamics*, vol. 85, no. 2, pp. 163–176, 2000.

[5] R. Kunz, I. A. Khatib, and N. Moussiopoulos, "Coupling of mesoscale and microscale models—an approach to simulate scale interaction," *Environmental Modelling and Software*, vol. 15, no. 6-7, pp. 597–602, 2000.

[6] R. P. Hosker, "Flow and diffusion near obstacles," in *Atmospheric Science and Power Production*, pp. 241–326, 1984.

[7] R. Bornstein, "Mean diurnal circulation and thermodynamic evolution of urban boundary layers," in *Modeling the Urban Boundary Layer*, pp. 52–94, American Meteorological Society, Boston, Mass, USA, 1987.

[8] H. Kusaka, H. Kondo, Y. Kikegawa, and F. Kimura, "A simple single-layer urban canopy model for atmospheric models: comparison with multi-layer and slab models," *Boundary-Layer Meteorology*, vol. 101, no. 3, pp. 329–358, 2001.

[9] H. Kondo, Y. Genchi, Y. Kikegawa, Y. Ohashi, H. Yoshikado, and H. Komiyama, "Development of a Multi-Layer Urban Canopy Model for the analysis of energy consumption in a big city: structure of the Urban Canopy Model and its basic performance," *Boundary-Layer Meteorology*, vol. 116, no. 3, pp. 395–421, 2005.

[10] A. Martilli and J. L. Santiago, "CFD simulation of air flow over a regular array of cubes. Part II: analysis of spatial average properties," *Boundary-Layer Meteorology*, vol. 122, no. 3, pp. 635–654, 2007.

[11] J. L. Santiago, O. Coceal, A. Martilli, and S. E. Belcher, "Variation of the sectional drag coefficient of a group of buildings with packing density," *Boundary-Layer Meteorology*, vol. 128, no. 3, pp. 445–457, 2008.

[12] Y. Cheng, F. S. Lien, E. Yee, and R. Sinclair, "A comparison of large Eddy simulations with a standard k-ε Reynolds-averaged Navier-Stokes model for the prediction of a fully developed turbulent flow over a matrix of cubes," *Journal of Wind Engineering and Industrial Aerodynamics*, vol. 91, no. 11, pp. 1301–1328, 2003.

[13] E. Meinders, *Experimental study of heat transfer in turbulent flows over wall-mounted cubes [Ph.D. thesis]*, Faculty of Applied Sciences, Delft University of Technology, Delft, The Netherlands, 1998.

[14] E. Meinders and K. Hanjalić, "Vortex structure and heat transfer in turbulent flow over a wall-mounted matrix of cubes," *International Journal of Heat and Fluid Flow*, vol. 20, no. 3, pp. 255–267, 1999.

[15] TransAT, Transport Phenomena Analysis Tool, 2009, http://ascomp.ch/.

[16] J. Smagorinsky, "General circulation experiments with the primitive equations," *Monthly Weather Review*, vol. 91, no. 3, pp. 99–164, 1963.

[17] M. Germano, U. Piomelli, P. Moin, and W. H. Cabot, "A dynamic subgrid-scale eddy viscosity model," *Physics of Fluids A*, vol. 3, no. 7, pp. 1760–1765, 1991.

[18] D. Lilly, "A proposed modification of the Germano subgrid-scale closure method," *Physics of Fluids A*, vol. 4, no. 3, pp. 633–635, 1992.

[19] E. Driest, "On turbulent flow near a wall," *Journal of the Aeronautical Sciences*, vol. 23, no. 11, pp. 1007–1011, 1956.

[20] M. W. Rotach, "Turbulence close to a rough urban surface. Part I: reynolds stress," *Boundary-Layer Meteorology*, vol. 65, no. 1-2, pp. 1–28, 1993.

10

Evaluation of Parameterization Schemes in the WRF Model for Estimation of Mixing Height

R. Shrivastava,[1] **S. K. Dash,**[2] **R. B. Oza,**[1] **and D. N. Sharma**[3]

[1] *Radiation Safety Systems Division, Bhabha Atomic Research Centre, Mumbai 400 085, India*
[2] *Centre for Atmospheric Sciences, Indian Institute of Technology Delhi, New Delhi 110 016, India*
[3] *Health Safety and Environment Group, Bhabha Atomic Research Centre, Mumbai 400 085, India*

Correspondence should be addressed to R. Shrivastava; roopa@barc.gov.in

Academic Editor: Daiwen Kang

This paper deals with the evaluation of parameterization schemes in the WRF model for estimation of mixing height. Numerical experiments were performed using various combinations of parameterization schemes and the results were compared with the mixing height estimated using the radiosonde observations taken by the India Meteorological Department (IMD) at Mangalore site for selected days of the warm and cold season in the years 2004–2007. The results indicate that there is a large variation in the mixing heights estimated by the model using various combinations of parameterization schemes. It was seen that the physics option consisting of Mellor Yamada Janjic (Eta) as the PBL scheme, Monin Obukhov Janjic (Eta) as the surface layer scheme, and Noah land surface model performs reasonably well in reproducing the observed mixing height at this site for both the seasons as compared to the other combinations tested. This study also showed that the choice of the land surface model can have a significant impact on the simulation of mixing height by a prognostic model.

1. Introduction

Prognostic atmospheric models are used as meteorological drivers to air pollution models in the absence of representative measured meteorological data for a site. These models generally provide wind speed, wind direction, temperature, humidity, rainfall, and mixing height values to the air pollution models. Many times, the resolution at which these models are integrated is too coarse to resolve the exchanges of heat, momentum, and moisture taking place at the air soil interface and hence these exchanges have to be parameterized in atmospheric models. Parameterization schemes may also be included in an atmospheric model for the representation of atmospheric phenomena whose explicit treatment may become too prohibitive due to cost and computer limitations. A weather model includes parameterizations for radiation, surface layer fluxes, turbulence, cumulus convection, and clouds. Generally there are six to seven schemes available for representation of each of these processes with its own merits and demerits depending upon the terrain, geography, and climate of the area under consideration. Mixing height is an important input to air pollution models since the transport and extent of mixing of pollutants depend on it. The mixing in the atmosphere primarily takes place through convective and mechanical processes. During the daytime, differential heating due to solar radiation sets up strong thermals in the atmosphere and the convective processes dominate whereas, during the nighttime, mechanical processes are responsible for the turbulent mixing. The variation of mixing height with varying parameterization schemes was studied by other investigators also. For example, Shin and Hong [1] intercompared five Planetary Boundary Layer (PBL) schemes in the WRF model for a single day from the Cooperative Atmosphere Surface Exchange Study (CASES-99) field program. They reported a large variation in the mixing height values computed by the five schemes in the daytime and nighttime. Similarly Han et al. [2] had also evaluated five PBL schemes in the MM5 model for the East Asian domain for March 2001 and reported a large difference in the mixing heights predicted by the various combinations of parameterization schemes.

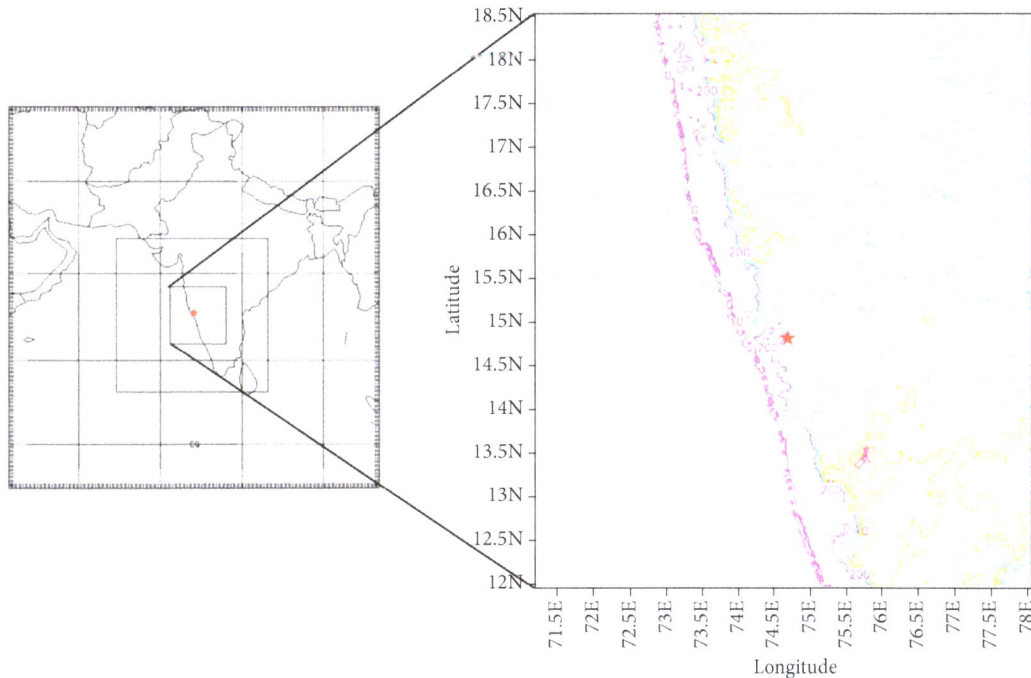

FIGURE 1: Model domains and the topography of the inner domain (red star denotes the plant site).

TABLE 1: Description of parameterization schemes.

Land surface	Surface layer	Planetary boundary layer	Combination
Thermal diffusion	Monin Obukhov	YSU	A
	Monin Obukhov	ACM2 (Pleim)	B
	Monin Obukhov (Janjic Eta)	Mellor Yamada Janjic (Eta)	D
	QNSE	QNSE	F
Noah land surface model	Monin Obukhov	YSU	C
	Monin Obukhov (Janjic Eta)	Mellor Yamada Janjic (Eta)	E

YSU: Yonsei University; ACM2: Asymmetric Convective Model version 2; QNSE: Quasi-Normal Scale Elimination.

Hu et al. [3] compared the 3-month mean diurnal variations of the simulated mixing heights using different combination of parameterization schemes in the WRF model with the observed values for South East Texas.

The present study focuses on the evaluation of parameterization schemes in the WRF model for estimation of mixing height for Kaiga site. Kaiga is one of the sites where nuclear power plants are operated for generation of electricity by Nuclear Power Corporation of India Ltd (NPCIL). It is a complex site with tall hills, evergreen forests, a reservoir, and so forth. Weather forecasting is important in the nuclear industry because of the aid it can provide in handling an emergency situation. However the weather forecast model needs to be validated/tuned with proper selection of parameterization schemes for it to be used in case of an emergency situation. Also many times a weather model is coupled with an offline atmospheric dispersion model for prediction of concentration. In order to validate the mixing height estimated by the WRF model, the radiosonde observations taken by the India Meteorological Department (IMD) are utilized. Since radiosonde data for Kaiga site are not available, the radiosonde data from the nearby radiosonde observation station, Mangalore, are utilized for this purpose. The Mangalore station is at a distance of ~300 km from Kaiga site. It should be noted that the model values were also extracted at this location for comparison with observed values. The cases selected for simulation correspond to selected days in the warm and cool season. Description of the model domain, numerical experiments, WRF model set-up, and physics parameterizations used is given in Section 2 and the results are presented in Section 3.

2. Model and Methodology

The study domain covers south west India centered at the Kaiga Generating Station (KGS) ($14°51'48''$N, $74°26'31''$ E) as shown in Figure 1. The WRF model version 3.1.1 (Skamarock et al. [4]; Wei et al. [5]) is integrated using three nested domains with grid spacing varying from 27 km, 9 km, and

TABLE 2: Description of experiments.

Simulation period	Case	Experiment
600Z 26 December 2004–00Z 29 December 2004		TMN1
00Z 17 January 2005–00Z 20 January 2005	Minimum temperature	TMN2
00Z 26 January 2006–00Z 29 January 06		TMN3
00Z 14 November 2007–00Z 17 November 07		TMN4
00Z 13 March 2004 –00Z 16 March 2004		TMX1
00Z 04 April 2005–00Z 07 April 2005	Maximum temperature	TMX2
00Z 23 February 2006–00Z 26 February 2006		TMX3
00Z 24 April 2007–00Z 27 April 2007		TMX4

TABLE 3: Comparison of model generated mixing height with the observed values for cases TMN1–TMN4.

	A (m)	B (m)	C (m)	D (m)	E (m)	F (m)	O (m)	Case study
12Z 26 December 2004	77	136	112	145	**146**	800	300	
00Z 27 December 2004	**248**	77	339	391	392	565	150	
12Z 27 December 2004	227	169	107	145	**398**	1069	300	TMN1
00Z 28 December 2004	405	**159**	546	395	393	2577	150	
12Z 28 December 2004	34	164	118	146	**399**	400	400	
00Z 29 December 2004	305	103	375	**144**	396	1064	100	
12Z 17 January 2005	256	293	493	401	**808**	580	900	
00Z 18 January 2005	29	33	30	145	**145**	144	350	
12Z 18 January 2005	112	189	297	1568	**580**	2073	600	TMN2
00Z 19 January 2005	30	38	32	145	**144**	144	200	
12Z 19 January 2005	215	213	360	400	**806**	807	700	
00Z 20 January 2005	159	50	30	144	**145**	144	100	
12Z 26 January 2006	350	258	417	146	**575**	575	550	
00Z 27 January 2006	133	53	**115**	142	142	253	100	
12Z 27 January 2006	237	192	236	146	**399**	258	650	TMN3
00Z 28 January 2006	**121**	39	234	394	143	1540	100	
00Z 29 January 2006	**232**	84	269	142	252	2041	200	
00Z 14 November 2007	309	**203**	347	145	398	397	200	
12Z 14 November 2007	137	54	171	142	**143**	253	100	
00Z 15 November 2007	133	206	**105**	146	146	258	100	TMN4
12Z 15 November 2007	101	57	29	142	143	**565**	480	
00Z 16 November 2007	35	231	202	145	**145**	395	100	
12Z 16 November 2007	40	37	29	141	141	**787**	400	
	3/23	3/23	1/23	1/23	13/23	2/23		

The bold values refer to the best match between model and observation.

3 km. The terrain and land use data for the innermost domain with 3 km grid resolution were taken from 30′ data, for domain with 9 km resolution from 5′ data and for the domain with 27 km resolution from 10′ data available from the United States Geological Survey (USGS). The parameterization schemes tested in this study are defined in Table 1. The PBL schemes tested are Yonsei University, (YSU, Hong et al. [6]), Mellor Yamada Janjic Eta (MYJ, Janjic [7, 8]), Asymmetric Convective Model, (ACM2, Pleim [9]), and Quasi Normal Scale Elimination (QNSE, Sukoriansky et al. [10, 11]). The surface layer (SL) schemes used are Monin Obukhov (Monin and Obukhov [12]), Monin Obukhov Janjic Eta (MYJ, Monin and Obukhov [12], Janjic [13]), and Quasi-Normal Scale

Elimination (QNSE, Sukoriansky et al. [10, 11]) and the land surface models (LSM) used are thermal diffusion (Dudhia [14]) and the Noah land surface model (Chen and Dudhia [15]). The other physical parameterizations are identical for all experiments which are Rapid radiation Transfer Model for long-wave radiation (RRTM, Mlawer et al. [16]), Dudhia (Dudhia [17]) for short-wave radiation, Kain Fritsch scheme for cumulus parameterization (Kain and Fritsch [18], J. S. Kain [19]) and Ferrier new Eta (Ferrier et al. [20]) scheme for representing microphysical processes in the clouds. A total of six combinations of parameterization schemes are considered which are defined in the Table 1. The National Centers for Environmental Prediction (NCEP) Final Analysis data

TABLE 4: Comparison of model generated mixing height with the observed values for cases TMX1–TMX4.

	A (m)	B (m)	C (m)	D (m)	E (m)	F (m)	O (m)	Case study
12Z March 2004	256	293	493	401	**808**	580	900	
00Z 14 March 2004	29	33	30	145	**145**	144	350	
12Z 14 March 2004	112	189	297	1568	**580**	2073	900	TMX1
00Z 15 March 2004	30	38	32	145	**144**	144	200	
12Z 15 March 2004	215	213	360	400	**806**	807	700	
00Z 16 March 2004	159	50	30	144	**145**	144	100	
12Z 04 April 2005	257	365	399	580	**809**	809	900	
00Z 05 April 2005	30	39	30	145	**146**	145	150	
12Z 05 April 2005	86	237	338	1575	**581**	2084	900	TMX2
00Z 06 April 2005	30	59	30	145	**146**	145	350	
12Z 06 April 2005	117	272	226	147	403	**587**	800	
00Z 07 April 2005	30	39	30	146	147	**259**	300	
12Z 23 February 2006	117	153	276	146	**577**	580	600	
00Z 24 February 2006	89	35	42	144	**143**	1062	100	
12Z 24 February 2006	84	1017	206	146	**576**	2058	400	TMX3
00Z 25 February 2006	134	**86**	29	143	571	254	100	
12Z 25 February 2006	331	223	355	398	**575**	398	900	
00Z 26 February 2006	198	52	29	395	**144**	2049	100	
12Z 24 April 2007	429	409	533	811	**809**	1083	800	
00Z 25 April 2007	30	53	41	**145**	579	258	150	
12Z 25 April 2007	214	263	350	588	**584**	819	700	TMX4
00Z 26 April 2007	58	**54**	30	147	146	261	50	
12Z 26 April 2007	422	436	443	585	585	**1088**	900	
00Z 27 April 2007	30	57	30	146	**146**	260	100	
No. of successes	—	2/24	—	1/24	18/24	3/24		

The bold values refer to the best match between model and observation.

available at 6 hourly interval on a $1° \times 1°$ resolution are used to supply initial and boundary conditions. The model uses two-way nested boundary conditions and they are updated every 6 hours. A total of eight simulation cases are considered in the present study which are based on either maxima or minima in temperature and these are described in Table 2. This study focuses on the comparison of mixing height estimated by the model in the various combinations of parameterization schemes with the observed values. Each simulation is for a three-day period one on the day of occurrence, one day prior, and one day later to it. The mixing height computation by the model differs according to the PBL scheme used. The schemes YSU and ACM2 estimate mixing height based on the Richardson number whereas the remaining two PBL schemes namely MYJ and QNSE diagnose the mixing height on the basis of the turbulent kinetic energy (TKE). While both the approaches may be theoretically correct, the one which produces values of mixing height which compare well with the observation can be considered to be the better approach. In addition to the model's own diagnosis of mixing height, the temperature profile given by the model was also used to estimate the mixing height based on the Holtzworth [21] method. In this case, the variation between different combinations of schemes due to varying methodologies is eliminated and the differences observed are only due to those in the temperature profile of the model. Both the methods of estimating mixing height are in turn compared with the observation.

3. Results and Discussion

The mixing heights obtained in all combinations of parameterization schemes tested are shown in Tables 3 and 4 for the experiments TMN1–TMN4 and TMX1–TMX4, respectively, with the best performing combinations presented in bold font. The observed mixing heights are calculated from the radiosonde temperature profile following the method of Holtzworth [21]. For a three-day simulation, six radiosonde profiles are available for comparison (one each at 00Z and 12Z). However, sometimes, a particular sounding may not be available and hence that time is not considered in the analysis for example, 12Z on January 28, 2006. In order to remove the dependence of mixing height on parameterization schemes, the mixing height values are also generated by using Holtzworth method on the model derived temperature profiles and the results are presented in Table 5 for experiments TMN1–TMN4 and in Table 6 for experiments TMX1–TMX4.

TABLE 5: Comparison of model generated mixing height (using Holtzworth's method) with the observed values for experiments TMN1–TMN4.

	A (m)	B (m)	C (m)	D (m)	E (m)	F (m)	O (m)	Case study
12Z 26 December 2004	1300	1400	550	1200	**450**	1250	300	
00Z 27 December 2004	100	100	—	**150**	50	200	150	
12Z 27 December 2004	1350	1300	650	1300	**400**	1450	300	TMN1
00Z 28 December 2004	—	80	—	50	**80**	—	150	
12Z 28 December 2004	1100	800	500	1200	**500**	1300	400	
00Z 29 December 2004	—	80	—	70	**80**	50	100	
12Z 17 January 2005	600	600	600	580	**550**	625	900	
00Z 18 January 2005	200	180	**200**	150	100	150	350	
12Z 18 January 2005	1300	1200	1000	1400	**950**	1350	600	TMN2
00Z 19 January 2005	250	300	150	**200**	100	250	200	
12Z 19 January 2005	1100	1050	900	1150	**850**	1150	700	
00Z 20 January 2005	200	200	80	150	**80**	200	100	
12Z 26 January 2006	1400	1800	1050	2000	**900**	2000	550	
00Z 27 January 2006	50	200	50	200	**100**	200	100	
12Z 27 January 2006	1450	950	750	1400	**700**	1400	650	TMN3
00Z 28 January 2006	50	50	—	50	**100**	100	100	
00Z 29 January 2006	100	80	—	100	80	**120**	200	
00Z 14 November 2007	**150**	—	—	—	—	—	200	
12Z 14 November 2007	**550**	650	750	1250	600	1300	100	
00Z 15 November 2007	50	120	—	120	**100**	180	100	TMN4
12Z 15 November 2007	500	**490**	400	900	700	950	480	
00Z 16 November 2007	110	120	50	150	**100**	200	100	
12Z 16 November 2007	1250	1100	1000	1200	**900**	1200	400	
No. of successes	2/23	1/23	1/23	2/23	16/23	1/23		

The bold values refer to the best match between model and observation.

Here also, the best performing combinations are presented in bold font. The Holtzworth method works on the basis of intersection of the temperature profile with the dry adiabatic line drawn from the surface with the maximum temperature of the day for the 12Z mixing height and with the minimum temperature +5°C for the 00Z mixing height. In cases where the two curves do not intersect, the mixing height cannot be determined. These are represented by a "—" in Tables 5 and 6. The model derived mixing heights are shown in columns A–F and the observed mixing height is presented in column O. From Tables 3 and 4, one can estimate the magnitude of variation among the different parameterization schemes for estimation of mixing height at a given time making a parametric study so very essential. The differences in the model estimates of mixing height arise because of the various methods of calculation used in the schemes. For example at 12Z 26th December 2004, the observed mixing height from the radiosonde data was 300 m. The model generated values range from 77 m to 800 m indicating a large variation as compared to the observation. Similar variation is noted at other times also. It is also important to note the role of the land surface model in estimation of mixing height. In an atmospheric model, the land surface model is used to calculate the exchange coefficients of heat, momentum, and moisture at the air soil interface which is used in the estimation of heat, momentum, and moisture fluxes. These

fluxes provide the lower boundary condition for vertical transport in the PBL. The five-layer thermal diffusion model is a simple soil temperature model. The layers are 1 cm, 2 cm, 4 cm, 8 cm, and 16 cm thick. Below this the temperature is fixed at a deep layer average value. The soil moisture values are season dependent constant values based on land use categorization without considering the effects of vegetation. The Noah land surface model on the other hand predicts soil temperature and moisture in 4 layers extending 10 cm, 30 cm, 60 cm, and 100 cm from the surface and summing up to 2 m below the surface. It includes land use, monthly vegetation fraction, and evapotranspiration in the estimation of sensible and latent heat fluxes. In general, it is seen that the values of mixing height obtained using the Noah land surface model are better than those obtained with the thermal diffusion model for identical combinations of PBL/SL schemes (i.e., E is better than D and C is better than A) due to better representation of physical processes leading to transfer of heat, momentum, and moisture at the air soil interface. In fact Stensrud [22] says that "if variations in surface conditions occur over small scales, then the resulting rapid horizontal changes in the values of surface fluxes can lead to the development of non-classical mesoscale circulations, such as vegetation or inland sea breezes." The model's representation of such processes are important for weather prediction for a complex terrain sites. For a given land surface model,

Table 6: Comparison of model generated mixing height (using Holtzworth method) with the observed values for experiments TMX1–TMX4.

	A (m)	B (m)	C (m)	D (m)	E (m)	F (m)	O (m)	Case study
12Z 13 March 2004	1200	950	950	1200	**920**	1450	900	
00Z 14 March 2004	220	150	250	**250**	200	200	350	
12Z 14 March 2004	350	300	500	**900**	500	950	900	TMX1
00Z 15 March 2004	150	180	180	200	**200**	250	200	
12Z 15 March 2004	1150	950	800	1450	**750**	1400	700	
00Z 16 March 2004	300	200	100	300	**100**	400	100	
12Z 04 April 2005	950	950	950	1300	**900**	1300	900	
00Z 05 April 2005	**100**	200	350	250	300	250	150	
12Z 05 April 2005	100	—	500	700	550	**800**	900	TMX2
00Z 06 April 2005	**250**	180	100	200	180	200	350	
12Z 06 April 2005	300	**400**	200	250	300	250	800	
00Z 07 April 2005	200	**300**	200	200	150	200	300	
12Z 23 February 2006	1050	1150	750	1600	**700**	1600	600	
00Z 24 February 2006	—	**100**	80	80	—	100	100	
12Z 24 February 2006	1400	1300	700	1500	**600**	1500	400	TMX3
00Z 25 February 2006	80	150	50	80	**80**	150	100	
12Z 25 February 2006	1300	1350	950	1450	**850**	1450	900	
00Z 26 February 2006	—	50	50	80	**50**	150	100	
12Z 24 April 2007	700	700	750	700	**750**	700	800	
00Z 25 April 2007	100	120	200	150	**150**	200	150	
12Z 25 April 2007	400	350	450	400	450	**500**	700	TMX4
00Z 26 April 2007	50	100	80	80	**50**	150	50	
12Z 26 April 2007	800	700	700	1050	**700**	1100	900	
00Z 27 April 2007	120	100	120	100	**120**	180	100	
No. of successes	2/24	3/24	—	2/24	15/24	2/24	—	

The bold values refer to the best match between model and observation.

Table 7: Statistical analysis.

Statistic	TMN_E	TMN_HE	TMX_E	TMX_HE
AVG_WRF (m)	318	385	423	415
AVG_O (m)	320	320	498	493
MAE (m)	123	159	179	125
IOA	0.86	0.83	0.84	0.91
CORR	0.74	0.76	0.75	0.86
RMSE (m)	160	222	229	186

AVG_WRF: average of model, AVG_O: average of observation, MAE: mean absolute error, IOA: index of agreement, CORR: correlation coefficient, RMSE: root mean square error, TMN_E: experiments TMN1–TMN4, TMN_HE: experiments TMN1–TMN4 but using Holtzworth method, and TMX_E, TMX_HE: same as TMN_E, TMN_HE for the case studies TMX1–TMX4.

say the Noah LSM, the results with the Monin Obukhov (Janjic Eta) PBL scheme and the Mellor Yamada Janjic (Eta) surface layer scheme are better than those with the YSU PBL and Monin Obukhov surface layer scheme (i.e., E is better than C). Hence it is seen that it is a combined effect of all parameterization schemes which translates into a good simulation of mixing height by an atmospheric model. As already mentioned, the Holtzworth method was also used to estimate the mixing height based on the model derived temperature profile under various combinations of parameterization schemes. For a comparison of the two methods, the scatter plots of the model generated mixing heights with the observations are shown in Figures 2 and 3 for the case studies TMN1–TMN4 and TMX1–TMX4, respectively. Also, many statistical parameters like average of model, average of observation, mean absolute error, index of agreement, and root mean square error were computed and are presented in Table 7. From Figure 2, it is seen that the model's own diagnosis of mixing height is underpredicted whereas the same obtained using Holtzworth method on the model generated temperature profile is overpredicted. However the Holtzworth method used on the model derived temperature profile has a smaller intercept as compared to that of the model's own diagnosis of mixing height. The correlation coefficient and index of agreement obtained in both the cases are comparable. From Table 7, referring to the errors, namely,

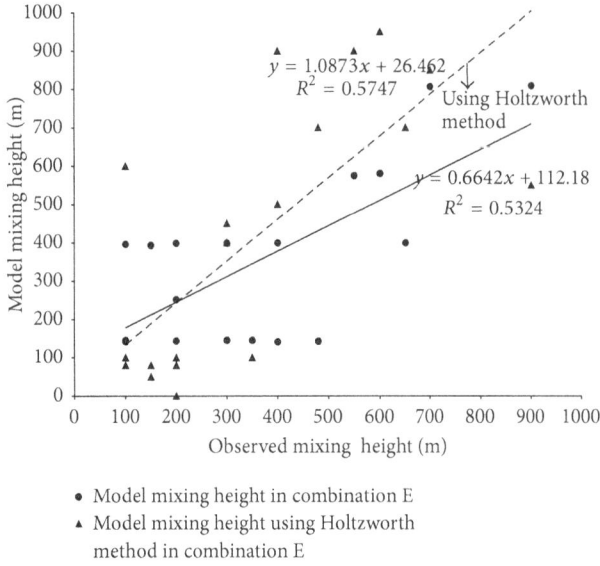

$y = 1.0873x + 26.462$
$R^2 = 0.5747$

Using Holtzworth method

$y = 0.6642x + 112.18$
$R^2 = 0.5324$

• Model mixing height in combination E
▲ Model mixing height using Holtzworth method in combination E

FIGURE 2: Comparison of model generated (in combination E) mixing height with the observed values for experiments TMN1–TMN4.

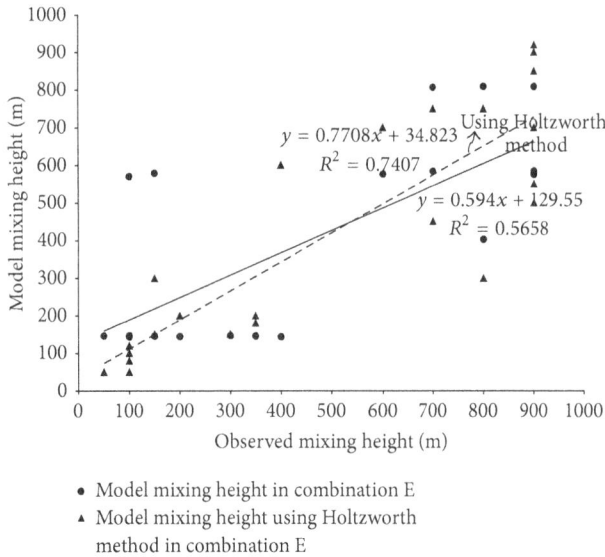

$y = 0.7708x + 34.823$
$R^2 = 0.7407$

Using Holtzworth method

$y = 0.594x + 129.55$
$R^2 = 0.5658$

• Model mixing height in combination E
▲ Model mixing height using Holtzworth method in combination E

FIGURE 3: Comparison of model generated (in combination E) mixing height with the observed values for experiments TMX1–TMX4.

mean absolute and root mean square error, it is seen that the model's own estimation of mixing height is better than that obtained using the Holtzworth method on the model generated temperature profile on the basis of lower values of errors. The difference in the mean absolute error between the two methods is ~12% in an average mixing height of 320 m. Hence making a definite statement about any of the two methods being superior is not possible for the case studies TMN1–TMN4 based on the current simulations carried out. Similar analysis was done for the case studies TMX1–TMX4 also, with the scatter plot being shown in Figure 3. From the

plot and the statistical analysis in Table 7, it can be concluded that for these cases, the mixing height computed using the Holtzworth method on the model generated temperature profile is better than the model's own estimation of mixing height based on the turbulent kinetic energy. All the statistical parameters computed for the case TMX_HE (referring to Holtzworth method on the model derived temperature profile for experiments TMX1–TMX4) are superior to the case TMX_E (referring to experiments TMX1–TXM4). From Table 4, it is seen that on many occasions the mixing height at 12Z is underestimated by the model. Even if the land surface model chosen accounts for the physical processes occurring in the atmosphere, the coefficients/constants being used in the respective algorithms may require further refinement. When the Holtzworth method is applied on the model derived temperature profile in combination E, this underestimation is eliminated. Hence for the experiments TMX1–TXM4 corresponding to selected days in the warm season, the mixing height estimated using the Holtzworth method on the model derived temperature profile is better than the model's own estimation of mixing height. It is also to be noted that the application of the Holtzworth method on other combinations of parameterization schemes A, B, D, and F in this study leads to deterioration of model results and should be avoided. Finally to summarize, this study has shown that the set of parameterizations schemes defined by the combination E gives reasonable results for 13 out of 23 case studies for the cool season and 18 out of 24 case studies for the warm season mainly due to the Noah land surface model used. Nevertheless, this mixing height should still be verified by using the Holtzworth method on the model derived temperature profile obtained using the same combination of parameterization schemes for those locations where radiosonde data are available. This study also shows that the state-of-the-art parameterizations schemes in the WRF model still have scope for improvement.

Index of agreement

$$IOA = 1 - \frac{\sum_{i=1}^{N} (P_i - O_i)^2}{\sum_{i=1}^{N} (|P_i - O_{mean}| + |O_i - O_{mean}|)^2}. \quad (1)$$

Pearson correlation coefficient

$$CORR = \left(N \left(\sum_{i=1}^{N} O_i P_i \right) - \left(\sum_{i=1}^{N} O_i \right) \left(\sum_{i=1}^{N} P_i \right) \right)$$

$$\times \left(\left[N \left(\sum_{i=1}^{N} O_i^2 \right) - \left(\sum_{i=1}^{N} O_i \right)^2 \right] \right.$$

$$\left. \times \left[N \left(\sum_{i=1}^{N} P_i^2 \right) - \left(\sum_{i=1}^{N} P_i \right)^2 \right] \right)^{-1/2}. \quad (2)$$

Mean absolute error

$$MAE = \frac{1}{N} \sum_{i=1}^{N} |P_i - O_i|. \quad (3)$$

Root mean square error

$$\text{RMSE} = \sqrt{\frac{1}{N}\sum_{i=1}^{N}\left(P_i - O_i\right)^2}. \qquad (4)$$

4. Conclusions

The numerical experiments carried out for the selection of the optimum combination of parameterization schemes for estimation of mixing height have shown that the physics option consisting of Mellor Yamada Janjic (Eta) as the PBL scheme, Monin Obukhov Janjic (Eta) as the Surface Layer scheme, and Noah land surface model performs reasonably well in reproducing the observed mixing height for almost all the cases considered for simulation as compared to the other combinations tested. The role played by the land surface model for a good simulation of mixing height was also emphasized. The application of the Holtzworth method on the model derived temperature profile to estimate mixing height as opposed to the model diagnosed mixing height suggests that during the warm season, the model derived mixing height based on the Holtzworth method matches reasonably well with the observations as opposed to the model diagnosed mixing height. However, for the cold season no such conclusive statement can be made. Such studies on the proper choice of parameterization schemes at a power plant site are useful when a prognostic weather model like WRF is to be coupled with an air quality model for atmospheric dispersion studies.

Conflict of Interests

The authors declare that there is no conflict of interests regarding the publication of this paper.

Acknowledgments

The authors express their gratitude to Dr. K. S. Pradeepkumar, Head Radiation Safety Systems Division, BARC, and Dr. R. N. Nair, (former Head), Environmental Modeling Section, BARC, for constant encouragement and fruitful discussions during the period of study. The authors also wish to thank the anonymous reviewers for useful suggestions and comments which have improved the quality of the paper. The FNL data for this study are from the Research Data Archive (RDA) which is maintained by the Computational and Information Systems Laboratory (CISL) at the National Center for Atmospheric Research (NCAR). NCAR is sponsored by the National Science Foundation (NSF). The original data are available from the RDA (http://rda.ucar.edu/) in dataset no. ds083.2.

References

[1] H. H. Shin and S. Y. Hong, "Intercomparison of planetary boundary-layer parameterizations in the WRF model for a single day from CASES-99," *Boundary-Layer Meteorology*, vol. 139, no. 2, pp. 261–281, 2011.

[2] Z. Han, H. Ueda, and J. An, "Evaluation and intercomparison of meteorological predictions by five MM5-PBL parameterizations in combination with three land-surface models," *Atmospheric Environment*, vol. 42, no. 2, pp. 233–249, 2008.

[3] X. M. Hu, J. W. Nielsen-Gammon, and F. Zhang, "Evaluation of three planetary boundary layer schemes in the WRF model," *Journal of Applied Meteorology and Climatology*, vol. 49, no. 9, pp. 1831–1844, 2010.

[4] W. C. Skamarock, J. B. Klemp, J. Dudhia et al., "A description of the advanced research WRF version 3," NCAR Technical Note, NCAR, Boulder, Colo, USA, 2008.

[5] W. Wei, B. Cindy, D. Michael et al., *ARW Version 3 Modeling System User's Guide*, 2009.

[6] S. Y. Hong, Y. Noh, and J. Dudhia, "A new vertical diffusion package with an explicit treatment of entrainment processes," *Monthly Weather Review*, vol. 134, no. 9, pp. 2318–2341, 2006.

[7] Z. I. Janjic, "The step-mountain coordinate: physical package," *Monthly Weather Review*, vol. 118, no. 7, pp. 1429–1443, 1990.

[8] Z. I. Janjic, "Nonsingular implementation of the Mellor Yamada level 2.5 Scheme in the NCEP Meso model," NCEP Office Note 437, 2002.

[9] J. E. Pleim, "A combined local and non-local closure model for the atmospheric boundary layer. Part I: model description and testing," *Journal of Applied Meteorology and Climatology*, vol. 46, no. 9, pp. 1383–1395, 2007.

[10] S. Sukoriansky, B. Galperin, and V. Perov, "Application of a new spectral theory of stably stratified turbulence to the atmospheric boundary layer over sea ice," *Boundary-Layer Meteorology*, vol. 117, no. 2, pp. 231–257, 2005.

[11] S. Sukoriansky, B. Galperin, and V. Perov, "A quasi-normal scale elimination model of turbulence and its application to stably stratified flows," *Nonlinear Processes in Geophysics*, vol. 13, no. 1, pp. 9–22, 2006.

[12] A. S. Monin and A. M. Obukhov, "Basic laws of turbulent mixing in the surface layer of the atmosphere," *Contributions of the Geophysical Institute of the Slovak Academy of Sciences*, vol. 24, no. 151, pp. 163–187, 1954.

[13] Z. I. Janjic, "The surface layer in the NCEP Eta Model," in *Proceedings of the 11th Conference on Numerical Weather Prediction*, pp. 354–355, American Meteorological Society, Norfolk, Va, USA, August 1996.

[14] J. Dudhia, "A multi layer soil temperature model for MM5," in *Proceedings of the 6th PSU/NCAR Mesoscale Model Users' Workshop*, pp. 49–50, Boulder, Colo, USA, July 1996.

[15] F. Chen and J. Dudhia, "Coupling and advanced land surface-hydrology model with the Penn State-NCAR MM5 modeling system. Part I: model implementation and sensitivity," *Monthly Weather Review*, vol. 129, no. 4, pp. 569–585, 2001.

[16] E. J. Mlawer, S. J. Taubman, P. D. Brown, M. J. Iacono, and S. A. Clough, "Radiative transfer for inhomogeneous atmospheres: RRTM, a validated correlated-k model for the longwave," *Journal of Geophysical Research D*, vol. 102, no. 14, pp. 16663–16682, 1997.

[17] J. Dudhia, "Numerical study of convection observed during the Winter Monsoon Experiment using a mesoscale two-dimensional model," *Journal of the Atmospheric Sciences*, vol. 46, no. 20, pp. 3077–3107, 1989.

[18] J. S. Kain and J. M. Fritsch, "A one-dimensional entraining/detraining plume model and its application in convective parameterization," *Journal of the Atmospheric Sciences*, vol. 47, no. 23, pp. 2784–2802, 1990.

[19] J. S. Kain, "The Kain Fritsch convective parameterization: an update," *Journal of Applied Meteorology*, vol. 43, no. 1, pp. 170–181, 2004.

[20] B. S. Ferrier, Y. Lin, T. Black, E. Rogers, and G. DiMego, "Implementation of a new grid scale cloud and precipitation scheme in the NCEP Eta model," in *Proceedings of the 15th Conference on Numerical Weather Prediction*, pp. 280–283, American Meteorological Society, San Antonio, Tex, USA, 2002.

[21] G. C. Holtzworth, "Mixing depths, wind speeds and air pollution potential for selected locations in the United States," *Journal of Applied Meteorology*, vol. 6, no. 6, pp. 1039–1044, 1967.

[22] J. Stensrud David, *Parameterization Schemes Keys to Understanding Numerical Weather Prediction Models*, Cambridge University Press, 2007.

11

An Advanced Review of the Relationships between Sahel Precipitation and Climate Indices: A Wavelet Approach

Churchill Okonkwo

Beltsville Center for Climate System Observation, Atmospheric Science Program, Howard University, Washington, DC 20059, USA

Correspondence should be addressed to Churchill Okonkwo; churchill.okonkwo@bison.howard.edu

Academic Editor: Helena A. Flocas

The interannual and decadal to multidecadal variability of precipitation in western Sahel region was examined using wavelet transform and coherency analysis. The aim was to identify the major climate index that has a robust relationship with Sahel precipitation (drought). The results show that ENSO, North Atlantic Oscillation (NAO), Atlantic Multidecadal Oscillation (AMO), and Indian Ocean Dipole (IOD) all have some relationship with precipitation at different time scales which is in agreement with recent studies. There is an antiphase relationship between Sahel precipitation and ENSO at the 3-4-year band localized around 1982/83 El Niño episode. This indicates a cause and effect relationship between the droughts of 1983 and 1982/83 El Niño. In addition, wavelet transform coherence analysis also revealed a relatively antiphase relationship between AMO and precipitation signifying cause and effect. The wavelet analyses indicate that IOD control on rainfall variability in Sahel is limited to the east (15°E–35°E). Advancing this understanding of variability in rainfall and climate forcing could improve the accuracy of rainfall forecast.

1. Introduction

Several authors have reported marked interannual variability in rainfall across Africa [1–3]. Since economic development in the region is highly dependent on water availability [4], the effect of climate variability on rainfall is critical [5]. Western Sahel region (latitudes 14°N and 18°N—longitude –18°W to 10°W) is the semiarid transition zone between the Sahara desert and humid tropical Africa that is prone to drought [6, 7]. The Disaster Management Center (DMC) [8] reported that more than 900,000 people were severely affected by the devastating drought of the 1970s across the Sahel. The associated social and economic consequence of drought such as failure in crop yield, destruction of pasture, and famine has led to a series of studies exploring the interactions and dynamics that control precipitation within the region.

Over the past three decades, studies on the possible causes of drought in Sahel have focused on forcing by either sea surface temperature (SST) or land-atmosphere interaction. Simulations of hydrological impact of land-atmosphere interactions include [9–12] which all attributed reduced rainfall to degradation of land surface at least in part. Li et al. [12] confirmed the impact of land surface changes on the

regional climate through a feedback mechanism that sustains drought. The contribution of these mechanisms has however been exaggerated [13, 14] especially the characterization of desertification in the Sahel as irreversible.

There have also been several studies that examined the teleconnection between rainfall variability in Sahel and variation in SST over the tropical Pacific [15, 16]. While it has been concluded that SST patterns play a significant role in rainfall variability in West Africa [17, 18], there is still a debate regarding the major drivers [19]. Several results found that regional weather patterns forced by North Atlantic Oscillation (NAO) have more influence on the local climate in Sahel region [20–23]. Zhang and Delworth [24] and Delworth et al. [25] have linked the Atlantic Multidecadal Oscillation (AMO) to low frequency variations extending back to the nineteenth century.

In addition to the above teleconnection, the role of the Mediterranean sea as having a strong influence on precipitation across the Sahel has been suggested [26]. On the other hand, the influence of the Indian Ocean Dipole (IOD) on both Sahel rainfall and Indian monsoon has been suggested [27, 28]. Studies on the role of El Niño-Southern Oscillation (ENSO) and Pacific Ocean in modulating Sahel precipitation include those of [29–31]. A strong link between the decaying phase of La Niña and developing phase of El Niño has been established by Joly and Voldoire [32]. The limiting role of ENSO on rainfall variability only on eastern Sahel has also been suggested [33]. One of the conclusions that came out of these multiple studies is that the interannual to decadal variability in precipitation over West Africa is controlled by competing physical mechanisms [34].

Despite these advances on the possible causes and trends of persistent drought in Sahel region, there are still projection uncertainties and consensus that models cannot reliably predict future climates in the region [17, 18, 30, 34, 35]. This had led to big spreads in projections [17, 36].

The spread in projections in this semiarid region with high interseasonal, interannual, and interdecadal variability in rainfall [22, 37, 38] could lead to some serious repercussions on the local economy, farming, and livestock production. More accurate modeling is therefore not only essential in improving our understanding of circulation across the Sahel region but also crucial in planning for future impact of climate change and variability. Accurate modeling and better projections however depend on use of appropriate climate forcing and understanding of the complex dynamics in play in this region.

The major aim of this advanced review is to identify the major climate index that has a robust relationship with Sahel precipitation (drought) using a wavelet approach. Most of the studies reviewed above have employed either coupled ocean atmosphere general circulation model [24] or dynamical modeling [21]. The specific nature of rainfall time series (variability) and drought outbreaks in the Sahel region makes its analysis with wavelet potentially advantageous. The most important advantage of wavelet analysis is that, unlike classical spectral analysis that requires the restrictive assumption of stationarity, wavelet approach focuses on time series that change with time [39, 40]. Wavelet analysis is also a better tool for extracting features locally. It achieves this by decomposing patterns while preserving and displaying locational information that makes analysis of dependencies between two signals easier. Finally, wavelet cross coherence analysis between ENSO-and LC level is of high importance with respect to long-term water resources problem. One disadvantage of continuous wavelet approach is the edge effect at the beginning and end of the time series. This makes information within the cone of influence less accurate.

This advanced review differs from previous studies by applying wavelet statistical analysis to climate indices to study how they modulate precipitation in the Sahel region. The datasets and preparation will be described in Section 2. The rainfall characteristics and climatology of the Sahel region are provided in Section Three. Description of the wavelet approach and results will be given in Section 4 together with the difference between rainfall variability in eastern and western Sahel in relation to these climate indices. Section 5 provided summary and conclusions.

2. Data and Preparation

2.1. Dataset. There are two primary precipitation datasets used in this study. One is the Global Precipitation Clima-

tology Project (GPCP v.2) 2.5-degree global grid monthly estimate of precipitation, which is a combination of merged satellite data (from infrared and microwave imagers) and gauge observations [41]. These observed data are daily data resolved to monthly time steps. The second precipitation dataset used in the study is the gridded station monthly rainfall anomalies (cm) taken from the National Oceanic and Atmospheric Administration (NOAA) and Global Historical Climatology Network (GHCN) [42]. The gridded data points were produced on a 5°-by-5° basis and averaging for Sahel region to address inhomogeneity was based on a rotated principal component analysis of African precipitation by Janowiak [42]. Also, stations with at least 20 years of data between 1961 and 1990 period were used. Further detail for the data is described at http://www.ncdc.noaa.gov/temp-and-precip/ghcn-gridded-products.php.

In this study, we focused on the North Atlantic Multi-decadal Oscillation (AMO), North Atlantic Oscillation index (NAO), the Indian Ocean Dipole (IOD), and the Nino 3.4 indices. These are the most studied climate indices linked to modulation of precipitation in the Sahel region. AMO uses gridded global sea surface temperature (SST) and anomalies from 1856 till present derived from UK Met Office SST data [43].

El Niño-Southern Oscillation (ENSO) is represented in this study by SST data from Niño 3.4 region (5°S–5°N/120°–170°W). The Niño-3.4 SST period used in this study is based on a new strategy of updating by Climate Prediction Center (CPC) of National Oceanic and Atmospheric Administration (NOAA). In this new approach, multiple centered 30-year base periods are used to calculate anomalies for successive 5-year periods in the historical record [44]. This has the advantage of defining El Niño and La Niña episodes based by their contemporary climatology while previous classification will mostly remain fixed over historical period. Also, warning from longer-term trends that do not reflect interannual ENSO variability defined by a single fixed 30-year base period is removed [44].

IOD was first identified by [45] and is an interannual climate pattern across tropical Indian Ocean. Cooler than normal water in tropical eastern Indian Ocean and warmer than normal water in tropical western Indian Ocean characterizes the positive IOD period [45]. The trend is reversed during the negative IOD period.

The NAO is a large-scale mode of natural climate variability that is dominant in winter with important impact across the North Atlantic region [46]. According to Jones et al. (1997) [47], NAO measures the sea level pressure difference between the Azores High and Icelandic Low [48]. Positive NAO is linked with strong westerlies while negative NAO is linked with weakened Atlantic storm track [49].

2.2. Data Preparation. As part of the data preparation we aggregated the monthly climate indices into seasons in order to reduce noise following Brown et al. [50]. We applied area weighted averaging to the GPCP gridded monthly time series. This has the advantage of minimizing the spatial data gaps in a semiarid region. Also, Huffman et al. [51] reported that high mean absolute error for individual grid points can

(a)

(b)

(c)

(d)

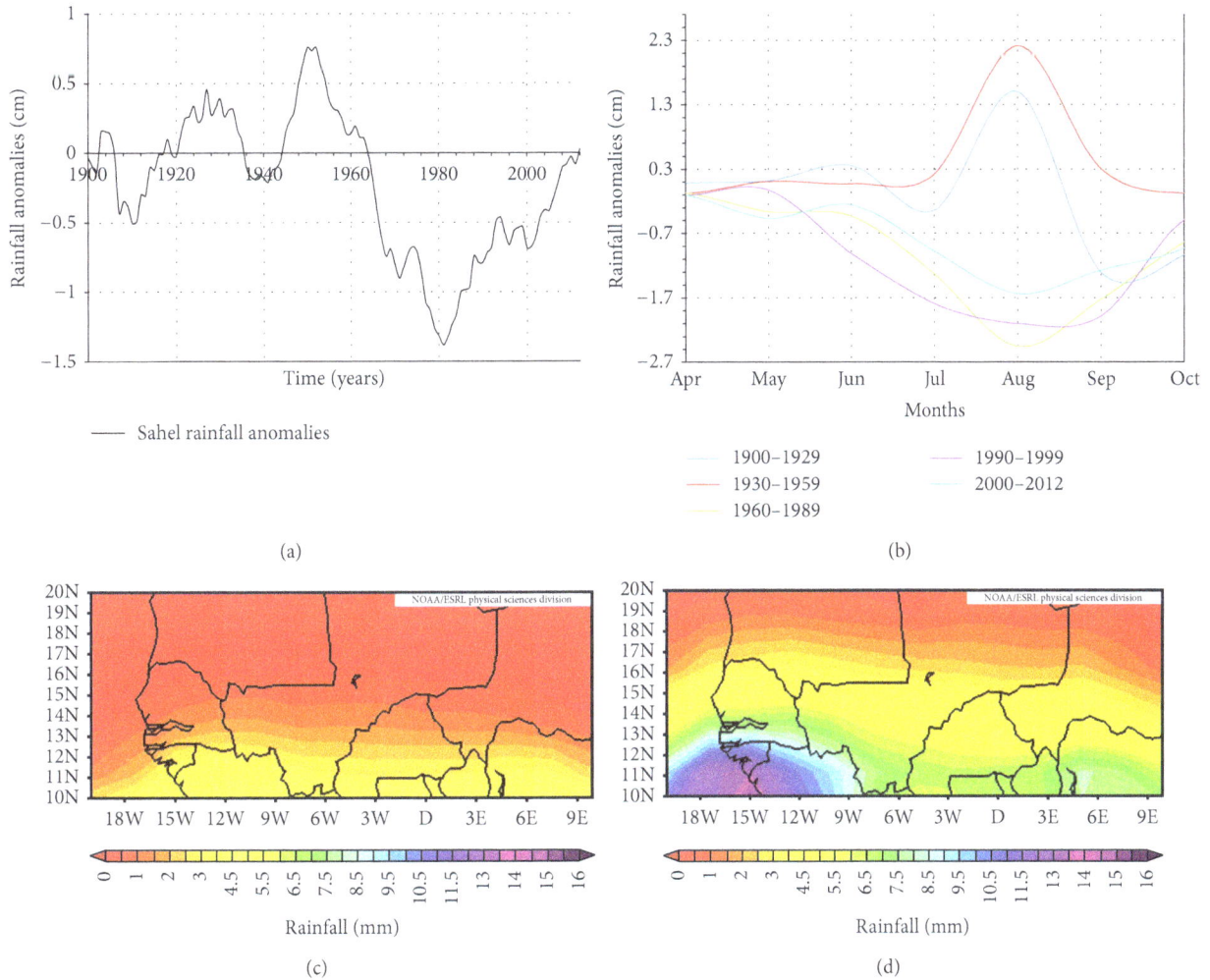

FIGURE 1: (a) Eight-year running mean of observed Sahel precipitation anomalies (cm). (b) April–October monthly cycle of rainfall anomalies (cm). The rainfall anomalies are relative to 1950–1979. Spatial distribution of western Sahel seasonal precipitation climatology (GPCP) 1981–2010 for (b) AMJ and (c) JAS.

be minimized by spatial and temporal averaging. We used JAS (July, August, and September) average index correlated with aggregated monthly rainfall totals of JAS corresponding to rainfall peak of raining season in Sahel. The aggregated JAS rainfall allowed us to focus on periods of considerable precipitation, thereby reducing the noise from long periods of dry spells characteristic of this region. The criteria for selecting the GHCN dataset (1900–2012) were based on the suitability for correlation analysis with Atlantic Multidecadal Oscillation (AMO). The climate indices analyzed in this study are mainly archived as a monthly dataset and thus the suitability of GPCP monthly precipitation. The different time frames for climate indices were based on the available time series for each dataset.

There are conflicting definitions of the extent of Sahel region domain in the literature. While there is a consensus on the latitudinal domain (14°N to 18°N), the east-west bounds have varied. Nicholson [52] generally used the 30° to 35°E as the eastern bound while Lamb [53–55] used eastern bound to 10°E. Some studies have found significant difference in rainfall variability between different sections of Sahel [55].

In this advanced review, we divided the Sahel into east (10°E to 35°E) and west (Atlantic to 10°E) for the purpose of comparing rainfall variability in relation to these climate indices.

3. Spatial and Temporal Characterization of Western Sahel Rainfall

We first characterize the monthly and seasonal (spatial) climatology of rainfall in Sahel region using observational data set. The rainfall time series (Figure 1(a)) in Sahel shows large variability with negative rainfall anomalies since early 1960s. The negative rainfall anomaly of 1.4 cm around 1983-1984 corresponds to the period of one of the severe droughts that have plagued the region. This drought condition is characteristic of this region and has been documented by several authors (e.g., [19]). Also of note is the gradual recovery in rainfall (Figure 1(a)) from the year 2000 corresponding to onset of recovery in Sahel rainfall as reported by Nicholson [3].

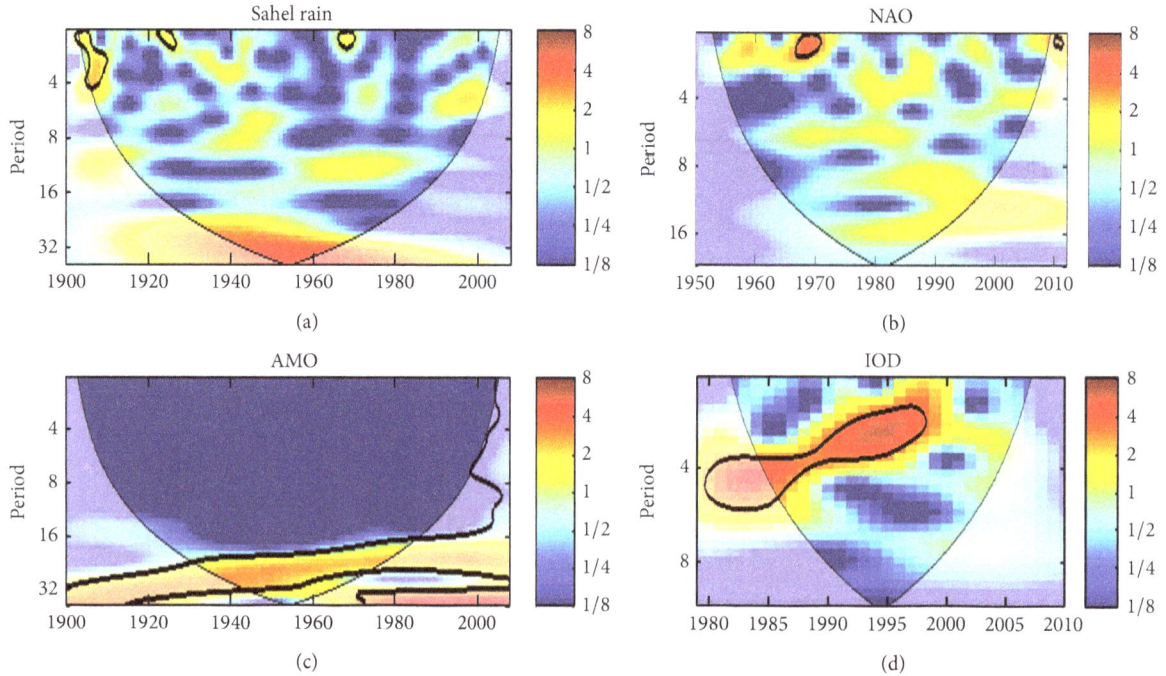

FIGURE 2: Continuous wavelet transform for (a) western Sahel rainfall, (b) North Atlantic Oscillation (NAO), (c) Atlantic Multidecadal Oscillation (AMO), and (d) Indian Ocean Dipole (IOD). The thick contour enclosed regions are greater than 95% confidence for a red-noise process. The thin solid line indicates the "cone of influence," where edge effects become important.

Figure 1(b) shows the monthly (average) evolution of 1900–2012 April to October western Sahel rainfall (cm) anomalies with respect to 1950–1979. There is short period of rainfall in the region (boreal summer) with maximum rainfall in August. The "wettest" period (1930–1959) has a positive rainfall anomaly of about 2.3 cm in August. In the 2013 State of the Climate report (Sima et al. 2013) [56], the 2012 extensive flooding in the Sahel was reported as a pointer to a full return to "wet" periods across the region. Our analysis however shows negative rainfall anomalies of −2.1 cm and −1.7 cm in August for 1990–2000 and 2000–2012 periods, respectively (Figure 1(b)). This difference could be explained by the use of 1981–2010 rainfall climatology in the State of the Climate report, a period characterized by severe drought in Sahel region, whereas our anomaly analysis was based on 1950–1979 rainfall anomaly, a relatively wetter period.

Figure 1(c) (April–June (AMJ)) and Figure 1(d) (July–September (JAS)) show the spatial distribution of seasonal rainfall in the region. Predominantly, heavy rainfall occurs south of latitude 20°N while dry conditions are common northwards at the proximity of Sahara desert. Also, east-west uniformity and south-north gradient described by Nicholson (2013) are evident. There is a characteristic rain-band at the southwest region off the coast of Guinea in the JAS seasonal rainfall distribution (Figure 1(d)). The evidence of the limitation of rainfall in this region can also be seen in the difference in the spatial distribution of rainfall between boreal spring (Figure 1(c)) and boreal summer (Figure 1(d)). There is heavier rainfall (between 5.5 and 7.5 mm/day) in JAS south of latitude 14°N. Conversely, there is only about 4 mm/day rainfall for the same latitude band in AMJ.

4. Wavelet Analysis

Using wavelet analysis, we examined the interannual variability of precipitation and climate indices by decomposing their (a multiscale nonstationary process) time series into frequency space following the program developed by Torrence and Compo [57]. A summary of the basic theory of continuous wavelet transform (CWT), cross wavelet transform (XWT), and wavelet transform coherency (WTC) following [57, 58] is given as

$$\mathcal{W}(\tau, s) = \frac{1}{(s)^{1/2}} \int_{-\infty}^{+\infty} X(t)\, \psi^* \left(\frac{t-\tau}{s}\right) dt, \qquad (1)$$

where $\psi(t)$ is the mother wavelet defined by τ, the transition parameter corresponding to the position of the wavelet, and s is the scale dilation parameter that determines the width of the wavelet. The variability of the dominant mode with time was determined using the Morlet wavelet with a wavenumber $w0 = 6$ as the mother wavelet. The choice of Morlet wavelet is based on its localization in time and frequency making it a good tool in extracting features [58]. Information about the periodicity of the time series data can be extracted from the CWT while cross wavelet transform (XWT) helps in determining whether the two time series are statistically significant by Pearson's correlation coefficient. Detailed methodology of wavelet transform can be found in [57, 58].

4.1. Continuous Wavelet Transform. Figure 2 shows the CWT of rainfall in the Sahel and the climate indices. The major periodicity can be seen in the power spectrum near the 2-

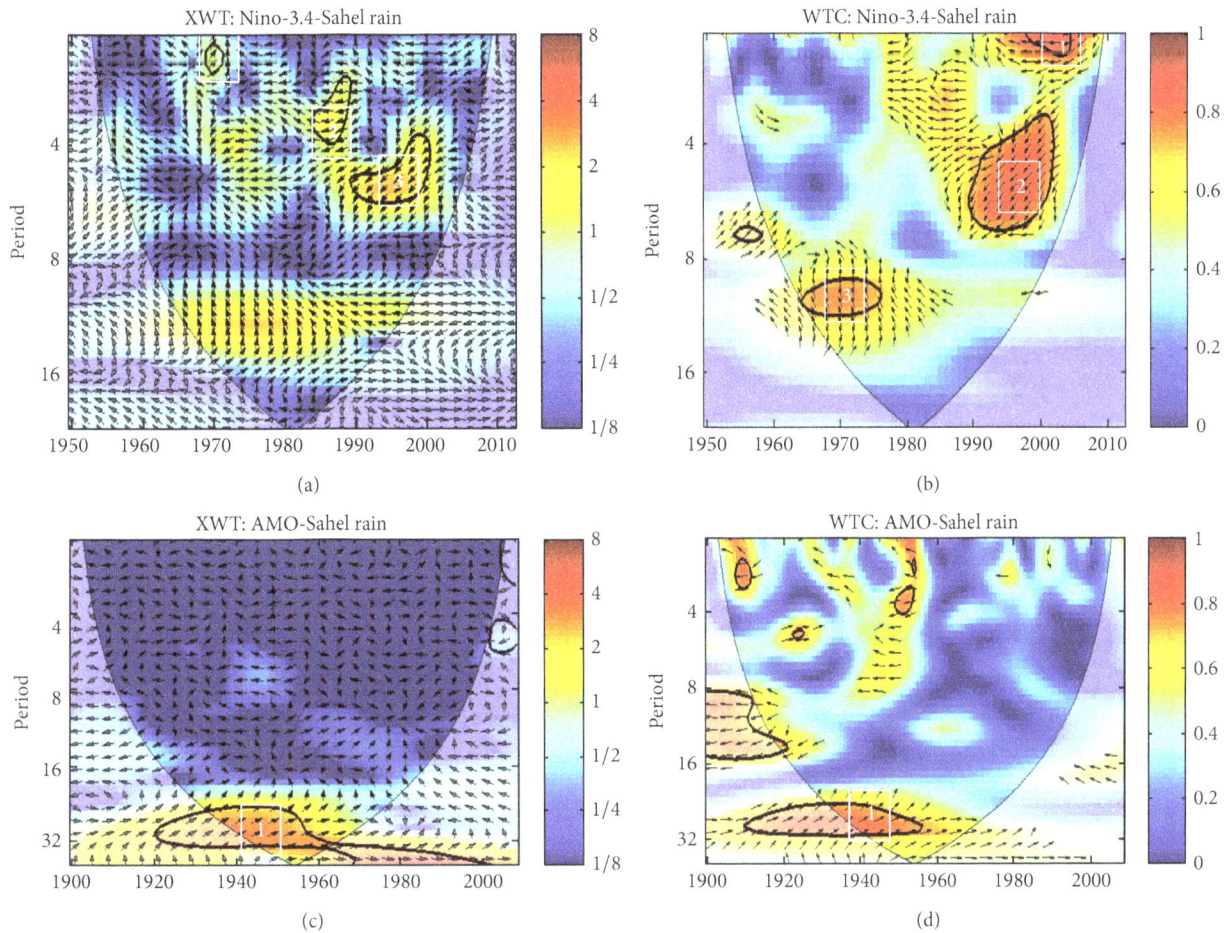

FIGURE 3: Cross wavelet spectrum. (a) Nino 3.4, (b) Nino 3.4, (c) Atlantic Multidecadal Oscillation (AMO) and wavelet coherence, and (d) Atlantic Multi-decadal Oscillation (AMO). (In-phase, pointing right; anti-phase, pointing left; and leading by 90° pointing straight down; see text for details and interpretation). The thick contour enclosed regions are greater than 95% confidence for a red-noise process. The thin solid line indicates the "cone of influence," where edge effects become important.

year band, 1–4-year band, 1-2-year band, and 2–5-year band for ENSO, NAO, and IOD, respectively. These interannual cycles have time scales of less than 8 years. In addition, the region that is statistically significant at 95% confidence for IOD is important with reference to drought of 1980s in the region. Also, multidecadal variability of AMO shows regions of statistical significance at 38–32-year period around the early 1900s. The statistically significant periodicity has however gradually decreased from 40-year to 16-year cycle in 2000s. The next step is to expand these time series into frequency space by applying CWT as a band filter to the time series. This will help in finding localized periodicities and in feature extraction. Cross wavelet transform will determine if these associations are merely coincidences by obtaining the frequency component of the hydroclimate variability as a function of time. Wavelet transform coherency (WTC) between two CWT will address the statistical significance of the coherences as well as confidence level against noise. In this analysis, we applied the 95% confidence level.

4.2. Cross Wavelet Transform and Wavelet Coherence. The XWT and WTC between rainfall and the four climate indices

are shown in Figures 3 and 4. According to Torrence and Compo [57], XWT is a representation of cross correlation between rainfall and each climate index as a function of time and frequency. The phase difference between the climate indices and precipitation is represented by the vectors while the locally significant power of the red-noise spectrum at significance level of $\alpha = 0.1$ is shown by the bold solid contour line [57]. The lighter black contour line is the cone of influence (COI), where edge effects are not negligible. The coherence power between two series is shown in the color code, red to blue (strong to weak).

ENSO's role in modulating Sahel precipitation shows high common power (good correlation) at 3 bands, (1) 1-2-year band localized around 1970, (2) 3-4-year band localized between 1983 and 1986, and (3) 4–6-year band localized between 1990 and 2000 years (Figure 3(a)). According to Grinsted [58], cause and effect relationship in XWT is indicated by phase lock oscillation. The 3-4-year band is pointing left (antiphase) indicating a cause and effect relationship between the drought of 1983 and 1982/83 El Niño episode. The coherency between precipitation and ENSO (Figure 3(b)) also shows three high common powers that are

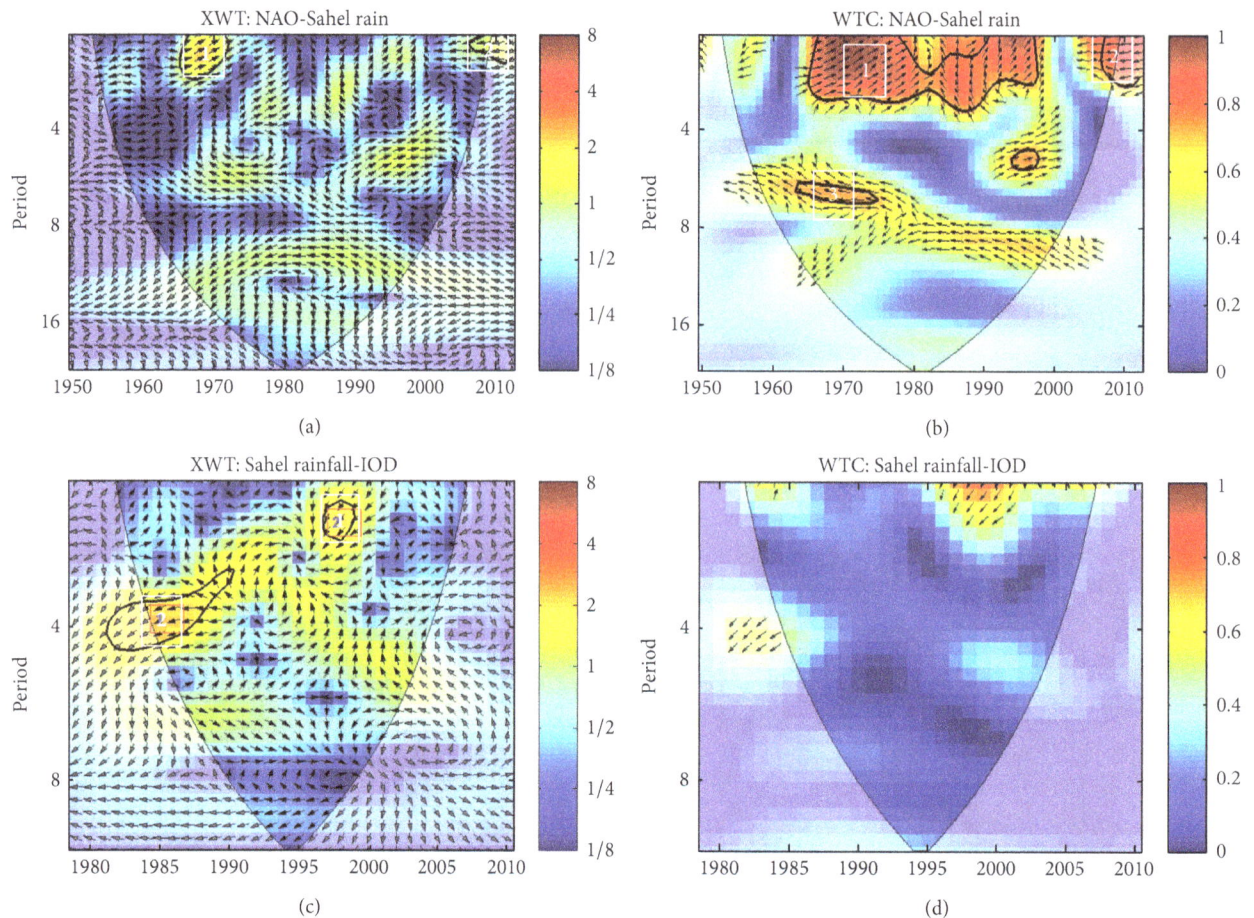

FIGURE 4: The same as Figure 3 but for (a-b) North Atlantic Oscillation and (c-d) Indian Ocean Dipole.

statistically significant, (1) 1-2-year band localized around 2000, (2) 4–6-year band localized between 1990 and 2000, and (3) 9-10-year band localized between 1965 and early 1970s. The spectral coherency in the 4–6-year band is very strong and noteworthy since it coincides with the onset of rain recovery in the Sahel region [59]. However, because the oscillation is not phase-locked, we can only speculate that there is statistically significant association between the increasing rainfall and Nino 3.4. ENSO, according to Ward [60], influenced high frequency rainfall variability in Sahel while [32] linked strong ENSO events teleconnection to Sahel precipitation to developing and decaying phase of El Niño and La Niña, respectively.

In the XWT between rainfall and AMO shown in Figure 3(c), notice the statistically significant cross wavelet power and consistent phase angle stretching from early 1920s to 2000 which is associated with signal in the 25–38-year band. Many studies have shown that precipitation in the Sahel region of Africa is influenced by oceanic conditions impact on atmospheric circulations especially Atlantic Multidecadal Oscillation AMO [23] and El Niño-Southern Oscillation ENSO [16]. The coherency spectra that are significant at the 25–32-year band from late 1910s to late 1950s correspond to the years of positive rainfall anomalies shown in Figure 1(a). This is significant in relation to the reported increase in

precipitation associated with warm phase of AMO [24, 25]. Wavelet transform coherence analysis also revealed a relatively antiphase relationship between AMO and precipitation signifying cause and effect.

As seen in Figure 4(a), there is a strong agreement between NAO and Sahel rainfall indicated in the XWT between 1- and 2-year band around late 1970 and late 2000s. While the second band (2) is phase-locked, the first band (1) however indicates a nonlinear relationship between ANO and Sahel precipitation. A corresponding strong coherency between NAO and precipitation is evident in the 1-2-year band of late 2000s (Figure 4(b)). In addition, NAO shows a significant spectral coherence with Sahel rainfall between 1- and 3-year band localized from 1960 to 2000 (Figure 4(b)). The phase angle is however chaotic. There is antiphase relationship between IOD and GPCP precipitation for period of 3-4 years around 1983–1990 (Figure 4(c)). There is also a significant cross wavelet power for the period of 1997-1998 associated with 2-year band (Figure 4(d)). However, there is no significant coherency between rainfall and IOD for most of the years within COI. The power is also very weak, except for the 1-year band around 1995 to 1998.

These results confirm the conclusions by [34] and Nicholson [19] that the interannual to decadal variability in precipitation over in this region is not controlled by any dominant mechanism. The 1983–1987 years corresponding to the drought of the late 1970s to late 1980s have been described as intense [19]. While Bader and Latif [33] attributed the 1983 drought to Indian Ocean SSTs, our results show that AMO, ENSO, NAO, and IOD all have some correlation with precipitation around the same year though on different bands. There is however strong coherency with ENSO, AMO, and NAO unlike IOD. This suggests that, contrary to [33], AMO, NAO, and ENSO may have influenced events during those drought years [26].

4.3. East versus West Sahel. In this section, we compared the influence of the climatic indices on rainfall variability in east and west Sahel as defined in our data preparation section. The XWT and coherence plots of GPCP rainfall in west and east Sahel for two selected climate indices (ENSO and IOD) with significant difference are shown in Figure 5. As seen in Figures 5(a) and 5(c), there are some significant differences in ENSO association with rainfall variability in eastern and western Sahel. ENSO relationship with precipitation in the east (Figure 5(c)) stretches over longer period and years while showing more variability compared to rainfall-ENSO relationship in western Sahel (Figure 5(a)).

ENSO teleconnection with precipitation in the east shows coherency that is localized from 1990 to 2004 (Figure 5(d)). The coherency for western Sahel is however only localized from 1995 to 2004 (Figure 5(b)). We can thus argue that ENSO teleconnection effect on precipitation is stronger in east Sahel, which is in agreement with [33] who had suggested that ENSO effect on rainfall variability is limited to eastern Sahel.

The results also show some evidence of difference in climatic control of precipitation in west and east Sahel. From Figures 5(e) and 5(g), it appears that IOD control on rainfall variability in Sahel is limited to the east. The XWT for IOD-rainfall in the east shows 2 regions with statistically significant correlation: (1) in the 2–4-year band localized between late 1980s and early 1990s (Figure 5(g)) and (2) in the 5-6-year band localized between 1986 and 1990. In the west however there is relatively small band with statistically significant correlation around the same band (Figure 5(e)). The spectral coherency plots show a very weak association except a thin spectrum at the 1-year band in the early 2000s (Figures 5(f) and 5(h)). The coherency plot for east Sahel on the other hand has a similar pattern of the weak band as the west. In addition, there is a relatively strong power in the 3-4-year band in the early 1990 (Figure 5(g)), though none of these coherencies are statistically significant.

5. Summary and Discussion

With the establishment of a statistical connection between these climate indices and Sahel precipitation, we will next discuss the dynamical mechanism of the connection. The physical mechanisms influencing the multidecadal variability in the Sahel region have been linked to AMO [24, 25] while the interannual variability at different time scales are linked to ENSO [30], IOD [33], and NAO [61]. The warm (cold)

phase of AMO is associated with enhancement (weakening) of Sahel precipitation [24]. The increased precipitation during the warm phase has been attributed to increased African easterly wave activity and northward displacement of Intertropical Convergence Zone (ITCZ) [62]. On the other hand, El Niño/La Niña Southern Oscillation (ENSO/LNSO) [63], a coupled cycle of atmosphere and ocean [64], has been linked to the devastating droughts of 1970s and 1980s in the Sahel of Africa [15]. Also, Okonkwo et al. [65] characterized the relationship between West African jet streams and ENSO and found enhanced variability of African Easterly Jet (AEJ), Tropical Easterly Jet (TEJ), and low-level African Westerly Jet (AWJ) which are coupled with (ENSO/LNSO). Further analysis by [65] suggests a statistically significant association between TEJ and the El Niño events of the 1980s that led to intense drought in the Sahel region of West Africa. El Nino results in the weakening of West African Monsoon (WAM) flow, creating a dry condition across the Sahel region [66]. La Niña on the other hand creates a wet condition through the enhancement of Walker circulation [59].

This advanced review had focused on identifying and illuminating the nature of interannual and decadal to multidecadal variability of precipitation in Sahel region. It should however be pointed out that the Sahelian climate under future climate warming is still complex and controlled by one dominant ocean dynamics. The analysis is however a useful tool in identifying the major climate index with a robust relationship with Sahel precipitation (drought) variability on annual to interannual timescale.

The findings are summarized as follows.

(1) ENSO, NAO, AMO, and IOD all have strong relationship with precipitation at periodicity.

(2) There is an antiphase relationship between Sahel precipitation and ENSO, the 3-4-year band localized around 1982/83 El Niño episode.

(3) This indicates a cause and effect relationship between the droughts of 1983 and 1982/83 El Niño.

(4) The interrelationship between ENSO composite of 1983–1987, cold phase of AMO, and warm phase of NAO explained the drought of the early 1980s.

While this advanced review does not resolve all the outstanding issues on rainfall variability in Sahel region, it however points to some climate indices that have significant control to current improvement in precipitation as well as the most recent severe drought in the region. This improved knowledge is a first step in better planning and management of water resources in Sahel region. One limitation of the study is that it is based on the different time scales for different climate indices considered. This study however has provided some key aspects of SST forcing on rainfall variability across the Sahel Region. Advancing this understanding of variability in rainfall and climate forcing could therefore improve the accuracy of rainfall forecast.

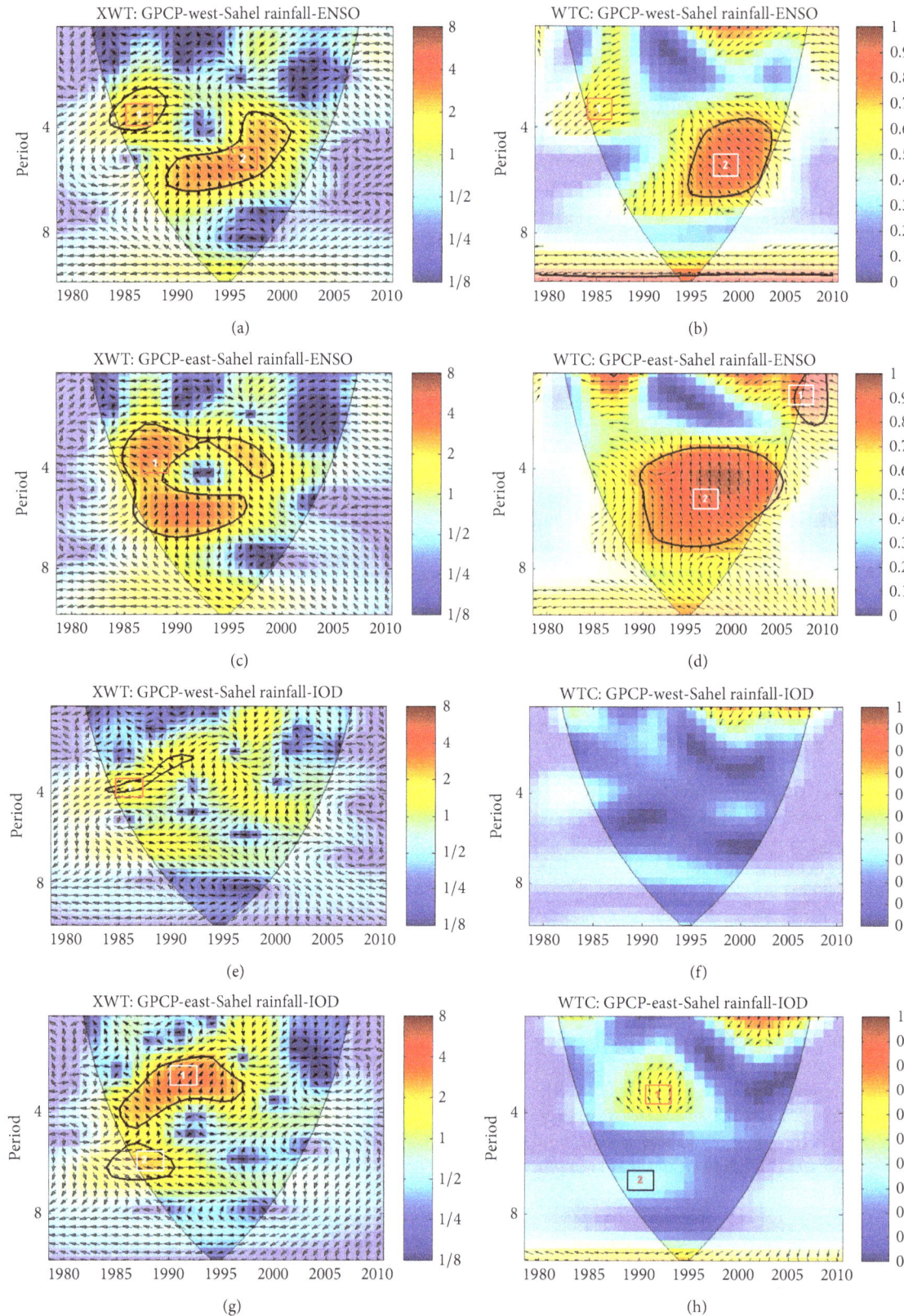

FIGURE 5: Cross wavelet spectrum (left) and wavelet coherence (right) between GPCP Sahel precipitation (east and west) and selected climate indices (ENSO and IOD). (a)-(b) Western Sahel rainfall and ENSO, (c)-(d) eastern Sahel Rainfall and ENSO, (e)-(f) western Sahel rainfall and IOD, and (g)-(h) eastern Sahel rainfall and IOD. In-phase, pointing right; anti-phase, pointing left; and leading by 90° pointing straight down; (see text for details and interpretation). The thick contour enclosed regions are greater than 95% confidence for a red-noise process. The thin solid line indicates the "cone of influence," where edge effects become important.

Conflict of Interests

The author declares that there is no conflict of interests regarding the publication of this paper.

Acknowledgments

The author would like to acknowledge the support of Beltsville Center for Climate Systems Observations (BCCSO) and the National Oceanic and Atmospheric Administration (NOAA) Center for Atmospheric Sciences (NCAS) both at Howard University.

References

[1] Z. T. Segele and P. J. Lamb, "Characterization and variability of Kiremt rainy season over Ethiopia," *Meteorology and Atmospheric Physics*, vol. 89, no. 1–4, pp. 153–180, 2005.

[2] C. Cook, C. J. C. Reason, and B. C. Hewitson, "Wet and dry spells within particularly wet and dry summers in the South African summer rainfall region," *Climate Research*, vol. 26, no. 1, pp. 17–31, 2004.

[3] S. E. Nicholson, "On the factors modulating the intensity of the tropical rainbelt over West Africa," *International Journal of Climatology*, vol. 29, no. 5, pp. 673–689, 2009.

[4] P. Desanker and C. Magadza, *Africa Climate Change 2001: Impacts, Adaptation and Vulnerability*, Cambridge University Press, New York, NY, USA, J. J. McCarthy, O. F. Canziani, N. A. Leary, D. J. Dokken, K. S. White, Eds., 2001.

[5] R. Washington, M. Harrison, D. Conway et al., "African climate change: taking the shorter route," *Bulletin of the American Meteorological Society*, vol. 87, no. 10, pp. 1355–1366, 2006.

[6] S. E. Nicholson, C. J. Tucker, and M. B. Ba, "Desertification, drought, and surface vegetation: an example from the West African Sahel," *Bulletin of the American Meteorological Society*, vol. 79, no. 5, pp. 815–829, 1998.

[7] A. Tarhule and P. J. Lamb, "Climate research and seasonal forecasting for West Africans," *Bulletin of the American Meteorological Society*, vol. 84, no. 12, pp. 1741–1759, 2003.

[8] Disaster Management Centre D. M. C., *Natural Disasters: Causes and Effects. Lesson 7: Drought*, Disaster Management Centre, University of Wisconsin, Madison, Wis, USA, 1995.

[9] J. G. Charney, "Dynamics of deserts and drought in the Sahel," *Quarterly Journal of the Royal Meteorological Society*, vol. 101, no. 428, pp. 193–202, 1975.

[10] Y. Xue and J. Shukla, "The influence of land surface properties on Sahel climate—part I: desertification," *Journal of Climate*, vol. 6, no. 12, pp. 2232–2245, 1993.

[11] C. M. Taylor, E. F. Lambin, N. Stephenne, R. J. Harding, and R. L. H. Essery, "The influence of land use change on climate in the Sahel," *Journal of Climate*, vol. 15, no. 24, pp. 3615–3629, 2002.

[12] K. Y. Li, M. T. Coe, N. Ramankutty, and R. D. Jong, "Modeling the hydrological impact of land-use change in West Africa," *Journal of Hydrology*, vol. 337, no. 3-4, pp. 258–268, 2007.

[13] J. F. Reynolds and S. D. M. Stafford, *Global Desertification: Do Humans Cause Deserts?* Dahlem University Press, Berlin, Germany, 2002.

[14] S. D. Prince, *Spatial and Temporal Scales For Detection of Desertification in Global Desertification: Do Humans Cause Deserts?* Dahlem University Press, Berlin, Germany, J. F. Reynolds, D. M. S. Smith, Eds., 2002.

[15] A. Giannini, R. Saravanan, and P. Chang, "Oceanic forcing of Sahel rainfall on interannual to interdecadal time scales," *Science*, vol. 302, no. 5647, pp. 1027–1030, 2003.

[16] C. Caminade and L. Terray, "Twentieth century Sahel rainfall variability as simulated by the ARPEGE AGCM, and future changes," *Climate Dynamics*, vol. 35, no. 1, pp. 75–94, 2010.

[17] M. Biasutti, I. M. Held, A. H. Sobel, and A. Giannini, "SST forcings and Sahel rainfall variability in simulations of the twentieth and twenty-first centuries," *Journal of Climate*, vol. 21, no. 14, pp. 3471–3486, 2008.

[18] A. Giannini, M. Biasutti, I. M. Held, and A. H. Sobel, "A global perspective on African climate," *Climatic Change*, vol. 90, no. 4, pp. 359–383, 2008.

[19] S. E. Nicholson, "The West African Sahel: a review of recent studies on the rainfall regime and its interannual variability," *ISRN Meteorology*, vol. 2013, Article ID 453521, 32 pages, 2013.

[20] C. K. Folland, T. N. Palmer, and D. E. Parker, "Sahel rainfall and worldwide sea temperatures, 1901–85," *Nature*, vol. 320, no. 6063, pp. 602–607, 1986.

[21] J. Lu and T. L. Delworth, "Oceanic forcing of the late 20th century Sahel drought," *Geophysical Research Letters*, vol. 32, no. 22, Article ID L22706, pp. 1–5, 2005.

[22] M. Hoerling, J. Hurrell, J. Eischeid, and A. Phillips, "Detection and attribution of twentieth-century northern and southern African rainfall change," *Journal of Climate*, vol. 19, no. 16, pp. 3989–4008, 2006.

[23] I. Polo, A. Ullmann, P. Roucou, and B. Fontaine, "Weather regimes in the Euro-Atlantic and Mediterranean sector, and relationship with West African rainfall over the 1989–2008 period from a self-organizing maps approach," *Journal of Climate*, vol. 24, no. 13, pp. 3423–3432, 2011.

[24] R. Zhang and T. L. Delworth, "Impact of Atlantic multidecadal oscillations on India/Sahel rainfall and Atlantic hurricanes," *Geophysical Research Letters*, vol. 33, no. 17, Article ID L17712, 2006.

[25] T. Delworth, L. R. Zhang, and M. E. Mann, "Decadal to centennial variability of the Atlantic from observations and models. Ocean circulation: mechanisms and impacts," *Geophysical Monograph Series*, vol. 173, pp. 121–148, 2007.

[26] I. Polo, B. Rodríguez-Fonseca, T. Losada, and J. García-Serrano, "Tropical atlantic variability modes (1979–2002)—part I: time-evolving SST modes related to West African rainfall," *Journal of Climate*, vol. 21, no. 24, pp. 6457–6475, 2008.

[27] F. Raicich, N. Pinardi, and A. Navarra, "Teleconnections between Indian monsoon and Sahel rainfall and the Mediterranean," *International Journal of Climatology*, vol. 23, no. 2, pp. 173–186, 2003.

[28] C. E. Chung and V. Ramanathan, "Weakening of north Indian SST gradients and the monsoon rainfall in India and the Sahel," *Journal of Climate*, vol. 19, no. 10, pp. 2036–2045, 2006.

[29] D. P. Rowell, "Teleconnections between the tropical Pacific and the Sahel," *Quarterly Journal of the Royal Meteorological Society*, vol. 127, no. 575, pp. 1683–1706, 2001.

[30] M. Joly, A. Voldoire, H. Douville, P. Terray, and J.-F. Royer, "African monsoon teleconnections with tropical SSTs: validation and evolution in a set of IPCC4 simulations," *Climate Dynamics*, vol. 29, no. 1, pp. 1–20, 2007.

[31] S. Janicot and B. Sultan, "Intra-seasonal modulation of convection in the West African monsoon," *Geophysical Research Letters*, vol. 28, no. 3, pp. 523–526, 2001.

[32] M. Joly and A. Voldoire, "Role of the Gulf of Guinea in the interannual variability of the West African monsoon: what do we learn from CMIP3 coupled simulations?" *International Journal of Climatology*, vol. 30, no. 12, pp. 1843–1856, 2010.

[33] J. Bader and M. Latif, "The impact of decadal-scale Indian ocean sea surface temperature anomalies on Sahelian rainfall and the North Atlantic oscillation," *Geophysical Research Letters*, vol. 30, no. 22, pp. 2169–2173, 2003.

[34] L. M. Druyan, "Studies of 21st-century precipitation trends over West Africa," *International Journal of Climatology*, vol. 31, no. 10, pp. 1415–1424, 2011.

[35] S.-Y. Wang and R. R. Gillies, "Observed change in Sahel rainfall, circulations, African easterly waves, and Atlantic hurricanes since 1979," *International Journal of Geophysics*, vol. 2011, Article ID 259529, 14 pages, 2011.

[36] J. H. Christensen, T. R. Carter, M. Rummukainen, and G. Amanatidis, "Evaluating the performance and utility of regional climate models: the PRUDENCE project," *Climatic Change*, vol. 81, no. 1, pp. 1–6, 2007.

[37] G. S. Jenkins, A. T. Gaye, and B. Sylla, "Late 20th century attribution of drying trends in the Sahel from the Regional Climate Model (RegCM3)," *Geophysical Research Letters*, vol. 32, no. 22, Article ID L22705, pp. 1–4, 2005.

[38] G. Wang and E. A. B. Eltahir, "Ecosystem dynamics and the Sahel drought," *Geophysical Research Letters*, vol. 27, no. 6, pp. 795–798, 2000.

[39] B. Cazelles, M. Chavez, D. Berteaux et al., "Wavelet analysis of ecological time series," *Oecologia*, vol. 156, no. 2, pp. 287–304, 2008.

[40] M. À. Rodríguez-Arias and X. Rodó, "A primer on the study of transitory dynamics in ecological series using the scale-dependent correlation analysis," *Oecologia*, vol. 138, no. 4, pp. 485–504, 2004.

[41] R. F. Adler, G. J. Huffman, A. Chang et al., "The version-2 global precipitation climatology project (GPCP) monthly precipitation analysis (1979–present)," *Journal of Hydrometeorology*, vol. 4, no. 6, pp. 1147–1167, 2003.

[42] J. E. Janowiak, "An investigation of interannual rainfall variability in Africa," *Journal of Climate*, vol. 1, no. 3, pp. 240–255, 1988.

[43] A. Kaplan, M. A. Cane, Y. Kushnir, A. C. Clement, M. B. Blumenthal, and B. Rajagopalan, "Analyses of global sea surface temperature 1856–1991," *Journal of Geophysical Research C: Oceans*, vol. 103, no. 9, pp. 18567–18589, 1998.

[44] Climate Prediction Center (CPC), "Description of changes to oceanic Niño indexes," 2013, http://www.cpc.ncep.noaa.gov/products/analysis_monitoring/ensostuff/ONI_change.shtml.

[45] N. H. Saji, B. N. Goswami, P. N. Vinayachandran, and T. Yamagata, "A dipole mode in the tropical Indian ocean," *Nature*, vol. 401, no. 6751, pp. 360–363, 1999.

[46] J. W. Hurrell, "Decadal trends in the North Atlantic oscillation: regional temperatures and precipitation," *Science*, vol. 269, no. 5224, pp. 676–679, 1995.

[47] P. D. Jones, T. Jonsson, and D. Wheeler, "Extension to the North Atlantic oscillation using early instrumental pressure observations from gibraltar and south-west Iceland," *International Journal of Climatology*, vol. 17, no. 13, pp. 1433–1450, 1997.

[48] J. C. Rogers, "The association between the North Atlantic oscillation and the Southern oscillation in the Northern Hemisphere," *Monthly Weather Review*, vol. 112, no. 10, pp. 1999–2015, 1984.

[49] UCAR, *Persistent Patterns That Shape Weather and Climate Variability: A Glossary*, NCAR & UCAR, Boulder, Colo, USA, 2009, B. H. K.Trenberth, Z. Gallon, Eds.

[50] M. E. Brown, K. de Beurs, and A. Vrieling, "The response of African land surface phenology to large scale climate oscillations," *Remote Sensing of Environment*, vol. 114, no. 10, pp. 2286–2296, 2010.

[51] G. J. Huffman, R. F. Adler, M. M. Morrissey et al., "Global precipitation at one-degree daily resolution from multisatellite observations," *Journal of Hydrometeorology*, vol. 2, no. 1, pp. 36–50, 2001.

[52] S. Nicholson, "On the question of the "recovery" of the rains in the West African Sahel," *Journal of Arid Environments*, vol. 63, no. 3, pp. 615–641, 2005.

[53] P. J. Lamb, "Sub-Saharan rainfall update for 1982: continued drought," *Journal of Climatology*, vol. 3, no. 4, pp. 419–422, 1983.

[54] P. J. Lamb and R. A. Peppler, "Further case studies of tropical Atlantic surface atmospheric and oceanic patterns associated with sub-Saharan drought," *Journal of Climate*, vol. 5, no. 5, pp. 476–488, 1992.

[55] T. Lebel and A. Ali, "Recent trends in the Central and Western Sahel rainfall regime (1990–2007)," *Journal of Hydrology*, vol. 375, no. 1-2, pp. 52–64, 2009.

[56] F. Sima, A. Kamga, I. Raiva, F. S. Dekaa, and A. I. James, "State of the climate 2012: [Africa] West Africa," *Bulletin of the American Meteorological Society*, vol. 94, no. 8, pp. S161–S166, 2013.

[57] C. G. Torrence and P. Compo, "A practical guide to wavelet analysis," *Bulletin of the American Meteorological Society*, vol. 79, no. 1, pp. 61–78, 1998.

[58] A. Grinsted, J. C. Moore, and S. Jevrejeva, "Application of the cross wavelet transform and wavelet coherence to geophysical times series," *Nonlinear Processes in Geophysics*, vol. 11, no. 5-6, pp. 561–566, 2004.

[59] S. E. Nicholson, B. Some, and B. Kone, "An analysis of recent rainfall conditions in West Africa, including the rainy seasons of the 1997 El Nino and the 1998 La Nina years," *Journal of Climate*, vol. 13, no. 14, pp. 2628–2640, 2000.

[60] M. N. Ward, "Diagnosis and short-lead time prediction of summer rainfall in tropical North Africa at interannual and multidecadal timescales," *Journal of Climate*, vol. 11, no. 12, pp. 3167–3191, 1998.

[61] C. K. Folland, J. Knight, H. W. Linderholm, D. Fereday, S. Ineson, and J. W. Hurrel, "The summer North Atlantic oscillation: past, present, and future," *Journal of Climate*, vol. 22, no. 5, pp. 1082–1103, 2009.

[62] E. R. Martin and C. D. Thorncroft, "The impact of the AMO on the West African monsoon annual cycle," *Quarterly Journal of the Royal Meteorological Society*, vol. 140, no. 678, pp. 31–46, 2014.

[63] K. E. Trenberth and D. P. Stepaniak, "Indices of El Niño evolution," *Journal of Climate*, vol. 14, no. 8, pp. 1697–1701, 2001.

[64] J. Bjerknes, "Atmospheric teleconnections from the equatorial Pacific," *Monthly Weather Review*, vol. 97, no. 3, pp. 163–172, 1969.

[65] C. Okonkwo, B. Demoz, and S. Tesfai, "Characterization of West African jet streams and their association to ENSO events and

rainfall in ERA-interim 1979–2011," *Advances in Meteorology*, vol. 2014, Article ID 405617, 12 pages, 2014.

[66] S. Janicot, F. Mounier, S. Gervois, B. Sultan, and G. N. Kiladis, "The dynamics of the West African monsoon—part V: the detection and role of the dominant modes of convectively coupled equatorial Rossby waves," *Journal of Climate*, vol. 23, no. 14, pp. 4005–4024, 2010.

12

Solitary Rossby Waves in the Lower Tropical Troposphere

Andre Lenouo[1] and Francois Kamga Nkankam[2]

[1] Department of Physics, Faculty of Science, University of Douala, P.O. Box 24157, Douala, Cameroon
[2] Laboratory for Environmental Modelling and Atmospheric Physics, Department of Physics, University of Yaounde 1, Yaoundé, Cameroon

Correspondence should be addressed to Andre Lenouo; lenouo@yahoo.fr

Academic Editors: L. Ahrens, M. N. Lorenzo, and R. Singh

Weakly nonlinear approximation is used to study the theoretical comportment of large-scale disturbances around the intertropical midtropospheric jet. We show here that the Korteweg de Vries (KdV) theory is appropriated to describe the structure of the streamlines around the African easterly jet (AEJ) region. The introduction of the additional velocity of the soliton C_1 permits to search the stage where the configuration of the wave structures is going to emerge out of specified initial conditions and this is the direct and inverse cascade method. It was also shown that the configurations of disturbances can be influenced by this parameter so that we can look if the disturbances are in the control or not of their dispersive effects. This permits to explain the evolution of initial conditions of the Tropical Storm (TS) Debby over West Africa from 20 to 24 August 2006.

1. Introduction

The Rossby waves are the most important in large-scale atmospheric flow processes [1]. For their analysis, it is usually sufficient to study the horizontal structure of waves. Most theories treating the structure of these waves are based on linear models which only take into account their dispersive behaviour. Nonlinear processes are more interesting because they can help to explain, for example, the hurricane spiral bands observed in the tropical zone [2] and energy exchanges between different modes of the waves [3]. Solitary Rossby waves in a zonal flow appear to have been discovered (analytically) by [4] and have been studied subsequently by [5–10]. All invoke Rossby's β-plane model, in which the northerly gradient of the vertical component of the Earth's rotation is constant.

Many studies have dealt with nonlinear waves and particularly solitary waves in the atmosphere, stating with works by [11, 12]. On the theoretical level, nonlinear waves have been examined by [3] in the midatmosphere where the African easterly waves (AEWs) are propagated. In the same region, [13] showed that the Korteweg de Vries (KdV) theory is appropriate to describe the Rossby solitary waves. But the physical interpretation of the results in terms of the Rossby solitary waves is not evident and the roles that these waves could influence the structure and energy of these waves have not been examined. Moreover, [14] established that the propagation of the Rossby solitary wave has behaviour closer to those of ridges and troughs. They therefore showed that these waves can travel long distances in the northern hemisphere without a change both in speed or structure, and for any hour.

The first well-known studies of [15] have helped to identify solitary wave connected to internal gravity waves in the atmosphere. They showed that these waves are described by the KdV equations when they move in the upper atmosphere and by the Benjamin-David-Ono (BDO) equation when they appear in the lower level. They therefore correctly analysed the observations of [16] by using the KdV model, whereas the observation of [17] were better explained by the BDO model. This was the first evidence of these types of wave observed in the atmosphere and their comparison with theoretical models. The solution of the three-dimensional nonlinear Charney-Obukhov equation describing solitary pancake Rossby vortices was found by [18]. Its solution was represented in the form of an axially symmetric cylindrical monopole (anticyclonic) vortical structure moving with constant velocity. However, the role of westward-travelling planetary (Rossby) waves in the block onset and the deformation of eddies during the interaction between synoptic-scale eddies and an incipient block was examined by [10]. This author has constructed an incipient block that consists of a stationary dipole wave for zonal wavenumber and a westward-travelling monopole wave with constant amplitude for zonal wavenumber.

The role of nonlinear wave was also being studied in oceanography. Hence, time series observations of nonlinear internal waves in the deep basin of the South China Sea are used to evaluate mechanisms for their generation and evolution by [19]. They showed that internal tides are generated by tidal currents over ridges in the Luzon Strait and steeper as they travel west, subsequently generating high-frequency nonlinear waves. Although nonlinear internal waves appear repeatedly on the western slopes of the South China Sea, their appearance in the deep basin is intermittent and more closely related to the amplitude of the semidiurnal than the predominant diurnal tidal current in the Luzon Strait.

In the present study, we will use the weakly nonlinear theory to examine the behaviour of the large-scale waves around the midtropospheric African easterly jet (AEJ), where the wave is more intensely specified in the case of Tropical Storm (TS) Debby observed over West Africa from 20 to 24 August 2006. Considering that the vertical extent of this jet is smaller than its horizontal extent [3], we admit in first approximation that the motion of the air in this region is dominated by the effects of the rotation of the earth. Under these assumptions, we will look how the nonlinear disturbance could influence the structure and the energy of solitary waves in the midtroposphere where AEWs are propagated. This study is organised as follows. In Section 2, we will present method used to examine the nonlinear vorticity equation. In Section 3, the linear and nonlinear solutions are discussed; a case study of intense AEW of TS Debby is examined whereas conclusion is presented in Section 4.

2. Methodology and Data

2.1. Basics Equations. Rossby solitary waves are sought by using a nonlinear vorticity equation in a barotropic model [13]. This equation integrates the horizontal shear in mean zonal wind. We define a coordinate system (x, y, t), where t is the time component and the space components x and y are along the direction of wave propagation in east and north direction, respectively. When the flow is no divergent, zonal and meridional components of the velocity can be written as a function of a streamfunction perturbation ψ as

$$u = -\frac{\partial \psi}{\partial y}, \qquad v = \frac{\partial \psi}{\partial x}. \tag{1}$$

Otherwise, in a barotropic model, the evolution of streamline is described by the equation [1]

$$\left[\frac{\partial}{\partial t} + U\frac{\partial}{\partial x}\right]\nabla^2\psi + J\left(\psi, \nabla^2\psi\right) + \left[\beta - \frac{d^2U}{dy^2}\right]\frac{\partial \psi}{\partial x} = 0 \tag{2}$$

with the following boundary conditions:

$$\psi = 0 \quad \text{at } y = 0, \quad y = L. \tag{3}$$

In (2), J is the Jacobian operator $J(a, b) = (\partial a/\partial x)(\partial b/\partial y) - (\partial a/\partial y)(\partial b/\partial x)$, $\nabla^2 = \partial^2/\partial x^2 + \partial^2/\partial y^2$ the Laplacian operator, U the mean zonal wind component, and β the meridional gradient of the coriolis parameter.

The effects of nonlinearity are introduced through the Jacobian term, which is nonlinear. In the case of weak-amplitude waves, the individual oscillations can be represented in the form of linear or nonlinear wave superposition.

2.2. Theory. A soliton is localised wave, solution to a nonlinear partial derivatives equation without change of velocity or profile in a weakly dispersive area. By using a multiple scale method, we can write a stream function ψ in the form of a power expansion in a small parameter ϵ so that

$$\psi = \epsilon\psi_1 + \epsilon^2\psi_2 + \cdots. \tag{4}$$

It is necessary to introduce a convenient space and time variables ζ and τ, adapted to describe a weakly dispersive nonlinear system [15]. In this new Galilean reference frame, the transformations of [15] are given by

$$\zeta = \epsilon^{1/2}(x - C_0 t), \qquad \tau = \epsilon^{3/2}t, \tag{5}$$

where C_0 is the phase velocity of the eastward wave. The procedure consists in rewriting (2) using ζ and τ and then seeks a solution in a power series expansion in the amplitude parameter ϵ. Then by collecting terms of order $O(\epsilon^{3/2})$, we obtain the following linear equation in ψ_1:

$$(U - C_0)\frac{\partial}{\partial \zeta}\left(\frac{\partial^2\psi_1}{\partial y^2}\right) + \left[\beta - \frac{d^2U}{dy^2}\right]\frac{\partial\psi_1}{\partial \zeta} = 0. \tag{6}$$

Also the terms of order $O(\epsilon^{5/2})$ give:

$$\frac{\partial}{\partial \tau}\left(\frac{\partial^2\psi_1}{\partial y^2}\right) + (U - C_0)\frac{\partial^3\psi_1}{\partial \zeta^3} + \frac{\partial\psi_1}{\partial \zeta}\frac{\partial^3\psi_1}{\partial y^3} - \frac{\partial\psi_1}{\partial y}\frac{\partial}{\partial \zeta}\left(\frac{\partial^2\psi_1}{\partial y^2}\right) = 0. \tag{7}$$

Solutions of (6) and (7) are sought in the amplitude $A(\zeta, \tau)$ and modulated by the meridional function $\varphi(y)$, as given by the relation

$$\Psi(\zeta, y, \tau) = A(\zeta, \tau)\varphi(y). \tag{8}$$

(*a*) *Determination of* $\varphi(y)$. By substituting relation (8) into (6), we set the following eigenvalue equation for $\varphi(y)$:

$$\varphi''(y) + \frac{\beta - U''}{U - C_0}\varphi(y) = 0, \qquad (9)$$

where the prime denotes differentiation with respect to y. The boundary conditions are the same as those given by relation (3)

$$\varphi = 0 \quad \text{at } y = 0, \quad y = L. \qquad (10)$$

Equation (9) is solved numerically by using GAUSS-SEIDEL's relaxation methods. The shape of horizontal shear is chosen such that the wind is zero at the boundaries. Using centred-difference differentiation, (9) is rewritten as

$$\varphi_i = \frac{[\varphi_{i+1} + \varphi_{i-1}]}{[2 - F_i(\Delta y)^2]}, \qquad (11)$$

where Δy is the grid size, $i = 1, 2, \ldots, N$, and $F_i = [(\beta - U'')/(U - C_0)]_i$.

(*b*) *Nonlinear Waves.* By substituting relation (8) into (7), we can obtain a nonlinear equation as the KdV equation in the form

$$\frac{\partial A}{\partial \tau} + a_n A \frac{\partial A}{\partial \zeta} + b_n \frac{\partial^3 A}{\partial \zeta^3} = 0, \qquad (12)$$

where parameters a_n and b_n are determined by eigenfunctions φ and depend on the profile of $U(y)$ (see Appendix A for more details). They are given by the following expression:

$$a_n = \frac{\int_0^L \left[\varphi\varphi''' - \varphi'\varphi''\right] dy}{\int_0^L \varphi'' dy}, \qquad (13)$$

$$b_n = \frac{\int_0^L (U - C_0)\,\varphi\, dy}{\int_0^L \varphi'' dy}. \qquad (14)$$

We now examine solutions of (12) in the form of nonlinear progressive waves (soliton) $A = A(\zeta + C_1\tau)$, where C_1 is the phase velocity of the soliton which is a weak contribution to the principal phase velocity C_0. Thus, the total velocity of the system is

$$C = C_0 + \epsilon C_1. \qquad (15)$$

The solitary waves, solution of (12) is given by the following relation (see Appendix B for more details):

$$A(\zeta, \tau) = A_0 \operatorname{Sech}^2\left[\kappa\left(\zeta - C_1\tau\right)\right], \qquad (16)$$

where $A_0 = 3C_1/a_n$ is the nonlinear wave amplitude and $\kappa = (1/2)(C_1/b_n)^{1/2}$. Going back to the original variable, we finally have

$$A(x, t) = A_0 \operatorname{Sech}^2\left(\frac{x - Ct}{\Delta}\right), \qquad (17)$$

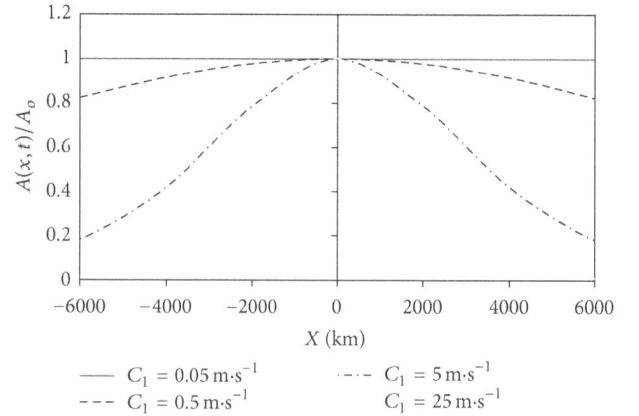

FIGURE 1: Variations of $A(x, t)/A_0$ in the propagation direction for different values of the additional wind velocity C_1.

where $\Delta = \sqrt{b_n/\epsilon C_1}$ is the soliton characteristic width.

The product of the soliton amplitude and the square of its characteristic width are independent of the soliton phase velocity C_1. It is however proportional to b_n and inversely proportional to a_n or ϵ and is expressed as

$$\Delta^2 A_0 = \frac{3b_n}{\epsilon a_n}. \qquad (18)$$

Since a_n and b_n are given by relation (13) and (14), respectively, the only known remaining is ϵ. Figure 1 shows some profiles of $A(x, t)/A_0$ for different values of additional phase velocity C_1 and $\epsilon = 0.01$. For small C_1, the figure shows that $A(x, t)$ is nearly constant in space but takes the form of a pulse when this parameter becomes important. It is to be noticed that this parameter appears explicitly in the expression of Δ. Thus, when C_1 is set to zero, Δ is even larger and the wave becomes evanescent. But as C_1 increases, for example, $C_1 = 25\,\mathrm{m}\cdot\mathrm{s}^{-1}$, the wave propagates symmetrically below the plan which passed trough the origin where $x = Ct$. Hence, we can say that the wave amplitude A_0 grows with the additional phase velocity C_1 (Figure 2). This shows that the wave moves faster as its amplitude becomes larger. Figure 2 also shows that the soliton characteristic width Δ decreases with C_1.

2.3. Data. During the African Monsoon Multidisciplinary Analysis (AMMA) SOP3 experimentation period, observations were conducted over West Africa from August 15 to September 15, 2006 [20]. The climatology data are from AMMA SOP3 operational database. From the briefing made during this campaign, the convective activity was very high and data for the entire West African region was available from August 15 to September 15, 2006, when wave 2 on the 7 waves was observed over this period. For more details about the 7 AEWs observed during this period, see the paper of [20]. Wave 2 was initiated during the period August 20 to 24, 2006. The satellite data used for this study are digitized images from Meteosat 8 calibrated from the MSG channel IR10.8, representing equivalent blackbody temperatures emitted by clouds. Apart from snow and ice artefacts, contiguous areas

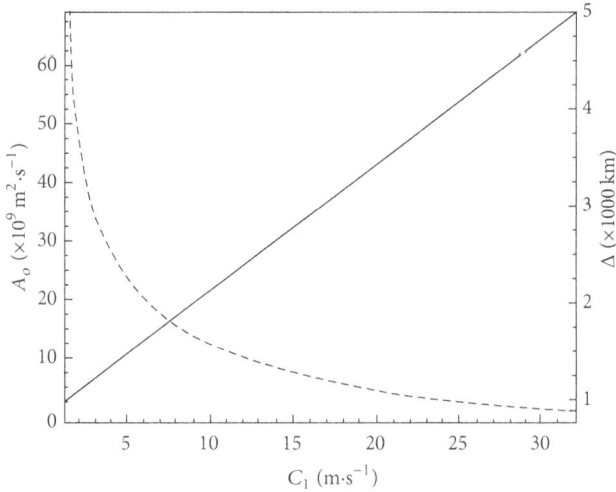

FIGURE 2: Variations of amplitude A_0 (in solid line) and the width Δ (in dash line) of the soliton as function of the additional wind velocity C_1.

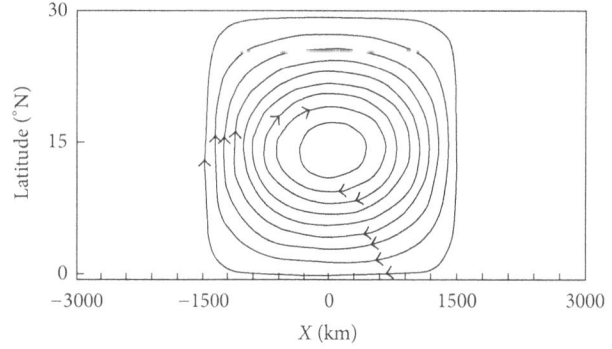

FIGURE 3: Configuration of streamlines in the case of linear approximation with $C_0 = 10 \, \text{m} \cdot \text{s}^{-1}$.

of depressed brightness temperatures in 85 GHz imagery indicate the presence of deep convection. The magnitudes of the brightness temperatures are inversely correlated with the strength of the convective updraughts producing the ice hydrometeors. Unlike infrared imagery, 85 GHz imagery can be used to analyse quantitatively the intensity of deep convection. This allows it to be used to study the properties of convective systems at a variety of spatial scales.

3. Results and Discussions

As we know, the solitary wave can be considered only when we define a coordinates system which moves with the wave at the velocity C, for the soliton to appear stationary. At the origin $X = 0$, where $X = x - Ct$, the amplitude solitary wave is maximal. This is due to the fact that in the translation $X = x - Ct$ of a nonfixed distance Ct, the maximum wave amplitude, initially at $x = 0$, stays until we are at $X = 0$.

Explicitly, the streamline depends on the soliton phase velocity C_1 and on the form of the mean zonal wind $U(y)$. We adopt in this work a basic flow with a horizontal shear as proposed by [21] to describe the midtropospheric jet in the West African Tropical zone. Based on observations the zonal wind can readily be represented in the functional form

$$U(y) = -U_0 \sin^2\left(\frac{\pi}{L}y\right), \qquad (19)$$

where $U_0 = 20 \, \text{m/s}$ is the maximal value of wind at the centre of jet (15°N) and L the distance between the equator ($y = 0$) and 30°N latitude ($y = L$). This jet corresponds to the one seen in the atmosphere during summer at an altitude of around 3000 m in the Northern African troposphere. The principal phase velocity is found to be 7.0 m/s [22–24].

Before examining the influence of the additional wind velocity C_1 in the present theory, let us first consider the case of linear waves.

3.1. Analysis of Linear Effects.

In the linear theory case, the solution of (2) without the Jacobian term is sought in the normal mode:

$$\Psi(x, y, t) = Y(y) \exp\left[ik(x - Ct)\right], \qquad (20)$$

where k is the zonal wavenumber, $k = 2\pi/\lambda$, λ the zonal wavelength, and $Y(y)$ is the amplitude function which depends only on y and solution to the following equation:

$$Y'' + \left[\frac{\beta - U''}{U - C_0} - k^2\right] Y = 0. \qquad (21)$$

This equation differs from (9) by the presence of the k^2 term, but it still must verify the boundary conditions $Y = 0$ at $y = 0$ and $y = L$. The numerical solution of (21) is found as earlier by GAUSS-SEIDEL's relaxation methods.

Figure 3 illustrates the configurations of streamlines in the (X, y) plane obtained from this approximation. We note that the region of instability corresponds to the depression centred along the principal axes of the jet. These streamlines have a quasi-concentric form, on the one hand symmetric respect of the plan passing $X = 0$ and on the other hand to the jet axis.

3.2. Nonlinear Effects.

The streamlines in the case of the weakly nonlinear approximation are presented in Figure 4 for different values of the additional wind velocity C_1. We see that this velocity has a predominant role in the configuration of patterns in the domain under consideration.

For $C_1 = 0.05 \, \text{m} \cdot \text{s}^{-1}$ (Figure 4(a)), the patterns are essentially parallel to the zonal direction. Here, the perturbations are swamped by the mean flow and this explains why for weak value of C_1, one cannot observe the track of the wave. The air flow can be assimilated in this case to the displacement of a solid that presents an axis of symmetric.

When the value of the parameter C_1 is increased, the streamlines have a new configuration as shown in Figure 4(b). We can deduce that the streamlines tend to unclose and stretch in the zonal direction isolating a depression centred at the region of maximum shear. The presence of a depression characterises the linear effects in the system. The fact that it is presently limited to the maximum disturbance region

(a)

(b)

(c)

(d)

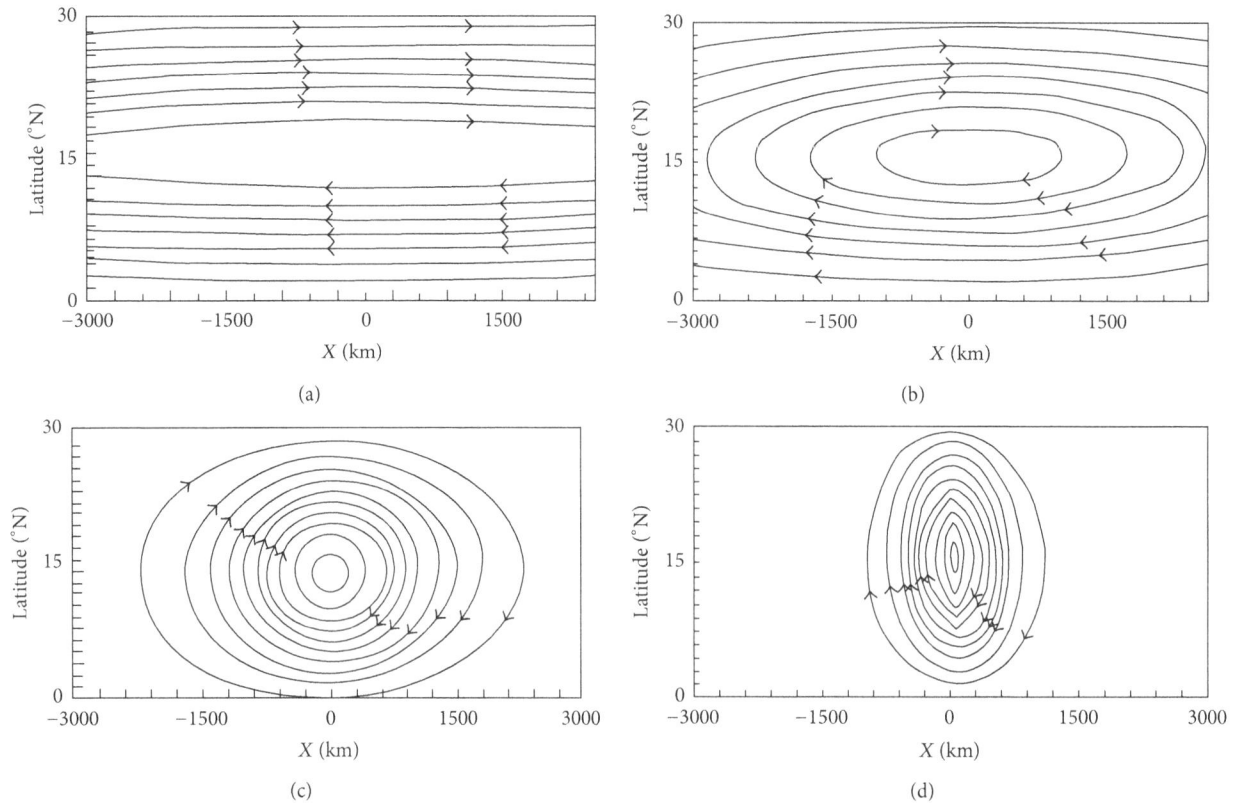

FIGURE 4: Configuration of streamlines in the case of weakly nonlinear approximation with $C_0 = 10\,\mathrm{m \cdot s^{-1}}$ and $\epsilon = 0.01$ for (a) $C_1 = 0.05\,\mathrm{m \cdot s^{-1}}$; (b) $C_1 = 0.5\,\mathrm{m \cdot s^{-1}}$; (c) $C_1 = 5\,\mathrm{m \cdot s^{-1}}$; and (d) $C_1 = 25\,\mathrm{m \cdot s^{-1}}$.

shows that these effects are in their early stages. This is why little is felt at the boundary of the domain, where only weak deformations of streamlines are observed.

We can continue to increase the value of the phase velocity of the soliton in order to determine a value that for which the weakly nonlinear theory leads to the same structure of linear waves as given by normal mode theory. Figure 4(c) presents the streamlines for $C_1 = 5\,\mathrm{m \cdot s^{-1}}$. We note that all the patterns have concentric ovoidal form around the region of maximal instability region where the amplitude of the nonlinear wave is high. The difference from previous configurations is the fact that these patterns are zonally limited at $X = \pm 2100\,\mathrm{km}$. Since our interest is to study the behaviour of wave around the jet, this result is not in contradiction with those obtained in the case of the linear approximation but matter confirms that the maximum instability of the jet is located in depressionary area. As also shown in Figure 4(c), this streamline can be superposed to those given by linear approximation (Figure 3). Hence the nonlinear wave is strongly governed by its linear effects. The first manifestation of the nonlinearity effects appears here, as noted by [25]; the presence of a weak nonlinearity in the system can produce important effects capable of countering those due to the dispersion. According to the weakly nonlinear approximation, the soliton results from a balance between linear and nonlinear effects. In other term, it is for the value of C_1 equal to $5\,\mathrm{m \cdot s^{-1}}$ that the Rossby soliton may be observed. Its profile described by relation (17) is represented in Figure 1.

As the parameter C_1 increases, the nonlinear effects grow and the wave patterns are concentrated around the region where their amplitudes grow (Figure 4(d)). Here, the streamlines tend to stretch along the meridional direction while being confined in a small zonal domain. This shows that the nonlinear waves became strongly localised.

3.3. Case of TS Debby. For energy consideration, we admit the principle that the energy of the perturbation is proportional to the square of amplitude of the wave. So, for weak values of C_1, the energy of the wave is dissipated in the space $x - Ct$. We found that the energy spreads in the space $x - Ct$ but in a reasonable interval compared to the purely linear case. However, Figure 4(c) shows that nonlinearity, though weak, leads to live for the perturbation. The value of C_1 can be found in a mosaic of satellite images for the period August 20–24, 2006, taken every six hours as shown in Figure 5. The total velocity of the wave is around 15 m/s where the classical value of AEW velocity is 10 m/s [21, 22, 24]; then we deduce that the additional wave (soliton) velocity is about 5 m/s.

This figure presents a very distinct signal that can be tracked from an initial region of convection, developing through several stages and moving off the African coast. The convection is more active at 0000 and 1800 UTC showing diurnal cycle of deep convection. The diurnal cycle of convection is important in many aspects of climate studies, in particular, through its strong modulation of the radiative budget and precipitation and its control of surface temperature [26]. Diurnal changes in deep convective clouds

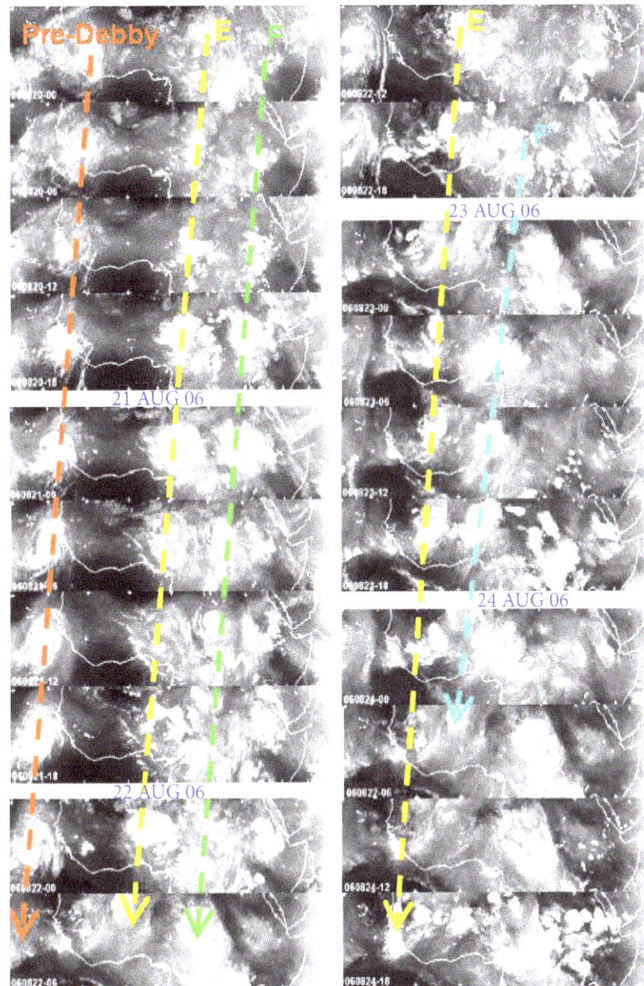

FIGURE 5: The situation on August 20–24, 2006, over West Africa corresponds first to the AEW ridge between the two troughs associated with "pre-Debby" and "F" (August 21 00 UTC to August 22 06 UTC) with weak convective development over east Guinea, then to the slow and rather inefficient propagation of AEW trough "E" over West Africa. There is no indication of triggering of an AEW (which can be traced back further east) by convection, nor strong development of convection induced by the propagating AEW "F" trough (except for the small thunderstorms on 22nd evening ahead of "E" and the short-lived MCS associated with "E").

influence incoming short wave and outgoing longwave radiations thereby affecting the Earth's radiation balance [27]. Two initial convection cells occur around 0000 UTC on August 21, 2006: the first at the west of the Senegalese coast and the second covers the northeast of Nigeria, south of Niger, and near the Jos Mountain around 15°N where the AEW is located. This continental region is favourable for the initiation of long-lived mesoscale convective weather systems which is a manifestation of nonlinear disturbance [28].

Interestingly, this convection begins at a time of the day when the mean MCS genesis frequency over Africa has been shown to be at a minimum [28]. On these images, the brightness of the pixels indicates that the initial convection is located east of and earlier than the first location of the AEW2 (around 20°W). This suggests that convection acts as a precursor for the AEW that develops downstream. Subsequent satellite images show that a convective zone rapidly develops, with high clouds covering a region of 1000 km square at 0600 UTC on August 21. The brightness of

the pixels indicates how the amount of very deep convection begins to propagate westward, crossing Senegal and reaching the eastern Atlantic at approximately 0600 UTC. The "pre-Debby" cyclonic part of the AEW is over West Africa and propagates westward over the warm waters off the coast of Guinea then becomes TS Debby on August 22, 2006. The cell at the west of the Senegalese coast persists until August 23, 2006, at 0600 UTC. The average speed of the cold cloud (<190 K) was about 15 m/s as noted earlier.

The convection over Jos Mountain remains whereas a new cell appears over the east of Central African Republic. The deep convection does not decrease within the convective region for the following 6 h as it moves over the Jos Mountain. Deep convection occurs on the leading edge of the remnants of the first convective burst along the coast of the Gulf of Guinea and over Central Africa Republic on August 21, 2006, at 1800 UTC. This convection persists over the entire West African region until August 23, 2006, at 0000 UTC. During the three days, the deep convection which develops over central Nigeria, near the Jos Plateau, is stationary, whereas the deep convection over the Senegalese coast marks the beginning of the second convective growth cycle over the Atlantic Ocean. On August 23, 2006, at 1800 UTC, the deep convection initiated over Central Africa Republic at 0600 UTC moves westward. In addition to this continuously developing region of convection propagating westward, the IR imagery shows another interesting feature: persistent localized convection occurs between 10° and 15°N. As shown by [7], a nonlinear evolution of the KdV equation can describe the interaction of solitary wave propagation in a zonal shear flow as AEJ.

4. Conclusion

We have presented a nonlinear theory to study the evolution of perturbation due to the shear mean wind in the midtropospheric African jet. Its formulation is necessarily complicated, but we have carefully described all the stages which permit to obtain the final result, so that one can use, in some conditions, these results which seem applicable to description of Rossby solitary waves. This requires the choice of additional wind speed C_1. The use of the Gardner and Morikawa transformations helps us to introduce the phase velocity of the soliton C_1. Its influence on the structure and the amplitude for streamline is important. Hence, the weakly nonlinear approximation through the KdV theory explains how the solitary Rossby waves are propagated over West Africa. The weakly nonlinear theory leads to explain the evolution of AEW2 which becomes TS Debby on August 22, 2006, when the soliton velocity was $C_1 = 5\,\mathrm{m}\cdot\mathrm{s}^{-1}$. The presence of a weak nonlinearity in the system can produce important effects capable of countering those due to the dispersion.

In the tropical zone, the cyclones are some time presented as the soliton. It is interesting to confront this theory with other observations since the nonlinear amplitude function $A(x,t)$, which is proportional to pressure, presents a maximum when the eastern wave propagation arrives at $X = 0$. These results have to be compared to other weakly nonlinear theory models, for example, with model of solitary pancakes Rossby vortices.

Appendices

A. Determination of Parameters a_n and b_n

By substituting relation (8) in (7), we obtain

$$\varphi''\frac{\partial A}{\partial \tau} + (U - C_0)\varphi\frac{\partial^3 A}{\partial \zeta^3} + \varphi\frac{\partial A}{\partial \zeta}A\varphi''' - \varphi'A\varphi''\frac{\partial A}{\partial \zeta} = 0 \tag{A.1}$$

or

$$\varphi''\frac{\partial A}{\partial \tau} + (U - C_0)\varphi\frac{\partial^3 A}{\partial \zeta^3} + \left(\varphi\varphi''' - \varphi'\varphi''\right)A\frac{\partial A}{\partial \zeta} = 0 \tag{A.2}$$

and if we integrate this equation into meridional domain, we have

$$\int_0^L \varphi''\frac{\partial A}{\partial \tau}dy + \int_0^L (U - C_0)\varphi\frac{\partial^3 A}{\partial \zeta^3}dy$$
$$+ \int_0^L \left(\varphi\varphi''' - \varphi'\varphi''\right)A\frac{\partial A}{\partial \zeta}dy = 0. \tag{A.3}$$

By setting $\int_0^L \varphi''dy$, (A.3) can be rewritten at last as follows:

$$\frac{\partial A}{\partial \tau} + a_n A\frac{\partial A}{\partial \zeta} + b_n\frac{\partial^3 A}{\partial \zeta^3} = 0 \tag{A.4}$$

with

$$a_n = \frac{\int_0^L \left[\varphi\varphi''' - \varphi'\varphi''\right]dy}{\int_0^L \varphi''dy},$$
$$b_n = \frac{\int_0^L (U - C_0)\varphi\,dy}{\int_0^L \varphi''dy}. \tag{A.5}$$

B. Solution of the KdV-Equation

To solve (12), we introduce the following Galilee transformation:

$$s = \zeta - C_1\tau. \tag{B.1}$$

In this new referential where the wave is propagated with the velocity C_1, (12) becomes

$$-C_1\frac{\partial A}{\partial s} + a_n A\frac{\partial A}{\partial s} + b_n\frac{\partial^3 A}{\partial s^3} = 0. \tag{B.2}$$

The integration of (B.2) with respect to s gives

$$-C_1 A + a_n\frac{A^2}{2} + b_n\frac{\partial^2 A}{\partial s^2} = 0. \tag{B.3}$$

By multiplying this last equation by dA/ds and integrating them, we have

$$-C_1\frac{A^2}{2} + a_n\frac{A^3}{6} + \frac{b_n}{2}\left(\frac{dA}{ds}\right)^2 = 0 \tag{B.4}$$

or

$$\frac{dA}{ds} = \left(\frac{C_1}{b_n}\right)^{1/2}A\sqrt{1 - \frac{a_n}{3C_1}A}. \tag{B.5}$$

By setting $\widetilde{A} = a_n A/(3C_1)$, we obtain the integral

$$\int \frac{d\widetilde{A}}{\widetilde{A}\left(1 - \widetilde{A}\right)^{1/2}} = \left(\frac{C_1}{b_n}\right)^{1/2}(s - s_0). \tag{B.6}$$

If we assume $s_0 = 0$, (B.6) can be written as follows:

$$\text{Arg Sech } \widetilde{A} = \left(\frac{C_1}{b_n}\right)^{1/2}s. \tag{B.7}$$

And at last

$$A = \frac{3C_1}{a_n}\text{Sech}^2\left[\left(\frac{C_1}{b_n}\right)^{1/2}s\right]. \tag{B.8}$$

Acknowledgment

This research is supported by ICTP, Trieste, Italy, through the Associate and Federation Schemes Program (Reference 431).

References

[1] J. R. Holton, *An Introduction to Dynamic Meteorology*, vol. 16 of *International Geophysics Series*, 4th edition, 2004.

[2] T. A. Guinn and W. H. Schubert, "Hurricane spiral bands," *Journal of the Atmospheric Sciences*, vol. 50, no. 20, pp. 3380–3403, 1993.

[3] A. Lenouo, F. N. Kamga, and E. Yepdjuo, "Weak interaction in the African easterly jet," *Annales Geophysicae*, vol. 23, no. 5, pp. 1637–1643, 2005.

[4] R. R. Long, "Solitary waves in the westerlies," *Journal of the Atmospheric Sciences*, vol. 21, no. 2, pp. 197–200, 1964.

[5] L. H. Larsen, "Comments on solitary waves in the westerlies," *Journal of the Atmospheric Sciences*, vol. 22, no. 2, pp. 222–224, 1965.

[6] D. J. Benney, "Long nonlinear waves in fluid flows," *Journal of Mathematical Physics*, vol. 45, pp. 52–63, 1966.

[7] L. G. Redekopp and P. D. Weidman, "Solitary Rossby waves in zonal shear flows and their interactions," *Journal of the Atmospheric Sciences*, vol. 35, pp. 790–804, 1978.

[8] J. W. Miles, "On solitary Rossby waves," *Journal of the Atmospheric Sciences*, vol. 36, no. 7, pp. 1236–1238, 1979.

[9] B. J. Hoskins and T. Ambrizzi, "Rossby wave propagation on a realistic longitudinally varying flow," *Journal of the Atmospheric Sciences*, vol. 50, no. 12, pp. 1661–1671, 1993.

[10] D. Luo, "A barotropic envelope Rossby soliton model for block-eddy interaction. Part II: role of westward-travelling planetary waves," *Journal of the Atmospheric Sciences*, vol. 62, no. 1, pp. 22–40, 2005.

[11] M. Tepper, "A proposed mechanism of squall line. The pressure jump line," *Journal of Meteorology*, vol. 7, no. 1, pp. 21–29, 1950.

[12] A. J. Abdullah, "The atmospheric solitary wave," *Bulletin of the American Meteorological Society*, vol. 10, pp. 511–518, 1955.

[13] E. M. Dobryshman, "Theoretical studies of tropical waves," *GATE*, vol. 25, pp. 121–177, 1982.

[14] H. S. Huang Sixun and Z. Ming, "Periodic, solitary and discontinuous periodic solution of nonlinear waves in the atmosphere and their existence. Part I and II.," *Scientia Sinica B*, vol. 31, no. 12, pp. 1489–1502, 1988.

[15] J. W. Rottman and F. Einaudi, "Solitary waves in the atmosphere," *Journal of the Atmospheric Sciences*, vol. 50, no. 14, pp. 2116–2136, 1993.

[16] Y. L. Yuh-Lang Lin and R. C. Goff, "A study of a mesoscale solitary wave in the atmosphere originating near a region of deep convection," *Journal of the Atmospheric Sciences*, vol. 45, no. 2, pp. 194–205, 1988.

[17] R. K. Smith and B. R. Morton, "An observational study of northeasterly "morning glory" wind surges," *Australian Meteorological Magazine*, vol. 32, pp. 155–175, 1984.

[18] T. D. Kaladze, "New solution for nonlinear pancake solitary Rossby vortices," *Physics Letters A*, vol. 270, no. 1-2, pp. 93–95, 2000.

[19] Q. Li and D. M. Farmer, "The generation and evolution of nonlinear internal waves in the deep basin of the South China Sea," *Journal of Physical Oceanography*, vol. 41, no. 7, pp. 1345–1363, 2011.

[20] E. J. Zipser and Coauthors, "The Saharan air layer and the fate of African easterly waves: the NAMMA field program," *Bulletin of the American Meteorological Society*, vol. 90, pp. 1137–1156, 2009.

[21] M. A. Rennick, "The generation of African wave," *Journal of the Atmospheric Sciences*, vol. 33, no. 10, pp. 1955–1969, 1976.

[22] R. W. Burpee, "The origin and structure of easterly waves in the lower troposphere of North Africa," *Journal of the Atmospheric Sciences*, vol. 29, no. 1, pp. 77–90, 1972.

[23] C. Mass, "A linear equation model of African wave disturbances," *Journal of the Atmospheric Sciences*, vol. 36, pp. 2075–2092, 1979.

[24] A. Lenouo and F. Mkankam Kamga, "Sensitivity of African easterly waves to boundary layer conditions," *Annales Geophysicae*, vol. 26, no. 6, pp. 1355–1363, 2008.

[25] B. B. Kadomtsev, *Phénomènes Collectifs dans le Plasmas*, Editions Mir, Moscou, Russia, 1979.

[26] R. A. Houze Jr., "Mesoscale convective systems," *Reviews of Geophysics*, vol. 42, no. 4, 2004.

[27] L. A. T. Machado, H. Laurent, and A. A. Lima, "Diurnal march of the convection observed during TRMM-WETAMC/LBA," *Journal of Geophysical Research D*, vol. 107, no. 20, pp. 31.1–31.15, 2002.

[28] K. I. Hodges and C. D. Thorncroft, "Distribution and statistics of african mesoscale convective weather systems based on the ISCCP meteosat imagery," *Monthly Weather Review*, vol. 125, no. 11, pp. 2821–2837, 1997.

13

Temporal Patterns of the Two-Dimensional Spatial Trends in Summer Temperature and Monsoon Precipitation of Bangladesh

Avit Kumar Bhowmik

Institute for Environmental Sciences, Quantitative Landscape Ecology, University of Koblenz-Landau, Fortstraße 7, 76829 Landau (Pfalz), Germany

Correspondence should be addressed to Avit Kumar Bhowmik; bhowmik@uni-landau.de

Academic Editors: E. Tagaris and F. Yang

Two climate indices, TXx and PRCPTOT, representing the summer maximum temperature and annual total monsoon precipitation, respectively, in Bangladesh were computed. The temperature and precipitation measurements from 34 meteorological stations during the temporal extent of 1948–2007 were applied for indices' computation under thorough quality control. The spatial trends of the indices were analyzed by applying two-dimensional least square approach along latitudes and longitudes of the observation points. The temporal patterns of the spatial trends were identified by temporally interpolating them applying thin plate smoothing spline method. The analyses of TXx identified regional scale spatial trends in the east-west and south-north directions, which were increasing between 1948 and 1980s. After the 1980s the spatial trends started decreasing, and after 2000 the spatial trend along the south-north changed its direction to the north-south and continued until present. The analyses of the PRCPTOT identified spatial trends in the west-east and north-south directions, which were decreasing between 1948 and 1980s and thereafter increasing until present. About half of the spatial trends were significant in F-statistics at or more than 90% confidence level. Thus, the obtained results indicated a significant climatic shift within the regional scale of the country during the study period.

1. Introduction

In light of the recent climatic change concern, evidence has been presented that the indices for seasonal temperature and precipitation show augmented responses to the actual mean climatic trend [1, 2]. The global multidimensional trends of such climate indices have already been analyzed by incorporating several anthropogenic and natural factors, to predict the pattern of climate change [3–5]. As IPCC [6, 7] put emphasis on the need for detailed information of regional patterns of climate change, trends in seasonal climatic events and their statistical significance have also been analyzed in different regions of the world, that is, in Nigeria [8], Australia [9], Asia and central Pacific [10], UK [11], and some parts of India [12]. These studies represent that the regional spatiotemporal trends of temperature and precipitation are more complex and significant than the global trends; they are particularly significant for the regions that have already been under climate change stress, such as Bangladesh.

Bangladesh, situated in south-east Asia, is one of the most vulnerable countries of the world regarding the adverse impacts of anthropogenic climate change [13–17]. The total area of the country is 147,570 square kilometer [18], approximately one-fifth of which consists of low-lying coastal zones within one meter of the high water mark [7]. Threats of sea level rise, droughts, floods, and seasonal shifts due to the global warming have been presented in many recent studies on the country [13–17, 19–21]. The mean annual temperature increased during the period of 1895–1980 by 0.31°C [19], and the annual maximum temperature is predicted to be increased by 0.4°C and 0.73°C by the year of 2050 and 2100, respectively [20, 21]. Monsoon precipitation is forecasted to be increased, and at the same time winter precipitation is forecasted to be decreased in the coming decades [22, 23].

The temperature and precipitation fields are highly variable in both space and time, and therefore they should include both the two-dimensional space and one-dimensional time into their trend analysis [24, 25]. To obtain an accurate

TABLE 1: Name of the meteorological stations in Bangladesh with their longitudes, latitudes, altitudes, and temporal records of TXx and PRCPTOT.

NIR	Stations	Longitude (dd)	Latitude (dd)	Altitude (m)	TXx record	PRCPTOT record
1	Barisal	90.4	22.667	4	1949–1951, 1953-1954, 1956–1963, 1966–2007	1949–1951, 1953-1954, 1956–2007
2	Bhola	90.667	22.333	5	1966–1978, 1979–2007	1966–2007
3	Bogra	89.416	24.833	20	1948-1949, 1952, 1954, 1956–2007	1948–2007
4	Chandpur	90.666	23.2	7	1964, 1966–1971, 1973–1977, 1979, 1981–2007	1964, 1966–1970, 1973–1977, 1979, 1981–2007
5	Chittagong	91.866	22.316	6	1949–2007	1949–2007
6	Chuadanga	88.866	23.666	12	1989–2007	1989–2007
7	Comilla	91.2	23.45	10	1948–1962, 1964–1966, 1969–2007	1948–1962, 1964–1977, 1979–2007
8	Cox's Bazar	92.033	21.417	4	1948–2007	1948–2007
9	Dhaka	90.4	23.683	9	1953–1973, 1975–2007	1953–1973, 1975–2007
10	Dinajpur	88.616	25.583	37	1948–1972, 1981–2007	1948–1972, 1981–2007
11	Faridpur	89.866	23.583	9	1948–2007	1948–2007
12	Feni	91.4	23	8	1973–2007	1974–2007
13	Hatiya	91.1	22.43	4	1966–1971, 1973–1980, 1982–1994, 2000–2007	1966–1971, 1973–1980, 1981–1994, 1999–2007
14	Ishurdi	89.166	24.166	14	1961–2007	1961–1966, 1968–2007
15	Jessore	89.233	23.166	7	1948–1977, 1979–2007	1948–1977, 1979–2007
16	Khepupara	90.233	21.933	9	1975–2007	1974–2007
17	Khulna	89.583	22.833	4	1948–1955, 1957–1966, 1968–1974, 1976–2007	No record
18	Kutubdia	91.833	21.833	7	1985–2007	1977, 1979-1980, 1985–2007
19	Madaripur	90.216	23.183	13	1977–2007	1977–1978, 1980–2007
20	Maijdee Court	91.133	22.833	6	1951–1956, 1958–1975, 1978–2007	1951–1975, 1978–2007
21	Mongla	89.666	22.417	2	1989–2007	1991–2007
22	Mymensingh	90.416	24.766	19	1948, 1950–2007	1948, 1951–2007
23	Patuakhali	90.417	22.333	3	1973, 1975–1979, 1981–2007	1973, 1975, 1977–1979, 1981–2007
24	Rajshahi	88.583	24.333	20	1964–1969, 1971–2007	1964–1968, 1971–2007
25	Rangamati	92.216	22.583	17	1957–2007	1957–1966, 1969–2007
26	Rangpur	89.3	25.7	34	1957–1967, 1969–1972, 1978–2007	1954–1967, 1969–1975, 1976–2007
27	Swandip	91.416	22.416	6	1966–1974, 1976–2002, 2005–2007	1966–1974, 1976–2002, 2004–2007
28	Satkhira	89.166	22.683	6	1948–1954, 1955–1967, 1969–2007	
29	Sayedpur	88.833	25.766	6	1991–2007	1991–2007
30	Sitakunda	91.583	22.533	23	1977–2007	1977–2007
31	Srimongol	91.666	24.3	44	1948–1979, 1982–2007	1948–1980, 1982–2007
32	Sylhet	91.833	24.9	10	1956–1972, 1974–2007	1956–1972, 1974–2007
33	Tangail	89.833	24.166	20	1987–2007	1987–2007
34	Teknaf	92.333	20.833	6	1977–2007	1977–2007

dd: decimal degree.

prediction of the trends in climate phenomena conformant to the real earth scenario, a three-dimensional spatiotemporal trend analysis is very important [26, 27]. All the previous studies on Bangladesh climate basically dealt with the one-dimensional temporal trend of the climate phenomena and spatial distribution of these temporal trends. No study has so far dealt with the two-dimensional spatial trends and the temporal distribution of the spatial trends of climate.

In addition, Shahid [23] analyzed the one-dimensional temporal trends and their spatial distribution for 15 different precipitation indices, but so far no study has been carried out to analyze the trends of temperature indices. On one hand, the annual mean rainfall varies from 1400 mm in the west to more than 4300 mm in the east of the country [22, 23]. On the other hand, the north-west division of Bangladesh-Rajshahi experiences the highest temperature in summer which causes

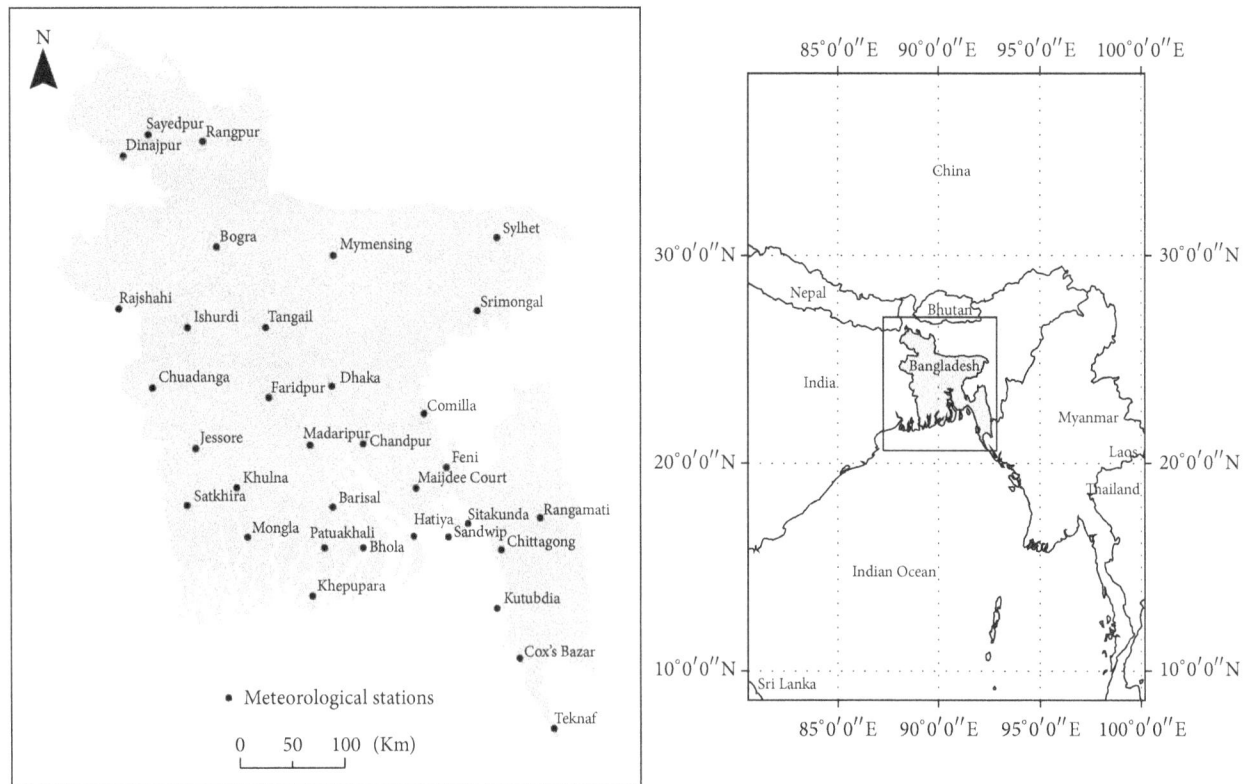

FIGURE 1: Maps of the part of Southeast Asia with the location of Bangladesh and neighboring countries and Bangladesh with the location of 34 meteorological stations for which the data is available in 2007. In general, the time series used for these stations is 1948–2007.

long-term droughts and famine in this region. The southern coastal regions in contrast experience the lowest moderate temperature in summer [21]. These study results provide the evidences of spatial trends in temperature and precipitation phenomena over the country. Therefore, a two-dimensional spatial trend analysis of the temperature and precipitation indices, which represent the seasonal climatic phenomena in Bangladesh, and the temporal pattern analysis of the spatial trends are of utmost importance for climate change mitigation and climate resilience activities. More importantly, the analyses of the spatial trends will provide more detailed knowledge of the climate variability within the region over time. A two-dimensional spatial trend surface provides indicators (slopes), which are spatial characterizations of the climatic fields in each time step. Thus the temporal interpolation of the spatial trends brings important information on regional climatic shifts, since fields of climate anomalies, once the background climate means are removed, tend to be more spatially coherent than the raw observations themselves [8–12]. It is also urgent to compare between the spatial trends of precipitation and temperature indices to discover the status of vulnerability of the country and people with respect to the change in both climatic phenomena.

This study, in spite of the discussion above, attempts to provide new information on the spatial trends of the climatic phenomena in the regional scale of Bangladesh, using long-term observations of daily temperature and precipitation

from 34 meteorological stations during 1948–2007. It analyses and compares between the two-dimensional spatial trends of the indices for summer maximum temperature and annual total monsoon precipitation events by fitting the two-dimensional least square surfaces along latitudes and longitudes of the observation points. The annual spatial trends are further interpolated in time to analyze the temporal pattern of the spatial trends of the two indices within the extent of the study period. Two climate indices have been selected for the spatial trend analysis and comparison, which are recommended by the Joint Project Commission for Climatology/Climate Variability and Predictability (CLIVAR) and Joint WMO/IOC Technical Commission for Oceanography and Marine Meteorology Expert Team on Climate Change Detection and Indices, namely, TXx and PRCPTOT [28, 29]. In case of Bangladesh, according to the definition of the indices, TXx represents the annual summer maximum temperature, and PRCPTOT represents the annual total monsoon precipitation. The representative trend surfaces of the two indices from each decade during the study period have been presented, and the mean trend surfaces have been computed based on the mean spatial trends. Finally, the change in the regional variation of climate in Bangladesh has been analyzed, and corresponding consequences on crop production and other resources have been predicted. These analyses are very important for Bangladesh since any change in the regional scale of climate can have large impacts on the daily life of the population that is dependent on agriculture and threatened by devastating floods every year.

2. Material and Methods

2.1. Data, Tools, and Quality Control. The daily maximum temperatures and daily precipitation data obtained from the 34 meteorological stations across Bangladesh for the period between 1948 and 2007 have been utilized in this study. The data are not available from all stations in every year because all the stations did not start working and providing data at the same time [30]. On one hand, temperature data from 10 stations are available for 1948 which gradually increases to the data from 34 meteorological stations by 2007. On the other hand, precipitation data from 8 stations are available for 1948 which gradually increases to the data from 32 meteorological stations by 2007. Therefore, the 1948–2007 period has been chosen because it characterizes a long-term dataset for each station and fosters the annual spatial trend analysis. The station locations are shown in Figure 1, and their latitudes, longitudes, and altitudes are provided in Table 1. The dataset is provided by the Bangladesh Meteorological Department (BMD), which is the authorized government organization for all meteorological activities in the country [30].

The open source statistical software "RStudio" [31] with its particular packages, "spacetime," "intamap," "fields," "gstat," "scatterplot3d," "maptools," and "sm," has been used in this study for the climate indices computation, their spatial trend analyses, temporal interpolation and visualization. The R-module especially designed for climate indices computation and data quality control entitled "RClimdex" [32] has been adapted for this study. ASCII format for raster grids and CSV or TXT formats for the temperature and precipitation records have been used as the input data file as *R* requires. The temperature units are degrees Celsius (°C), and the precipitation units are millimeters (mm).

An exhaustive data quality control has been conducted being aware of the fact that indices are sensitive to changes in station location, exposure, equipment, and observer practice [32]. The procedures that have been performed are: (1) replacing all missing values to an internal format that the software recognizes, that is, NA, not available, and (2) replacing all unreasonable values into NA, which include (a) daily maximum temperature more than 60°C (only one observation has been identified which is unlikely for a subtropical climate like Bangladesh [14–16]; this has been considered as a systematic bias) and (b) daily precipitation less than zero. The count of NA values for each month was also recorded to ensure rational computation of the climate indices based on their definition [32, 33]. This reduces any systematic bias caused by changes of precipitation gauges and temperature measurement instruments over the 60 years.

2.2. Methodological Framework

2.2.1. Indices for Summer Temperature and Monsoon Precipitation. TXx refers to the yearly maximum value of the daily maximum temperature [28, 34]. Bangladesh has a clearly defined summer season that is recognized in March–June every year [14, 15]. The maximum temperature of every year is observed during this summer season which characterizes the magnitude of corresponding summer. Therefore, TXx for Bangladesh stands for the maximum summer temperature recorded in the country, and it is the most representative of the summer temperature anomaly. The formula for calculating TXx is if TXx is the daily maximum temperatures in period *j*, then the maximum daily temperature for each period is as (1) [28]. The period *j* is set to March–June for each year.

Therefore,

$$TXx_j = \max\left(TXx_j\right). \tag{1}$$

Considering that a single observation can be affected by a particular synoptic episode over a limited area and may not represent the actual overall warming or cooling tendency in that year; the top 20% TXx over the summer period have been averaged for each station.

PRCPTOT refers to the annual total precipitation in wet days [28, 35]. Bangladesh has clearly defined wet days in each year, this particular precipitation phenomenon is known as "Monsoon" and is present in June–September every year [7, 14, 36–38]. Therefore, PRCPTOT for Bangladesh stands for the total monsoon precipitation in each year and it is the most representative of the mean precipitation anomaly. The formula for calculating PRCPTOT is if RR_{ij} is the daily precipitation amount on day *i* in wet period *j* and if *l* represents the number of days in *j*, then PRCPTOT is as (2) [28]. The period *j* for PRCPTOT calculation is set to June–September for each year.

Hence,

$$PRCPTOT_j = \sum_{i=1}^{l} RR_{ij}. \tag{2}$$

These two indices have been computed from the available temperature and precipitation data for each year of 1948–2007 and for each station. Following the monthly NA count for the observations, the threshold has been set to 80% for the designated periods. Therefore the indices have only been calculated when 80% of the daily observations of maximum temperature and precipitation are available for the summer (March–June) and monsoon (June–September), respectively; otherwise the indices have been set to NA. Table 1 demonstrates the producible TXx and PRCPTOT records within the temporal extent of the study with the available temperature and precipitation data.

2.2.2. Spatial Trends of the Indices. The resulting series of the summer temperature and monsoon precipitation indices have been analyzed through spatial trends. Significant correlations of the computed indices to the latitudes and longitudes justify the spatial trend analysis (will be discussed in Section 3). The spatial trend analysis includes fitting a two-dimensional least square surface along latitudes and longitudes as (3).

Consider,

$$B = Y + P_1 A_1 + P_2 A_2, \tag{3}$$

where B represents the least square value of the particular index and A_1 and A_2 represent the two independent variables—longitude and latitude in two different dimensions—along which the least square trend surface is computed. A third dimension, altitude, could have been included, but the correlation of the indices to altitude has resulted insignificant. This is an artifact of the altitudes of the stations being less than 50 m. The intercept Y and slopes P_1 and P_2 of the annual spatial trends of the climate indices have been calculated based on the fitted least square surfaces according to

$$Y = B - P_1 A_1 - P_2 A_2,$$

$$P_i = \frac{\sum (A_i - A_{av})(B_i - B_{av})}{\sum (A_i - A_{av})^2}, \tag{4}$$

where A_{av} and B_{av} indicate the average value of the A and B, respectively. The intercept y calculates the point at which the least square surface of a particular index intersects the index axis by using existing index values, latitude values and longitude values. It determines the index surface value when there is no spatial trend along the latitude and longitude. The slope values P_1 and P_2 represent the trends of the index along latitude and longitude, respectively. Trends are obtained for each index at 60 years, the statistical significance p of the trends were assessed through applying the F-statistics [39], and the number of degrees of freedom was obtained based on the length of the dataset, that is, 8–34 for the varying number of spatial data points [32, 39]. According to the F-statistics, for 8–34 degrees of freedom, statistically significant trends at the 90% ($0.05 < p \leq 0.1$), 95% ($0.01 < p \leq 0.05$), and 99% ($p \leq 0.01$) confidence levels have been identified.

2.2.3. Temporal Interpolation of the Spatial Trends. The spatial trends of the indices have been interpolated in time to analyze the temporal change in regional pattern of climate variability. Thin plate smoothing spline method described by Hutchinson and Hancock [40, 41] has been applied to interpolate the spatial trend values within the temporal extent of 1948–2007. The spatial trends of the indices are considered as observations $(z_i, x_{1i}, x_{2i}, \ldots, x_{di})$ measuring a dependent variable z and predictor variables $x_1, x_2, \ldots, x_D$ which are included in a set of time domain D. If z has both continuous long-term variation as well as discontinuous and random short-term variation, then the data model can be expressed as (5)

$$z_i = g(x_1, x_2, \ldots, x_D) + \epsilon_i \quad (i = 1, \ldots, n), \tag{5}$$

where n is the number of data observations, g is a slowly varying continuous function, and ϵ_i is the realization of a random variable ϵ. The function g represents the temporally continuous long range variation in the process measured by z_i. The errors of ϵ_i are assumed to be independent with

mean zero and variance σ^2. The thin plate smoothing spline predicts the process g in (5) by a suitably continuous function f that is able to separate the continuous signal g from the discontinuous noise ϵ_i. This function can be estimated by minimizing

$$\frac{1}{n} \sum_{i=1}^{n} (z_i - f_i)^2 + \lambda J_m^D(f) \tag{6}$$

over functions $f \in X$, where X is a time function whose partial derivatives of total order m are in $L^2(E^d)$. The f_i are values of the fitted function at the ith measurement, λ is a fixed smoothing parameter, and $J_m^D(f)$ is a measure of the roughness of the function f in terms of mth order partial derivatives. The thin plate smoothing spline represents a smooth gradual increase or decrease of the spatial trend values along the time series.

3. Results

3.1. Spatial Trends of the Temperature and Precipitation Indices. The computed values of the climate indices—TXx and PRCPTOT—vary between 30°C–46°C and 750 mm–4516 mm, respectively. Figure 2 represents the distribution of the computed TXx and PRCPTOT values at every station in every year. A general spatial trend of the indices over the study area during the study period can be observed. On one hand, in almost all the years, the maximum TXx values ranging between 38°C and 46°C are observed at the stations situated in the north-western, whereas the minimum TXx values ranging between 30°C and 32°C are observed at the stations situated in the south-eastern parts of the country. There is a gradual increase in the TXx values from the southeast to the north-west of the country in all years which provides evidence that spatial trend in the TXx index exists in the same direction. On the other hand, the calculated maximum PRCPTOT values which range 2000 mm–4500 mm are observed at the stations situated in the south-eastern and the minimum PRCPTOT values which range 750 mm–1500 mm at the stations situated in the north-western parts of the country. There is also a gradual increase in the PRCPTOT values from the north-west to the south-east of the country out of the precipitation noise in all the years, which clearly indicates a spatial trend in the PRCPTOT index from the north-west to south-east.

Moreover, the calculated correlation-coefficients of the TXx with the stations' latitudes and longitudes are 0.34 and −0.52, respectively, which indicates that the spatial dependence of the index is more dominant in the west-east direction than in the south-north direction. Similarly, the correlation-coefficients of PRCPTOT with the stations' latitudes and longitudes are −0.42 and 0.55, respectively, which indicates that the spatial dependence of the index is more dominant in the east-west direction than in the north-south direction. In parallel, the correlation of both TXx and PRCPTOT with altitude has also been computed, which are found insignificant, and therefore the decision is made that altitude does not affect any of the indices in their spatial dependence.

TABLE 2: Yearly spatial trends and intercepts of the indices for the temperature and precipitation over Bangladesh during 1948–2007.

Years	Intercept (°C)	TXx Trend-longitude (°C dd⁻¹)	Trend-latitude (°C dd⁻¹)	Intercept (mm)	PRCPTOT Trend-longitude (mm dd⁻¹)	Trend-latitude (mm dd⁻¹)
1948	103.8016	-0.7537^{***}	0.0625	-29088.7	376.9	-143
1949	68.3107	-0.2333^{***}	-0.5117	-23148.8	333.5	-223.8^{**}
1950	73.4924	-0.4364	0.1456	-4596.5	195.8	-483.2^{*}
1951	125.5522	-0.9277^{***}	-0.2017	-29394.81	350.62	-38.25
1952	17.360687	-0.007369	0.911620	-15961.6	238.2	-161.9^{***}
1953	78.5432	-0.5917^{**}	0.4979^{*}	-15137.5	245	-221.8^{*}
1954	161.1209	-1.5983	0.9162	-10420.3	197.2	-233.4^{**}
1955	198.2763	-1.6914^{**}	-0.3202	-37526.6	382.6^{***}	196.2
1956	222.6913	-1.9503^{*}	-0.2872	-29687.48	348.799^{*}	5.367^{**}
1957	144.29750	-1.14795	-0.09892	-17896.22	206.47	21.77
1958	161.4466	-1.5386^{*}	0.7119^{*}	-30510.06	336.95^{**}	51.96
1959	195.1917	-1.7594^{*}	0.1079^{**}	-23323.1	288.6^{***}	-45.7^{***}
1960	25.3299	-0.1297	1.1547^{*}	-2568.8	93.29	-178.83
1961	179.60017	-1.56272	0.04544^{**}	-18565.7	323.8^{***}	-383.1^{*}
1962	137.363	-1.240	0.557	-31848.51	360.01^{*}	29.38^{**}
1963	165.88467	-1.42300^{*}	0.03316	-33962.83	380.80^{*}	45.02^{***}
1964	159.5418	-1.2443	-0.4053	-44464	465.7^{*}	169.9
1965	165.0051	-1.5313^{*}	0.5039^{*}	-25449.01	316.44^{***}	-54.99
1966	101.2386	-0.8977	0.7845^{**}	-38080.81	451.77^{*}	-49.72^{*}
1967	-4.64955	0.42860	0.06532	-31650.7	396.2^{**}	-110.6^{**}
1968	104.3254	-0.9106^{**}	0.6626^{*}	-47098.03	534.94^{*}	20.13^{**}
1969	192.1136	-1.6794^{*}	-0.1081	-27956.9	369.6^{**}	-157.6^{**}
1970	162.1099	-1.5462^{*}	0.6616^{*}	-35447.65	395.05^{*}	59.78
1971	30.6352	-0.0764	0.5080^{**}	-7052.3	200.9	-402
1972	173.6323	-1.6547^{*}	0.5991^{*}	-25619.05	291.13^{**}	22.49
1973	156.567	-1.598	1.098	-5934.26	96.83	-59.97
1974	160.5881	-1.4366	0.2716	-34744.4	441.6^{**}	-145.7^{**}
1975	168.8692	-1.6862^{*}	0.9412^{*}	-11241.5	194.5	-215.0^{*}
1976	254.41470	-2.39811	0.04196^{***}	-47562.3	490.9^{*}	210.3
1977	112.5683	-0.9820	0.5351	-10141.5	188.2	-228.3^{*}
1978	72.3471	-0.5746^{***}	0.7181	-16762.9	240	-143.4^{**}
1979	215.8671	-2.0008	0.1767	-10600.1	102.9	121.4
1980	176.1498	-1.5847	0.2362^{*}	-2900.498	46.222	1.344
1981	98.0694	-0.7403	0.2140	-21583.3	296.3^{**}	-155.8^{*}
1982	108.0663	-0.8164^{*}	0.1579^{*}	-30514.4	413.8^{**}	-215.5
1983	123.6575	-0.9997	0.1674^{*}	-35315.24	411.807^{*}	-8.191^{**}
1984	122.0941	-1.0714	0.5498	-12669.5	193.4	-115.2^{***}
1985	198.261425	-1.775920	-0.007286^{*}	-22804.02	283.93^{**}	-55.01^{**}
1986	159.2983	-1.4260	0.3307	-12779.4	195.0^{**}	-134.3
1987	200.0741	-1.7634	-0.1075	-33677.72	401.94^{**}	-23.91^{**}
1988	156.13074	-1.33725	0.08981^{*}	-44693.04	502.69	46.42^{**}
1989	195.5940	-1.8502	0.4334	-49170.7	517.8	161.7
1990	90.2426	-0.6675^{**}	0.2680^{*}	-25302.23	312.52^{*}	-58.26^{*}
1991	108.60294	-0.81179^{*}	0.08409^{*}	-29569.9	381.8	-129.7
1992	166.5412	-1.4896	0.2765	-20994.75	257.81^{*}	-40.51^{*}
1993	100.70048	-0.72120^{*}	0.03924^{**}	-45265.5	478.7	163.3

TABLE 2: Continued.

		TXx			PRCPTOT	
Years	Intercept ($^\circ$C)	Trend-longitude ($^\circ$C dd^{-1})	Trend-latitude ($^\circ$C dd^{-1})	Intercept (mm)	Trend-longitude (mm dd^{-1})	Trend-latitude (mm dd^{-1})
1994	151.8259	−1.3179	0.2347	−14942.1	243.8**	−246.7
1995	183.0792	−1.5888	−0.0298	−24787.1	266.1**	107.5
1996	155.78969	−1.32219	0.07176	−18987.7	258.6**	−119.9
1997	102.0541	−0.7728	0.2109	−29502.38	355.16	−34.13
1998	127.94642	−1.00331	0.02429*	−41110.5	480.8	−19.3*
1999	97.3587	−0.7064**	0.1831**	−16258.9	247.4**	−176.6
2000	97.24909	−0.65306	−0.04446**	−31375.2	399.6*	−135.3
2001	83.09171	−0.50906**	0.03273***	−23168.2	324.1**	−188.6*
2002	160.745	−1.208	−0.618	−17521.24	219.80***	−18.59***
2003	247.777	−2.013**	−1.180	−31519.6	396.3*	−116.5
2004	1.8089	0.1002	1.1987***	−26830.19	330.83*	−42.72*
2005	165.6108	−1.3359	−0.2931**	−19607.49	260.30**	−96.51*
2006	104.1688	−0.6952*	−0.1426	−11883.7	205.3**	−218.5
2007	121.1362	−0.9814	0.2382	−29451	357.95	−40.97

*99% level of significance ($p \leq 0.01$); **95% level of significance ($0.01 < p \leq 0.05$); ***90% level of significance ($0.05 < p \leq 0.1$); dd: decimal degree.

These significant observations lead to the detailed analyses of the spatial trends in the climate indices. Representative examples of the approaches are presented in Figure 3. Since space is two-dimensional, the spatial trend analysis of the indices requires fitting a two-dimensional least square surface along latitudes and longitudes, which have been represented by the dotted regular grids. Least square surface of TXx computed for 2007 represents the increasing spatial trend with the increasing latitudes and decreasing longitudes, whereas the least square surface of PRCPTOT for 2007 represents the increasing spatial trend with the decreasing latitudes and increasing longitudes (Figure 3). The corresponding trend (slope) values with the intercepts represent the attitudes of the spatial trends.

Similar approach, as presented in Figure 3, has been applied to analyze the spatial trends of the two indices in every year of 1948–2007. The intercepts and the spatial trend values of the indices with their F-statistical significances at the three particular confidence levels are presented in Table 2. As observed in Figure 2, in most of the years, TXx shows a spatial trend increasing along latitudes and decreasing along longitudes, whereas PRCPTOT shows a spatial trend increasing along longitudes and decreasing along latitudes. Statistically significant spatial trend values have been resulted which represents a significant two-dimensional spatial trend in the climate variability over Bangladesh in regional scale.

An in-depth analysis has been performed on the obtained spatial trend values from Table 2. In spite of the F-statistical significance, as presented in Figure 4(a), on an average 22.92% of the spatial trend values show statistical significance at 99% confidence level. Another 18.75% and 7.08% of the spatial trend values on average show statistical significance at 95% and 90% confidence levels, respectively. Therefore 48.75% of the spatial trend values on average show statistical significance at 90% or more than 90% confidence level,

which represents that the spatial trends in the indices and thereby the regional climate variability over years are very significant in Bangladesh. In this circumstance, some parts of Bangladesh represent more vulnerability due to climate change than the others.

As presented in Figure 4(b), the comparison of the relative values of the spatial trends depicts that, in 75% of the years, TXx shows an increasing trend along latitudes and, in around 90% of the years, the spatial trends of TXx along latitudes are lower than the spatial trends of TXx along longitudes. In contrast, in around 96.67% of the years, TXx shows a decreasing spatial trend along longitudes, when around 90% of the spatial trends of TXx along longitudes are higher than the ones along latitudes. These results clearly illustrate that the spatial trend of TXx in Bangladesh increases along latitudes and decreases along longitudes, and the trend along longitudes is more dominant than the trend along latitudes. This proves that the relative western regions of the country are experiencing the steadily higher temperatures in summers than the eastern regions of the country. The relative northern regions of the country are also experiencing the steadily higher temperatures in summers than the relative southern regions of the country, but not as severely as the western regions of the country.

Figure 4(b) also depicts that the spatial trends of PRCPTOT along latitudes in 71.67% of the years show decreasing tendency, and 83.33% of the spatial trends of PRCPTOT along latitudes are lower than the ones along longitudes. In contrast, in all the years (100%), the spatial trends of PRCPTOT along longitudes show increasing tendency when 83.33% of the trends are higher than the ones along latitudes. This proves the fact that the spatial trend of PRCPTOT in Bangladesh increases along longitudes and decreases along latitudes, and it is more dominant along longitudes than the one along latitudes. Therefore, the relative eastern regions

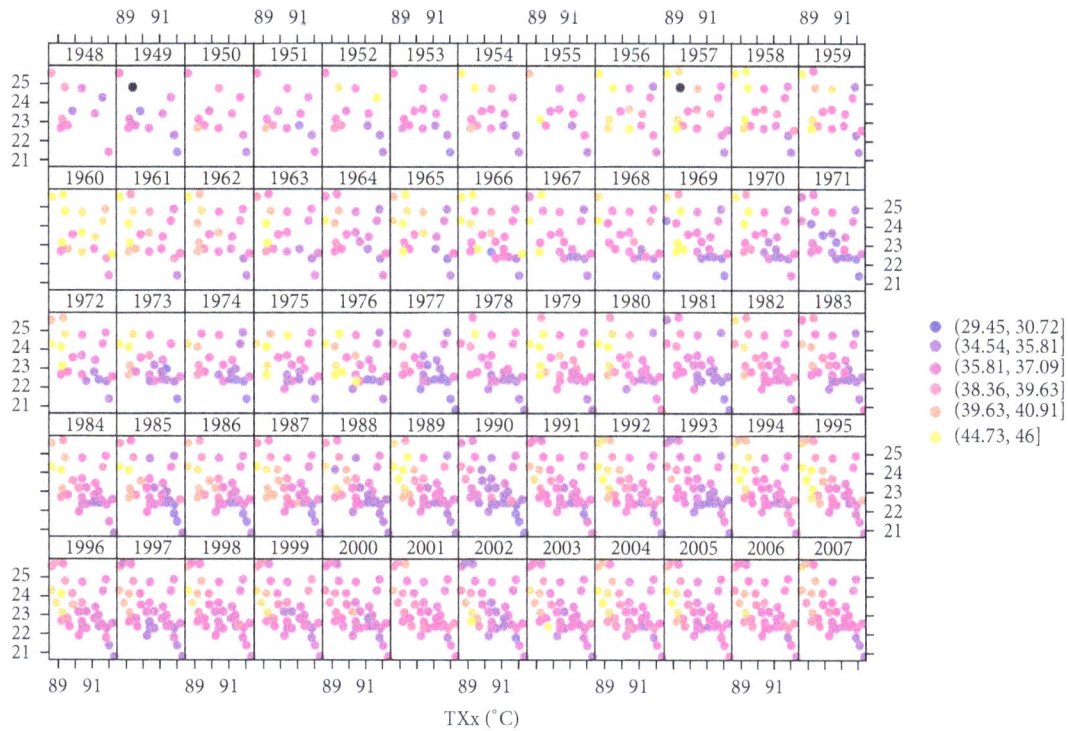

(a)

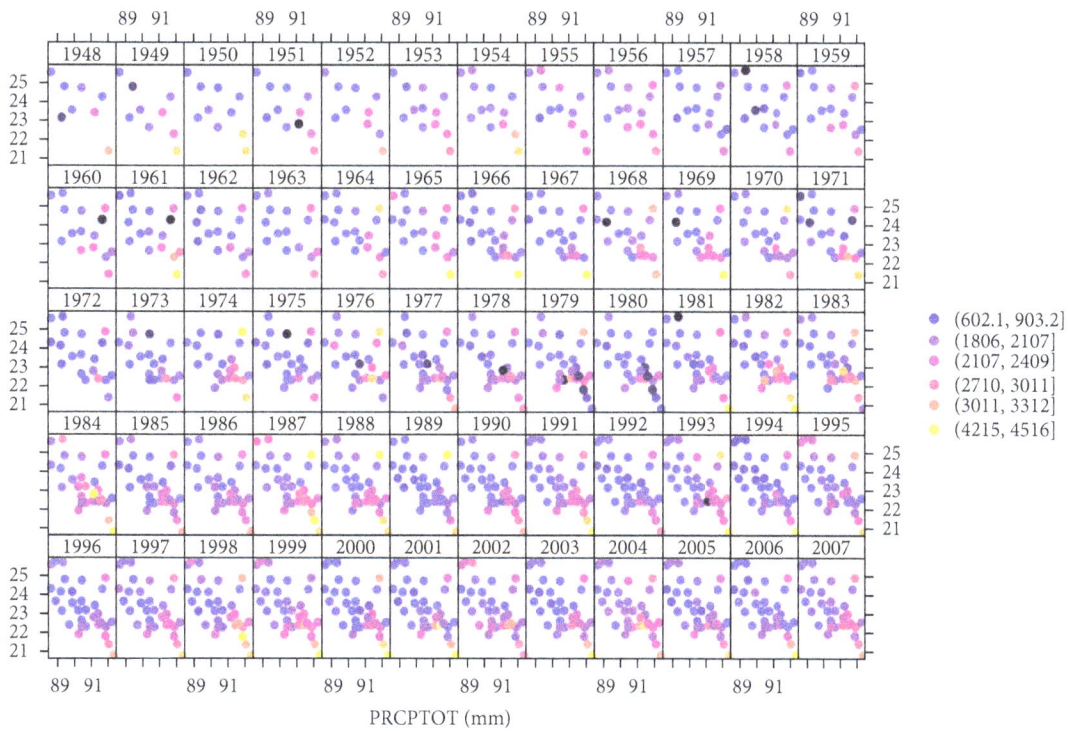

(b)

FIGURE 2: Computed (a) TXx and (b) PRCPTOT at every station and its spatial dependence over Bangladesh in every year during 1948–2007. The computed TXx and PRCPTOT values are in °C and mm, respectively.

TXx 2007 = 121.14 − 0.98*longitude + 0.24* latitude

(a)

PRCPTOT 2007 = −29451 + 357.95*longitude − 40.97* latitude

(b)

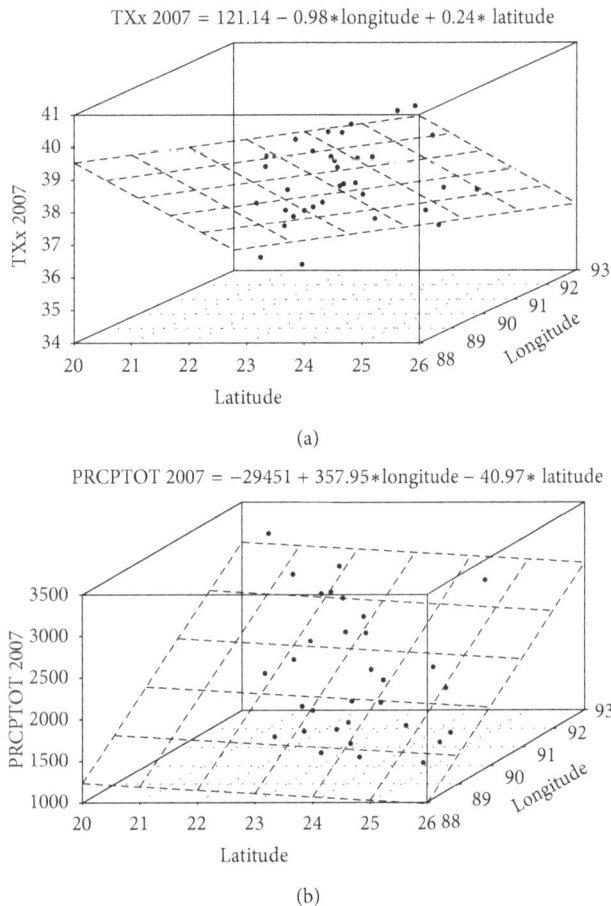

FIGURE 3: Representative least square surfaces, presented by the dotted regular grids, of (a) TXx and (b) PRCPTOT for 2007 applying corresponding intercepts and trends (slopes), fitted along the longitudes and latitudes.

of the country are experiencing steadily higher amount of monsoon precipitation than the regions situated in the relative western part of the country. In a similar fashion, the southern coastal regions of the country are also experiencing higher monsoon precipitation than the northern regions but not as higher as the eastern regions. Nevertheless, a few deviant tendencies of the spatial trends contrary to the usual tendencies have also been observed from the analyses as represented in Figure 4(c). 3.33% of the spatial trends of TXx show increasing tendencies along both latitudes and longitudes, whereas around 25% of the trends show decreasing tendencies along both dimensions.

28.33% of the spatial trends of PRCPTOT show increasing tendency along both latitudes and longitudes, but none of the trends shows decreasing tendency along both dimensions. After all, majority of the trends' tendencies have been considered as the mean actual spatial trends of the indices for the summer temperature and monsoon precipitation over Bangladesh during 1948–2007. The temporal period of 1948–2007 consists of six decades, and the least square trend surfaces of the TXx and PRCPTOT indices from each decade are presented in Figure 5. The surfaces show the indices values

in regular grids calculated by fitting the least squares using the corresponding intercepts and trends (slopes) along the latitudes and longitudes as shown in Figure 3.

The decadal least square surfaces of TXx (Figure 5(a)) represent the actual dominant spatial trend from the eastern regions to the western regions of the country. The surface of 1978 represents the spatial trend from the south east to north-west regions of the country. In 1958, the midwestern regions and north-western regions experienced the summer maximum temperature up to 50°C. Generally, these regions experience the summer maximum temperature of 40°C–45°C. In contrast, the eastern regions of the country generally experienced the summer maximum temperature around 35°C, though in 1968 and 1998 a few hill tracts regions of the south-eastern region experienced the maximum temperature slightly towards 40°C in summer time.

In contrast to the least square surfaces of TXx, the decadal least square surfaces of PRCPTOT (Figure 5(b)) represent the dominant spatial trend from the western regions to the eastern regions of the country. In 1968 and 1988 onwards, the north-eastern and south-eastern regions of the country experienced the annual monsoon precipitation of 4000 mm–4500 mm. In 1958 and 1978, the tendency is 2000 mm–3000 mm for these regions. The northern regions of the country, violating the traditional trend, also experienced the monsoon precipitation of around 2500 mm in 1958, 1988, and 1998, which is basically experienced in the south and midwestern regions of the country generally. Following the typical spatial trends, the western regions of the country always experienced the monsoon precipitation less than 1500 mm.

3.2. Temporal Interpolation of the Spatial Trends in Climate Indices. The spatial trend values of the TXx and PRCPTOT have been interpolated in time applying the thin plate smoothing spline with several iterations. The interpolated splines of TXx and PRCPTOT are presented in Figure 6. The number of iterations used to best fit the smoothing spline in spite of the general cross-validations is 11–13, and the resulted smoothing parameters (λ) range 0.71–1.02. The interpolated splines of the spatial trends of TXx are very similar and used the same number of iterations and almost the same λ. In contrast, the interpolated splines of the spatial trends of PRCPTOT are very different which have been reflected in the different number of iterations required and the different smoothing parameters calculated. The interpolated splines depict the continuous change in the spatial trends of the indices during 1948–2007, which is very important to take into account for regional climate variability analyses of Bangladesh.

Figure 6(a) shows that almost throughout the study period, TXx represents increasing spatial trend along latitude and decreasing spatial trend along longitude. The increasing trend values (+ slopes) along latitudes have been increasing from around 0.1 to around 0.35 within 1948–1972, afterwards the increasing trend values have been decreasing and after 2000 they have been representing decreasing trends (− slopes) below 0. In a similar way and as have been

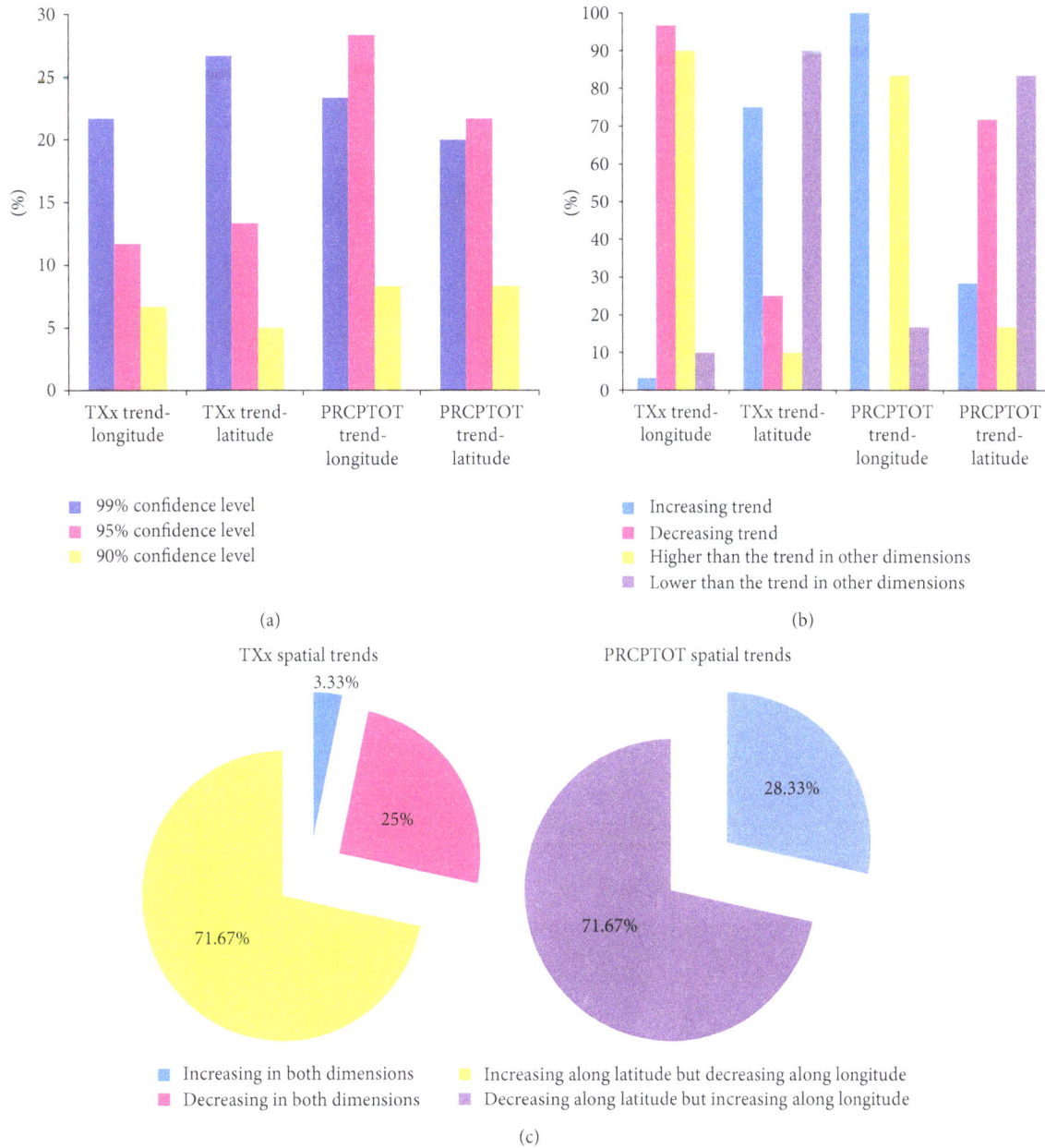

Figure 4: Percentage of the (a) F-statistically significant spatial trends at 99%, 95%, and 90% confidence levels; the (b) increasing and decreasing spatial trends of TXx and PRCPTOT and their comparative values; (c) comparative increasing and decreasing spatial trends of the TXx and PRCPTOT along latitudes and longitudes.

analyzed before, the decreasing trend values (− slopes) along longitudes have been increasing within 1948–1980 from −0.85 to −1.25. Afterwards, the decreasing trends (− slopes) have been decreasing again till 2007. These results illustrate that the spatial variability of the annual maximum temperature in Bangladesh has been increasing from 1948 until around 1980, and since then it has been decreasing again. The tendency of the variability is from the southern and eastern regions towards the northern and western regions, though after 2000 the tendency has been changed from south-north to north-south. If the tendency continues, there will be a complete change in the usual spatial trend of TXx over Bangladesh. This is an evidence of the regional shift of climate in Bangladesh.

Figure 6(b) depicts that throughout the study period PRCPTOT represents decreasing spatial trend along latitude and increasing spatial trend along longitude. The decreasing trend values (− slopes) along latitudes have been decreasing from around −130 to around −60 within 1948–1985, afterwards the decreasing trend values (− slopes) have been increasing until 2007, and most likely they will continue increasing. In a sharp contrast, the increasing spatial trends (+ slopes) of the PRCPTOT along longitudes display cyclic order in their change with time. In 1955–1965 and 1979–1990, the increasing trend values (+ slopes) along longitudes have been increasing from around 260 to 360. In the remaining years, they display a decreasing tendency in a similar

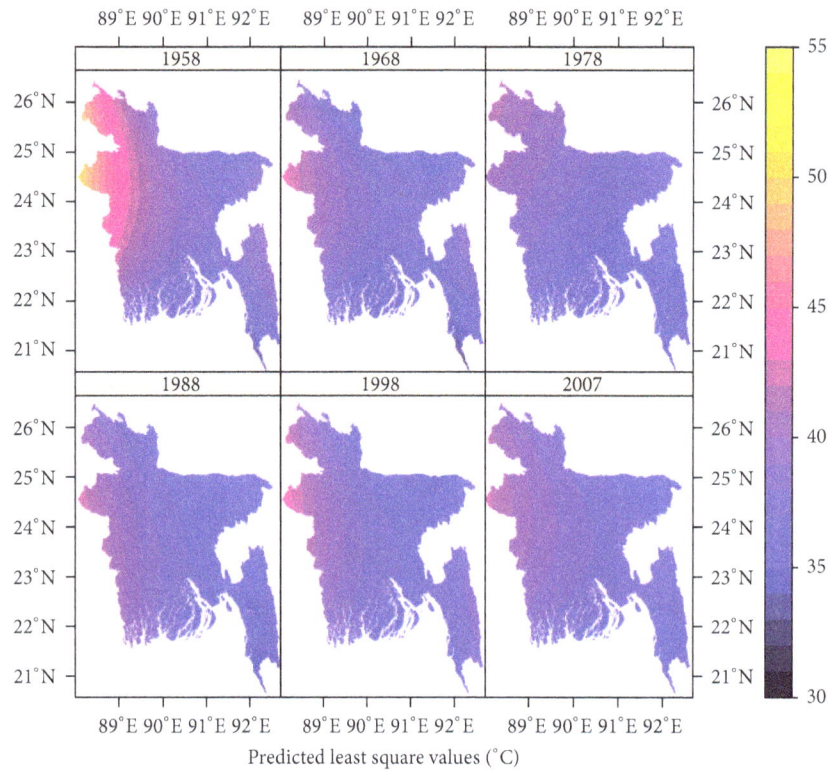

(a)

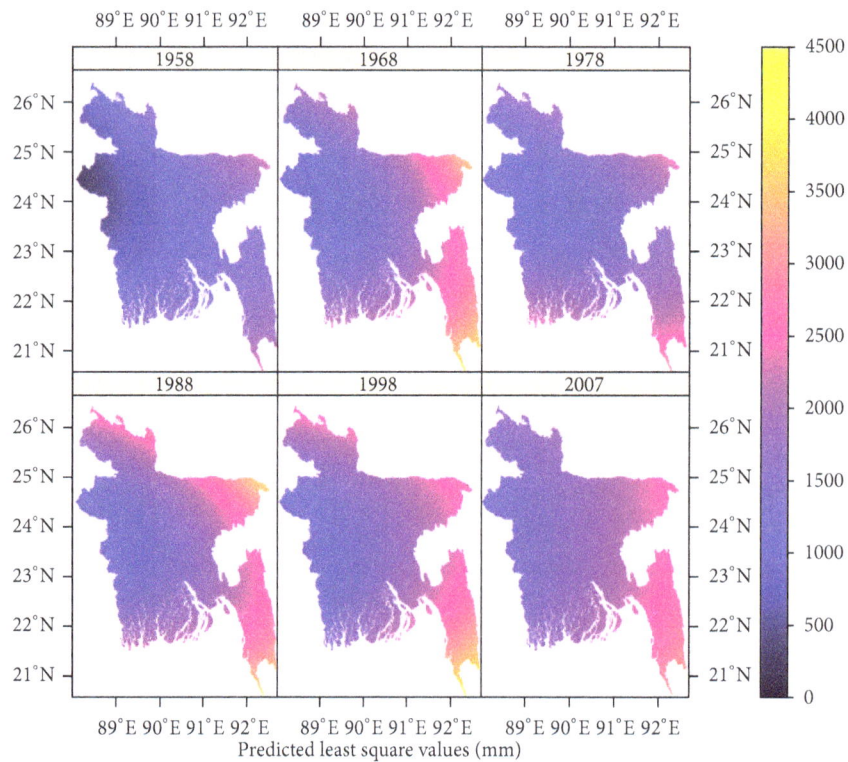

(b)

FIGURE 5: Representative trend surfaces of (a) TXx and (b) PRCPTOT from each decade, fitted using the least square approach with corresponding intercepts and trend (slope) values from Table 2.

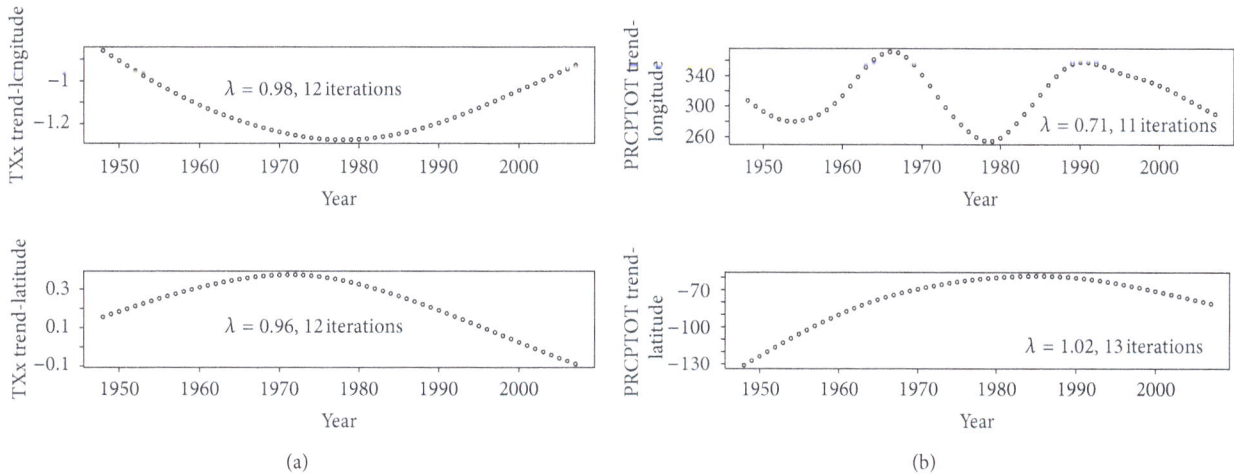

FIGURE 6: Temporally interpolated spatial trend values of (a) TXx and (b) PRCPTOT along longitudes and latitudes in the extent of 1948–2007, fitted by the thin plate smoothing spline.

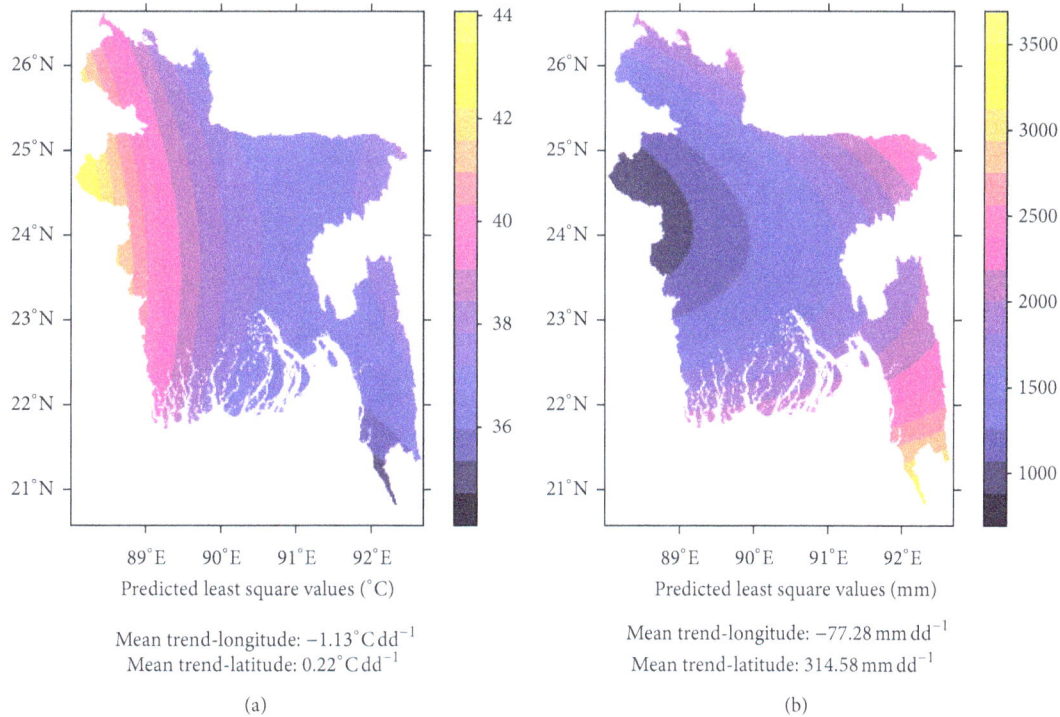

FIGURE 7: Mean trend surfaces of (a) TXx and (b) PRCPTOT with gridded mean indices values, fitted by the two-dimensional least square approach with the mean intercepts and mean trend (slope) values.

range. Since 1990 and until 2007, the increasing trends (+ slopes) have been decreasing, and most likely it will continue decreasing in the next decade. These results illustrate that the spatial variability of the monsoon precipitation in Bangladesh is highly discontinuous despite of the spatial trends, but no evidence of the shift of the precipitation regime has been observed. Precipitation by nature is a very noisy phenomenon, and therefore high variation in the tendency of the spatial trend of the PRCPTOT is expected. But it is also important to consider that the rain-fed agriculture of

Bangladesh is highly and adversely affected by the intense spatial variation of the monsoon precipitation. This also creates dilemma in the flood mitigation planning.

The mean least square surfaces of the TXx and PRCPTOT considering the trends along latitudes and longitudes during the time period of 1948–2007 have been computed and presented in Figure 7. The surfaces represent the actual mean trend behavior of the indices and thereby the expected climate regime of summer temperature and monsoon precipitation in Bangladesh. According to the computed surfaces, the summer maximum temperature in Bangladesh varies from 44°C to 32°C in the spatial direction from the west to the east of the country. The monsoon precipitation of the country varies from 700 mm to 3800 mm in the spatial direction from the midwest to the north-east and south-east of the country. Therefore, both of the indices show high spatial dependence in their behavior. The western regions of the country are the warmest, and the south-eastern and eastern regions of the country are the coolest in summer. Heat waves and droughts are most likely to occur in the western regions from this point of view. In addition, these regions experience the least monsoon precipitation, which will decrease the rain-fed crop production potentials in those regions and will create severe scarcity of water resources. But the combined phenomena increase the tea production potentials in the north-eastern regions of the country, though the observed shift in the temperature regime which resulted in the temporal interpolation might decrease this potential.

3.3. *Discussion of the Results.* According to the results obtained, there is a clear significant spatial trend in both of the climatic indices' tendency and variability over Bangladesh. The summer maximum temperature decreases, and the monsoon precipitation increases towards the southern coastal zone and the eastern hill tract zone of the country. These particular spatial trends of the indices are describable from the geographic location and terrain distribution in Bangladesh and in the neighboring countries, as presented in Figure 8. The southern periphery of the country is bordered by the Bay of Bengal, which mingles with the Indian Ocean at the far south. The eastern periphery of the country is characterized by the hills, which are further extended to the small mountains of Assam and Mizoram regions of India. The expansion of the ocean in the south and hills at the east of Bangladesh are presented in blue and green legends, respectively (Figure 8). Another important terrain characteristic is the location of Mount Himalaya in Nepal, close to the north-western periphery of the country. From the perspective of geographic location, Bangladesh is situated in the windward part of the monsoon wind. The funnel-shaped exposure to the ocean allows plenty of rain-wind to propagate over the country and finally obstructed by the elevated mountains causing heavy rainfall in monsoon.

Therefore the highest monsoon precipitation occurs in the eastern part of the country because of geographical closeness to the mountainous area. Mount Himalaya has also an impact, which is visible in the relative higher monsoon precipitation in the neighboring areas at the typical north of Bangladesh from Figures 5 and 7. Closeness to the ocean leads to the higher monsoon precipitation events in the southern coastal zones than in the northern zones. And due to the less precipitation events and higher distance from the mountains, the western and northern parts of the country experience higher summer maximum temperatures. This can also be explained by the lapse rate of temperature phenomenon [42], which increases the temperature with the decreasing elevation. So, despite Bangladesh being a flat region, altitudes have a clear impact on the precipitation and temperature

although the correlation of the station altitudes with the indices has been resulted insignificant. This contradiction has resulted because of the fact that the precise altitudes of the meteorological station locations do not represent the altitudes of the corresponding geographical areas and are below 50 m. More importantly, none of the stations is located in the high elevated hills which could represent the correlated values of the indices with altitudes. The bounded continuum of precipitation and lapse rate of temperature should be taken into account for such analyses, which are subject to further research on the distribution of the climate indices in Bangladesh. The number of meteorological stations available in the country is not enough to predict spatiotemporal climatic trend with acceptable accuracy [43]. Innovative and complex models have to be adapted to design climate variability in Bangladesh more accurately. The observations from the stations of neighboring countries can be taken into account to predict the climate trend into larger spatial extent, which might bring interesting results in the spatial trends inside the boundary of Bangladesh.

An increase in the spatial trends of TXx and a decrease in the spatial trends of PRCPTOT until the 1980s have been observed from the temporal interpolation. Since then, until the current stage, the spatial trends of TXx and PRCPTOT are in the decreasing and increasing stages, respectively. As previous studies on Bangladesh represent the increasing temporal trend in temperatures and monsoon precipitation, the results from this study represent an alarming stage of climate change in the country. In light of the increasing temporal trend and decreasing spatial trend of TXx, it is comprehensible that the summer maximum temperature is increasing in the relative cooler regions and decreasing in the warmest regions of the country. This will entail a clear climatic shift in terms of temperature in Bangladesh, which is devastating for such an agriculturally dependent economy. While, in light of the increasing temporal and spatial trend of PRCPTOT, it can be clearly stated that the monsoon precipitation is increasing more and more in the wet regions and decreasing in the dry regions of the country. This will entail the increasing flood risk in the wet regions and drought risk in the dry regions and put the underground water table and soil moisture into risk since they are highly dependent on the monsoon precipitation. The case might be that the TXx and PRCPTOT are increasing in time in all the regions of the country, which depicts the much higher increase in TXx and PRCPTOT in the relative cooler and wet southern and eastern regions of the country. But the situation of climate change and threats due to the climatic shift remain the same.

The mean spatial trends of TXx observed in the country are $-1.13°C\,dd^{-1}$ along longitude and $0.22°C\,dd^{-1}$ along latitude, whereas the mean spatial trends of PRCPTOT observed are $314.58\,mm\,dd^{-1}$ along longitude and $-77.28\,mm\,dd^{-1}$ along latitude. These trends represent the complete opposite continuous spatial variation of these climate phenomena along the same axes. But the spatial trend of TXx along latitude has shifted to the opposite direction since 2000. This indicates that the southern coastal regions of the country have been experiencing higher summer maximum temperature than the northern regions since 2000. Generally, the northern

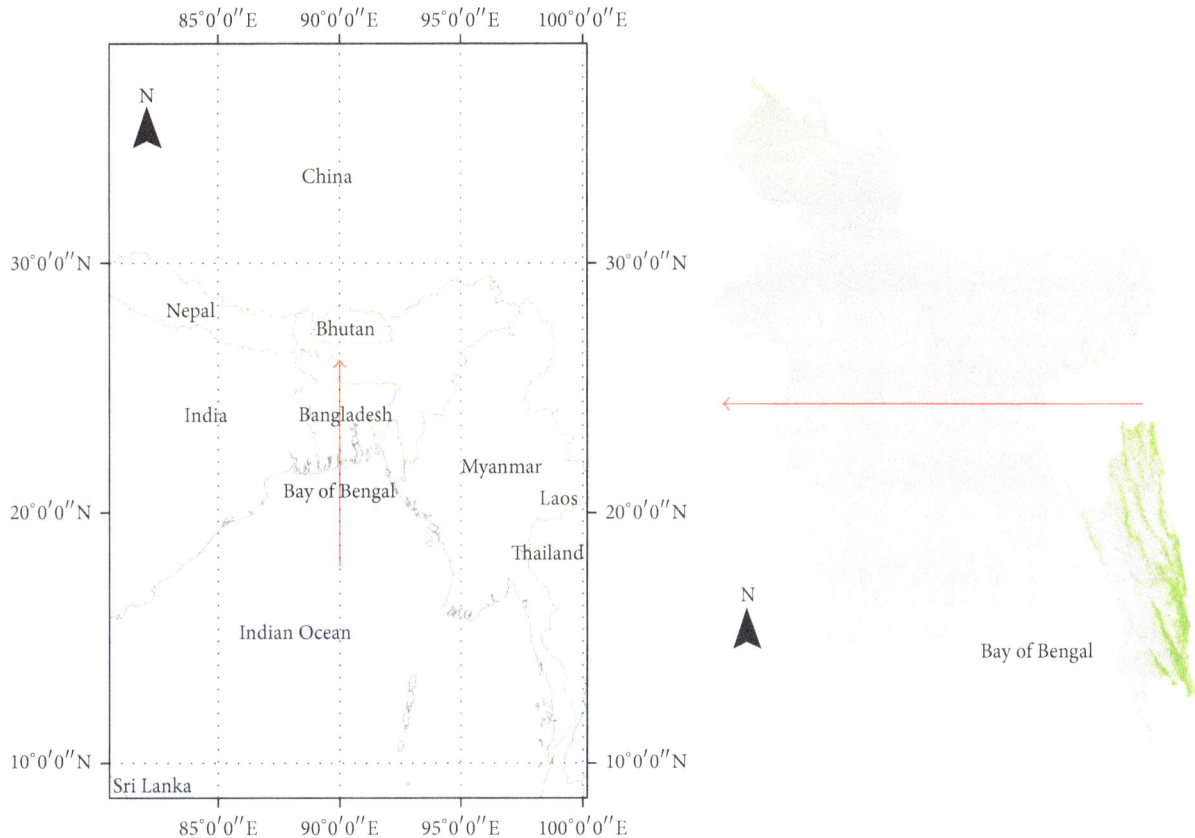

FIGURE 8: Geographic coastal and windward location of Bangladesh in relation to the Indian Ocean, Mount Himalaya, and other small mountains in India in the eastern border. The ocean is represented by blue, and hills inside Bangladesh boundary at the southeastern part are represented by green legends. The directions of both arrows represent the increase in TXx and decrease in PRCPTOT.

regions of the country represent the warmer regions; therefore there has been a clear shift in the temperature regime of Bangladesh. The study shows conformant results in the change of temperature and precipitation regimes to Bhowmik and Cabral's [44], Shahid's [22], and the other studies' results discussed in Section 1.

4. Conclusion

Previous studies have proved that one of the most important questions regarding climatic events is whether their occurrence is changing over space and time, and can be characterized by their frequency and statistical significance. This paper presents analyses of the spatial trends in two different annual indices representing summer temperature and monsoon precipitation for Bangladesh. The analyses have been conducted using long-term datasets for 34 meteorological stations under quality control, in general, for a temporal extent between 1948 and 2007.

As predicted by IPCC [7], increases in climate stress events such as floods, droughts, and heat waves will pose further challenges to South-Asian farmers. IPCC [7] also reveals a decrease of about 20% in precipitation at the high latitudes in most subtropical land regions. Bangladesh, being a subtropical South-Asian country, entails similar risk of

climate change to the farmers, who are contributing significantly to the national GDP. In light of the results from this study, many regions of the country will lose their particular crop productivities. Bhowmik and Costa [45] have already discovered the decrease in the nonrain-fed irrigated crop production with the decrease in precipitation in Bangladesh. Further threats are most likely to take place in the near future, which must be taken into account in the climate resilience activities in the country.

There are other indices available for temperature events than TXx, which are urgent to analyze for Bangladesh to track the actual change in temperature. Nevertheless, it is difficult to forecast regional variability of climate phenomena and their potential effects on agriculture by using global climate models. Such regional studies on Bangladesh using the regional datasets will bring benefit to the climate variability analyses for South Asia and eventually for the world.

Acknowledgments

This study has been carried out in the framework of the European Commission, Erasmus Mundus Program, Master of Science in Geospatial Technologies, project no. 2007-0064. The author acknowledges Ana Cristina Costa, Edzer Pebesma, Jorge Mateu, and Pedro Cabral for their continuous support to shape this research work. The author would like to thank ISRN Atmospheric Sciences for providing with the opportunity to share this research.

References

[1] C. A. C. dos Santos, C. M. U. Neale, T. V. R. Rao, and B. B. da Silva, "Trends in indices for extremes in daily temperature and precipitation over Utah, USA," *International Journal of Climatology*, vol. 31, no. 12, pp. 1813–1822, 2011.

[2] C. K. Folland, T. R. Karl, J. R. Christy et al., *Observed Climate Variability and Change. Climate Change 2001: The Scientific Basis. Contribution of Working Group I to the Third Assessment Report of the Intergovernmental Panel on Climate Change*, Cambridge University Press, Cambridge, UK, 2001.

[3] L. V. Alexander, P. Hope, D. Collins, B. Trewin, A. Lynch, and N. Nicholls, "Trends in Australia's climate means and extremes: a global context," *Australian Meteorological Magazine*, vol. 56, no. 1, pp. 1–18, 2007.

[4] M. Hulme, T. J. Osborn, and T. C. Johns, "Precipitation sensitivity to global warming: comparison of observations with Had CM2 simulations," *Geophysical Research Letters*, vol. 25, no. 17, pp. 3379–3382, 1998.

[5] F. Lambert, P. Stott, and M. Allen, "Detection and attribution of changes in global terrestrial precipitation," *Geophysical Research Abstracts*, vol. 5, Article ID 06140, 2003.

[6] Intergovermental Panel on Climate Change (IPCC), "Climate change 2001: impacts, adaptation and vulnerability," in *Contribution of Working Group II to the Third Assessment Report of the Intergovernmental Panel on Climate Change*, Cambridge University Press, Cambridge, Mass, USA, 2001.

[7] Intergovermental Panel on Climate Change (IPCC), "Climate change 2007: impacts, adaptation and vulnerability," in *Contribution of Working Group II to the Fourth Assessment Report of the Intergovernmental Panel on Climate Change*, M. L. Parry, O. F. Canziani, J. P. Palutikof, P. J. van der Linden, and C. E. Hanson, Eds., p. 976, Cambridge University Press, Cambridge, UK, 2007.

[8] A. Tarhule and M. K. Woo, "Changes in rainfall characteristics in northern Nigeria," *International Journal of Climatology*, vol. 18, no. 11, pp. 1261–1271, 1998.

[9] M. Haylock and N. Nicholls, "Trends in extreme rainfall indices for an updated high quality data set for Australia, 1910–1998," *International Journal of Climatology*, vol. 20, pp. 1533–1541, 2000.

[10] G. M. Griffiths, M. J. Salinger, and I. Leleu, "Trends in extreme daily rainfall across the South Pacific and relationship to the South Pacific convergence zone," *International Journal of Climatology*, vol. 23, no. 8, pp. 847–869, 2003.

[11] T. J. Osbom, M. Hulme, P. D. Jones, and T. A. Basnett, "Observed trends in the daily intensity of United Kingdom precipitation," *International Journal of Climatology*, vol. 20, no. 4, pp. 347–264, 2000.

[12] S. S. Roy and R. C. Balling, "Trends in extreme daily precipitation indices in India," *International Journal of Climatology*, vol. 24, no. 4, pp. 457–466, 2004.

[13] M. H. K. Chowdhury and S. K. Debsharma, "Climate change in Bangladesh—a statistical review," in *Report on the IOC-UNEP Workshop on the Impacts of Sea Level Rise due to Global Warming*, NOAMI, 1992.

[14] D. Braun, *Bangladesh, India Most Threatened by Climate Change, Risk Study Finds*, National Geographic, 2010.

[15] C. Yao, W. Qian, S. Yang, and Z. Lin, "Regional features of precipitation over Asia and summer extreme precipitation over Southeast Asia and their associations with atmospheric-oceanic conditions," *Meteorology and Atmospheric Physics*, vol. 106, no. 1, pp. 57–73, 2010.

[16] A. M. G. Klein Tank, T. C. Peterson, D. A. Quadir et al., "Changes in daily temperature and precipitation extremes in central and south Asia," *Journal of Geophysical Research D*, vol. 111, no. D16, Article ID D16105, 2006.

[17] A. M. Hussain and N. Sultana, "Rainfall distribution over Bangladesh stations during the monsoon months in the absence of depressions and cyclonic storms," *Mausam*, vol. 47, pp. 339–348, 1996.

[18] Bangladesh Bureau of Statistics (BBS), *Statistical Pocket Book Bangladesh*, 2009.

[19] B. Parthasarathy, N. A. Sontakke, A. A. Monot, and D. R. Kothawale, "Droughts/floods in the summer monsoon season over different meteorological subdivisions of India for the period 1871–1984," *Journal of Climatology*, vol. 7, no. 1, pp. 57–70, 1987.

[20] S. Karmakar and M. L. Shrestha, *Recent Climate Change in Bangladesh*, SMRC no.4, SAARC Meteorological Research Center, Dhaka, Bangladesh, 2000.

[21] N. M. Mia, "Variations of temperature of Bangladesh," in *Proceedings of SAARC Seminars on Climate Variability in the South Asian Region and Its Impacts*, SMRC, Dhaka, Bangladesh, 2003.

[22] S. Shahid, "Spatio-temporal variability of rainfall over Bangladesh during the time period 1969–2003," *Asia-Pacific Journal of Atmospheric Science*, vol. 45, pp. 375–389, 2009.

[23] S. Shahid, "Trends in extreme rainfall events of Bangladesh," *Theoretical and Applied Climatology*, vol. 104, no. 3-4, pp. 489–499, 2011.

[24] W. Wang, X. Chen, P. Shi, and P. H. A. J. M. van Gelder, "Detecting changes in extreme precipitation and extreme streamflow in the Dongjiang River Basin in southern China," *Hydrology and Earth System Sciences*, vol. 12, no. 1, pp. 207–221, 2008.

[25] M. E. Mann, R. S. Bradley, and M. K. Hughes, "Global-scale temperature patterns and climate forcing over the past six centuries," *Nature*, vol. 392, no. 6678, pp. 779–787, 1998.

[26] J. Chen, A. D. Del Genio, B. E. Carlson, and M. G. Bosilovich, "The spatiotemporal structure of twentieth-century climate variations in observations and reanalyses—part I: long-term trend," *Journal of Climate*, vol. 21, no. 11, pp. 2611–2633, 2008.

[27] D. C. Norton and S. J. Bolsenga, "Spatiotemporal trends in lake effect and continental snowfall in the Laurentian Great Lakes, 1951–1980," *Journal of Climate*, vol. 6, no. 10, pp. 1943–1956, 1993.

[28] T. C. Peterson, C. Folland, G. Gruza, W. Hogg, A. Mokssit, and N. Plummer, "Report on the activities of the Working Group on Climate Change Detection and Related Rapporteurs 1998–2001," Tech. Rep. WCDMP-47, WMO-TD, 1071, World Meteorological Organization, Geneva, Switzerland, 2001.

[29] X. Zhang, Climate Research Division, Environment Canada under the auspices of ETCCDI, ETCCDI/CRD Climate Change Indices, August 2011, http://cccma.seos.uvic.ca/etccdi/index.shtml.

[30] Disaster Management Information Center (DMIC) of Comprehensive Disaster Management Program (CDMP), Bangladesh Meteorological Department, January 2012, http://www.bmd.gov.bd/index.php.

[31] Institute for Statistics and Mathematics, WU Wien, (ISMWUWW), "The R Project for Statistical Computing," January 2012, http://www.r-project.org/.

[32] M. R. Haylock, T. C. Peterson, L. M. Alves et al., "Trends in total and extreme South American rainfall in 1960–2000 and links with sea surface temperature," *Journal of Climate*, vol. 19, no. 8, pp. 1490–1512, 2006.

[33] X. Zhang and F. Yang, *RClimDex (1.0) User Guide*, Climate Research Branch Environment, Ontario, Canada, 2004.

[34] N. Plummer, M. J. Salinger, N. Nicholls et al., "Changes in climate extremes over the Australian region and New Zealand during the twentieth century," in *Weather and Climate Extremes: Changes, Variations and a Perspective from the Insurance Industry*, T. R. Karl, N. Nicholls, and A. Ghazi, Eds., pp. 183–202, Kluwer Academic publishers, 1999.

[35] Q. You, S. Kang, E. Aguilar et al., "Changes in daily climate extremes in China and their connection to the large scale atmospheric circulation during 1961–2003," *Climate Dynamics*, vol. 36, no. 11-12, pp. 2399–2417, 2011.

[36] E. Alexander, *The Third World Natural Disasters*, Kluwer Academic Publishers, Dordrecht, The Netherlands, 1999.

[37] Ministry of Environment and Forests Government of the People's Republic of Bangladesh (MEF), *Bangladesh Climate Change Strategy and Action Plan*, Ministry of Environment and Forests, Government of the People's Republic of Bangladesh, Dhaka, Bangladesh, 2008.

[38] The World Bank (WB), "Bangladesh: Economics of Adaptation to Climate Change Study," January 2012, http://climate change.worldbank.org/content/bangladesh-economics-adaptation-climate-change-study/.

[39] S. W. Greenhouse and S. Geisser, "On methods in the analysis of profile data," *Psychometrika*, vol. 24, no. 2, pp. 95–112, 1959.

[40] M. F. Hutchinson, "Interpolating mean rainfall using thin plate smoothing splines," *International Journal of Geographical Information Systems*, vol. 9, no. 4, pp. 385–403, 1995.

[41] P. A. Hancock and M. F. Hutchinson, "Spatial interpolation of large climate data sets using bivariate thin plate smoothing splines," *Environmental Modelling and Software*, vol. 21, no. 12, pp. 1684–1694, 2006.

[42] C. J. Willmott, C. M. Rowe, and W. D. Philpot, "Small-scale climate maps: a sensitivity analysis of some common assumptions associated with grid-point interpolation and contouring," *American Cartographer*, vol. 12, no. 1, pp. 5–16, 1985.

[43] A. K. Bhowmik, "A comparison of Bangladesh climate surfaces from the geostatistical point of view," *ISRN Meteorology*, vol. 2012, Article ID 353408, 20 pages, 2012.

[44] A. K. Bhowmik and P. Cabral, "Statistical evaluation of spatial interpolation methods for small-sampled region: a case study of temperature change phenomenon in Bangladesh," in *Computational Science and Its Applications—ICCSA*, B. Mugante, O. Gervasi, A. Iglesias, D. Taniar, and B. O. Apduhan, Eds., Lecture Notes in Computer Science, pp. 44–59, Springer, New York, NY, USA, 2011.

[45] A. K. Bhowmik and A. C. Costa, "A geostatistical approach to the seasonal precipitation effect on boro rice production in Bangladesh," *International Journal of Geosciences*, vol. 3, pp. 443–462, 2012.

Evaluation of Regional Climatic Model Simulated Aerosol Optical Properties over South Africa Using Ground-Based and Satellite Observations

M. Tesfaye,[1,2] J. Botai,[1] V. Sivakumar,[1,3] and G. Mengistu Tsidu[4]

[1] Department of Geography, Geoinformatics and Meteorology, University of Pretoria, Lynwood Road, Pretoria 0002, South Africa
[2] National Laser Centre, Council for Scientific and Industrial Research, P.O. Box 395, Pretoria 0001, South Africa
[3] Discipline of Physics, School of Chemistry and Physics, University of KwaZulu Natal, Westville, Durban 4000, South Africa
[4] Department of Physics, Addis Ababa University, P.O. Box 1176, Addis Ababa, Ethiopia

Correspondence should be addressed to M. Tesfaye; mela_20062@yahoo.com

Academic Editors: G. A. Gerosa and K. Schaefer

The present study evaluates the aerosol optical property computing performance of the Regional Climate Model (RegCM4) which is interactively coupled with anthropogenic-desert dust schemes, in South Africa. The validation was carried out by comparing RegCM4 estimated: aerosol extinction coefficient profile, Aerosol Optical Depth (AOD), and Single Scattering Albedo (SSA) with AERONET, LIDAR, and MISR observations. The results showed that the magnitudes of simulated AOD at the Skukuza station (24°S, 31°E) are within the standard deviation of AERONET and ±25% of MISR observations. Within the latitudinal range of 26.5°S to 24.5°S, simulated AOD and SSA values are within the standard deviation of MISR retrievals. However, within the latitude range of 33.5°S to 27°S, the model exhibited enhanced AOD and SSA values when compared with MISR observations. This is primarily associated with the dry bias in simulated precipitation that leads to the overestimation of dust emission and underestimation of aerosol wet deposition. With respect to LIDAR, the model performed well in capturing the major aerosol extinction profiles. Overall, the results showed that RegCM4 has a good ability in reproducing the major observational features of aerosol optical fields over the area of interest.

1. Introduction

Atmospheric aerosols which originate from different natural events (e.g., wind-blown dust and sea salt particles) and human activities such as combustion of biomass and fossil fuels, as well as various industrial processes (e.g., sulfates, nitrates, ammonium, and carbonaceous aerosols) are ubiquitous in the Earth's atmosphere [1]. Relative to the well mixed and long-lived greenhouse gases, one of the main typical properties of atmospheric aerosols is their immense diversity, not only with respect to their physicochemical and optical properties, but also with regards to their spatial and temporal distributions (e.g., [2]). This is attributed to their diverse local source mechanisms, rapid aging, and chemical transformation processes, as well as short lifetime [3]. Though, owing to these heterogeneous properties of aerosols, the quantification of their climatic roles remains with large uncertainties; they are increasingly reported as one of the crucial components of the atmosphere for multiclimatic issues ([4] and references therein).

Primarily, atmospheric aerosols play an important role in modulating the regional radiation budget either through scattering or absorption of radiation (direct effects) (e.g., [5]). The perturbation of the radiation balance of the Earth through scattering of the incoming solar radiation back to space cools the Earth's surface as well as certain portions of the troposphere, but it induces stratospheric warming (e.g., [6]). The absorption of short and long wave radiation predominantly prompts atmospheric heating effects; nevertheless, depending on the underlying surface as well as the atmospheric situations, it might also result to surface cooling (e.g., [7, 8] and references therein). Particulates that

Evaluation of Regional Climatic Model Simulated Aerosol Optical Properties over South Africa Using
Ground-Based and Satellite Observations

137

are highly absorbing solar radiation such as black carbon and mineral dust particles have a substantial influence in converting the solar energy into heat; this radiative heating in turn creates the semidirect effect of aerosols. The semi-direct effect is the response of thermal, hydrological, and dynamical variables of the climate to atmospheric heating induced by light-absorbing aerosols (e.g., [7, 9]). For instance, the warming influence of aerosols in lower troposphere often enhances the low-level cloud evaporation and atmospheric stability, which consecutively results in the reduction of cloudiness as well as the slowing of the hydrological cycle and the suppression of convection processes (e.g., [1, 9–14]). Additionally, the strong heating effects of absorbing aerosols in the lower troposphere will produce alterations in atmospheric circulation (e.g., [14, 15]). Furthermore, as discussed by different studies, depending on the relative position of the absorbing aerosol layer with respect to clouds (e.g., [13, 16–18]) as well as the underlying surface properties [19, 20], the semidirect effect may also result in instability of the atmosphere and an increase in cloud water.

In general, attributed to the involvement of various atmospheric, surface, and other variables, the computation of semi-direct effects of aerosols is quite complicated and highly variable [19, 20]. Moreover, aerosols enhance the cloud number droplets and decrease its mean droplet size by acting as cloud condensation nuclei. This results in the change in cloud albedo and radiative properties, reducing precipitation efficiency which might influence the cloud lifetime as well as its formation processes and coverage (indirect effects) ([21–23] and references therein). Likewise, different reports point to the involvement of atmospheric aerosols in several climatic system topics: in a range of tropospheric chemistry variations [24], stratospheric ozone depletion [25], and in several ecological concerns (e.g., [26–28]).

Once aerosols are released into (formed in) the atmosphere, they will be transported to fields far away from the areas of their origin. However, during their transportation they will be subjected to numerous physicochemical transformations and removal processes such as dry and wet deposition and gravitational settling [29–31]. Thus, as aerosols travel further away from their source regions, their concentration and impact will decline drastically [3, 32]. As a result of this declination, the impact of aerosols on climate must be understood and quantified on a regional scale (i.e., in and around their source regions) rather than on a global-average basis (e.g., [1, 33, 34]). Due to the extreme heterogeneity of aerosol space-time distribution, as well as physicochemical properties, the quantitative assessments of certain puzzling climatic roles and different aspects of aerosols through observations (field experiments) are prohibitively expensive and highly constrained by various factors (e.g., [4, 35–39]).

Therefore, studying the climatic effects of aerosols using chemistry/aerosol models which are radiatively active and coupled with the meteorological models with online feedback on the radiation and climatic schemes (e.g., [40–45]) is crucial. In addition, models are also indispensable tools for estimating the past and projecting the future climatic role of aerosol forcing (e.g., [46, 47]). Since the late 1980s, interactively coupled climate-aerosol models for global scale (e.g.,

[48, 49] and references therein) and regional scale ([50] and references therein) simulations have been developed. Global and regional models are now becoming more complex as they incorporate new parameterizations of aerosol properties and processes. Global-scale models, due to their frequently implemented coarse grid resolution, do not accurately simulate the regional-scale spatiotemporal distributions of atmospheric aerosols, as well as meteorological processes that govern the aerosol-atmosphere-radiation-chemistry interactions (e.g., [30, 51–54]). As a result of this and other aspects, predictions of aerosol optical properties and climatic forcings employing global-scale models are exposed to remarkable uncertainties (e.g., [8, 30, 53, 55–58] and references therein).

On the other hand, the high-resolution climatic system (i.e., the surface, ocean as well as atmospheric processes) representations of Regional Climate Models (RCMs) offer enhanced advantages in assessing the downscaled meteorological processes as well as different climatic information and patterns (e.g., [59–61] and references therein). Furthermore, interactively coupled high-resolution regional climate-chemistry/aerosol models progressively turn out to be a suitable tool in assessing the regional scale distribution and complex climatic roles of aerosols with a much better computational cost, relative to global climatic models (e.g., [29, 62–69]). In addition, the results from high-resolution RCMs are well suited for comparison with measurements of individual events at selected sites/areas. Therefore, to evaluate the regional scale aerosol distributions, along with their radiative and climatic impact with improved accuracy, simulations utilizing high resolution RCMs are vital (e.g., [29, 69]).

The literature on aerosol studies employing the art of RCM over Africa is not exhaustive. Most studies which have been reported over this continent such as Solmon et al. [67], Konare et al. [70], and Malavelle et al. [71] have focused on the effect of mineral dust and biomass burning particles over the West African regions. On the other hand, South Africa, which has plenty of industries and mining sectors, is the most industrialized country in the continent. These human activities, along with the wide usage of coal for electricity generation, make South Africa one of the remarkable spots in the globe, which contributes several types of aerosols via anthropogenic activities (e.g., [72–75]). Further, many space- and ground-based observational studies (e.g., [2, 76–80] and references therein) and modeling studies ([69, 80, 81] and references therein) identify South Africa as a major source of anthropogenic aerosols in the subcontinent. Different intensive field-campaign observations such as Southern African Regional Science Initiative (SAFARI) (e.g., [82–84]) and aerosol climatology studies (e.g., [2, 85]) indicate that during the dry seasons, South Africa experiences a drastic burden of aerosols from biomass burning activities. In addition, the dust blowing from the arid/semiarid regions of South Africa and its neighboring countries [2, 76, 86], along with marine aerosols—which are induced from the surrounding oceans [2, 85]—is another main component of natural aerosols over South Africa.

In overall, due to various natural/anthropogenic events, the South African atmosphere is burdened by almost all major types of aerosols. As aforementioned, the impact of aerosols is considerably substantial near to their source regions; therefore, the regional scale distributions as well as climatic impact of aerosols—which are induced in and around South African regions—need to be assessed and reported separately. To date, only a single study has been reported employing interactively coupled regional climate-chemistry/aerosol model (ICTP RegCM4-aerosol model) over southern Africa [69]. This study focused on the direct and semidirect radiative effects of biomass burning and dust aerosols on southern Africa's regional climate during the dry winter season only. A study devoted to compressively examine the seasonal distributions and long-term climatic signals of individual/combined aerosol components using the above model over South Africa is still lacking. Therefore, using the ICTP RegCM4-aerosol model [50], studies that follow this paper and are reported elsewhere will compressively examine the seasonal distributions, as well as the direct and semi-direct effects of different components of aerosols over South Africa. However, before employing the model for the investigation of atmospheric aerosol radiative and climatic effects, its performance in computing the magnitudes as well as the spatiotemporal evolution of the optical properties of aerosols needs to be evaluated via comparing with a range of remote sensing/in-situ observations.

Modeling the direct influence of aerosols on the earth's radiation balance by solving the radiative transfer equation needs the following spectral aerosol optical parameters: (a) aerosol extinction optical depth (AOD), (b) asymmetry parameter, and (c) single scattering albedo (SSA) (e.g., [5, 29]). These optical parameters are significantly dependent on the aerosol's composition (complex refractive index), particle size (particle size distribution), shape, wavelength, and relative humidity [87–90]. AOD is the vertical integral of fraction of solar/terrestrial radiation either scattered or absorbed by airborne particles (i.e., the sum of aerosol scattering and absorption optical depths) (e.g., [87]). The asymmetry parameter is the intensity-weighted mean value of the cosine of the scattering angle (e.g., [91]); it determines the net angular distribution of aerosol scattered light. The SSA (i.e., the ratio of the extinction due to scattering to the total extinction due to scattering plus absorption) is an important parameter that governs the relative efficiency of particles to scatter solar/thermal radiation compared to absorption (e.g., [92]). Depending on the underlying surface albedo, these optical properties of aerosols are the key parameters driving the magnitude, as well as a sign of aerosol direct radiative forcing (i.e., in driving the aerosol radiative cooling/heating roles) (e.g., [7, 93]). Therefore, to understand and evaluate various aspects of atmospheric aerosols, a reasonable quantification of these optical parameters is crucial. Additionally, these aerosol optical properties are the most comprehensive standard quantities that link the observations with the outcomes of the model.

Concerning aerosol microphysical and optical property inquiries, field measurements provide more detailed information with better accuracies (e.g., [87, 94]); however, they are confined within temporal or spatial coverage (e.g., [95]). Satellite observations provide the requisite aerosol optical property distributions with extensive temporal and spatial coverage (e.g., [95]). Nonetheless, due to high variability of the earth's system reflectance, both in space and time, as well as aerosol physicochemical properties, satellite retrievals are exposed to some accuracy limitations and constraints to deliver some essential aerosol quantities such as aerosol compositions [37, 38, 96–100]. As a result, neither field measurements nor satellite observations, alone, would be sufficient to fully describe the total regional scale aerosol distributions as well as its physical, chemical, and optical properties. Alternatively, interactively coupled regional climate-chemistry/aerosol models (RC-aerosol models), which comprises a suite of major atmospheric aerosols, with their detailed parameterizations, are essential in delivering various parameters which are related to aerosols and their climatic roles, with high-temporal and spatial resolutions. To this end, the RC-aerosol models have a capability of providing important information about the complex aerosols-radiation-climatic interactions and the physicochemical production/transformation rate of particles as well as their concentration and optical parameters. Some of this information gained from modeling studies cannot be easily addressed from either satellite retrievals or field measurements.

Notwithstanding the contribution of RC-aerosol models, modeling the entire complex aerosol processes (i.e., emission, transportation, physicochemical transformation, and removal, as well as wavelength and climatic condition dependent aerosol optical properties) is fundamentally a challenging task [29]. Further, errors in simulated meteorological fields, insufficiently understood physico-chemical processes of aerosols, inaccurately estimated precursor-gas/particulate emissions by the inventories used (e.g., [101]), and many other factors will impose significant uncertainties in simulating optical characteristics of aerosols [30, 34, 57]. The inaccuracies in model estimated optical parameters of aerosols will also propagate a substantial uncertainty on the computation of aerosol's direct radiative forcing and its consequential semidirect influences (e.g., [4, 34]). Thus, to compensate the deficiencies of one technique via another and to reduce inaccuracies in model predicted optical properties of aerosols, an hybrid research effort that integrates observational records and numerical modeling techniques is essential (e.g., [40, 81, 95, 102–105]).

The present study will evaluate the ICTP RegCM4-aerosol model capability of simulating the magnitude as well as the spatiotemporal evolution of optical properties of aerosols via comparing with different field observations, in South Africa. Such studies would also contribute important role to pointing-out the shortcomings of the model's parameterizations (e.g., [29]). In this study, including different types of aerosols which is induced from natural processes and distinct emission sectors in and around South Africa (see our simulation domain in Figure 1), a long-term regional climate/aerosol simulation has been carried out using RegCM4-aerosol model (see Sections 2.2 and 2.4). To estimate particulates/precursor-gases which are emitted from different anthropogenic/biomass burning sectors, recently updated emission inventories have been used (see Section 2.3). Subsequent to these, the evaluation of simulated

Evaluation of Regional Climatic Model Simulated Aerosol Optical Properties over South Africa Using
Ground-Based and Satellite Observations

139

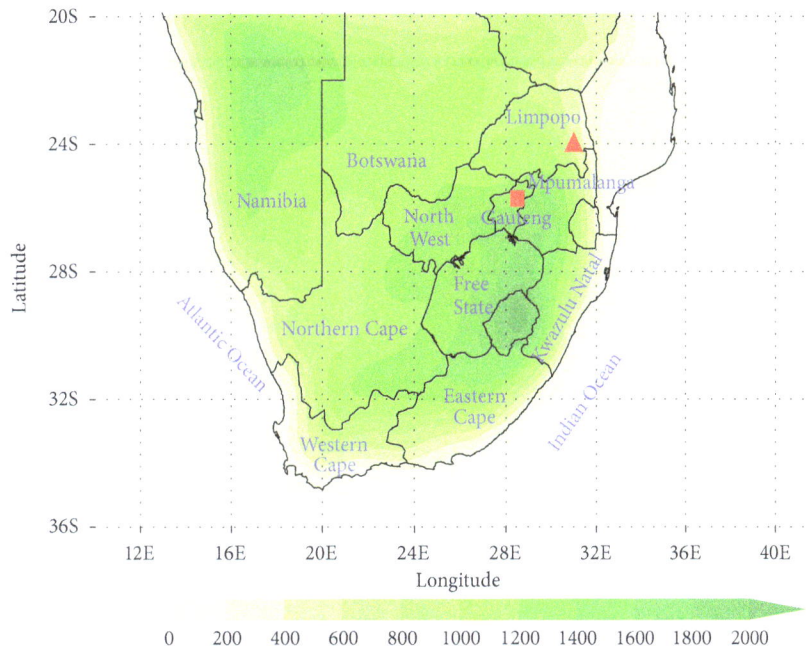

FIGURE 1: Model domain and topography (unit: m). The red triangle and square, respectively, indicate the geographical location of AERONET (Skukuza; 24°S, 31°E) and LIDAR (University of Pretoria; 25.7°S; 28.2°E) surface observation sites.

aerosol optical fields over South Africa (22°S to 34°S and 16°E to 32°E (see Figure 1)) has been carried out by comparing with values obtained from ground (sun-photometer and LIDAR) and spaceborne (MISR) observations. The paper is organized as follows. Section 2 will provide a brief description about the ICTP RegCM4-aerosol model along with the employed model physics parameterization, emission inventories used in the model, and the experimental design. In addition, the different surface/satellite products used for evaluating model outputs will be addressed in this section. In Section 3, comparing with different remote sensing products, we present the evaluation of the model's performance in simulating the magnitude and spatio-temporal evolution of column integrated aerosol optical properties, as well as their vertical distribution. Along with the validation, different rationale aspects which might be accountable for the biases of the simulated aerosol optical fields are discussed. A summary, concluding remarks, and future perspectives are given in Section 4.

2. Methodology

2.1. Model Description. In this study, for the regional climate/aerosol simulation, interactively coupled regional climate-aerosol model is used. The climate component of the coupled model is the Regional Climate Model (RegCM) version 4.0 (RegCM4.0), developed at the International Centre for Theoretical Physics (ICTP). RegCM4 is a hydrostatic, compressible, sigma vertical coordinate model, which is an upgraded version of RegCM3 with the similar basic model dynamics but with certain improvements in various physics representations and software code [50]. For a more detailed description of RegCM and its substantial evolution starting

from the first generation (RegCM1; [106]) to the current version, the reader is referred to ([50, 61] and references therein).

Among different model physics parameterization schemes of RegCM4, this study employs the following scheme: for the radiative transfer computation—the Community Climate Model-CCM3 radiative transfer package [107] is used. This radiative transfer package takes into account the radiative effect of different greenhouse gases, atmospheric aerosols and cloud water-ice, in different spectral bands. The radiative flux calculations include 18 spectral intervals which are within a wavelength range of 0.2 to 4.5 μm. Among these 18 spectral bands, seven of them are situated in the ultraviolet spectral interval (0.2–0.35 μm), one is in the visible band (0.35–0.64 μm) and the remaining spectral bands cover the infrared/special absorption windows [107]. The ocean surface fluxes are computed according to the scheme of Zeng et al. [108], and the land surface physics, which describe the transfer of energy, mass, and momentum between the atmosphere and the biosphere, is described by the biosphere-atmosphere transfer scheme (BATS; [109, 110]). The planetary boundary layer processes are characterized according to the nonlocal parameterization of Holtslag et al. [111]. The convective precipitation is represented by the mass flux scheme of Grell [112] with the Fritsch and Chappell [113] closure assumption, while the large-scale cloud and non-convective precipitation computations follow the Sub-grid explicit moisture scheme (SUBEX) of Pal et al. [114].

2.2. RegCM4-Aerosol Model. The RegCM4-aerosol model is interactively coupled model between RegCM4 and radiatively active simplified anthropogenic and dust aerosol models, which can be used to examine the two-way aerosol-climate feedback [50]. The RegCM4-aerosol model allows the simulation of major tropospheric aerosols, which originate from anthropogenic and biomass burning activities [29], as well as wind eroded desert dust particles [63, 67]. The anthropogenic and biomass burning aerosol schemes account for sulfur dioxide (SO_2), sulfate (SO_4^{-2}), hydrophobic and hydrophilic components of black carbon (BC), and organic carbon (OC) particles [29]. Following Qian et al. [42] the model takes into account the chemical conversion of SO_2 to SO_4^{-2} through both gaseous-phase and aqueous-phase pathways.

The atmospheric processes of these aerosols: surface emission, transportation (via advection by atmospheric winds, turbulent diffusion and deep convection), physico-chemical transformations, and removal processes (via wet and dry depositions) are described by the tracer transport equation of Solmon et al. [29]. The essential steps and mechanisms which are considered for developing and implementing the online dynamical dust production scheme, together with the parameterizations of several factors which influence the dust emission processes, are described in detail by Zakey et al. [63]. The dust scheme of RegCM4 represents the dry dust particle size distribution through size bin approach. The whole-size spectrum of dust particles covers a diameter range of 0.01 to 20.0 μm, divided into 4 size-bins, that is, the fine (0.01–1.0 μm), accumulation (1.0–2.5 μm), coarse (2.5–5.0 μm), and giant (5.0–20.0 μm) particle size modes [63]. As described in Zakey et al. [63], during the inclusion of dust module in RegCM framework, some new parameterizations of the dustatmospheric process (such as, size-dependent gravitational settling processes of dust particles) are incorporated into the tracer transport equation of Solmon et al. [29].

For each wavelength of the RegCM4 radiation scheme and for each aerosol species, the aerosol size distribution and refractive index dependent optical properties (i.e., asymmetry factor, single scattering albedo, and mass extinction coefficient) are computed using the Mie theory and employed in the model [29]. Using prognostic dust bin concentrations, long-wave refractive indices, and absorption cross sections, the dust particles long-wave emissivity/absorptivity influences are implemented based on Solmon et al. [67]. The relative humidity influences on optical properties of hydrophilic aerosols are specified according to Solmon et al. [29]. Accordingly, the model computes the shortwave radiative influences of all the above aerosol types using these optical properties, along with the long-wave effects of dust particles [50]. More information on different aspects of RegCM4-aerosol model is described in Giorgi et al. [50]. Over the years, this model has been widely used to examine the regional-scale direct radiative forcing of aerosols and their climatic effects in different parts of the globe (e.g., [43, 67, 69–71, 115–117]). In this study, also employing two recently updated emission inventories in the model (which are described in Section 2.3), a long term regional climate/aerosol simulation has been carried out.

Before we carry on to the next section, we would like to highlight some of the limitations which are associated with our simulation, as well as the aerosol schemes of RegCM4. This, further than assisting this study, it will also

avoid the repetition of information in studies that follow this contribution [118–121]. Most of the earlier studies, for example, several studies reported in IPCC [4] clearly stated the importance of the indirect effects of aerosols along with its current uncertainties. Nonetheless, one of the main caveats of RegCM4-aerosol model is that it does not include this aerosol effect. The next limitation is associated with the assumption used regarding the complex mixing state of the particles. Throughout our studies, the external mixing assumption is used for the computation of the total optical parameters of aerosols. This is based on previous studies which reported that nearby the aerosol source region, the extinction cross section is slightly sensitive to the mixing hypothesis (e.g., [93, 122]). However, there are several processes in the atmosphere that will alter an external mixture of particles into an internal mixture (e.g., [123]). In fact, different studies show a high sensitivity of the bulk aerosol optical and microphysical properties as well as its radiative and climatic influences to the aerosol mixing state assumptions (e.g., [41, 124–128]). Accordingly, the effort of understanding the internal/external mixture assumptions, along with their consequential aspects (via modifying the RegCM4-aerosol model parameterizations), is currently in progress [129], while the sea salt particles contribution is primarily noted only within limited coastal areas of South Africa, following the rise of southeasterly wind speed for a short period of time (e.g., [2]). Thus, for meanwhile our simulations do not encompass marine aerosols. Additionally, the simulation does not take into account the long-wave effects of carbonaceous and sulfate particles. In fact, the particle interaction with thermal infrared radiation is significant if the aerosols are large in size such as for dust and sea salt particles. For smaller aerosols, the extinction coefficient decreases rapidly with increasing wavelength (e.g., [90, 130]). Therefore, omitting the long-wave radiation effects of carbonaceous and sulfate particles will not impose a significant error in long-wave radiation computation, as well as in assessing the direct and semi-direct effects of these particles.

2.3. Emission Datasets. In this study, the emission estimates of black carbon (BC) and organic carbon (OC) particles, and sulfur dioxide (SO_2) which are induced from anthropogenic and biomass burning activities are derived from two recently updated emission inventories: MACCity [131] and Global Fire Emissions Database, version 3 (GFED3) [132], respectively. Both inventories are provided at spatial resolution of 0.5 × 0.5 degree over the globe with a monthly temporal resolution. The MACCity inventory is the upgraded extension of ACCMIP (Atmospheric Chemistry and Climate-Model Intercomparison Project) emissions dataset [131]. Covering a period from 1990 to 2010, this emission database provides various gases and aerosol species, which are contributed from different anthropogenic sectors such as power plant, industrial activities, and transportations. The GFED3 inventory applies global scale satellite-derived products such as vegetation characteristics, productivity, and burned area estimates, along with a fire module of a biogeochemical model, as well as several conditions to estimate emissions

Evaluation of Regional Climatic Model Simulated Aerosol Optical Properties over South Africa Using
Ground-Based and Satellite Observations

141

from biomass burning activities (i.e., from grass, peat, woodland, forest, open savanna, deforestation fires, and agricultural waste burnings; see Giglio et al. [133] and van der Werf et al. [132, 134]. Further details of GFED3 approaches and spatio-temporal emission variability are compressively described in van der Werf et al. [132] and references therein. This dataset contains several trace gases and particulate matter which emitted from these open biomass burning activities, for the period from 1997 to 2009.

2.4. Experimental Design.

For the purpose of evaluating the coupled model's (i.e., RegCM4-aerosol model) capability of simulating the total atmospheric aerosol optical parameters, as well as extracting the aerosol-induced climate-impact signal from the underlying noise, a long term simulation is essential. Implementing the above recently updated emission inventories we have conducted a series of simulations, which extends from January 1997 to December 2008 and analyses the recent 11 years' results. To eliminate boundary effects (e.g., [135, 136]), the simulation domain is designed to be larger than the area of interest; that is, our simulation domain (from 5.5°E to 47.1°E and from 37.5°S to 18.6°S) encompasses the entire South Africa and its surrounding land/ocean regions (Figure 1) at a horizontal grid resolution of 60 km with 18 vertical layers. Though, the finest horizontal resolution of RegCM4 can be set to 10 km; considering our domain size, the horizontal spacing used in this study is fairly adequate to make a multiyear simulation with a reasonable computational time. Taking into account the aerosol influence on radiation, as well as the two-way aerosol-climate feedbacks, all the ten tracers available in RegCM4 are included in our simulation under external mixture assumption. The European Centre for Medium-Range Weather Forecasts (ECMWF) reanalysis ERA interim (ERAIN: [137, 138]) and the weekly mean product of National Ocean and Atmosphere Administration's (NOAA) Optimum Interpolated sea surface temperature (OISST) [139] are implemented for limited-area model that required time-dependent initial and lateral boundary conditions. The simulations presented here use a time setup of 10 minutes for surface parameter files (topography, land use, vegetation, soil type, etc.), along with the dynamical model time step of 150 seconds and 6 hours updating lateral boundary conditions. Most of the model's climatic schemes, as well as meteorological lateral boundary condition selections, are based on Tummon [80] sensitivity and performance studies of RegCM schemes over Southern Africa. Since this work is the first step towards applying the RegCM4-aerosol model for the investigation of radiative and climatic effects of different types of aerosol over South Africa, we have tested and discussed the model's performance in terms of computing different optical fields of aerosols over this region only.

2.5. Observational Data

2.5.1. AERONET Surface Observation.

The Aerosol Robotic Network (AERONET) is a global ground-based network of automated sun-photometer measurements [140]. AERONET provides aerosol optical depth and surface solar flux, as well as

employing improved retrieval algorithms; it delivers different sets of atmospheric column aerosol optical and microphysical parameters (e.g., [141]). Even if there are different AERONET stations in South Africa, only one site has continuous AOD observations (from 1998 to 2008), that is, at Skukuza (24°S, 31°E; see Figure 1). In the present work, quality-assured dataset (Level 2.0) of AERONET aerosol optical depth (AOD at 500 nm) from this site is used for comparison with the AOD derived from the model simulations.

2.5.2. MISR Satellite Observation.

Multiangle Imaging SpectroRadiometer (MISR) was launched by the National Aeronautics and Space Administration (NASA) on December 18, 1999 and has been in operation since February 2000. The device consists of nine push broom cameras arranged to view at nominal zenith angles relative to the surface reference ellipsoid of 0.0°, ±26.1°, ±45.6°, 18 ± 60.0°, and ±70.5° and measures upwelling short wave radiance in each camera at four spectral bands, centered at 446, 558, 672, and 866 nm [142]. With high spatial resolution and a better radiometrical and geometrical accuracy, the multiple angle-band observations of MISR allow the retrieval of a number of aerosol optical and microphysical properties over land (including a bright desert surfaces) and ocean [36–38, 100, 143–145]. Furthermore, MISR-AOD retrievals (level-3 data) have a higher grid resolution ($0.5° × 0.5°$) in comparison to the Moderate Resolution Imaging Spectroradiometer (MODIS) of Level-3 which has the resolution of $1° × 1°$. In the present study from MISR-level-3-monthly-averaged datasets (version 31, which are available from 2000 onward), the AOD and SSA at 558 nm are utilized to evaluate the simulated results.

2.5.3. LIDAR Observation.

A mobile LIDAR system was developed at the Council for Scientific and Industrial Research (CSIR), National Laser Centre (NLC), Pretoria (25°5′S; 28°2′E), South Africa [146, 147]. At present, the CSIR-NLC mobile LIDAR can provide aerosol backscatter measurements at 532 nm for the altitude region from ground to 40 km with a height resolution of 10 m [148–150]. For a better understanding of the atmospheric boundary layer evolution and aerosol concentrations, during October 2008 the LIDAR experiment has been performed at the University of Pretoria (25.7°S; 28.2°E). The experiment has been made continuous for 23 hour measurement, that is, from 16 October, 16 h00 to 17 October, 15 h00. To assess the model's performance in simulating the vertical distribution of aerosols, LIDAR retrieved extinction coefficient profiles from this experiment are compared with the corresponding model-simulated results.

3. Results and Discussion

In the following subsections, we present the RegCM4-aerosol model's estimated aerosol optical field evaluation results, over South Africa. The magnitude and temporal variability of simulated columnar AOD comparison with AERONET and MISR observations at Skukuza (24S, 31E) is given in Section 3.1. The latitudinal variations of simulated AODs and SSAs values within South Africa, in comparison with their corresponding column-integrated MISR retrievals, are provided in Section 3.2. Further, the simulated aerosol extinction coefficient profiles with respect to ground-based CSIR-mobile LIDAR retrievals are presented in Section 3.3. The geographical location of the surface observation sites, that is, for AERONET and LIDAR, is shown in Figure 1. One of the important factors that propagate a bias in simulated aerosol concentration in model predicted aerosol optical fields is the model's insufficiency in simulating meteorological fields. In this context, even though the evaluation of model estimated meteorological parameters is beyond the scope and the aim of this study, to clarify some of the disparities between simulated and measured optical parameters, the bias in model estimated meteorological parameter, in comparison with specific meteorological sites, is presented in Appendix.

3.1. Comparisons with AERONET and MISR. The comparison of model-simulated monthly-averaged AOD (at 550 nm) with the available data from AERONET (at 500 nm) and MISR (at 558 nm) observations at Skukuza (24°S, 31°E), South Africa, is shown in Figure 2. Skukuza is situated in a region of the northeastern area of the Mpumalanga province (see Figure 1), which is relatively close to the major industrial areas of South Africa (i.e., Gauteng and western areas of Mpumalanga). This site is influenced by several aerosols, which are induced from various local activities, for instance, primary/secondary aerosols from frequently occurring local biomass burning activities (e.g., [69, 80]), a variety of agricultural practices, and natural-resource based industrial activities, such as coal production (e.g., [72, 77]). Besides, the conveyance of aerosols from the main industrial Highveld regions of South Africa and the slight contribution of long-range transported particles are the additional sources of aerosols for this site (e.g., [85, 151, 152]). In reproducing the magnitude of AERONET and MISR AOD values, the model relatively performs well. At least the magnitudes of simulated AODs are within the standard deviation of AERONET and ±25% of MISR observations. The other important aspect in studying the climatic role of aerosols is the model's performance in capturing the temporal evolution of aerosol loading. In this regard, the model shows a good performance of capturing the seasonal and interannual variability of AOD (i.e., the temporal pattern of simulated AOD exhibits a temporal correlation coefficient of ~0.6 when compared with both observations).

Under cloud-free conditions, the AERONET AOD uncertainties for wavelengths > 400 nm are quite small (< ±0.01, [94]); therefore, taking AERONET measurements as a reference in our evaluation during some years one/two month advance/late predictions of maximal/minimal values of AOD by the model has been noted. Nevertheless, similar levels of temporal pattern inconsistencies, between AERONET and MISR measurements, are also seen. Such arbitrarily occurring temporal biases on simulated AODs are most likely related (at least partially associated) to three factors: (1) since the air quality around this site is strongly influenced by biomass burning events and different

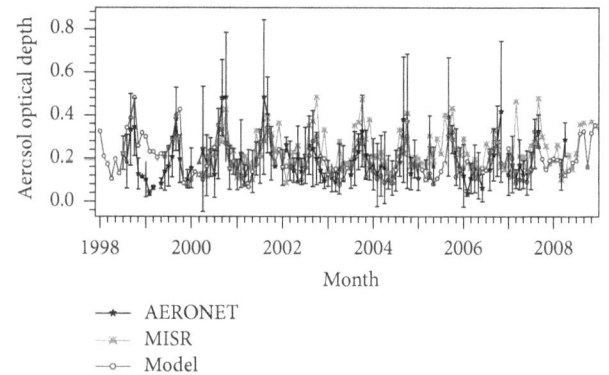

FIGURE 2: The comparison of RegCM4-aerosol model simulated monthly-averaged column-integrated aerosol extinction optical depth (at 550 nm—blue solid line) with its corresponding observations derived from MISR (at 558 nm—gray dotted line) and AERONET (at 500 nm—black solid line denoting the values and standard deviation), at Skukuza, South Africa.

industrial activities, uncertainties associated with biomass burning/anthropogenic/sporadic urban emissions estimates by the emission inventories used in our simulations will impose such temporal biases (e.g., [80, 101]); (2) it is also likely that the bias in simulated meteorological fields could be a contributing factor in inducing temporal evolution biases on simulated aerosol concentrations, in turn in model predicted AOD patterns (e.g., [30, 34, 57, 153]), and (3) biases that propagate from the interpolation scheme used to get model AOD at the AERONET site.

3.2. Comparison of Simulated and MISR Satellite-Observed AOD and SSA Latitudinal Variations. In order to evaluate the model performance in capturing the spatial variability of column-integrated AODs and SSAs values, the simulated results are compared with MISR observations. The 9-year (i.e., from 2000 to 2008) simulated AOD and SSA values (at 550 nm) are averaged over the longitudinal range, which encompasses only South Africa; subsequently their latitudinal variations (i.e., from the lower tip of Western Cape to the upper end of Limpopo, see Figure 1) in comparison with the corresponding MISR retrieved data (at 558 nm) are presented in Figures 3(a) and 3(b), respectively. Over most of the latitudinal locations which correspond to Gauteng, Mpumalanga, North West, and Western Cape regions, the simulated columnar AOD values are within the standard deviation of MISR observations. However, over the latitudinal locations which comprise the Northern Cape, KwaZulu Natal, Free State, and most areas of the Eastern Cape regions, the model slightly overestimates the AOD values with respect to the MISR observations. In contrast, relative to MISR retrieval, the model underestimates the AOD signal in Limpopo province (i.e., within latitudinal range of 25.5°S to 22°S). Within latitudinal ranges that extend from the southern tip of the Western Cape to central areas of Gauteng (except at 34.5°S), the model's estimated SSA values are within the standard deviation of MISR observations. However, this comparison shows a slight positive bias that varies from +0.6

Evaluation of Regional Climatic Model Simulated Aerosol Optical Properties over South Africa Using
Ground-Based and Satellite Observations

143

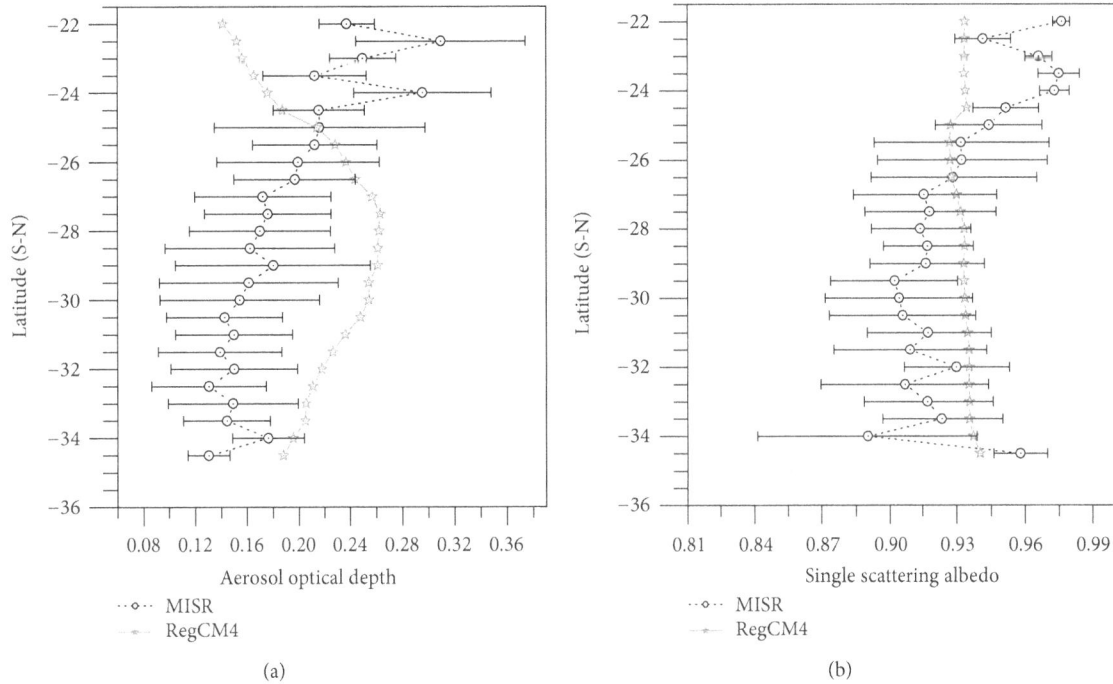

(a)

(b)

FIGURE 3: The latitudinal variation of the 9 years averaged column-integrated (a) AOD and (b) SSA values over South Africa, which are derived from MISR (at 558 nm—black dot line denoting the values and its standard deviation) and the model (at 550 nm—gray solid line).

to +4% relative to MISR mean values. For latitudinal locations which are above 25°S the model's predicted SSA values are lower than their corresponding MISR retrieved mean values.

The aerosol characteristics extracted from model simulations such as the distributions of atmospheric aerosol concentrations (in turn, their optical properties), the rate of production, and removal of particles are mostly controlled by meteorological parameters, which are highly variable both temporally and spatially (e.g., [2, 30, 31, 34, 153]). The model's performance in computing the actual precipitation values and patterns has a central role in determining the removal of aerosols from the atmosphere via wet deposition processes, as well as in regulating the soil moisture which in turn influences the dust production (e.g., [30, 31, 63]). As a result, the biases in the model's estimated precipitation values, in turn the aerosol concentrations, will significantly affect the robustness of simulated aerosol optical properties (e.g., [30, 57, 80, 154]). Even though it is not the intention of the current work to evaluate the simulated meteorological parameters, in order to decipher their contribution to the accuracy of model's estimated values of AOD and SSA, within a latitudinal range of 33.5°S to 27°S, it is valuable to assess the bias in the model's estimated precipitation values. For this purpose, a single South African weather service (SAWS) metrological station's datasets (only one representative station per Province) are used as representative of these provinces and are compared with simulated results (see Appendix, Figures 4 and 5). The comparison period covers 108 months (from 2000 to 2008; please see Appendix). From the total compared months in Bloemfontein (29.1°S, 26.3°E) which is in Free State and Upington (28.4°S, 21.3°E) which is in Northern Cape, ~80%

of simulated precipitation values are in a negative bias (see Appendix, Figure 5).

The processes that control the dust emission through wind erosion are quite different to those involved in the anthropogenic/biomass burning emissions [29, 63]. The computation of naturally emitted dust production involves numerous criteria of land surface characterization. It also depends upon the model's capability of simulating different meteorological fields and land surface conditions. One of the surface parameters that determine the dust production is soil moisture (e.g., [63, 155]) The model's predicted precipitation values exhibit a dry bias, which in turn will cause the surface to become dry and favorable for excessive dust particles emission depending on the wind intensity, [63, 156, 157]. Concurrently, the dry bias in simulated precipitation values will elongate the atmospheric residence time of the particles by reducing the wet removal rate (e.g., [30, 31, 158]). Soil dust particles do not readily dissolve in water ([159] and references therein); thus the negative bias in simulated precipitation will especially extend the atmospheric life-time of hydrophilic aerosols such as sulfate aerosols [30, 80].

Within the simulation domain of the present work, the arid/semiarid regions of the Northern Cape, as well as Namibia and Botswana, are major sources of wind-induced desert dust particles over South Africa ([2, 160] and references therein). Moreover, as presented in Tesfaye et al. [121] the largest contribution of sulfate aerosols to the total AOD is found in the Free State province of South Africa. In the visible part of the spectrum, excluding the slight absorption influence of larger dust particles [161], both dust and sulfate aerosols have a prevailing role of scattering (i.e., these aerosols

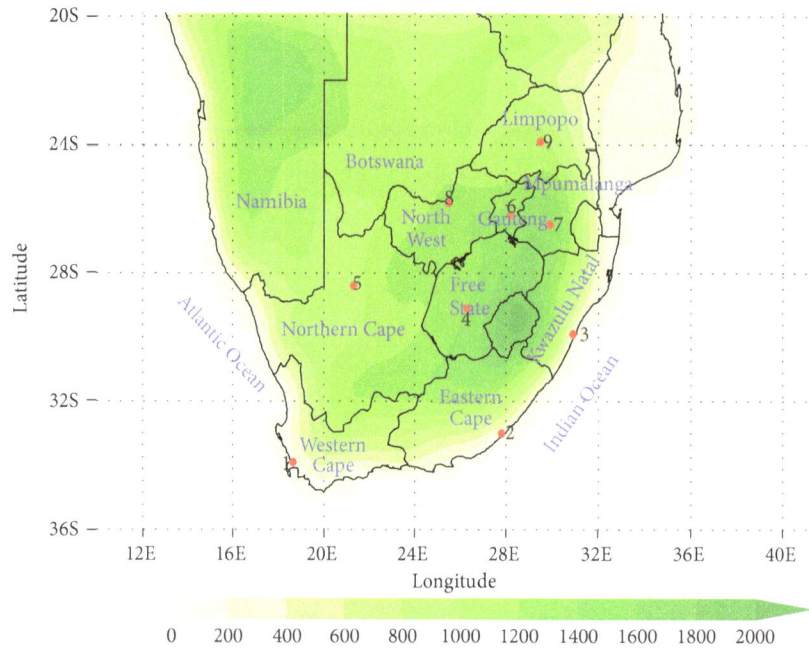

FIGURE 4: Model domain and topography (unit: m). The selected SAWS station locations are numbered as (1) Cape Town (33.9°S, 18.6°E), (2) East London (33.0°S, 27.8°E), (3) Durban South (29.9°S, 30.9°E), (4) Bloemfontein (29.1°S, 26.3°E), (5) Upington (28.4°S, 21.3°E), (6) Johannesburg (26.2°S, 28.2°E), (7) Ermelo (26.5°S, 29.9°E), (8) Mafikeng (25.8°S, 25.5°E), and (9) Polokwane (23.9°S, 29.5°E).

have higher values of SSA) (e.g., [88, 90, 92]). Connecting all the aforementioned interrelated actualities, it is evident that within latitudinal range of 33.5°S to 27°S (i.e., which includes the Northern Cape and Free State provinces of South Africa as well as the nearby regions; see Figure 1), the model's estimated higher values of AOD and SSA (relative to MISR retrieval, Figure 3) may have been caused by the overestimated dust and sulfate aerosol atmospheric concentrations. This is also ensued due to the prevailing dry bias incidences around the primary source region of these aerosols (see Appendix, Figure 5).

Even though the arid/semiarid surfaces are the main source areas of dust particles, anthropogenic activities induced land surface degradations that are related to agricultural use, mining activities, and many other events result in an increment in wind-generated dust production (e.g., [162]). The areas of South Africa which are bounded within a latitudinal range of 25°S to 22°S (i.e., the Limpopo province) are highly populated with different mining and agricultural practices. Consequently, these activities will raise the dust emission in local and regional scales. However, the RegCM4 dust emission parameterizations are effective for cells which are dominated by desert and semidesert land cover only [63]. Therefore, primarily owing to the lack of cooperating anthropogenic activity-related dust production in Limpopo province (i.e., 25°S to 22°S), the model underestimates the AOD, as well as SSA values with respect to MISR observations (Figure 3). Such bias, in turn, will affect the accuracy of the model's estimated direct and semi-direct radiative effects of total aerosols.

Due to the high dynamical nature of human activities, it is challenging to estimate anthropogenic events triggered by dust load. The above results notify the necessity of cooperating these particles for a better representation of bulk aerosol and its climatic roles in South Africa. Besides the above primary reason, the shortage of accounting the long-range transported particles in the model—especially from mining industries of Zambian to Limpopo areas (25°S to 22°S)—will have a slight contribution to the observed essential differences in Figure 3 (e.g., [163]). As a final point, as indicated in Section 2, our simulation did not take into account marine aerosols. In the meanwhile, at Cape Town's (33.9°S, 18.6°E) metrological site, largely positive biases in simulated precipitation values were noted (see Appendix, Figure 5), this will result in an overestimation of aerosol wet deposition (e.g., [30, 31]). Therefore, most likely related to these two factors, the model's estimated SSA values around the Cape Point (i.e., 34.5°S) are slightly underestimated with respect to MISR observations.

3.3. Comparisons with LIDAR. The aerosol extinction coefficient profiles that are retrieved from the LIDAR experiment (532 nm; see Section 2.5.3) on 16 October observation at 18:00 and 17 October at 00:00 and 06:00 were compared to their corresponding model results (550 nm) and provided in Figures 6(a)–6(c), respectively. In all comparisons, except at 18:00, both experimental and simulated profiles exhibit larger extinction coefficients below the altitude of 6 km. Above 6 km (i.e., above ~490 hPa), the simulated extinction profiles display a more rapid decline than the experimental observation. Especially on 16 October (at 18:00) and 17 October (at 06:00), considerable discrepancies between the model's and LIDAR's extinction profiles, above the height region of

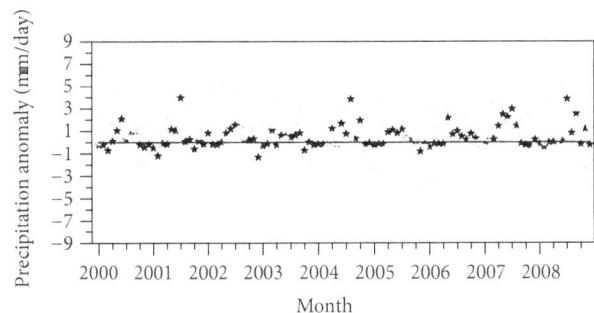

★ Cape Town: NB ~ 42%
Durban South: NB ~ 33%

(a)

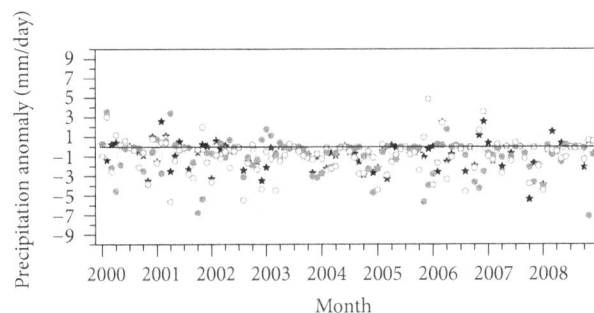

★ Bloemfontein: NB ~ 76%   Upington: NB ~ 82%
● Mafikeng: NB ~ 80%   ○ Polokwane: NB ~ 81%

(b)

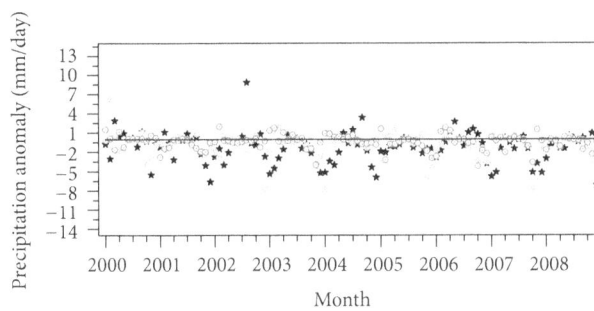

★ East London: NB ~ 70%
Johannesburg: NB ~ 70%
○ Ermelo: NB ~ 69%

(c)

FIGURE 5: The bias in RegCM4 estimated total daily precipitation (mm/day) in comparison with the specific metrological station's data. From the total compared months (108 months, i.e., 2000 to 2008) the percentage of the number of points that exhibit a negative bias (NB) is given in each plot.

6 km, have been noted. Using the same LIDAR datasets and air mass trajectory analysis, Tesfaye et al. [150] showed that particles above the height region of 6 km were particularly donated from long-range transportation processes. However, related with our future concern of investigating different types of aerosol impacts on South African climate—especially those which are originated in and around South Africa, we have configured the domain (see Figure 1), as well as the

inflow/outflow boundary conditions, in order to account only for the natural and anthropogenic aerosol sources in and around South Africa. These configurations neglect the contribution of aerosols from external sources (which are outside of the domain) to the regional aerosol budget. This may partially attribute for the differences between experimental and simulated extinction profiles which are noted above the altitude profile of 6 km.

In overall, to regulate the complex semidirect effect of aerosols—especially their role in cloud cover, studies have shown the importance of the relative position of the aerosol layer with respect to the cloud position (e.g., [16, 17]). Nonetheless, there are several factors that will impose substantial inaccuracies on simulated profiles of aerosol optical properties such as, the model's deficiency in representing convective processes (e.g., [57, 65, 68]), the number of aerosol components which are cooperated in the model along with their mixing state hypothesis (e.g., [5, 124]), and the various aspects which are mentioned in Section 1. Considering the presence of these all confining circumstances which will inflict discrepancies between simulated and LIDAR profiles, the model exhibits quite satisfactory performance in capturing the major aerosol extinction profiles. Although in our discussion several factors that will impose a bias on simulated aerosol optical signals are highlighted; at this scope of the study it is difficult to accurately assess the contribution of each factors in-depth and to point out which one is more responsible for enforcing these biases.

4. Summary and Concluding Remarks

Before we employ the Regional Climate Model-RegCM4 for the investigation of direct and semi-direct effects of aerosols over South Africa, in this study its performance to capture the observed aerosol optical properties has been evaluated and discussed. The evaluations were performed by comparing the simulated columnar Aerosol Optical Depth (AOD) and Single Scattering Albedo (SSA) against ground-based (AERONET) and satellite (Multiangle Imaging SpectroRadiometer: MISR) observations. Additionally, the simulated aerosol extinction profiles were compared with ground-based LIDAR retrieval. In our current contribution, the following conclusions can be drawn.

(i) At Skukuza (24°S, 31°E), the values of simulated AOD were within the standard deviation of AERONET and ±25% of MISR observations. Occasionally, the model-estimated maximum/minimum AOD values displayed a slight temporal shift with respect to AERONET observations. Nonetheless, such irregularly occurring temporal biases, as well as magnitude differences, are also noted among the MISR and AERONET platform estimates. In this frame, the model's simulated AOD climatology at Skukuza site is legitimately acceptable.

(ii) Considering the longitudinal range which includes only South Africa—the 9-year averaged values of simulated AOD and SSA latitudinal variations were also compared with the corresponding values retrieved

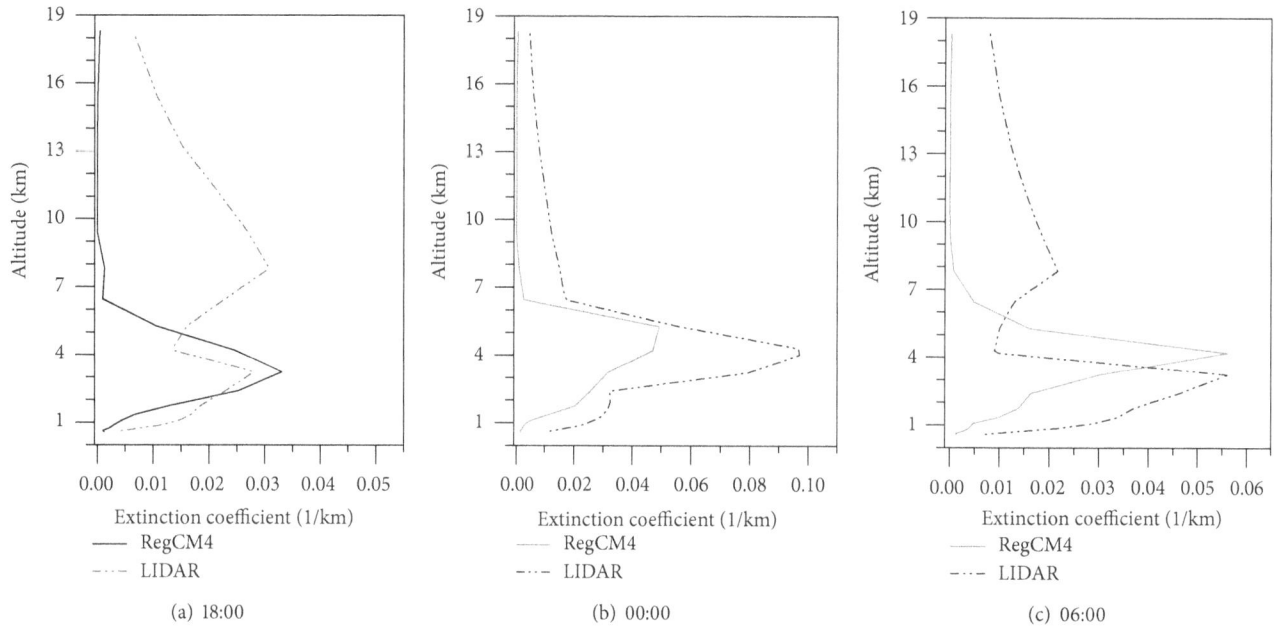

FIGURE 6: Aerosol extinction coefficient profiles, retrieved from LIDAR measurements at Pretoria (black dash-dot line) and its corresponding profiles provided by RegCM4 (gray line) (a) on 16 October at 18:00, (b) and (c) on 17 October at 00:00 and at 06:00, respectively.

from MISR observations. Within a latitudinal range of 26.5°S to 24.5°S, the simulated columnar AOD values were within the standard deviation of MISR. However, within the regions of 33.5°S to 27°S, the model tends to predict slightly higher values of AOD relative to MISR observation. This was predominantly caused by the negative (dry) bias in simulated precipitation that leads to the overestimation of dust and sulfate aerosol loads over these regions.

(iii) With respect to MISR, over the latitudinal range that corresponds to the Limpopo province (i.e., 25°S to 22°S), the model underestimates both AOD and SSA signals. This was primarily due to the model's shortage in cooperating the anthropogenic activities prompted dust loads, such as dust from agricultural and mining practices. Therefore, in our region of interest (i.e., South Africa), this is considered as the main deficiency of RegCM4. Besides, in most of the latitudinal ranges, the simulated SSA values were within standard deviations of MISR observations.

(iv) Excluding the model's underestimated extinction coefficient profiles above 6 km, RegCM4 performs very well in capturing the main aerosol extinction profiles, relative to LIDAR measurements. The aerosol extinction signals at higher altitude (>6 km) were donated by long-range transportation of particles from regions which were outside our simulation domain.

(v) Relative to observational data, the model fairly reproduced optical properties of aerosols. This affirmed that sulfate and carbonaceous aerosols from both anthropogenic and biomass burning activities and

wind eroded desert dust particles were the main aerosol components in the South African atmosphere.

(vi) Overall, RegCM4 appeared to be a suitable tool for the examination of the direct and semi-direct effects of aerosols over the South African regional climate.

(vii) In the series of studies to follow, we will provide the distribution, direct and semi-direct effects of wind eroded desert dust particles, and the different species of aerosols which were induced from anthropogenic and biomass burning sectors, in South Africa [118–121]. Thus, we would like to remark here that, in the present study, a bit of an extended general introduction, as well as a brief description about RegCM4, was addressed. Parts of these sections will be used as the basic framework for those of our future studies.

Appendix

The meteorological parameters such as precipitation, relative humidity, temperature, and wind speed play a major role in determining aerosol dynamics, as well as the changes in aerosol optical and physico-chemical properties (e.g., [2, 34, 153]). The model's deficiency in representing the actual precipitation values and patterns will impose a significant bias simulated aerosol production (via influencing surface wetness), as well as their atmospheric lifetime and concentration (via influencing wet aerosol deposition) [2, 30, 31]. The biases in simulated aerosol concentrations will consequently affect the model's accuracy in computing the optical properties of aerosols. Therefore, comparing the simulated precipitation values, with respect to the South Africa Weather Service (SAWS) metrological station observations, the biases in model's predicted precipitation values are estimated. For this purpose, from each province of South Africa, a single weather service station is designated as a representative of these areas (Figure 4).

The bias in simulated total precipitation, while being compared to the datasets of the specific metrological sites, is shown in Figure 5. Generally, excluding the stations which are nearby the coastal areas of South Africa (i.e., Cape Town and Durban South stations, Figure 5(a)), the model predominantly exhibits a negative bias (Figures 5(b) and 5(c)); that is, out of total compared months in 75% ± 5% of the points RegCM4 tends to underestimate the total daily precipitation values. The dry bias over the west (i.e., at Upington station, Figures 4 and 5(b)) and central (i.e., at Bloemfontein station, Figures 4 and 5(b)) parts of South Africa will promote the over emission of wind eroded dust particles and reduced wet deposition of hydrophilic aerosols. The precipitation bias in these comparisons has to be considered under caution; this is not a complete indicator of the overall performance of the model.

The model might actually be a demonstration of the region, but may not match exactly with the precise site observations. Further, the bias in metrological parameters can be induced by numerous complex factors such as radiation balance and energy flux inaccuracies induced through aerosol processes and surface properties (such as surface albedo), temperature advection, and cloud process representations [30, 57, 68, 164, 165]. However, determining the actual cause of these metrological parameter biases requires a further deep examination of the model's physics in representing the atmospheric dynamics, cloud, and surface processes as well as several other factors, which is beyond the scope of this study. Nevertheless, the expressive correlation between the biases in simulated AOD and precipitation values strengthens the remarkable importance of interactive coupling of aerosol-climate interactions.

Acknowledgments

The authors are grateful to Addis Ababa University, Department of Physics, for providing computational facilities. For the accessibility of RegCM model, the authors are thankful to the International Centre for Theoretical Physics (ICTP). The authors would like to acknowledge the AERONET, MISR, and SAWS for providing an easy access to the datasets used in this study. They are also indebted to Teresa Faleschini, Tamene Mekonnen, Fiona Tummon, and Addisu Gezahegn, for their valuable assistances. This work was supported by the African Laser Centre and NRF bi-lateral research grant (UID: 68688/65086), in addition to CSIR National Laser Centre.

References

[1] J. E. Penner, M. Andreae, H. Annegarn et al., "Aerosols, their direct and indirect effects," in *Climate Change 2001: The Scientific Basis. Contribution of Working Group I to the Third Assessment Report of the Intergovernmental Panel on Climate Change*, J. T. Houghton, Y. Ding, D. J. Griggs et al., Eds., pp. 289–348, Cambridge University Press, Cambridge, UK, 2001.

[2] M. Tesfaye, V. Sivakumar, J. Botai, and G. M. Tsidu, "Aerosol climatology over South Africa based on 10 years of multiangle imaging spectroradiometer (MISR) data," *Journal of Geophysical Research D*, vol. 116, no. 20, Article ID D20216, 2011.

[3] D. T. Shindell, H. Levy II, M. D. Schwarzkopf, L. W. Horowitz, J.-F. Lamarque, and G. Faluvegi, "Multi-model projections of climate change from short-lived emissions due to human activities," *Journal of Geophysical Research*, vol. 113, Article ID D11109, 2008.

[4] IPCC, "IPCC fourth assessment report (AR4)," in *Climate Change 2007: The Physical Science Basis. Contribution of Working Group I to the Fourth Assessment Report of the Intergovernmental Panel on Climate Change*, S. Solomon, D. Qin, M. Manning et al., Eds., p. 996, Cambridge University Press, Cambridge, UK, 2007.

[5] J. Haywood and O. Boucher, "Estimates of the direct and indirect radiative forcing due to tropospheric aerosols: a review," *Reviews of Geophysics*, vol. 38, no. 4, pp. 513–543, 2000.

[6] J. Hansen, M. Sato, R. Ruedy et al., "Efficacy of climate forcings," *Journal of Geophysical Research*, vol. 110, Article ID D18104, 2005.

[7] J. Hansen, M. Sato, and R. Ruedy, "Radiative forcing and climate response," *Journal of Geophysical Research*, vol. 102, no. 6, pp. 6831–6864, 1997.

[8] P. Forster, V. Ramaswamy, P. Artaxo et al., "Changes in atmospheric constituents and in radiative forcing," in *Climate Change 2007: The Physical Science Basis. Contribution of Working Group I to the Fourth Assessment Report of the Intergovernmental Panel on Climate Change*, S. Solomon, D. Qin, M. Manning et al., Eds., Cambridge University Press, Cambridge, UK, 2007.

[9] A. S. Ackerman, O. B. Toon, D. E. Stevens, A. J. Heymsfield, V. Ramanathan, and E. J. Welton, "Reduction of tropical cloudiness by soot," *Science*, vol. 288, no. 5468, pp. 1042–1047, 2000.

[10] R. L. Miller and I. Tegen, "Climate response to soil dust aerosols," *Journal of Climate*, vol. 11, no. 12, pp. 3247–3267, 1998.

[11] M. Z. Jacobson, "Control of fossil-fuel particulate black carbon and organic matter, possibly the most effective method of slowing global warming," *Journal of Geophysical Research D*, vol. 107, no. D19, p. 4410, 2005.

[12] I. Koren, Y. J. Kaufman, L. A. Remer, and J. V. Martins, "Measurement of the effect of Amazon smoke on inhibition of cloud formation," *Science*, vol. 303, no. 5662, pp. 1342–1345, 2004.

[13] J. Cook and E. J. Highwood, "Climate response to tropospheric absorbing aerosols in an intermediate general-circulation model," *Quarterly Journal of the Royal Meteorological Society*, vol. 130, no. 596, pp. 175–191, 2004.

[14] Y. Zhang, *The radiative effect of aerosols from biomass burning on the transition from dry to wet season over the Amazon as tested by a regional climate model [Ph.D. thesis]*, Georgia Institute of Technology, Atlanta, Ga, USA, 2008.

[15] C. E. Chung, V. Ramanathan, and J. T. Kiehl, "Effects of the South Asian absorbing haze on the northeast monsoon and surface-air heat exchange," *Journal of Climate*, vol. 15, no. 17, pp. 2462–2476, 2002.

[16] B. T. Johnson, K. P. Shine, and P. M. Forster, "The semi-direct aerosol effect: impact of absorbing aerosols on marine stratocumulus," *Quarterly Journal of the Royal Meteorological Society*, vol. 130, no. 599, pp. 1407–1422, 2004.

[17] G. Feingold, H. Jiang, and J. Y. Harrington, "On smoke suppression of clouds in Amazonia," *Geophysical Research Letters*, vol. 32, no. 2, Article ID L02804, 2005.

[18] J. E. Penner, S. Y. Zhang, and C. C. Chuang, "Soot and smoke aerosol may not warm climate," *Journal of Geophysical Research D*, vol. 108, no. 21, pp. 1–9, 2003.

[19] D. Koch and A. D. Del Genio, "Black carbon semi-direct effects on cloud cover: review and synthesis," *Atmospheric Chemistry and Physics*, vol. 10, no. 16, pp. 7685–7696, 2010.

[20] R. J. Allen and S. C. Sherwood, "Aerosol-cloud semi-direct effect and land-sea temperature contrast in a GCM," *Geophysical Research Letters*, vol. 37, no. 7, Article ID L07702, 2010.

[21] S. Twomey, "The influence of pollution on the shortwave albedo of clouds," *Journal of the Atmospheric Sciences*, vol. 34, pp. 1149–1152, 1977.

[22] B. A. Albrecht, "Aerosols, cloud microphysics, and fractional cloudiness," *Science*, vol. 245, no. 4923, pp. 1227–1230, 1989.

[23] U. Lohmann and J. Feichter, "Global indirect aerosol effects: a review," *Atmospheric Chemistry and Physics*, vol. 5, no. 3, pp. 715–737, 2005.

[24] F. J. Dentener, G. R. Carmichael, Y. Zhang, J. Lelieveld, and P. J. Crutzen, "Role of mineral aerosol as a reactive surface in the global troposphere," *Journal of Geophysical Research D*, vol. 101, no. 17, pp. 22869–22889, 1996.

[25] R. R. Dickerson, S. Kondragunta, G. Stenchikov, K. L. Civerolo, B. G. Doddridge, and B. N. Holben, "The impact of aerosols on solar ultraviolet radiation and photochemical smog," *Science*, vol. 278, no. 5339, pp. 827–830, 1997.

[26] W. L. Chameides, H. Yu, S. C. Liu et al., "Case study of the effects of atmospheric aerosols and regional haze on agriculture: an opportunity to enhance crop yields in China through emission controls?" *Proceedings of the National Academy of Sciences of the United States of America*, vol. 96, no. 24, pp. 13626–13633, 1999.

[27] H. R. Anderson, "Differential epidemiology of ambient aerosols," *Philosophical Transactions of the Royal Society A*, vol. 358, no. 1775, pp. 2771–2785, 2000.

[28] G. Oberdorster, J. Finkelstein, J. Ferin et al., "Ultrafine particles as a potential environmental health hazard: studies with model particles," *Chest*, vol. 109, no. 3, 1996.

[29] F. Solmon, F. Giorgi, and C. Liousse, "Aerosol modelling for regional climate studies: application to anthropogenic particles and evaluation over a European/African domain," *Tellus, Series B*, vol. 58, no. 1, pp. 51–72, 2006.

[30] B. Croft, J. R. Pierce, R. V. Martin, C. Hoose, and U. Lohmann, "Uncertainty associated with convective wet removal of entrained aerosols in a global climate model," *Atmospheric Chemistry and Physics*, vol. 12, pp. 10725–10748, 2012.

[31] N. Oshima, Y. Kondo, N. Moteki et al., "Wet removal of black carbon in Asian outflow: Aerosol Radiative Forcing in East Asia (A-FORCE) aircraft campaign," *Journal of Geophysical Research D*, vol. 117, no. 3, Article ID D03204, 2012.

[32] D. T. Shindell, M. Chin, F. Dentener et al., "A multi-model assessment of pollution transport to the Arctic," *Atmospheric Chemistry and Physics*, vol. 8, no. 17, pp. 5353–5372, 2008.

[33] V. Ramanathan, P. J. Crutzen, J. Lelieveld et al., "Indian ocean experiment: an integrated analysis of the climate forcing and effects of the great Indo-Asian haze," *Journal of Geophysical Research D*, vol. 106, no. 22, pp. 28371–28398, 2001.

[34] D. Rind, M. Chin, G. Feingold et al., "Modeling the effects of aerosols on climate," in *Atmospheric Aerosol Properties and Impacts on Climate, A Report by the U.S. Climate Change Science Program and the Subcommittee on Global Change Research*, M. Chin, R. A. Kahn, and S. E. Schwartz, Eds., pp. 64–97, National Aeronautics and Space Administration, 2009.

[35] E. Hirst, P. H. Kaye, R. S. Greenaway, P. Field, and D. W. Johnson, "Discrimination of micrometre-sized ice and super-cooled droplets in mixed-phase cloud," *Atmospheric Environment*, vol. 35, no. 1, pp. 33–47, 2001.

[36] D. J. Diner, B. H. Braswell, R. Davies et al., "The value of multiangle measurements for retrieving structurally and radiatively consistent properties of clouds, aerosols, and surfaces," *Remote Sensing of Environment*, vol. 97, no. 4, pp. 495–518, 2005.

[37] R. Kahn, P. Banerjee, and D. McDonald, "Sensitivity of multiangle imaging to natural mixtures of aerosols over ocean," *Journal of Geophysical Research D*, vol. 106, no. 16, pp. 18219–18238, 2001.

[38] R. A. Kahn, M. J. Garay, D. L. Nelson et al., "Satellite-derived aerosol optical depth over dark water from MISR and MODIS: comparisons with AERONET and implications for climatological studies," *Journal of Geophysical Research D*, vol. 112, no. 18, Article ID D18205, 2007.

[39] A. S. Goudie and N. J. Middleton, *Desert Dust in the Global System*, Springer, Berlin, Germany, 2006.

[40] M. Chin, P. Ginoux, S. Kinne et al., "Tropospheric aerosol optical thickness from the GOCART model and comparisons with satellite and sun photometer measurements," *Journal of the Atmospheric Sciences*, vol. 59, no. 3, pp. 461–483, 2002.

[41] S. H. Chung and J. H. Seinfeld, "Global distribution and climate forcing of carbonaceous aerosols," *Journal of Geophysical Research D*, vol. 107, no. 19, pp. 14–33, 2002.

[42] Y. Qian, F. Giorgi, Y. Huang, W. Chameides, and C. Luo, "Regional simulation of anthropogenic sulfur over East Asia and its sensitivity to model parameters," *Tellus, Series B*, vol. 53, no. 2, pp. 171–191, 2001.

[43] F. Giorgi, X. Bi, and Y. Qian, "Direct radiative forcing and regional climatic effects of anthropogenic aerosols over East Asia: a regional coupled climate-chemistry/aerosol model study," *Journal of Geophysical Research D*, vol. 107, no. 20, pp. 7–18, 2002.

[44] F. Giorgi, X. Bi, and Y. Qian, "Indirect vs. direct effects of anthropogenic sulfate on the climate of east Asia as simulated with a regional coupled climate-chemistry/aerosol model," *Climatic Change*, vol. 58, no. 3, pp. 345–376, 2003.

[45] A. M. L. Ekman and H. Rodhe, "Regional temperature response due to indirect sulfate aerosol forcing: impact of model resolution," *Climate Dynamics*, vol. 21, no. 1, pp. 1–10, 2003.

[46] T. Takemura, T. Nakajima, T. Nozawa, and K. Aoki, "Simulation of future aerosol distribution, radiative forcing, and long-range transport in East Asia," *Journal of the Meteorological Society of Japan*, vol. 79, no. 6, pp. 1139–1155, 2001.

[47] I. Tegen, M. Werner, S. P. Harrison, and K. E. Kohfeld, "Relative importance of climate and land use in determining present and future global soil dust emission," *Geophysical Research Letters*, vol. 31, no. 5, pp. L05105–4, 2004.

[48] A. Baklanov, B. Fay, G. Weather et al., "Overview of existing integrated (off-line and on-line) mesoscale systems in Europe," report of Working Group 2, 2007, http://www.cost728.org/.

[49] A. Baklanov, A. Mahura, and R. Sokhi, *Integrated Systems of Meso-Meteorological and Chemical Transport Models*, Springer, New York, NY, USA, 2011.

[50] F. Giorgi, E. Coppola, F. Solmon et al., "RegCM4: model description and preliminary tests over multiple CORDEX domains," *Climate Research*, vol. 52, pp. 7–29, 2012.

Evaluation of Regional Climatic Model Simulated Aerosol Optical Properties over South Africa Using
Ground-Based and Satellite Observations

149

[51] M. Kanakidou, J. H. Seinfeld, S. N. Pandis et al., "Organic aerosol and global climate modelling: a review," *Atmospheric Chemistry and Physics*, vol. 5, no. 4, pp. 1053–1123, 2005.

[52] X. Liu, J. E. Penner, B. Das et al., "Uncertainties in global aerosol simulations: assessment using three meteorological data sets," *Journal of Geophysical Research D*, vol. 112, no. 11, Article ID D11212, 2007.

[53] E. Vignati, M. Karl, M. Krol, J. Wilson, P. Stier, and F. Cavalli, "Sources of uncertainties in modelling black carbon at the global scale," *Atmospheric Chemistry and Physics*, vol. 10, no. 6, pp. 2595–2611, 2010.

[54] L. Lee, K. J. Pringle, C. L. Reddington et al., "The magnitude and causes of uncertainty in global model simulations of cloud condensation nuclei," *Atmospheric Chemistry and Physics*, vol. 13, pp. 6295–6378, 2013.

[55] G. Myhre, F. Stordal, T. F. Berglen, J. K. Sundet, and I. S. A. Isaksen, "Uncertainties in the radiative forcing due to sulphate aerosols," *Journal of the Atmospheric Sciences*, vol. 61, no. 5, pp. 485–498, 2004.

[56] M. Schulz, C. Textor, S. Kinne et al., "Radiative forcing by aerosols as derived from the AeroCom present-day and pre-industrial simulations," *Atmospheric Chemistry and Physics*, vol. 6, no. 12, pp. 5225–5246, 2006.

[57] H. Tost, M. G. Lawrence, C. Brühl, and P. Jöckel, "Uncertainties in atmospheric chemistry modelling due to convection parameterisations and subsequent scavenging," *Atmospheric Chemistry and Physics*, vol. 10, no. 4, pp. 1931–1951, 2010.

[58] P. Stier, N. A. J. Schutgens, N. Bellouin et al., "Host model uncertainties in aerosol radiative forcing estimates: results from the AeroCom prescribed inter-comparison study," *Atmospheric Chemistry and Physics*, vol. 13, pp. 3245–3270, 2013.

[59] F. Giorgi and M. R. Marinucci, "An investigation of the sensitivity of simulated precipitation to model resolution and its implications for climate studies," *Monthly Weather Review*, vol. 124, no. 1, pp. 148–166, 1996.

[60] F. Giorgi and C. Shields, "Tests of precipitation parameterizations available in latest version of NCAR regional climate model (RegCM) over continental United States," *Journal of Geophysical Research D*, vol. 104, no. 6, pp. 6353–6375, 1999.

[61] F. Giorgi and R. O. Anyah, "Evolution of regional climate modeling: the road towards RegCM4," *Climate Research*, vol. 52, pp. 3–6, 2012.

[62] M. Z. Jacobson, *Developing, coupling, and applying a gas, aerosol, transport, and radiation model to study urban and regional air pollution [Ph.D. thesis]*, Department of Atmospheric Sciences, University of California, Los Angeles, Calif, USA, 1994.

[63] A. S. Zakey, F. Solmon, and F. Giorgi, "Implementation and testing of a desert dust module in a regional climate model," *Atmospheric Chemistry and Physics*, vol. 6, no. 12, pp. 4687–4704, 2006.

[64] A. S. Zakey, F. Giorgi, and X. Bi, "Modeling of sea salt in a regional climate model: fluxes and radiative forcing," *Journal of Geophysical Research D*, vol. 113, no. 14, Article ID D14221, 2008.

[65] P. Zanis, C. Douvis, I. Kapsomenakis, I. Kioutsioukis, D. Melas, and J. S. Pal, "A sensitivity study of the Regional Climate Model (RegCM3) to the convective scheme with emphasis in central eastern and southeastern Europe," *Theoretical and Applied Climatology*, vol. 97, no. 3-4, pp. 327–337, 2009.

[66] Y. Zhang, X.-Y. Wen, and C. J. Jang, "Simulating chemistry-aerosol-cloud-radiation-climate feedbacks over the continental U.S. using the online-coupled Weather Research Forecasting Model with chemistry (WRF/Chem)," *Atmospheric Environment*, vol. 44, no. 29, pp. 3568–3582, 2010.

[67] F. Solmon, M. Mallet, N. Elguindi, F. Giorgi, A. Zakey, and A. Konaré, "Dust aerosol impact on regional precipitation over western Africa, mechanisms and sensitivity to absorption properties," *Geophysical Research Letters*, vol. 35, no. 24, Article ID L24705, 2008.

[68] F. Solmon, N. Elguindi, and M. Mallet, "Radiative and climatic effects of dust over West Africa, as simulated by a regional climate model," *Climate Research*, vol. 52, pp. 97–113, 2012.

[69] F. Tummon, F. Solmon, C. Liousse, and M. Tadross, "Simulation of the direct and semidirect aerosol effects on the southern Africa regional climate during the biomass burning season," *Journal of Geophysical Research D*, vol. 115, no. 19, Article ID D19206, 2010.

[70] A. Konare, A. S. Zakey, F. Solmon et al., "A regional climate modeling study of the effect of desert dust on the West African monsoon," *Journal of Geophysical Research D*, vol. 113, no. 12, Article ID D12206, 2008.

[71] F. Malavelle, V. Pont, M. Mallet et al., "Simulation of aerosol radiative effects over West Africa during DABEX and AMMA SOP-0," *Journal of Geophysical Research D*, vol. 116, no. 8, Article ID D08205, 2011.

[72] Eskom, "Effects of atmospheric pollution on the Mpumalanga Highveld," Power Technology No. 70, Eskom Technology Group, Cleveland, South Africa, 1996.

[73] R. Spalding-Fecher and D. K. Matibe, "Electricity and externalities in South Africa," *Energy Policy*, vol. 31, no. 8, pp. 721–734, 2003.

[74] K. E. Ross, S. J. Piketh, R. T. Bruintjes, R. P. Burger, R. J. Swap, and H. J. Annegarn, "Spatial and aerosol variations in CCN distribution and the aerosol-CCN relationship over southern Africa," *Journal of Geophysical Research D*, vol. 108, no. 13, pp. 1–18, 2003.

[75] A. J. Mills, A. V. Milewski, C. Sirami et al., "Aerosol capture by small trees in savannas marginal to treeless grassland in South Africa," *Geoderma*, vol. 189-190, pp. 124–132, 2012.

[76] H. Winkler, P. Formenti, D. J. Esterhuyse et al., "Evidence for large-scale transport of biomass burning aerosols from sunphotometry at a remote South African site," *Atmospheric Environment*, vol. 42, no. 22, pp. 5569–5578, 2008.

[77] A. J. Queface, S. J. Piketh, T. F. Eck, S.-C. Tsay, and A. F. Mavume, "Climatology of aerosol optical properties in Southern Africa," *Atmospheric Environment*, vol. 45, no. 17, pp. 2910–2921, 2011.

[78] B. I. Magi, "Chemical apportionment of southern African aerosol mass and optical depth," *Atmospheric Chemistry and Physics*, vol. 9, no. 19, pp. 7643–7655, 2009.

[79] M. Tesfaye, V. Sivakumar, J. Botai, and G. Mengistu, "Latitudinal variations of aerosol optical parameters over South Africa based on MISR satellite data," in *Proceedings of the 26th Annual Conference of South African Society for Atmosphere Science*, pp. 105–106, September 2010.

[80] F. Tummon, *Direct and semi-direct aerosol effects on the southern African regional climate during the austral winter season [Ph.D. thesis]*, University of Cape Town, Cape Town, South Africa, 2011.

[81] B. I. Magi, P. Ginoux, Y. Ming, and V. Ramaswamy, "Evaluation of tropical and extratropical Southern Hemisphere African aerosol properties simulated by a climate model," *Journal of Geophysical Research D*, vol. 114, no. 14, Article ID D14204, 2009.

[82] T. F. Eck, B. N. Holben, D. E. Ward et al., "Variability of biomass burning aerosol optical characteristics in southern

Africa during the SAFARI 2000 dry season campaign and a comparison of single scattering albedo estimates from radiometric measurements," *Journal of Geophysical Research D*, vol. 108, no. 13, pp. 1–21, 2003.

[83] R. J. Swap, H. J. Annegarn, J. T. Suttles et al., "The Southern African Regional Science Initiative (SAFARI 2000): overview of the dry season field campaign," *South African Journal of Science*, vol. 98, no. 3-4, pp. 125–130, 2002.

[84] R. J. Swap, H. J. Annegarn, J. T. Suttles et al., "Africa burning: a thematic analysis of the Southern African Regional Science Initiative (SAFARI 2000)," *Journal of Geophysical Research D*, vol. 108, no. 13, pp. 1–15, 2003.

[85] S. J. Piketh, H. J. Annegarn, and P. D. Tyson, "Lower tropospheric aerosol loadings over South Africa: the relative contribution of aeolian dust, industrial emissions, and biomass burning," *Journal of Geophysical Research D*, vol. 104, no. 1, pp. 1597–1607, 1999.

[86] J. M. Prospero, P. Ginoux, O. Torres, S. E. Nicholson, and T. E. Gill, "Environmental characterization of global sources of atmospheric soil dust identified with the Nimbus 7 Total Ozone Mapping Spectrometer (TOMS) absorbing aerosol product," *Reviews of Geophysics*, vol. 40, no. 1, pp. -1–31, 2002.

[87] O. Dubovik and M. D. King, "A flexible inversion algorithm for retrieval of aerosol optical properties from Sun and sky radiance measurements," *Journal of Geophysical Research D*, vol. 105, no. 16, pp. 20673–20696, 2000.

[88] C. Levoni, M. Cervino, R. Guzzi, and F. Torricella, "Atmospheric aerosol optical properties: a database of radiative characteristics for different components and classes," *Applied Optics*, vol. 36, no. 30, pp. 8031–8041, 1997.

[89] M. I. Mishchenko and L. D. Travis, *Light Scattering by Non-Spherical Particles*, Academic Press, San-Diego, Calif, USA, 2000.

[90] M. Hess, P. Koepke, and I. Schult, "Optical properties of aerosols and clouds: the software package OPAC," *Bulletin of the American Meteorological Society*, vol. 79, no. 5, pp. 831–844, 1998.

[91] E. Andrews, P. J. Sheridan, M. Fiebig et al., "Comparison of methods for deriving aerosol asymmetry parameter," *Journal of Geophysical Research*, vol. 111, no. D5, Article ID D05S04, 2006.

[92] T. Takemura, T. Nakajima, O. Dubovik, B. N. Holben, and S. Kinne, "Single-scattering albedo and radiative forcing of various aerosol species with a global three-dimensional model," *Journal of Climate*, vol. 15, no. 4, pp. 333–352, 2002.

[93] J. H. Seinfeld and S. N. Pandis, *Atmospheric Chemistry and Physics: From Air Pollution to Climate*, Wiley, New York, NY, USA, 2006.

[94] O. Dubovik, A. Smirnov, B. N. Holben et al., "Accuracy assessments of aerosol optical properties retrieved from Aerosol Robotic Network (AERONET) Sun and sky radiance measurements," *Journal of Geophysical Research D*, vol. 105, no. 8, pp. 9791–9806, 2000.

[95] H. Yu, Y. J. Kaufman, M. Chin et al., "A review of measurement-based assessments of the aerosol direct radiative effect and forcing," *Atmospheric Chemistry and Physics*, vol. 6, no. 3, pp. 613–666, 2006.

[96] Y. J. Kaufman, D. Tanré, L. A. Remer, E. F. Vermote, A. Chu, and B. N. Holben, "Operational remote sensing of tropospheric aerosol over land from EOS moderate resolution imaging spectroradiometer," *Journal of Geophysical Research D*, vol. 102, no. 14, pp. 17051–17067, 1997.

[97] P. Chylek, B. Henderson, and M. Mishchenko, "Aerosol radiative forcing and the accuracy of satellite aerosol optical depth retrieval," *Journal of Geophysical Research D*, vol. 108, no. 24, pp. 4–8, 2003.

[98] D. Tanré, Y. J. Kaufman, M. Herman, and S. Mattoo, "Remote sensing of aerosol properties over oceans using the MODIS/EOS spectral radiances," *Journal of Geophysical Research D*, vol. 102, no. 14, pp. 16971–16988, 1997.

[99] R. A. Kahn, B. J. Gaitley, J. V. Martonchik, D. J. Diner, K. A. Crean, and B. Holben, "Multiangle Imaging Spectroradiometer (MISR) global aerosol optical depth validation based on 2 years of coincident Aerosol Robotic Network (AERONET) observations," *Journal of Geophysical Research D*, vol. 110, no. 10, pp. 1–16, 2005.

[100] O. V. Kalashnikova, R. Kahn, I. N. Sokolik, and W.-H. Li, "Ability of multiangle remote sensing observations to identify and distinguish mineral dust types: optical models and retrievals of optically thick plumes," *Journal of Geophysical Research D*, vol. 110, no. 18, Article ID D18S14, pp. 1–16, 2005.

[101] A. De Meij, M. Krol, F. Dentener, E. Vignati, C. Cuvelier, and P. Thunis, "The sensitivity of aerosol in Europe to two different emission inventories and temporal distribution of emissions," *Atmospheric Chemistry and Physics*, vol. 6, no. 12, pp. 4287–4309, 2006.

[102] N. Mahowald, K. Kohfeld, M. Hansson et al., "Dust sources and deposition during the last glacial maximum and current climate: a comparison of model results with paleodata from ice cores and marine sediments," *Journal of Geophysical Research D*, vol. 104, no. 13, pp. 15895–15916, 1999.

[103] Y. J. Kaufman, D. Tanré, and O. Boucher, "A satellite view of aerosols in the climate system," *Nature*, vol. 419, no. 6903, pp. 215–223, 2002.

[104] D. J. Diner, T. P. Ackerman, T. L. Anderson et al., "Paragon: an integrated approach for characterizing aerosol climate impacts and environmental interactions," *Bulletin of the American Meteorological Society*, vol. 85, pp. 14911–11501, 2004.

[105] J. E. Penner, S. Y. Zhang, M. Chin et al., "A comparison of model- and satellite-derived aerosol optical depth and reflectivity," *Journal of the Atmospheric Sciences*, vol. 59, no. 3, pp. 441–460, 2002.

[106] R. E. Dickinson, R. M. Errico, F. Giorgi, and G. T. Bates, "A regional climate model for the western United States," *Climatic Change*, vol. 15, no. 3, pp. 383–422, 1989.

[107] J. T. Kiehl, J. J. Hack, G. B. Bonan et al., "Description of the ncar community climate model (ccm3)," Tech. Rep. NCAR/TN-420+STR, National Center for Atmospheric Research, 1996.

[108] X. Zeng, M. Zhao, and R. E. Dickinson, "Intercomparison of bulk aerodynamic algorithms for the computation of sea surface fluxes using TOGA COARE and TAO data," *Journal of Climate*, vol. 11, no. 10, pp. 2628–2644, 1998.

[109] R. E. Dickinson, A. Henderson-Sellers, and P. J. Kennedy, "Biosphere-atmosphere transfer scheme (bats) version 1e as coupled to the ncar community climate model," Tech. Rep. NCAR/TN-387+STR, National Center for Atmospheric Research, 1993.

[110] F. Giorgi, R. Francisco, and J. Pal, "Effects of a subgrid-scale topography and land use scheme on the simulation of surface climate and hydrology. Part 1: effects of temperature and water vapor disaggregation," *Journal of Hydrometeorology*, vol. 4, pp. 317–333, 2003.

[111] A. A. M. Holtslag, E. I. F. De Bruijn, and H.-L. Pan, "A high resolution air mass transformation model for short-range

Evaluation of Regional Climatic Model Simulated Aerosol Optical Properties over South Africa Using
Ground-Based and Satellite Observations

151

weather forecasting," *Monthly Weather Review*, vol. 118, no. 8, pp. 1561–1575, 1990.

[112] G. A. Grell, "Prognostic evaluation of assumptions used by cumulus parameterizations," *Monthly Weather Review*, vol. 121, no. 3, pp. 764–787, 1993.

[113] J. M. Fritsch and C. F. Chappell, "Numerical prediction of convectively driven mesoscale pressure systems. Part I: convective parameterization," *Journal of the Atmospheric Sciences*, vol. 37, no. 8, pp. 1722–1733, 1980.

[114] J. S. Pal, E. E. Small, and E. A. B. Eltahir, "Simulation of regional-scale water and energy budgets: representation of subgrid cloud and precipitation processes within RegCM," *Journal of Geophysical Research D*, vol. 105, no. 24, pp. 29579–29594, 2000.

[115] Y. Huang, W. L. Chameides, Q. Tan, and R. E. Dickinson, "Characteristics of anthropogenic sulfate and carbonaceous aerosols over East Asia: regional modeling and observation," *Advances in Atmospheric Sciences*, vol. 25, no. 6, pp. 946–959, 2008.

[116] D. F. Zhang, A. S. Zakey, X. J. Gao, F. Giorgi, and F. Solmon, "Simulation of dust aerosol and its regional feedbacks over East Asia using a regional climate model," *Atmospheric Chemistry and Physics*, vol. 9, no. 4, pp. 1095–1110, 2009.

[117] M. Santese, M. R. Perrone, A. S. Zakey, F. De Tomasi, and F. Giorgi, "Modeling of Saharan dust outbreaks over the mediterranean by RegCM3: case studies," *Atmospheric Chemistry and Physics*, vol. 10, no. 1, pp. 133–156, 2010.

[118] M. Tesfaye et al., "Mineral dust aerosol distributions, its direct and semi-direct effects over South Africa based on regional climate model simulation," submitted to a peer-reviewed journal and currently, 2013, ftp://ftp.csir.co.za/NLC/EE/.

[119] M. Tesfaye et al., "Simulation of anthropogenic aerosols mass distributions and their direct and semi-direct effects over South Africa using RegCM4," submitted to a peer-reviewed journal and currently, 2013, ftp://ftp.csir.co.za/NLC/EE/.

[120] M. Tesfaye et al., "Simulation of biomass burning aerosols mass distributions and their direct and semi-direct effects over South Africa using a regional climate model," submitted to a peer-reviewed journal and currently, 2013, ftp://ftp.csir.co.za/NLC/EE/.

[121] M. Tesfaye et al., "Simulation of bulk aerosol direct andsemi-direct effects in South Africa using RegCM4," submitted to a peer-reviewed journal and currently, 2013, ftp://ftp.csir.co.za/NLC/EE/.

[122] C. Liousse, F. Dulac, H. Cachier, and D. Tanré, "Remote sensing of carbonaceous aerosol production by African savanna biomass burning," *Journal of Geophysical Research D*, vol. 102, no. 5, pp. 5895–5911, 1997.

[123] S. Fuzzi, M. O. Andreae, B. J. Huebert et al., "Critical assessment of the current state of scientific knowledge, terminology, and research needs concerning the role of organic aerosols in the atmosphere, climate, and global change," *Atmospheric Chemistry and Physics*, vol. 6, no. 7, pp. 2017–2038, 2006.

[124] G. Lesins, P. Chylek, and U. Lohmann, "A study of internal and external mixing scenarios and its effect on aerosol optical properties and direct radiative forcing," *Journal of Geophysical Research D*, vol. 107, no. 9-10, pp. 5–1, 2002.

[125] S. E. Bauer and D. Koch, "Impact of heterogeneous sulfate formation at mineral dust surfaces on aerosol loads and radiative forcing in the Goddard Institute for Space Studies general circulation model," *Journal of Geophysical Research D*, vol. 110, no. 17, Article ID D17202, pp. 91–105, 2005.

[126] M. Z. Jacobson, "Strong radiative heating due to the mixing state of black carbon in atmospheric aerosols," *Nature*, vol. 409, no. 6821, pp. 695–697, 2001.

[127] R. C. Moffet and K. A. Prather, "In-situ measurements of the mixing state and optical properties of soot with implications for radiative forcing estimates," *Proceedings of the National Academy of Sciences of the United States of America*, vol. 106, no. 29, pp. 11872–11877, 2009.

[128] G. McFiggans, P. Artaxo, U. Baltensperger et al., "The effect of physical and chemical aerosol properties on warm cloud droplet activation," *Atmospheric Chemistry and Physics*, vol. 6, no. 9, pp. 2593–2649, 2006.

[129] M. Tesfaye, M. Tsidu, V. Sivakumar, and J. Botai, "Effective single scattering albedo estimation using regional climate model," in *Proceedings of the 27th Annual Conference of the South African Society for Atmospheric Sciences: The Interdependent Atmosphere, Land and Ocean, Hartbeespoort*, pp. 53–54, September 2011.

[130] M. Tesfaye, *Retrival of atmospheric aerosol optical and microphysical parameters from ground base passive remote sensing measurement over Addis Ababa [M.S. thesis]*, Addis Ababa University, Addis Ababa, Ethiopia, 2009.

[131] J.-F. Lamarque, T. C. Bond, V. Eyring et al., "Historical (1850–2000) gridded anthropogenic and biomass burning emissions of reactive gases and aerosols: methodology and application," *Atmospheric Chemistry and Physics*, vol. 10, no. 15, pp. 7017–7039, 2010.

[132] G. R. van der Werf, J. T. Randerson, L. Giglio et al., "Global fire emissions and the contribution of deforestation, savanna, forest, agricultural, and peat fires (1997–2009)," *Atmospheric Chemistry and Physics*, vol. 10, no. 23, pp. 11707–11735, 2010.

[133] L. Giglio, J. T. Randerson, G. R. Van Der Werf et al., "Assessing variability and long-term trends in burned area by merging multiple satellite fire products," *Biogeosciences*, vol. 7, no. 3, pp. 1171–1186, 2010.

[134] G. R. van der Werf, J. T. Randerson, L. Giglio, G. J. Collatz, P. S. Kasibhatla, and A. F. Arellano Jr., "Interannual variability in global biomass burning emissions from 1997 to 2004," *Atmospheric Chemistry and Physics*, vol. 6, no. 11, pp. 3423–3441, 2006.

[135] A. Seth and F. Giorgi, "The effects of domain choice on summer precipitation simulation and sensitivity in a regional climate model," *Journal of Climate*, vol. 11, no. 10, pp. 2698–2712, 1998.

[136] X. Wang, Z. Zhong, Y. Hu, and H. Yuan, "Effect of lateral boundary scheme on the simulation of tropical cyclone track in regional climate model RegCM3," *Asia-Pacific Journal of Atmospheric Sciences*, vol. 46, no. 2, pp. 221–230, 2010.

[137] A. Simmons, S. Uppala, D. Dee, and S. Kobayashi, "ERA-Interim: new ECMWF reanalysis products from 1989 onwards," *ECMWF Newsletter*, vol. 110, pp. 25–35, 2007.

[138] D. P. Dee, S. M. Uppala, A. J. Simmons et al., "The ERA-Interim reanalysis: configuration and performance of the data assimilation system," *Quarterly Journal of the Royal Meteorological Society*, vol. 137, no. 656, pp. 553–597, 2011.

[139] R. W. Reynolds, N. A. Rayner, T. M. Smith, D. C. Stokes, and W. Wang, "An improved in situ and satellite SST analysis for climate," *Journal of Climate*, vol. 15, no. 13, pp. 1609–1625, 2002.

[140] B. N. Holben, T. F. Eck, I. Slutsker et al., "AERONET—a federated instrument network and data archive for aerosol characterization," *Remote Sensing of Environment*, vol. 66, no. 1, pp. 1–16, 1998.

[141] O. Dubovik, M. Herman, A. Holdak et al., "Statistically optimized inversion algorithm for enhanced retrieval of aerosol properties from spectral multi-angle polarimetric satellite observations," *Atmospheric Measurement Techniques*, vol. 4, no. 5, pp. 975–1018, 2011.

[142] D. J. Diner, G. P. Asner, R. Davies et al., "New directions in earth observing: scientific applications of multiangle remote sensing," *Bulletin of the American Meteorological Society*, vol. 80, no. 11, pp. 2209–2228, 1999.

[143] J. V. Martonchik, D. J. Diner, R. A. Kahn et al., "Techniques for the retrieval of aerosol properties over land and ocean using multiangle imaging," *IEEE Transactions on Geoscience and Remote Sensing*, vol. 36, no. 4, pp. 1212–1227, 1998.

[144] J. V. Martonchik, D. J. Diner, K. A. Crean, and M. A. Bull, "Regional aerosol retrieval results from MISR," *IEEE Transactions on Geoscience and Remote Sensing*, vol. 40, no. 7, pp. 1520–1531, 2002.

[145] O. V. Kalashnikova and R. Kahn, "Ability of multiangle remote sensing observations to identify and distinguish mineral dust types: 2. Sensitivity over dark water," *Journal of Geophysical Research D*, vol. 111, no. 11, Article ID D11207, 2006.

[146] A. Sharma, V. Sivakumar, C. Bollig, C. Van Der Westhuizen, and D. Moema, "System description of the mobile LIDAR of the CSIR, South Africa," *South African Journal of Science*, vol. 105, no. 11-12, pp. 456–462, 2009.

[147] V. Sivakumar, M. Tesfaye, W. Alemu et al., "CSIR South Africa mobile LIDAR-first scientific results: comparison with satellite, sun photometer and model simulations," *South African Journal of Science*, vol. 105, no. 11-12, pp. 449–455, 2009.

[148] V. Sivakumar, M. Tesfaye, W. Alemu, A. Sharma, C. Bollig, and G. Mengistu, "Aerosol measurements over South Africa using LIDAR, Satellite and Sun Photometer," in *Advances in Geosciences, 16, Atmospheric Science*, chapter 22, pp. 253–262, World Scientific, 2010.

[149] M. Tesfaye, V. Sivakumar, J. Botai et al., "Retrieval of relative humidity from CSIR-NLC mobile LIDAR backscatter measurements," in *Proceedings of the 25th Annual Conference of South African society for Atmosphere Science*, September 2009.

[150] M. Tesfaye, V. Sivakumar, G. Mengistu et al., "Atmospheric Aerosol load morphological classification and retrieved visibility based on lidar backscatter measurement," in *Proceedings of the 25th International Laser Radar Conference*, pp. 487–490, Saint Pietersburg, Russia, 2010.

[151] J. R. Campbell, E. J. Welton, J. D. Spinhirne et al., "Micropulse lidar observations of tropospheric aerosols over northeastern South Africa during the ARREX and SAFARI 2000 dry season experiments," *Journal of Geophysical Research D*, vol. 108, no. 13, pp. 1–33, 2003.

[152] D. E. Terblanche, M. P. Mittermaier, S. J. Piketh, R. T. Bruintjes, and R. P. Burger, "The Aerosol Recirculation and Rainfall Experiment (ARREX): an initial study on aerosol-cloud interactions over South Africa," *South African Journal of Science*, vol. 96, no. 1, pp. 15–21, 2000.

[153] G.-J. Roelofs, H. Ten Brink, A. Kiendler-Scharr et al., "Evaluation of simulated aerosol properties with the aerosol-climate model ECHAM5-HAM using observations from the IMPACT field campaign," *Atmospheric Chemistry and Physics*, vol. 10, no. 16, pp. 7709–7722, 2010.

[154] V. Ramanathan and M. V. Ramana, "Persistent, widespread, and strongly absorbing haze over the Himalayan foothills and the Indo-Gangetic Plains," *Pure and Applied Geophysics*, vol. 162, no. 8-9, pp. 1609–1626, 2005.

[155] P. Ginoux, M. Chin, I. Tegen et al., "Sources and distributions of dust aerosols simulated with the GOCART model," *Journal of Geophysical Research D*, vol. 106, no. 17, pp. 20255–20273, 2001.

[156] T. Y. Tanaka and M. Chiba, "A numerical study of the contributions of dust source regions to the global dust budget," *Global and Planetary Change*, vol. 52, no. 1-4, pp. 88–104, 2006.

[157] X. Yue, H. Wang, H. Liao, and K. Fan, "Simulation of dust aerosol radiative feedback using the GMOD: 2. Dust-climate interactions," *Journal of Geophysical Research D*, vol. 115, no. 4, Article ID D04201, 2010.

[158] E. M. Wilcox and V. Ramanathan, "The impact of observed precipitation upon the transport of aerosols from South Asia," *Tellus, Series B*, vol. 56, no. 5, pp. 435–450, 2004.

[159] Z. Shi, M. D. Krom, T. D. Jickells et al., "Impacts on iron solubility in the mineral dust by processes in the source region and the atmosphere: a review," *Aeolian Research*, vol. 5, pp. 21–42, 2012.

[160] A. Bhattachan, P. Dodorico, M. C. Baddock, T. M. Zobeck, G. S. Okin, and N. Cassar, "The Southern Kalahari: a potential new dust source in the Southern Hemisphere?" *Environmental Research Letters*, vol. 7, no. 2, Article ID 024001, 2012.

[161] Y. J. Kaufman, D. Tanré, O. Dubovik, A. Karnieli, and L. A. Remer, "Absorption of sunlight by dust as inferred from satellite and ground-based remote sensing," *Geophysical Research Letters*, vol. 28, no. 8, pp. 1479–1482, 2001.

[162] I. Tegen and I. Fung, "Contribution to the atmospheric mineral aerosol load from land surface modification," *Journal of Geophysical Research*, vol. 100, no. 9, pp. 18–726, 1995.

[163] M. T. Freiman and S. J. Piketh, "Air transport into and out of the industrial Highveld region of South Africa," *Journal of Applied Meteorology*, vol. 42, no. 7994, 1002 pages, 2003.

[164] Y. Wang, L. R. Leung, J. L. McGregor et al., "Regional climate modeling: progress, challenges, and prospects," *Journal of the Meteorological Society of Japan*, vol. 82, no. 6, pp. 1599–1628, 2004.

[165] J. Heintzenberg and R. J. Charlson, *Clouds in the Perturbed Climate System*, MIT Press, Cambridge, Mass, USA, 2009.

15

Volcanic Ash versus Mineral Dust: Atmospheric Processing and Environmental and Climate Impacts

Baerbel Langmann

Institute of Geophysics, University of Hamburg, Geomatikum, Office 1411, Bundesstraße 55, 20146 Hamburg, Germany

Correspondence should be addressed to Baerbel Langmann; baerbel.langmann@zmaw.de

Academic Editors: A. M. Siani and E. Tagaris

This review paper contrasts volcanic ash and mineral dust regarding their chemical and physical properties, sources, atmospheric load, deposition processes, atmospheric processing, and environmental and climate effects. Although there are substantial differences in the history of mineral dust and volcanic ash particles before they are released into the atmosphere, a number of similarities exist in atmospheric processing at ambient temperatures and environmental and climate impacts. By providing an overview on the differences and similarities between volcanic ash and mineral dust processes and effects, this review paper aims to appeal for future joint research strategies to extend our current knowledge through close cooperation between mineral dust and volcanic ash researchers.

1. Introduction

Volcanic ash represents a major product of volcanic eruptions [1–3]. It is formed by fragmentation processes of the magma and the surrounding rock material of volcanic vents [1, 4]. Depending on the strength of a volcanic eruption, volcanic ash is released into the free troposphere or even the stratosphere [1, 5], where it is transported by the prevailing winds until it is removed from the atmosphere by gravitational settling and wet deposition [6]. Volcanic ash is also known to be mobilised by wind from its deposits [7–12], which have accumulated after volcanic eruptions on land located along the main transport directions of the volcanic cloud, which spreads out over hundreds to thousands of kilometres, dependent on wind speed, ash size, ash density, and eruption magnitude. In contrast to atmospheric mineral dust, the importance of volcanic ash for climate has long been considered negligible [5].

The global mineral dust cycle and its interactions with the Earth's climate system have been studied widely [13–18]. Mineral dust aerosols affect the radiative forcing of the atmosphere directly [13, 19] and indirectly by acting as cloud condensation or ice nuclei [20, 21]. Furthermore, mineral dust aerosols influence ozone photochemistry [22, 23] and supply nutrients to marine [24] and terrestrial ecosystems

[25]. Vice versa, climate variability affects the mineral dust burden of the atmosphere through modifications of precipitation, vegetation cover, and wind [15].

This review contrasts the environmental and climatic effects of volcanic ash versus those of mineral dust. A stronger focus is put on the description of volcanic ash, whereas mineral dust effects are described in less detail, but with referencing the extensive literature. Similarities and differences will be emphasised (Figure 1) to facilitate the different scientific communities studying volcanic ash and mineral dust to learn from each other in an interdisciplinary way, to think about future joint research projects, and to address the important, challenging, and compelling questions, which are still open such as the following:

(i) which physical-chemical processes during long-range transport in the atmosphere affect the surface chemical composition of volcanic ash and mineral dust?

(ii) how important is resuspension of volcanic ash from deposits on land for posteruptive climate and environmental effects?

(iii) what are the reasons for the huge variability of nutrient and toxic element fluxes from volcanic ashes and mineral dust to the surface ocean?

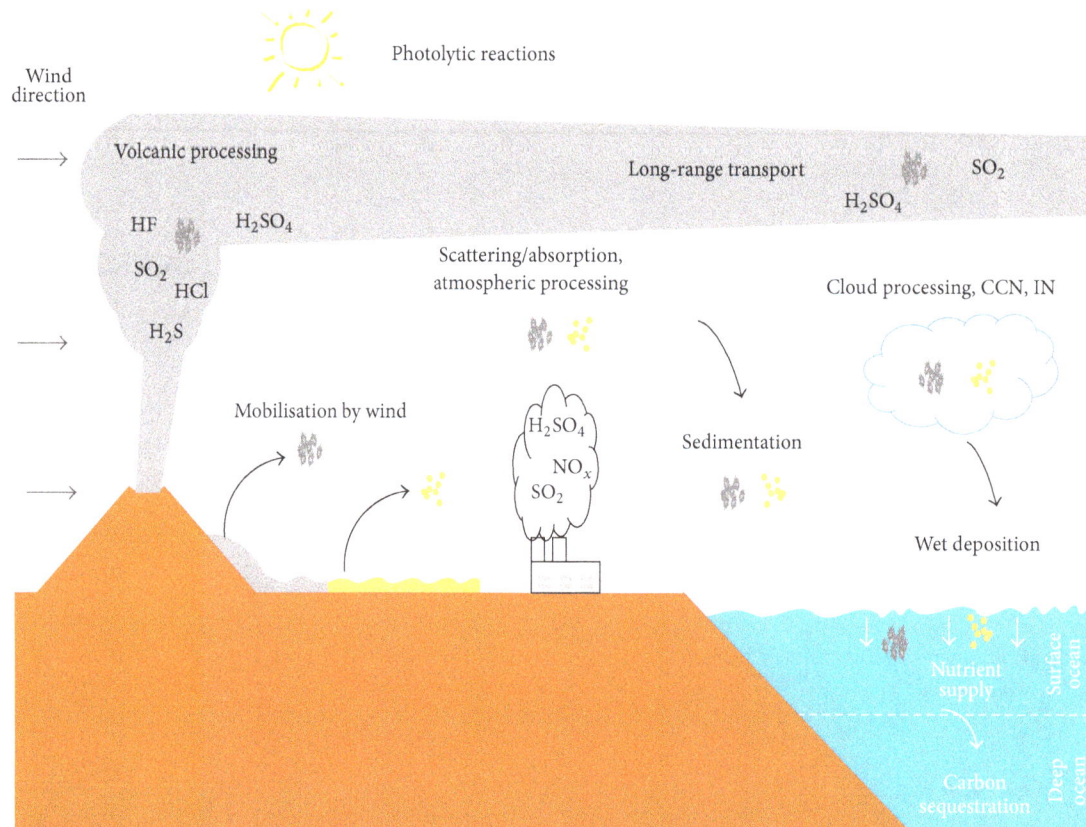

FIGURE 1: Schematic diagram showing the important processes controlling environmental and climate effects of volcanic ash (in grey) and mineral dust (in yellow). (CCN: cloud condensation nuclei; IN: ice nuclei).

(iv) how important is volcanic iron fertilisation of the surface ocean and the associated modifications of atmospheric CO_2 in comparison to that induced by mineral dust?

(v) how relevant is the role of volcanic ash and/or mineral dust for the Earth's climate?

Section 2 gives definitions for mineral dust and volcanic ash and provides information of the general chemical and physical properties. Sources, atmospheric load, and deposition processes are discussed in Section 3, atmospheric processing in Section 4, and environmental and climatic impacts in Section 5. The last section summarises needs for future research.

2. Definitions and Chemical and Physical Properties

2.1. General Definition. According to the "Glossary of Atmospheric Chemistry Terms" [26], dust consists of small, dry, and solid particles released into the atmosphere by natural forces, such as wind, volcanic eruptions, and by mechanical or man-made activities (e.g., crushing, milling, and shoveling). Dust particles are usually in the size range from about 1 to 100 μm in diameter and settle slowly from the atmosphere due to gravity [26]. Thus, mineral dust and volcanic ash may constitute a fraction of all recorded dust.

To distinguish between mineral dust and volcanic ash the following definitions are used: atmospheric mineral dust originates from a suspension of minerals constituting the soil, whereas volcanic ash is loose and unconsolidated material with particle diameters less than 2 mm [27] being either dispersed in the atmosphere or being deposited above the soil. Weathered mineral dust may originate from volcanic tephra (tephra is defined as any fragmental material produced by a volcanic eruption regardless of composition and fragment size; [28]); however, volcanic ash represents relatively fresh material produced during a recent volcanic eruption (recent in this context means no longer than about 100 years ago) and is therefore different form mineral dust.

2.2. Chemical Composition. The main chemical elements contained in mineral dust, as well as volcanic ash, are silicium and oxygen, which constitute the main components of minerals and rocks in the Earth's crust and mantle.

The chemical composition of the bulk volcanic ash is mainly determined by the magma from which it is generated. Generally, three types of magma are distinguished from each other (Table 1). These types of magma have different melting points, viscosities, and typical volatile contents. The mineral composition of volcanic ash consists of about 45–75 wt% of silica [27]. In addition, silicate is the main component of most minerals like feldspar, olivine, pyroxene, hornblende, and biotite [29]. These minerals are formed through successive

TABLE 1: Major types of magma.

Magma type	SiO$_2$ (wt%)	T_{melt} (°C)	Viscosity and gas content
Basaltic	45–55	1000–1200	Low
Andesitic	55–65	800–1000	Intermediate
Rhyolitic	65–75	650–1000	High

crystallisation during cooling and decompression when the magma ascends from the Earth's mantle through the Earth's crust into the conduit and subsequently into the volcanic plume [28]. During the crystallisation process, the composition of the melt is changing due to depletion of crystallised components and enrichment of remaining components driving the successive generation of different minerals, including those without silicate, such as magnetite or ilmenite.

A major difference in the mineral composition of mineral dust and volcanic ash results from chemical weathering of mineral dust, generally on geological time scales. Mineral composition changes under the influence of water, oxygen, and acids. For example, feldspar will weather to clay and iron-bearing minerals can form hematite and goethite. Therefore primary iron containing minerals (e.g., amphiboles and pyroxenes) of volcanic rocks are not identified in mineral dust [30]. Further multiphase chemical modifications at the surfaces of mineral dust and volcanic ash take place during atmospheric transport (see Section 4).

2.3. Physical Properties

2.3.1. Size Distribution. Close to the source regions, the size distribution of mineral dust particles varies from about 0.1 to over 100 μm in diameter, depending on soil characteristics and wind speed [18]. The particle size spectrum of emitted mineral dust largely controls the fraction that is transported over large distances. The coarser the particles, the faster they are deposited. However, to characterise long-range transported mineral dust, precise particle size measurements for less than about 10 μm are still a challenge [31].

The size distribution of volcanic ash is greatly dependent on the formation process, which is either an explosive volcanic eruption, a phreatomagmatic eruption, or a pyroclastic density current. In addition, secondary volcanic ash clouds result from the resuspension from volcanic ash depositions on land (see Section 3.1), which should not be mixed up with coignimbrite clouds as discussed later. Explosive volcanic eruptions occur when magma containing dissolved volatiles rises in the conduit. Thereby exsolution of volatiles forms gas bubbles that grow by diffusion, decompression, and coalescence. The further the magma-gas mixture rises, the more the pressure decreases, leading to an acceleration of the mixture against gravitational and friction forces, until a continuous gas stream with clots and clasts of magma (called pyroclasts) leaves the vent explosively [1]. The explosive character of a volcanic eruption depends considerably on the viscosity of the magma. In general, most efficient fragmentation occurs during explosive eruptions where magmas of rhyolitic composition are involved because of the higher volatile content (Table 1).

Phreatomagmatic eruptions [1] are triggered by the interaction of external water with magma, for example, from a glacier as during the early phase of the Eyjafjallajökull eruption [32]. External water may also be supplied from crater lakes or even the shallow ocean during seamount volcanic eruptions [33]. Highly efficient fragmentation is caused by thermal contraction of magma due to chilling on contact with water [4]. Water initially chills the magma at the interface, which then shatters. The water penetrates the mass of shattered hot glass and is transformed into high-pressure superheated steam by a runaway process of heat transfer and further magma fragmentation, until a violent explosion results. Violent phreatomagmatic eruptions produce especially fine-grained volcanic ash.

Pyroclastic flows occur when the eruption column or lava dome collapses leading to gas and tephra flows rushing down the flanks of a volcano at high speed, which thereby also contribute to the fragmentation process through milling by the collisional processes [1, 34]. Coignimbrite clouds can arise from pyroclastic flows when the material at the top of a pyroclastic flow gets more buoyant than the surrounding air. These convective clouds can form volcanic plumes as high as the original feeding plume and are a source of substantial amounts of fine volcanic ash as well.

As eruption conditions may be highly variable in time, all fragmentation processes can take place simultaneously. According to [2] volcanic ash particles with diameters smaller than 1 mm contribute about 55–97 wt% to the total ash content. Volcanic ash particles with diameters less than 30 μm make up only a few weight percent during basaltic eruptions, whereas they can contribute 30–50 wt% to the total ash content during rhyolitic eruptions. However, it is the even finer particle size fraction (PM$_{10}$ and PM$_{2.5}$) that may be carried for hundreds of kilometres before settling onto land or into the ocean.

2.3.2. Density and Surface Area. The morphology of mineral dust particles can be assumed to be spherical with a widely used particle density of 2650 kg/m^3 in mineral dust modelling, for example, [35]. The density of individual volcanic ash particles varies from one eruption to another and even during an eruption. Generally it lies between 2000 and 3000 kg/m^3 [28, 36] dependent on the basaltic or rhyolitic composition, the amount of crystallisation, and porosity [37]. Due to the expansion of magmatic gases like H$_2$O, CO$_2$, SO$_2$, H$_2$, CO, H$_2$S, HCl, and HF [38] during an explosive and phreatomatic volcanic eruption (see Section 2.3.1), volcanic ash generally consists of vesicular particles with an undifferentiated surface texture. Generally, the specific surface area of volcanic ash is smaller than 2 m^2/g [39]. However, values up to 10 m^2/g have also been reported [39].

3. Emissions into the Atmosphere, Atmospheric Load, and Subsequent Deposition

3.1. Emissions. The notation "emission" describes the release of material from outside the atmosphere into the atmosphere, where the location outside the atmosphere represents a source for the atmosphere. Mineral dust source areas are generally located in semiarid or arid areas where the surface is sparsely vegetated and dry. Here, fine grained material can accumulate and be mobilised into the atmosphere by wind. Numerical models for mineral dust mobilisation usually define dust emission areas based on, for example, soil moisture [40], soil texture [13], and vegetation effects [13, 41]. Mineral dust emissions into the atmosphere are a complex, nonlinear function of both soil surface properties (size distribution of the surface soils, roughness length of erodible and nonerodible particles, and soil moisture) and meteorological conditions (wind friction velocity and precipitation). Mineral dust emissions from an erodible surface occur when the wind friction velocity exceeds a threshold value, dependent on the soil properties [42]. However, by using different assumptions, for example, erodibility factors, numerical model estimates of global mineral dust emissions vary between 1500 and 1800 Tg/yr [16, 43, 44]. Injection heights are usually restricted to the planetary boundary layer (2–4 km) but may reach up to 6 km dependent on meteorological conditions [45]. Although mineral dust is usually considered of natural origin, it is estimated that about 30% of the mineral dust load in the atmosphere could be ascribed to human activities through desertification and land misuse. The Sahara desert is the major source of mineral dust, which subsequently spreads across the Mediterranean and European region, across the Caribbean Sea, and towards Central and North America. Additionally it plays a significant role in the nutrient inflow to the Amazon rainforest [25]. The Gobi Desert is another important source of dust in the atmosphere, which affects eastern Asia and western North America.

Volcanic ash is formed during explosive volcanic eruptions, phreatomagmatic eruptions, or pyroclastic density currents (see Section 2.3.1). On average about 20 volcanoes erupt at any given time worldwide, 50–70 volcanoes erupt throughout a year, and at least one large eruption with a Volcanic Explosivity Index (VEI, relative measure of the explosiveness of volcanic eruption; [46]) greater than 4 occurs annually [47]. The total emissions of volcanic ash into the troposphere by small volcanic eruptions with VEI < 4 (these eruptions make up the majority in number) is estimated to be 20 Tg/yr [48], equivalent to 10 km^3 when assuming a particle density of 2000 kg/m^3. However, these volcanic ash emissions are usually removed from the atmosphere quickly and are therefore only of local interest in the vicinity of the volcanoes up to a distance of about one hundred kilometers. Table 2 summarises ash emissions from major volcanic eruptions, lasting from a few hours to several weeks since 1900 with VEI ≥ 4 with a tephra release ranging from 0.1 to 100 km^3. It is obvious, that stronger volcanic eruptions are generally less frequent [28]. Note that before the satellite era starting about 1980, our knowledge on volcanic eruptions with VEI ≤ 4 is probably not complete due to limited observations of remote volcanic eruptions. Dependent on meteorological conditions and the injection height of the volcanic emissions, volcanic ash from eruptions with VEI ≥ 4 may be transported over thousands of kilometres in the atmosphere [49]. Therefore, volcanic ash, although released sporadically during volcanic eruptions, is an abundant atmospheric species.

It should be considered as well that fresh volcanic ash may be remobilised into the atmosphere from ash deposits on land, similar to what is observed for mineral dust, particularly in arid and semiarid regions. This contradicts with the general assumption that volcanic ash environmental and climate effects are restricted only to the duration of a volcanic eruption with time scales of days to weeks. Reference [9], for example, reports such posteruptive volcanic ash clouds being transported over the Patagonian desert for several months to years after the 1991 eruption of Mt. Hudson in Chile (Figure 2). Following the recent eruption of Eyjafjallajökull in Iceland during April/May 2010, volcanic ash remobilisation created poor air quality and health concerns for the local population for several months [10–12]. References [8, 9] also report volcanic ash from resuspensions events from the Katmai/Novarupta eruption in 1912, which typically occur in the fall before snowfall. The June 6–8, 1912 Katmai/Novarupta eruption was the largest volcanic eruption in the 20th Century and produced volcanic ash deposits of 1–10 meters around the volcano. A lack of snow combined with strong northerly winds is able to mobilise the hundred years old volcanic ash into the atmosphere even nowadays (Figure 2).

3.2. Atmospheric Load. Atmospheric concentrations of mineral dust and volcanic ash are subject to considerable temporal and spatial variability. Seasonal variability, for example, rainy and dry seasons, determines to a great extent the mineral dust load in the atmosphere, whereas volcanic ash atmospheric load is mainly dependent on the occurrences of sporadic and usually unpredictable volcanic eruptions.

Measurements in the Sahelian belt of West Africa [50] during 2006 to 2008 reveal median daily mineral dust concentrations of around 80 μg/m^3 with about 40% exceeding 100 μg/m^3 and less than 3% exceeding 500 μg/m^3. The maximum measured daily concentrations range is between 2250 and 4020 μg/m^3 [50]. The European standard for air quality (daily mean PM10 concentration of 50 μg/m^3 should not be exceeded for more than 35 days per year) is exceeded in this area by mineral dust aerosols about 200 days per year [50]. In the northern part of the Taklamakan desert a higher interannual variability with maximum daily mineral dust concentrations of 645–3800 μg/m^3 was measured between 2001 and 2004 [51]. An even higher variability is reported for the Inner Mongolia region [52] with PM10 concentrations of 190–9625 μg/m^3. Downwind from main mineral dust sources, concentrations are usually smaller, for example, on Cape Verde Islands between 65 and 264 μg/m^3 during mineral dust transport episodes [53].

During the eruption of Eyjafjallajökull on Iceland in 2010, maximum ash concentrations up to 4000 μg/m^3 are reported from measurements of the volcanic ash cloud spreading over Europe [54, 55], exceeding the threshold for safe aviation (2000 μg/m^3). Daily mean near surface concentrations reached up to 400 μg/m^3 in Scandinavia during the eruption [54]. Close to Eyjafjallajökull, the maximum daily average near surface concentration exceeded 1230 μg/m^3 during the ongoing eruption, but also after the eruption stopped, maximum daily average concentration reached

TABLE 2: Tephra mass release in DRE (Dense Rock Equivalent) of well-known volcanic eruptions since 1900 given for VEI values from 4 to 6 in three mass ranges: 0.1–1 km^3, 1–10 km^3, and 10–100 km^3 ([47]; http://www.volcano.si.edu/).

0.1 km^3	1 km^3	10 km^3	100 km^3
VEI = 4	VEI = 5	VEI = 6	
Nabro 2011 and Puyehue-Cordón Caulle 2011			
Grimsvötn 2011 and Merapi 2010			
Eyjafjallajökull 2010 and Sarychev Peak 2010			
Kasatochi 2008 and Chaiten 2008			
Reventator 2002 and Ulawun 2000			
Lascar 1993 and Mt. Spurr 1992			
Kelud 1990 and Kiluchevkoi 1987	Mt. Hudson 1991	Pinatubo 1991	
Chikurachki 1986 and Mount Augustine 1986			
Colo 1983 and Galunggung 1982	El Chichon 1982		
Pagan 1981 and Alaid 1981			
Mount Augustine 1976 and Tolbachik 1975	Mt. St. Helens 1980		
Volcan de Fuego 1974 and Tiatia 1973			
Fernandina 1968 and Mount Awu 1966			
Kelud 1966 and Taal 1965			
Shiveluch 1964 and Carran-Los Venados 1955	Agung 1963		
Mount Spurr 1953 and Bagana 1952	Bezymianny 1956		
Kelud 1951 and Mount Lamington 1951			
Ambrym 1950 and Hekla 1947			
Sarychev Paek 1946 and Avachinsky 1945			
Paricutin 1943–1952 and Suoh 1933			
Volcan De Fuego 1932 and Mont Aniakchak 1931	Kharimkotan 1933		
Kliucheskoi 1931 and Komagatake 1931	Cerro Azul 1932		
Komagatake 1929 and Avachinsky 1926			
Raikoko 1924 and Manam 1919			
Kelud 1919 and Agrhan 1917	Katla 1918		
Tungurahua 1916 and Sakurajima 1914			
Mount Lolobau 1911 and Grimsvötn 1903	Colima 1913	Katmai/Novarupta 1912	
Monut Pelee 1902	Ksudach 1907	St. Maria 1902	

more than 1000 μg/m^3 during resuspension events [12]. Volcanic ash concentrations within the volcanic eruption plume are expected to be even higher. However, measurements are difficult to obtain as saturation levels are reached by remote sensing instruments and direct measurements destroy measurement equipment and are too dangerous for humans to approach too close to the eruption. Altogether, the magnitudes of maximum atmospheric concentrations that may be reached during mineral dust storms and volcanic eruptions are relatively similar, although for volcanic eruptions with VEI $\geq$ 4 higher atmospheric ash concentrations are expected.

3.3. Deposition. Mineral dust and volcanic ash are removed from the atmosphere by gravitational settling, turbulent dry deposition, and wet scavenging by rain called wet deposition

[56]. Dry and wet aggregation of volcanic ash belongs per definition to the removal processes of dry and wet deposition [6]. However, because of their specialty they are discussed separately later. The ratio of dry-to-wet deposition differs considerably in mineral dust model estimates [19], with wet deposition over the ocean ranging from 30% to 95% of the total mineral dust deposition [18]. It should be noted that all uncertainties of the global mineral dust cycle, including emissions (see Section 3.1) and transport in the atmosphere, are summed up in the distribution and amount of deposition fluxes. Mineral dust deposition to ice cores reveals an up to factor 100 difference between cold and warm periods in the geological past [16], for example, much higher deposition during the Last Glacial Maximum, 21,000 years before present, compared to the present-day climate.

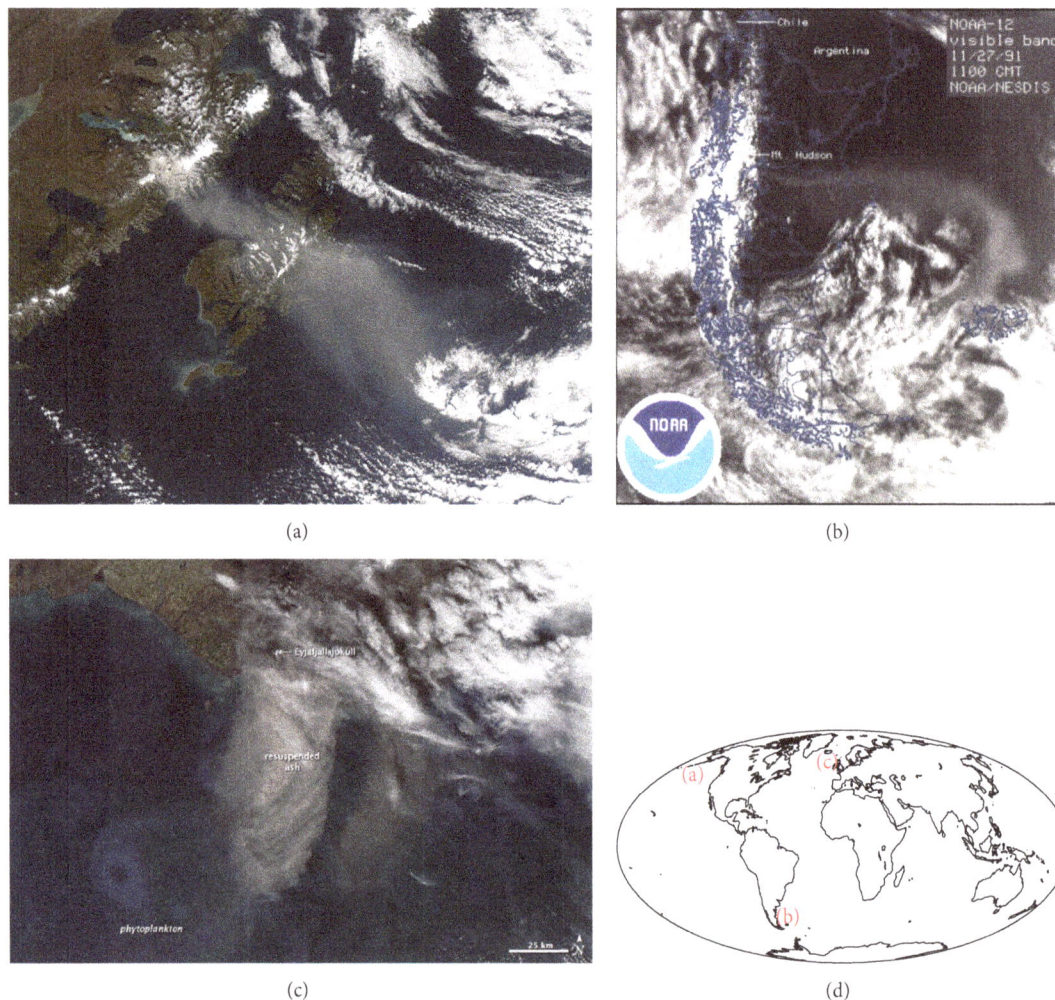

FIGURE 2: Volcanic ash resuspension event as seen from satellite: (a) September 21, 2003: Katmai/Novarupta, Alaska; (b) November 27, 1991: Cerro Hudson, Chile; (c) May 27, 2010: Eyjafjallajökull, Iceland; (d) approximate location of (a)–(c). Courtesy of NASA.

Reference [57] estimated the millennial scale flux of volcanic ash into the Pacific Ocean based on the marine sediment core data [58, 59] to be about 128–221 Tg/yr. This represents a conservative estimate, as already one volcanic eruption of VEI = 4 like Kasatochi in 2008 [60] can produce higher deposition fluxes. The estimated millennial volcanic ash flux is comparable to the mineral dust flux into the Pacific Ocean of around 100 Tg/yr [16]. So far, however, the importance of volcanic eruptions on climate, for example, by modifying the biogeochemistry of the surface ocean, has gained limited attention compared to the much better investigated effects of mineral dust. The amount of volcanic ash and bioavailable iron attached to the ash surface deposited into the ocean during episodic large volcanic eruptions may exceed the annual mineral dust flux by far. Reference [61], for example, estimated that iron deposition of the Mt. Hudson's volcanic eruption in Chile during August 12–15, 1991 is equivalent to ~500 years of Patagonian iron dust fallout.

Gravitational settling of volcanic ash has been observed to exceed the terminal settling velocity of single ash particles [6, 62]. This is explained by the formation of aggregates in the volcanic plume, as well as under more diluted conditions in the volcanic cloud [63–65]. Volcanic ash aggregation therefore represents a process that increases sedimentation and reduces atmospheric volcanic ash concentration during long-range transport. However, key aggregate formation processes and basic classifications are topics of ongoing debate. Several important questions remain unanswered up to now. Is aggregation driven primarily by hydrometeor formation? How does aggregation vary in time and space? What is the role of electrostatic charge and "secondary minerals"? How do instabilities (e.g., mammatus) change deposition rates? What proportion of fine-grained ash ends up in aggregates? Where does particle aggregation mainly occur (e.g., vertical plume, horizontal cloud, or during atmospheric sedimentation)? A better understanding of volcanic ash aggregation will be necessary to improve the modelling of volcanic ash dispersion and deposition.

4. Atmospheric Processing

4.1. Chemical Processing in Volcanic Plumes. Before atmospheric processing occurs at ambient temperatures, volcanic ash undergoes extreme temperature gradients (from about

$1000°C$ to less than $0°C$) in extreme short periods of time (few minutes) in the volcanic eruption plume [1, 66, 67], which is expanding vertically into the atmosphere from the vent to the level of neutral buoyancy (Figure 1). Besides fragmentation processes taking place here (see Section 2.3.1), quenching represents an important process in the production of the glass material contained in volcanic ash together with minerals formed by incomplete crystallisation reactions (see Section 2.2). Intensive lightning in the volcanic eruption plume is an often-observed phenomenon [68, 69] due to vertically separated regions of oppositely charged volcanic ash particles. Aggregation processes of volcanic ash (see Section 3.3) may be affected by the charged volcanic ash particles. Until now, the temperature and ionising effects of lighting strokes on volcanic ash chemical composition have not been investigated. A number of potentially important processes for physical-chemical modifications of volcanic ash surfaces, without considering lightning, are discussed in the literature, as summarised below. Volcanic eruption plumes cool significantly during rise from about $1000°C$ at the vent to ambient temperature leading to various homogeneous and heterogeneous chemical and microphysical modifications on the ash surfaces. During volcanic eruptions, large amounts of volatiles [1, 38, 66] are released into the atmosphere along with volcanic ash. Through the interaction between these gases and secondary aerosols produced from these gases with volcanic ash within the eruption plume, it is assumed that soluble compounds are produced on the volcanic ash surfaces [70–72], scavenging up to 30%–40% of the sulfur and 10%–20% of the chlorine released from volcanic eruptions [70, 73]. Reference [71] introduced the idea that scavenging of volatiles by ash within eruption plumes occurs in three temperature-dependent zones: (1) the "salt formation zone" representing the hot core of the eruption plume where sulfate and halide salt aerosols, which were formed at near magmatic temperatures, are adsorbed onto ash particles; (2) in the "surface adsorption zone" halogen gases react directly with the surface of ash during the cooling of the plume until temperatures of about $700°C$ are reached; (3) the "condensation zone" is characterised by the formation of sulfuric and halogen acids at temperatures below $338°C$.

Leaching experiments with pristine volcanic ash in water have been performed for decades [74] revealing the release of various sulfate and halide compounds in addition to biologically relevant elements such as N, P, Si, Fe, Cd, Co, Cu, Mn, Mo, Pb, and Zn [75–77]. Several studies suggest adsorption of volcanic salts on volcanic ash surfaces as the main mechanism for the production of soluble compounds on volcanic ash [78, 79]. Other studies emphasise that condensation of sulfuric acid onto volcanic ash may drive dissolution reactions, thereby providing the source of the soluble cations measured in ash leachates [70]. Different to these assumptions, measurements of [80] point to a rapid acid dissolution of the ash surface material within eruption plumes followed by precipitation of secondary minerals and salts at the ash-liquid interface. Reference [80] assumes adsorption of volcanic salts to represent a process of minor importance. Reference [81] reports about the high-temperature scavenging of volcanic SO_2 by volcanic ash

with potential important modifications of the ash surfaces under cooler conditions. Only a few of these aspects have been explored so far by using complex plume models, for example, [82]. Despite the progress made in recent years, the physical-chemical mechanisms, which govern the modification of the surface composition of volcanic ash during its transit through the volcanic eruption plume, where large temperature (from about $1000°C$ to ambient temperature) and chemical gradients prevail, are poorly understood [83] and require further investigations. A further understanding of these processes under such extreme conditions may also help to increase our knowledge of atmospheric processing of mineral dust, even though mineral dust modifications are restricted to ambient temperatures.

4.2. Chemical Processing at Ambient Atmospheric Temperatures. After a volcanic plume reaches neutral buoyancy conditions, the volcanic cloud spreads out more horizontally (Figure 1). In the volcanic cloud further chemical modifications of volcanic ash surfaces at ambient temperatures, for example, during long-range atmospheric transport at high SO_2/sulfate concentrations, may also occur. Although the volcanic emissions are already diluted in the volcanic cloud, the acidic environment in volcanic clouds, which is dependent on the volcanic release rate of SO_2 [84, 85] (Figure 3) and other acidic substances like HCl and HF, may exceed by far that for mineral dust acid processing in anthropogenic polluted air masses [86]. However, volcanic ash, volcanic gases, and their oxidation products (e.g., H_2SO_4) are not necessarily released into similar atmospheric altitudes during a volcanic eruption (e.g., 1991 eruption of Pinatubo; [87]), resulting in different altitudes for the major dispersion pathways of volcanic ash and gases and limited acid processing during atmospheric transport. Furthermore, due to rapid sedimentation, a separation of volcanic ash particles from volcanic gases and secondary aerosols like sulphate occurs within some hours to days (dependent on the volcanic ash size distribution), even when the eruption height for all volcanic emissions is very similar as what was observed during the eruption of Kasatochi in 2008 [49].

In contrast, volcanic ash, which is remobilised from ash deposits, can be assumed to undergo very similar atmospheric processing as mineral dust. The importance of photochemistry for mineral dust under atmospheric conditions is highlighted in several studies [22, 23, 88, 89]. Mineral dust particles can act as a sink for SO_2, thus enabling the formation of sulfate on the mineral dust surfaces [90]. When mineral dust concentrations are low, they may trigger the nucleation of new sulfate particles via a series of photochemical reactions involving the mineral dust surface [91]. Metal oxides present in mineral dust act as atmospheric photocatalysts promoting the formation of gaseous OH radicals, which initiate the conversion of SO_2 to H_2SO_4 in the vicinity of dust particles. Comparable results for volcanic ash have also been measured (Dupart, personal information, 2011).

At ambient temperature, when clouds are present in the atmosphere, cloud processing is assumed to provide the main mechanism for the uptake of acid gases in the atmosphere by aerosols in general [56], including mineral dust [92–94]

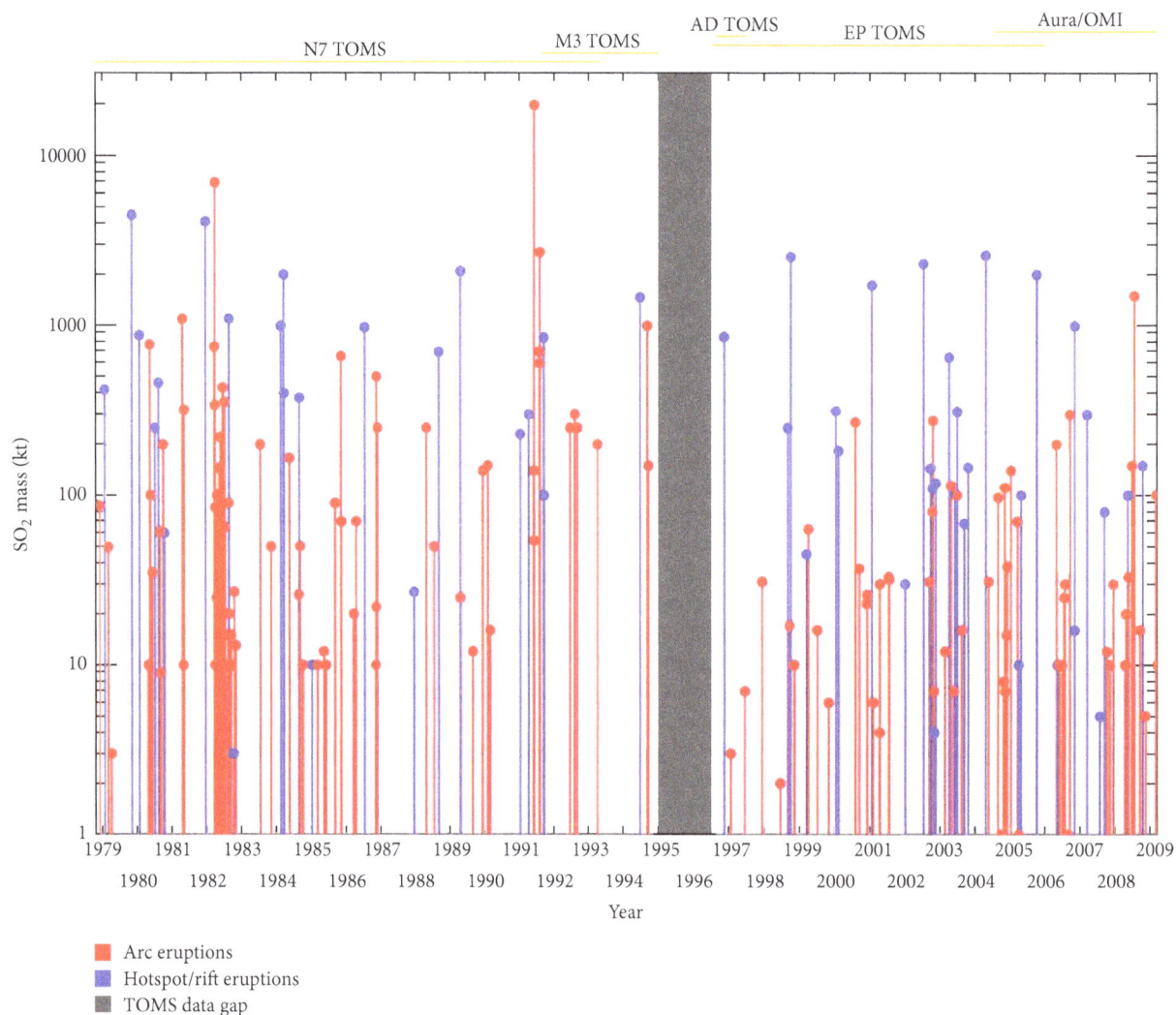

FIGURE 3: SO_2 emissions from major volcanic eruptions observed from satellite. Arc eruptions with SO_2 emissions exceeding 1000 kt: Mt. St. Helens 1980 (slightly less), Alaid 1981, El Chichon 1982, Mt. Pinatubo 1991, Mt. Hudson 1991, Raboul 1994, and Kasatochi 2008 (Courtesy of NASA).

and volcanic ash particles [95]. During atmospheric transport of mineral dust and volcanic ash particles, clouds often evaporate leaving only a thin film of aqueous electrolyte around each particle [56]. This film of aerosol water is very acidic compared to the cloud droplet, reaching pH values of 2 or even lower [86, 96]. As clouds can form and evaporate several times [56], five to ten cycles of pH alternation in the aqueous film around mineral dust and volcanic ash particles can occur before these particles are deposited via wet deposition or sedimentation out of the atmosphere [30]. Atmospheric processing is widely accepted to represent a key process for iron solubility in mineral dust [30, 97, 98]. Limited understanding of these processes hinders the development of accurate biogeochemical models predicting the impact of mineral dust and volcanic ash deposition on the chemistry of trace metals in the surface ocean and ultimately on the global carbon cycle (see Section 5.4.3).

Atmospheric processing of mineral dust and volcanic ash particles at freezing temperatures is a process, which

has been rarely studied. However, indications for increased iron solubility are presented [99]. Physical-chemical modifications of volcanic ash surfaces in the volcanic plume and/or volcanic cloud may also affect volcanic ash aggregation (see Section 3.3) [6], which in turn influences particle sedimentation from the plume and during long-range transport, thereby affecting the residence time of volcanic ash in the atmosphere.

Atmospheric and volcanic processing with modifications of the surface chemical composition of mineral dust and volcanic ash particles has implications on their behaviour to act as cloud condensation or ice nuclei (CCN or IN; see Section 5.4.2) and releasing nutrients (in particular iron) on contact with seawater (see Section 5.4.3). Observational studies in the field and the laboratory have confirmed that mineral dust and volcanic ash particles can act as CCN or IN, in particular, after being aged in the atmosphere [20, 21, 95, 100–102]. If the surface of volcanic ash has been partially dissolved, water uptake may be favoured [39], thereby explaining the ability of volcanic ash to act as cloud condensation and ice nuclei.

For the sake of completeness, mechanical and biogeo-chemical weathering that takes place at the Earth's surface under atmospheric conditions is mentioned here as well. These processes are important in the generation of mineral dust and decomposition of volcanic ash on geological time scales [103]. However, they are beyond the scope of this review. The interested reader is referred to the extensive litera-ture, for example, [104], which provides detailed information on weathering processes and important control variables, such as water, oxygen, and acids, as well as lithology, mor-phology, soil, ecosystems land use, temperature, and runoff.

5. Environmental and Climate Impacts

5.1. Human Health. Aeolian dust episodes represent a major health concern for humans due to elevated atmospheric concentrations (see Section 3.2). In urban centres in Asia, mineral dust episodes have become a growing concern for the health of populations and ecosystems due to mixing with pollution aerosols [105, 106]. With higher levels of exposure, widespread chronic respiratory and lung diseases, asthma, allergic alveolitis, and eye irritations are well-documented health effects [107]. Human exposure to high volcanic ash concentrations may create similar health effects as those related with mineral dust. However, due to its sharp surface structures, small ash particles have additional mechanical effects; for example, they can abrade the front of the eye under windy conditions [108] (Horwell and Baxter, 2006) and lead to silicosis in the worst case.

5.2. Aviation. Strong mineral dust storms mainly affect take-off and landing of aircraft due to poor visibility. However, aircraft can operate in environments with high mineral dust concentration without any engine problems. Mineral dust which typically melts at temperatures of around 1700°C does not melt when it is ingested in jet engines. However, jet engines may have problems with volcanic ash as the melting temperatures are at or below the operating temperatures of high-performance jet engines, which are around 1400°C [109]. The molten material deposited on the cooler parts of the engines can cause flame-outs, which may also result from glass shards when temperature exceeds their glass transition temperature. 129 flights in the last 60 years were affected by volcanic ash [110], for example, during the erup-tion of Galunggung, Indonesia (1982) or Redoubt, Alaska (1989/1990). In nine cases, one or more engines temporarily failed due to the melting of ash in the jet engine turbine [110]. To prevent dangerous flights in the presence of volcanic ash, Volcanic Ash Advisory Centres (VAACs) have been installed with the mandate to regularly publish warnings on the location of volcanic ash in the atmosphere. After the eruption of Eyjafjallakökull in Iceland in 2010, a growing demand developed, suggesting that VAACs should not only follow the zero tolerance rule, but also provide absolute volcanic ash concentrations [32, 111], with a preliminary threshold of 2 mg/m³ for safe aircraft operation.

5.3. Soil Fertilisation. Heavy volcanic ash or mineral dust deposition completely buries vegetation and soil. Plant sur-vival is dependent on deposit thickness, chemistry, com-paction, rainfall, and duration of burial. Slight deposition of volcanic ash and mineral dust can affect vegetation and soil positively and negatively. Although thin volcanic ash fall inhibits transpiration and photosynthesis and alters growth, buried plants may survive [37]. A positive effect is attributed to mineral dust originating from the Sahara desert, which is regularly transported across the Atlantic Ocean and is argued to represent the main mineral fertiliser of the Amazon region [25]. A positive effect of volcanic ash deposition is an increase in agricultural production due to mulching, for example, following the 1980 Mount St. Helens eruption [112]. Other positive effects are due to the leaching of nutrients released from volcanic ash [28], for example, following the 1995/96 Mt. Ruapehu eruption. In particular, nutrient-poor soils benefit from volcanic ash leaching; however, soluble toxic elements may also be washed-out [37].

5.4. Climate

5.4.1. Direct Radiative Effects. Short-term effects during mineral dust storms, volcanic eruptions, and volcanic ash resuspension events considerably reduce visibility and solar irradiation reaching the Earth's surface, whereas long-term effects occur in a more diluted environment. These effects can be measured by an increase in atmospheric optical depth, represented by enhanced absorption and/or scattering of solar and thermal radiation and modifications in surface temperature. During the first two days following the 1980 eruption of Mount St. Helens, [113] reported a decrease in daytime surface temperature of 8°C induced by volcanic ash absorption of solar radiation in the ash affected area downwind of Mt. St. Helens. Here, the colour of volcanic ash pays an important role; low-Si, high-Fe ash tends to develop a dark brown and black colouration, whereas high-Si ash with low Fe ash tends to appear pale and white coloured [37]. The colour of volcanic ash is also an important factor for the modification of the surface albedo for volcanic ash deposits covering the soil. Reference [112] estimated that the light-coloured volcanic ash from Mount St. Helens reflected two to three times more incoming solar radiation than ash-free soils. In contrast, dark-coloured ash will decrease the surface albedo. A modification of the surface albedo will also occur after volcanic ash deposition on snow or ice. In most cases a reduction is observed, although [114] argues that reflectivity measurements of dry volcanic ash can show albedo values as high as snow. Therefore, large areas covered by volcanic ash may have considerable implications for climate [114].

Direct radiative effects of mineral dust have been studied widely [13, 19, 31, 115–117]. Reference [111] compared optical properties of volcanic ash from the Eyjafjallajökull eruption in 2010 with Sahara dust. Although the optical properties of volcanic ash and mineral dust are relatively similar, the imag-inary part of the refractive index shows weaker absorption and drops stronger between the blue and red parts of the spectrum for mineral dust when compared with volcanic ash. Another difference in optical properties was observed for the polarisation properties, with slightly stronger forward and lower backward scattering for volcanic ash. These differences allow discrimination between mineral dust and volcanic ash during, for example, lidar measurements [118]. However, atmospheric processing (see Section 4) may modify the radiative properties of both, mineral dust [119] and volcanic ash, although further investigations are needed.

Due to the relatively short residence time of mineral dust and volcanic ash in the atmosphere (in the order of a few days), their direct radiative effects (and indirect radiative effects; see Section 5.4.2) exhibit a high temporal and spatial variability, with major effects around the main source regions and the main transport pathways from a few kilometers to some thousand kilometres. In these regions, however, the direct radiative effects of mineral dust and volcanic ash may be dominant.

5.4.2. Indirect Radiative Effects. With about 60% of the Earth being cloud covered, clouds represent an important factor in regulating the Earth's radiation budget [120]. It has been estimated that a 5% increase of the shortwave cloud forcing could compensate the radiative effect due to increased greenhouse gases between the years 1750–2000 [107]. Cloud formation, lifetime, and radiative properties are affected by aerosols. These so-called indirect aerosol effects are related to changes of the cloud droplet and ice spectrum.

In general, a greater quantity of cloud droplets are formed with typically smaller size if more aerosols are available to act as cloud condensation nuclei (CCN). High cloud droplet number concentrations (CDNC) reduce the diffusional droplet growth. Therefore droplet sizes cannot be reached which are large enough for an efficient growth by droplet collision. Thus, changes in CDNC can influence cloud albedo (first indirect aerosol effect) [121], cloud lifetime, and precipitation formation (second indirect aerosol effect) [122]. In the tropics, the modification of precipitation formation in deep convective clouds is of special importance. The local effect near the aerosol source regions may lead to suppression of precipitation. However, as more liquid water and water vapour stays in the atmosphere, precipitation will be formed elsewhere and the potential of flooding and erosion is increased. Global circulation can be influenced by the release of latent heat due to condensation of water vapour. Therefore, aerosols including mineral dust and volcanic ash, which act as cloud condensation [20, 123] and ice nuclei (see later), have the potential to affect the Earth's radiation budget, the hydrological cycle, and regional or even global circulation [124]. Satellite observations in the South Atlantic and North Pacific [123] show that either natural degassing or weakly explosive volcanoes affect low marine stratocumulus for up to 1300 km downwind by decreasing the effective radius of droplets and increasing visible brightness, which may add cloud cover in otherwise cloudless areas.

Super-cooled clouds are abundant in the atmosphere, which contain metastable water that freezes as soon as suitable ice nuclei are available. In the presence of particulate material, such as mineral dust, volcanic ash, or pollen [95, 100, 125–130], it may undergo heterogeneous nucleation, where freezing may be initiated at significantly lower

supersaturations and higher temperatures than in the case of homogeneous freezing [125]. Upon glaciation, the size distribution and the lifetime and radiative forcing of the clouds are modified. A negative correlation between super-cooled clouds (at $-20°C$) and the occurrence of mineral dust has been found by [131], likely due to glaciation by dust. Reference [100] measured ice formation in the presence of fine volcanic ash particles between about 250 and 260 K. During the Eyjafjallajökull volcanic eruption on Iceland in spring 2010, lidar measurements over Europe clearly observed volcanic ash having an impact on cloud glaciation [130], and in central Germany the highest IN number concentrations within a two year record of daily IN measurements were measured [95]. Also in Israel, about 5000 km away from Eyjafjallajökull, INs were as high in spring 2010 as during desert dust storms. Reference [95] showed that aging increased the ice nucleating ability of the volcanic ash during its transport in the atmosphere (see Section 4.2).

5.4.3. Ocean Fertilisation. In the past 20 years, iron-enrichment experiments ranging from bottle incubations to open-ocean amendment studies in regions of $50–100 \, km^2$ have demonstrated that iron supply stimulates phytoplankton growth in High-Nutrient-Low-Chlorophyll (HNLC) waters [132]. Thereby, the surface ocean is fertilised with iron, which affects marine primary production (MPP), phytoplankton community structures, and subsequently has an impact on higher trophic levels of the oceanic food-web (zooplankton, fish). Through the conversion of CO_2 to organic carbon and the sinking of parts of this organic matter into the deep ocean, the process referred to as the "biological pump" is activated and atmospheric CO_2 concentrations can be modified. However, it has been difficult to quantify export production via subsurface storage of carbon. The details of the "iron hypothesis" [24] and the possible magnitude of its effect on the global carbon cycle are subject of intense international debate, particularly in connection with climate engineering.

Ocean fertilisation by mineral dust has been studied extensively, as mineral dust has long been assumed to be the main component of atmospheric deposition of minerals into the open ocean [15]. A significant correlation of dust with climate indicators is found in paleorecords such as ice cores [133–135]. In the NE Pacific Ocean, the supply of iron from dust sources occurs episodically, for example, [136, 137], which is dependent on dust storm frequency and atmospheric circulation. For the Southern Ocean, an important multiproxy dataset was recently presented from a marine sediment core in the sub-Antarctic Atlantic [138]. A close correlation was observed between iron input and marine export production, implying that the process of ion fertilisation on marine biota was a recurred process operating in the sub-Antarctic region over the glacial/interglacial cycles of the last 1.1 Ma. A 25%–50% decrease in CO_2 observed during glacial maxima is attributed to mineral dust [132, 139].

Volcanic ash deposition into the ocean represents another external and largely neglected source of iron. However, its significance and impact on climate has long been considered negligible. The major climate forcing effect following volcanic eruptions is widely assumed to occur due to the reduction of solar radiation through volcanic sulfate aerosols [5]. In contrast to volcanic gases and aerosols, volcanic ash is removed from the atmosphere much faster after an eruption. Recent work, however, showed that volcanic ash modifies the biogeochemical processes in the surface ocean [57, 60, 76, 140, 141] thereby directly affecting climate. When airborne volcanic ash is deposited in the surface ocean, it may release trace species upon contact with seawater [75, 142]. Volcanic ash, though released sporadically, can therefore play a similar role as mineral dust. Other trace metals contained in volcanic ash such as zinc or copper may have both, fertilising or toxic effects on phytoplankton [140].

The first direct evidence for iron fertilisation in an HNLC ocean area by volcanic ash emerged after the eruption of the Kasatochi volcano, situated on the Aleutian Islands in August 2008. Atmospheric and oceanic conditions in the NE Pacific were ideal for generating a massive and large-scale phytoplankton bloom, which was observed by satellite instruments [60], confirmed by insitu measurements [143, 144] and ocean biogeochemical modelling [145]. In 2010, it was speculated that the population of sockeye salmon returning to the Fraser River in Canada which was the largest for decades was associated with the fertilisation of the NE Pacific Ocean by Kasatochi ash in 2008 [146]. However, the effect of volcanic ash on salmon populations is discussed controversially; for example, the analysis of [147] rejects the hypothesis of [146].

After the eruption of Kasatochi in 2008 on the Aleutian Islands, atmospheric CO_2 decreased slightly by ~ 0.01 Pg C as diatoms and mesozooplankton increased export of organic carbon from the surface to the deeper ocean [143]. This carbon sequestration was negligible compared to the rate at which fossil fuel emissions are rising (7–9 Pg C/yr; [107]). While the volcanic ash flux from Kasatochi of 0.2–0.3 km^3 [60] was relatively small, there is abundant evidence for regular volcanic ash emissions into the atmosphere (see Section 3.1, Table 2). Although strong volcanic eruptions with VEI $\geq$ 5 are rare and not necessarily close to an oceanic HNLC area, they are argued to have affected MPP and atmospheric CO_2 on geological time scales [148–152]. In addition, volcanic ocean fertilisation is not restricted to HNCL areas, as reported by [153] for the Mediterranean Sea or by [141] for the North Atlantic Ocean. However, all these effects are discussed controversially.

Reduced atmospheric CO_2 concentrations were observed in the years following the 1991 Pinatubo eruption [154, 155]. Reference [156] argued that this was the consequence of increased vegetation photosynthesis induced by the presence of a volcanic sulfate aerosol layer in the atmosphere. Notably, [154] Sarmiento (1993) suggested that the atmospheric CO_2 drawdown was the result of ocean fertilisation by Pinatubo ash. While the 1991 Pinatubo eruption released 5-6 km^3 of ash (about 30 times the volume of ash emitted by Kasatochi in 2008), a percentage limited amount fell into the iron-limited Southern Ocean. However, the eruption of Mt. Hudson around the same time deposited approximately 1.1 km^3 of ash into the iron-limited Atlantic sector of the Southern Ocean [157]. Surprisingly, this ash deposition event has never

been evoked to explain the decrease in atmospheric CO_2 concentration. Furthermore, the fertilisation potential of the Mt. Hudson ash deposited in Patagonia ($\sim$1.6 km^3), which was easily remobilised by the roaring forties during several months after the eruption has never been considered.

Another interesting event is the eruption of Huaynaputina in Peru in 1600, which produced more than 9.6 km^3 of volcanic ash [158], which is known to have settled into the tropical Pacific as well as the Southern Ocean, two large HNLC areas. An iron-fertilisation effect could partly explain the 10 ppm decrease in atmospheric CO_2 concentration measured in Antarctic ice cores after 1600 [159].

Ocean iron fertilisation may also affect the climate relevant exchange of trace gases between the ocean and the atmosphere. An increase of the MPP is accompanied by an increased contribution of organic carbon (OC) to submicron marine aerosols [160] and the release of dimethylsulfide (DMS) [161], oxidised to sulfate in the atmosphere. OC and sulfate aerosols can act as efficient cloud condensation nuclei and significantly influence cloud properties via the indirect aerosol effects (see Section 5.4.2; [120]), thereby further cooling the Earth's surface.

6. Future Research Needs

Although there are substantial differences in the history of mineral dust and volcanic ash particles before they are released into the atmosphere (see Sections 2 and 3), there are on the other hand a number of similarities in atmospheric processing at ambient temperatures (see Section 4.2) and environmental and climate impacts (see Section 5). Therefore, this review tries to trigger a closer cooperation between the research communities studying mineral dust and volcanic ash atmospheric chemical modifications and impacts.

Model parameterisations of volcanic ash remobilisation from its deposits on land build on mineral dust mobilisation schemes [10]. However, as the availability of ash in its deposits is limited, modified approaches will be necessary considering mass conserving parameterisations, where the migration of deposits is also included. Such parameterisations might be of interest for mineral dust researchers as well.

The extreme conditions for multiphase chemistry in volcanic plumes (see Section 4.1) regarding temperature and their associated gradients, acidity, lightning, and particle load represent an obstacle which hindered an overall understanding of the important processes up to now. Despite these difficulties, the multiphase volcanic plume chemistry under extreme conditions, however, offers the possibilities to illuminate processes which might also be important for mineral dust atmospheric chemical processing under less extreme conditions. Here, in particular, the formation of bioavailable iron on mineral dust and volcanic ash surfaces for ocean fertilisation (see Section 5.4.3) is emphasised. Joint experimental and modelling research projects between mineral dust [18] and volcanic ash researchers could substantially increase our incomplete understanding beyond what we know today, particularly from leaching experiments [74]. Although leaching experiments are extremely important for our current knowledge, they could be even more important if standard protocols would be defined and applied to allow a comparison between the experiments conducted at different laboratories [97]. Also particle size distributions and mineralogy for particle diameters substantially smaller than $2\,\mu m$ should be increasingly studied, particularly by volcanological researchers, as these particles are subject to long-range transport.

During a volcanic eruption, ash particles are easily injected into atmospheric regimes where freezing temperatures prevail, and therefore a better understanding of the processes affected by freezing temperatures, like IN formations or Fe mobilisation and their climate impacts [99], should be studied more systematically. CCN and IN formation is linked with wet deposition processes of mineral dust and volcanic ash out of the atmosphere. Regarding volcanic ash, an improved knowledge of aggregation (see Section 3.3), a deposition process reducing the amount of volcanic ash for long-range transport, is urgently needed. This process is insufficiently handled in all ash dispersion models [162]. Reference [62] requests continued interaction between the meteorological and volcanological communities to achieve advances in understanding the fundamentals of ash aggregation. Besides ash-ice aggregation processes, which considerably increase the terminal settling velocity of the aggregates in comparison to the single fine ash particle and thereby increase the removal rate of volcanic ash, the process of wet deposition of volcanic ash must also be considered as a nonlinear interaction process between volcanic ash and meteorological clouds. Even without the effects of volcanic ash, understanding of the fundamentals of cloud formation is challenging for atmospheric scientists. Interactions of aerosols, including mineral dust, with water and ice in atmospheric clouds and their influence on cloud formation, lifetime, and precipitation formation is one of the hot topics in climate research [120].

For paleoclimate research, the results from terrestrial and marine environmental archives, namely, ice, peat, sea, and ocean sediment cores for mineral dust [139] and volcanic ash deposition [58, 59], need to be assembled to better assess the climate impacts of volcanic ash versus mineral dust during the geological past. However, until we have a good understanding of present day processes, we will not be able to adequately address these processes either in the palaeo-records or with regard to the future impacts of mineral dust in contrast to volcanic ash on climate.

Acknowledgments

The financial support through the Cluster of Excellence "CliSAP" (EXC177), University of Hamburg, funded through the German Science Foundation (DFG) is gratefully acknowledged. The author thanks Michael Hemming, Gholamali Hoshyaripour, and Matthias Hort for their comments on the paper.

References

[1] R. Sparks, M. Bursik, J. Gilbert, L. Glaze, H. Sigurdsson, and A. Woods, *Volcanic Plumes*, John Wiley, Chichester, UK, 1997.

[2] W. I. Rose and A. J. Durant, "Fine ash content of explosive eruptions," *Journal of Volcanology and Geothermal Research*, vol. 186, no. 1-2, pp. 32–39, 2009.

[3] D. B. Dingwell, Y. Lavallée, and U. Kueppers, "Volcanic ash: a primary agent in the Earth system," *Physics and Chemistry of the Earth*, vol. 45-46, pp. 2–4, 2012.

[4] B. Zimanowski, K. Wohletz, P. Dellino, and R. Büttner, "The volcanic ash problem," *Journal of Volcanology and Geothermal Research*, vol. 122, no. 1-2, pp. 1–5, 2003.

[5] A. Robock, "Volcanic eruptions and climate," *Reviews of Geophysics*, vol. 38, no. 2, pp. 191–219, 2000.

[6] R. J. Brown, C. Bonadonna, and A. J. Durant, "A review of volcanic ash aggregation," *Physics and Chemistry of the Earth*, vol. 45-46, pp. 65–78, 2012.

[7] P. V. Hobbs, D. A. Hegg, and L. F. Radke, "Resuspension of volcanic ash from Mount St. Helens," *Journal of Geophysical Research*, vol. 88, no. 6, pp. 3919–3921, 1983.

[8] D. Hadley, G. L. Hufford, and J. J. Simpson, "Resuspension of relic volcanic ash and dust from katmai: still an aviation hazard," *Weather and Forecasting*, vol. 19, pp. 829–840, 2004.

[9] T. M. Wilson, J. W. Cole, C. Stewart, S. J. Cronin, and D. M. Johnston, "Ash storms: impacts of wind-remobilised volcanic ash on rural communities and agriculture following the 1991 Hudson eruption, southern Patagonia, Chile," *Bulletin of Volcanology*, vol. 73, no. 3, pp. 223–239, 2011.

[10] S. J. Leadbetter, M. C. Hort, S. Von Lwis, K. Weber, and C. S. Witham, "Modeling the resuspension of ash deposited during the eruption of Eyjafjallajökull in spring 2010," *Journal of Geophysical Research*, vol. 117, no. D20, 2012.

[11] T. Thorsteinsson, G. Gísladóttir, J. Bullard, and G. McTainsh, "Dust storm contributions to airborne particulate matter in Reykjavík, Iceland," *Atmospheric Environment*, vol. 45, no. 32, pp. 5924–5933, 2011.

[12] T. Thorsteinsson, T. Jóhannsson, A. Stohl, and N. I. Kristiansen, "High levels of particulate matter in Iceland due to direct ash emissions by the Eyjafjallajökull eruption and resuspension of deposited ash," *Journal of Geophysical Research B*, vol. 117, no. B9, 2012.

[13] I. Tegen and I. Fung, "Modeling of mineral dust in the atmosphere: sources, transport, and optical thickness," *Journal of Geophysical Research*, vol. 99, no. D11, pp. 22897–22914, 1994.

[14] J. M. Prospero, P. Ginoux, O. Torres, S. E. Nicholson, and T. E. Gill, "Environmental characterization of global sources of atmospheric soil dust identified with the Nimbus 7 Total Ozone Mapping Spectrometer (TOMS) absorbing aerosol product," *Reviews of Geophysics*, vol. 40, no. 1, pp. 2-1-2-31, 2002.

[15] T. D. Jickells, Z. S. An, K. K. Andersen et al., "Global iron connections between desert dust, ocean biogeochemistry, and climate," *Science*, vol. 308, no. 5718, pp. 67–71, 2005.

[16] N. M. Mahowald, A. R. Baker, G. Bergametti et al., "Atmospheric global dust cycle and iron inputs to the ocean," *Global Biogeochemical Cycles*, vol. 19, no. 4, 2005.

[17] Y. Shao, K.-H. Wyrwoll, A. Chappell et al., "Dust cycle: an emerging core theme in Earth system science," *Aeolian Research*, vol. 2, no. 4, pp. 181–204, 2011.

[18] M. Schulz, J. M. Prospero, A. R. Baker et al., "Atmospheric transport and deposition of mineral dust to the ocean: implications for research need," *Environmental Science and Technology*, vol. 46, pp. 10390–10404, 2012.

[19] N. Huneeus, M. Schulz, Y. Balkanski et al., "Global dust model intercomparison in AeroCom phase I," *Atmospheric Chemistry and Physics*, vol. 11, no. 15, pp. 7781–7816, 2011.

[20] D. Rosenfeld, Y. Rudich, and R. Lahav, "Desert dust suppressing precipitation: a possible desertification feedback loop," *Proceedings of the National Academy of Sciences of the United States of America*, vol. 98, no. 11, pp. 5975–5980, 2001.

[21] P. Kumar, I. N. Sokolik, and A. Nenes, "Measurements of cloud condensation nuclei activity and droplet activation kinetics of fresh unprocessed regional dust samples and minerals," *Atmospheric Chemistry and Physics*, vol. 11, no. 7, pp. 3527–3541, 2011.

[22] F. J. Dentener, G. R. Carmichael, Y. Zhang, J. Lelieveld, and P. J. Crutzen, "Role of mineral aerosol as a reactive surface in the global troposphere," *Journal of Geophysical Research*, vol. 101, no. D17, pp. 22869–22889, 1996.

[23] S. E. Bauer, Y. Balkanski, M. Schulz, D. A. Hauglustine, and F. Dentener, "Global modeling of heterogenous chemistry on mineral aerosol surfaces: influence on tropospheric ozone chemistry and comparison to observations," *Journal of Geophysical Research*, vol. 109, no. 2, p. D2, 2004.

[24] J. H. Martin, "Glacial-interglacial CO_2 change: the iron hypothesis," *Paleoceanography*, vol. 5, no. 1, pp. 1–13, 1990.

[25] I. Koren, Y. J. Kaufman, R. Washington et al., "The Bodélé depression: a single spot in the Sahara that provides most of the mineral dust to the Amazon forest," *Environmental Research Letters*, vol. 1, no. 1, Article ID 014005, 2006.

[26] J. G. Calvert, *Glossary of Atmospheric Chemistry Terms*, IUPAC, 1990.

[27] G. Heiken, "Morphology and Petrography of volcanic ashes," *Geological Society of America Bulletin*, vol. 83, pp. 1961–1988, 1972.

[28] H. U. Schmincke, *Volcanism*, Springer, Berlin, Germany, 2004.

[29] M. Nakagawa and T. Ohba, "Minerals in volcanic ash 1: primary minerals and volcanic glass," *Global Environmental Research*, vol. 6, pp. 41–51, 2003.

[30] Z. Shi, M. D. Krom, T. D. Jickels et al., "Impacts on iron solubility in the mineral dust by processes in the source region and the atmosphere: a review," *Aeolian Research*, vol. 5, pp. 21–42, 2012.

[31] P. Formenti, L. Schuetz, Y. Balkanski et al., "Recent progress in understanding physical and chemical properties of African and Asian mineral dust," *Atmospheric Chemistry and Physics*, vol. 11, pp. 8231–8256, 2011.

[32] B. Langmann, A. Folch, M. Hensch, and V. Matthias, "Volcanic ash over Europe during the eruption of Eyjafjallajökull on Iceland, April–May 2010," *Atmospheric Environment*, vol. 48, pp. 1–8, 2012.

[33] S. A. Colgate and T. Sigurgeirsson, "Dynamic mixing of water and lava," *Nature*, vol. 244, no. 5418, pp. 552–555, 1973.

[34] S. Dartevelle, G. G. J. Ernst, and A. Bernard, "Origin of the Mount Pinatubo climactic eruption cloud: implications for volcanic hazards and atmospheric impacts," *Geology*, vol. 30, no. 7, pp. 663–666, 2002.

[35] Y. H. Lee, K. Chen, and P. J. Adams, "Development of a global model of mineral dust aerosol microphysics," *Atmospheric Chemistry and Physics*, vol. 9, no. 7, pp. 2441–2458, 2009.

[36] T. M. Wilson, C. Stewart, V. Sword-Daniels et al., "Volcanic ash impacts on critical infrastructure," *Physics and Chemistry of the Earth*, vol. 45-46, no. 5, 23 pages, 2012.

[37] P. M. Ayris and P. Delmelle, "The immediate environmental effects of tephra emission," *Bulletin of Volcanology*, vol. 74, pp. 1905–1936, 2012.

[38] R. B. Symonds, W. I. Rose, G. J. S. Bluth, and T. M. Gerlach, "Volcanic gas studies: methods, results and applications," in *Volatiles in Magma*, M. R. Caroll and J. R. Holloway, Eds., vol. 30, pp. 1–66, Reviews in Mineralogy and Geochemistry, 1994.

[39] P. Delmelle, F. Villiéras, and M. Pelletier, "Surface area, porosity and water adsorption properties of fine volcanic ash particles," *Bulletin of Volcanology*, vol. 67, no. 2, pp. 160–169, 2005.

[40] S. Joussaume, "3-dimensional simulations of the atmospheric cycle of desert dust particles using a general-circulation model," *Journal of Geophysical Research*, vol. 95, pp. 1909–1941, 1990.

[41] N. Mahowald, K. Kohfeld, M. Hansson et al., "Dust sources and deposition during the last glacial maximum and current climate: a comparison of model results with paleodata from ice cores and marine sediments," *Journal of Geophysical Research*, vol. 104, no. D13, pp. 15895–15916, 1999.

[42] B. Marticorena and G. Bergametti, "Modeling the atmospheric dust cycle: 1. Design of a soil-derived dust emission scheme," *Journal of Geophysical Research*, vol. 100, no. 8, pp. 16415–16430, 1995.

[43] I. Tegen, M. Werner, S. P. Harrison, and K. E. Kohfeld, "Relative importance of climate and land use in determining present and future global soil dust emission," *Geophysical Research Letters*, vol. 31, no. 5, 2004.

[44] C. S. Zender, H. Bian, and D. Newman, "Mineral Dust Entrainment and Deposition (DEAD) model: description and 1990s dust climatology," *Journal of Geophysical Research*, vol. 108, no. D14, 2003.

[45] P. Formenti, J. L. Rajot, K. Desboeufs et al., "Airborne observations of mineral dust over western Africa in the summer Monsoon season: spatial and vertical variability of physicochemical and optical properties," *Atmospheric Chemistry and Physics*, vol. 11, no. 13, pp. 6387–6410, 2011.

[46] C. G. Newhall and S. Self, "The volcanic explosivity index (VEI): an estimate of explosive magnitude for historical volcanism," *Journal of Geophysical Research*, vol. 87, no. C2, pp. 1231–1238, 1982.

[47] T. Simkin and I. Siebert, *Volcanoes of the World*, Smithonian Institution; Geoscience Press, Misoula, Mont, USA, 1994.

[48] T. A. Mather, D. M. Pyle, and C. Oppenheimer, "Tropospheric volcanic aerosol," in *Volcanism and the Earth's Atmosphere, Geophysical Monograph*, vol. 139, pp. 89–211, 2003.

[49] B. Langmann, K. Zaksek, and M. Hort, "Atmospheric distribution and removal of volcanic ash after the eruption of Kasatochi volcano: a regional model study," *Journal of Geophysical Research*, vol. 115, no. D2, 2010.

[50] B. Marticorena, B. Chatenet, J. L. Rajot et al., "Temporal variability of mineral dust concentrations over West Africa: analyses of a pluriannual monitoring from the AMMA Sahelian Dust Transect," *Atmospheric Chemistry and Physics*, vol. 10, no. 18, pp. 8899–8915, 2010.

[51] M. Mikami, O. Abe, M. Du et al., "The impact of Aeolian dust on climate: sino-Japanese cooperative project ADEC," *Journal of Arid Land Studies*, vol. 11, pp. 211–222, 2002.

[52] C. Hoffmann, R. Funk, M. Sommer, and Y. Li, "Temporal variations in PM10 and particle size distribution during Asian dust storms in Inner Mongolia," *Atmospheric Environment*, vol. 42, no. 36, pp. 8422–8431, 2008.

[53] I. Chiapello, "Origins of African dust transported over the northeastern tropical Atlantic," *Journal of Geophysical Research*, vol. 102, no. D12, pp. 13701–13709, 1997.

[54] A. J. Prata and A. T. Prata, "Eyjafjallajökull volcanic ash concentrations determined using Spin enhanced visible and infrared imager measurements," *Journal of Geophysical Research*, vol. 117, no. D20, 2012.

[55] H. N. Webster, D. J. Thomson, B. T. Johnson et al., "Operational prediction of ash concentrations in the distal volcanic cloud from the 2010 Eyjafjallajökull eruption," *Journal of Geophysical Research D*, vol. 117, no. D20, 2012.

[56] J. H. Seinfeld and S. N. Pandis, *Atmospheric Chemistry and Physics: From Air Pollution to Climate Change*, John Wiley, New York, NY, USA, 2006.

[57] N. Olgun, S. Duggen, P. L. Croot et al., "Surface ocean iron fertilization: the role of airborne volcanic ash from subduction zone and hot spot volcanoes and related iron fluxes into the Pacific Ocean," *Global Biogeochemical Cycles*, vol. 25, no. 4, 2011.

[58] S. M. Straub and H. U. Schmincke, "Evaluating the tephra input into Pacific Ocean sediments: distribution in space and time," *Geologische Rundschau*, vol. 87, no. 3, pp. 461–476, 1998.

[59] S. Kutterolf, A. Freundt, U. Schacht et al., "Pacific offshore record of plinian arc volcanism in Central America: 3. Application to forearc geology," *Geochemistry, Geophysics, Geosystems*, vol. 9, no. 2, 2008.

[60] B. Langmann, K. Zakšek, M. Hort, and S. Duggen, "Volcanic ash as fertiliser for the surface ocean," *Atmospheric Chemistry and Physics*, vol. 10, no. 8, pp. 3891–3899, 2010.

[61] D. M. Gaiero, J.-L. Probst, P. J. Depetris, S. M. Bidart, and L. Leleyter, "Iron and other transition metals in Patagonian riverborne and windborne materials: geochemical control and transport to the southern South Atlantic Ocean," *Geochimica et Cosmochimica Acta*, vol. 67, no. 19, pp. 3603–3623, 2003.

[62] W. I. Rose and A. J. Durant, "Fate of volcanic ash: aggregation and fallout," *Geology*, vol. 39, no. 9, pp. 895–896, 2011.

[63] A. J. Durant, W. I. Rose, A. M. Sarna-Wojcicki, S. Carey, and A. C. M. Volentik, "Hydrometeor-enhanced tephra sedimentation: constraints from the 18 May 1980 eruption of Mount St. Helens," *Journal of Geophysical Research*, vol. 114, no. B3, 2009.

[64] C. Bonadonna, R. Genco, M. Gouhier et al., "Tephra sedimentation during the 2010 Eyjafjallajökull eruption (Iceland) from deposit, radar, and satellite observations," *Journal of Geophysical Research*, vol. 116, no. B12, 2011.

[65] A. Folch, "A review of tephra transport and dispersal models: evolution, current status and future perspectives," *Journal of Volcanology and Geothermal Research*, vol. 235, pp. 96–115, 2012.

[66] C. Textor, H. F. Graf, C. Timmreck, and A. Robock, "Emissions from volcanoes," in *Emissions of Chemical Compounds and Aerosols in the Atmosphere*, C. Granier, C. Reeves, and P. Artaxo, Eds., vol. 18 of *Advances in Global Change Research*, pp. 269–303, Kluwer, Dordrecht, The Netherlands, 2004.

[67] L. G. Mastin, "A user-friendly one-dimensional model for wet volcanic plumes," *Geochemistry, Geophysics, Geosystems*, vol. 8, no. 3, 2007.

[68] M. R. James, L. Wilson, S. J. Lane et al., "Electrical charging of volcanic plumes," *Space Science Reviews*, vol. 137, no. 1–4, pp. 399–418, 2008.

[69] S. R. McNutt and E. R. Williams, "Volcanic lightning: global observations and constraints on source mechanisms," *Bulletin of Volcanology*, vol. 72, no. 10, pp. 1153–1167, 2010.

[70] W. I. Rose, "Scavenging of volcanic aerosol by ash: atmospheric and volcanologic implications," *Geology*, vol. 5, pp. 621–624, 1977.

[71] N. Oskarsson, "The interaction between volcanic gases and tephra: fluorine adhering to tephra of the 1970 Hekla eruption," *Journal of Volcanology & Geothermal Research*, vol. 8, no. 2–4, pp. 251–266, 1980.

[72] E. Bagnato, A. Aiuppa, A. Bertagnini et al., "Scavenging of sulphur, halogens and trace metals by volcanic ash: the 2010 Eyjafjallajökull eruption," *Geochimica Et Cosmochimica Acta*, vol. 103, pp. 138–160, 2013.

[73] J. M. de Moor, T. P. Fischer, D. R. Hilton, E. Hauri, L. A. Jaffe, and J. T. Camacho, "Degassing at Anatahan volcano during the May 2003 eruption: implications from petrology, ash leachates, and SO_2 emissions," *Journal of Volcanology and Geothermal Research*, vol. 146, no. 1–3, pp. 117–138, 2005.

[74] C. S. Witham, C. Oppenheimer, and C. J. Horwell, "Volcanic ash-leachates: a review and recommendations for sampling methods," *Journal of Volcanology and Geothermal Research*, vol. 141, no. 3-4, pp. 299–326, 2005.

[75] P. Frogner, S. R. Gíslason, and N. Óskarsson, "Fertilizing potential of volcanic ash in ocean surface water," *Geology*, vol. 29, no. 6, pp. 487–490, 2001.

[76] S. Duggen, P. Croot, U. Schacht, and L. Hoffmann, "Subduction zone volcanic ash can fertilize the surface ocean and stimulate phytoplankton growth: evidence from biogeochemical experiments and satellite data," *Geophysical Research Letters*, vol. 34, no. 1, Article ID L01612, 2007.

[77] S. Duggen, N. Olgun, P. Croot et al., "The role of airborne volcanic ash for the surface ocean biogeochemical iron-cycle: a review," *Biogeosciences*, vol. 7, no. 3, pp. 827–844, 2010.

[78] P. S. Taylor and R. E. Stoiber, "Soluble material on ash from active Central American volcanoes," *Geological Society of America Bulletin*, vol. 84, pp. 1031–1042, 1973.

[79] D. B. Smith, R. A. Zielinski, W. I. Rose Jr., and B. J. Huebert, "Water-soluble material on aerosols collected within volcanic eruption clouds (Fuego, Pacaya, Santiaguito, Guatamala)," *Journal of Geophysical Research*, vol. 87, no. 7, pp. 4963–4972, 1982.

[80] P. Delmelle, M. Lambert, Y. Dufrêne, P. Gerin, and N. Óskarsson, "Gas/aerosol-ash interaction in volcanic plumes: new insights from surface analyses of fine ash particles," *Earth and Planetary Science Letters*, vol. 259, no. 1-2, pp. 159–170, 2007.

[81] P. M. Ayris, A. F. Lee, K. Wilson, U. Kueppers, D. B. Dingwell, and P. Delmelle, "SO_2 sequestration in large volcanic eruptions: high-temperature scavenging by tephra," *Geochimica Et Cosmochimica Acta*, vol. 110, pp. 58–69, 2013.

[82] C. Textor, H. F. Graf, M. Herzog, J. M. Oberhuber, W. I. Rose, and G. G. J. Ernst, "Volcanic particle aggregation in explosive eruption columns. Part I: parameterization of the microphysics of hydrometeors and ash," *Journal of Volcanology and Geothermal Research*, vol. 150, no. 4, pp. 359–377, 2006.

[83] P. Ayris and P. Delmelle, "Volcanic and atmospheric controls on ash iron solubility: a review," *Physics and Chemistry of the Earth*, vol. 45-46, pp. 103–112, 2012.

[84] M. M. Halmer, H.-U. Schmincke, and H.-F. Graf, "The annual volcanic gas input into the atmosphere, in particular into the stratosphere: a global data set for the past 100 years," *Journal of Volcanology and Geothermal Research*, vol. 115, no. 3-4, pp. 511–528, 2002.

[85] C. Gao, A. Robock, and C. Ammann, "Volcanic forcing of climate over the past 1500 years: an improved ice core-based index for climate models," *Journal of Geophysical Research*, vol. 113, no. D23, 2008.

[86] N. Meskhidze, W. L. Chameides, A. Nenes, and G. Chen, "Iron mobilization in mineral dust: can anthropogenic SO_2 emissions affect ocean productivity?" *Geophysical Research Letters*, vol. 30, no. 21, 2003.

[87] J. Fero, S. N. Carey, and J. T. Merrill, "Simulating the dispersal of tephra from the 1991 Pinatubo eruption: implications for the

formation of widespread ash layers," *Journal of Volcanology and Geothermal Research*, vol. 186, no. 1-2, pp. 120–131, 2009.

[88] M. Ullerstam, R. Vogt, S. Langer, and E. Ljungström, "The kinetics and mechanism of SO_2 oxidation by O_3 on mineral dust," *Physical Chemistry Chemical Physics*, vol. 4, no. 19, pp. 4694–4699, 2002.

[89] M. Ndour, P. Conchon, B. D'Anna, O. Ka, and C. George, "Photochemistry of mineral dust surface as a potential atmospheric renoxification process," *Geophysical Research Letters*, vol. 36, no. 5, 2009.

[90] R. C. Sullivan, S. A. Guazzotti, D. A. Sodeman, and K. A. Prather, "Direct observations of the atmospheric processing of Asian mineral dust," *Atmospheric Chemistry and Physics*, vol. 7, no. 5, pp. 1213–1236, 2007.

[91] Y. Dupart, S. M. King, B. Nekat et al., "Mineral dust photochemistry induces nucleation events in the presence of SO_2," *Proceedings of the National Academy of Science of the United States of America*, vol. 109, pp. 20842–20847, 2012.

[92] S. Wurzler, T. G. Reisin, and Z. Levin, "Modification of mineral dust particles by cloud processing and subsequent effects on drop size distributions," *Journal of Geophysical Research D*, vol. 105, no. D4, pp. 4501–4512, 2000.

[93] D. S. Mackie, P. W. Boyd, K. A. Hunter, and G. H. McTainsh, "Simulating the cloud processing of iron in Australian dust: pH and dust concentration," *Geophysical Research Letters*, vol. 32, no. 6, 2005.

[94] A. Matsuki, A. Schwarzenboeck, H. Venzac, P. Laj, S. Crumeyrolle, and L. Gomes, "Cloud processing of mineral dust: direct comparison of cloud residual and clear sky particles during AMMA aircraft campaign in summer 2006," *Atmospheric Chemistry and Physics*, vol. 10, no. 3, pp. 1057–1069, 2010.

[95] H. Bingemer, H. Klein, M. Ebert et al., "Atmospheric ice nuclei in the Eyjafjallajökull volcanic ash plume," *Atmospheric Chemistry and Physics*, vol. 12, pp. 857–867, 2012.

[96] F. Solmon, P. Y. Chuang, N. Meskhidze, and Y. Chen, "Acidic processing of mineral dust iron by anthropogenic compounds over the north Pacific Ocean," *Journal of Geophysical Research*, vol. 114, no. D2, 2009.

[97] A. R. Baker and P. L. Croot, "Atmospheric and marine controls on aerosol iron solubility in seawater," *Marine Chemistry*, vol. 120, no. 1–4, pp. 4–13, 2010.

[98] N. Meskhidze, W. L. Chameides, and A. Nenes, "Dust and pollution: a recipe for enhanced ocean fertilization?" *Journal of Geophysical Research D*, vol. 110, no. D3, 2005.

[99] D. Jeong, K. Kim, and W. Choi, "Accelerated dissolution of iron oxides in ice," *Atmospheric Chemistry and Physics*, vol. 12, pp. 11125–11133, 2012.

[100] A. J. Durant, R. A. Shaw, W. I. Rose, Y. Mi, and G. G. J. Ernst, "Ice nucleation and overseeding of ice in volcanic clouds," *Journal of Geophysical Research*, vol. 113, no. D9, 2008.

[101] Z. Shi, D. Zhang, M. Hayashi, H. Ogata, H. Ji, and W. Fujiie, "Influences of sulfate and nitrate on the hygroscopic behaviour of coarse dust particles," *Atmospheric Environment*, vol. 42, no. 4, pp. 822–827, 2008.

[102] T. L. Lathem, P. Kumar, A. Nenes et al., "Hygroscopic properties of volcanic ash," *Geophysical Research Letters*, vol. 38, no. 11, Article ID L11802, 2011.

[103] R. A. Dahlgren, F. C. Uoolim, and W. H. Casey, "Field weathering rates of Mt. St. Helens tephra," *Geochimica et Cosmochimica Acta*, vol. 63, no. 5, pp. 587–598, 1999.

[104] A. F. White and S. L. Brantley, "Chemical weathering rates of silicate minerals: an overview," *Reviews in Mineralogy and Geochemistry*, vol. 31, pp. 1–22, 1995.

[105] J.-I. Jeong and S.-U. Park, "Interaction of gaseous pollutants with aerosols in Asia during March 2002," *Science of the Total Environment*, vol. 392, no. 2-3, pp. 262–276, 2008.

[106] H. Yuan, G. Zhuang, J. Li, Z. Wang, and J. Li, "Mixing of mineral with pollution aerosols in dust season in Beijing: revealed by source apportionment study," *Atmospheric Environment*, vol. 42, no. 9, pp. 2141–2157, 2008.

[107] IPCC (International Panel of Climate Change), *Fourth Assessment Report: Climate Change*, 2007.

[108] C. J. Horwell and P. J. Baxter, "The respiratory health hazards of volcanic ash: a review for volcanic risk mitigation," *Bulletin of Volcanology*, vol. 69, no. 1, pp. 1–24, 2006.

[109] M. G. Dunn, A. J. Baran, and J. Miatech, "Operation of gas turbine engines in volcanic ash clouds," *Journal of Engineering for Gas Turbines and Power*, vol. 118, no. 4, pp. 724–731, 1996.

[110] M. Guffanti, T. J. Casadevall, and K. Budding, "1953–2009: Encounters of Aircraft with Volcanic Ash Clouds, A Compilation of Known Incidents," U.S. Geological Survey Data Series 545, 12 p., Plus 4 Appendixes Including the Compilation Database, 2010, http://pubs.usgs.gov/ds/545/.

[111] B. Weinzierl, D. Sauer, A. Minikin et al., "On the visibility of airborne volcanic ash and mineral dust from the pilot's perspective in flight," *Physics and Chemistry of the Earth*, vol. 45-46, pp. 87–102, 2012.

[112] R. J. Cook, J. C. Barron, R. I. Papendick, and G. J. Williams III, "Impact on agriculture of the Mount St. Helens eruptions," *Science*, vol. 211, no. 4477, pp. 16–22, 1981.

[113] C. Mass and A. Robock, "The short-term influence of the Mount St. Helens volcanic eruption on surface temperature in the northwest United States (Idaho, Montana)," *Monthly Weather Review*, vol. 110, no. 6, pp. 614–622, 1982.

[114] M. T. Jones, R. S. J. Sparks, and P. J. Valdes, "The climatic impact of supervolcanic ash blankets," *Climate Dynamics*, vol. 29, no. 6, pp. 553–564, 2007.

[115] Y. J. Kaufman, D. Tanré, O. Dubovik, A. Karnieli, and L. A. Remer, "Absorption of sunlight by dust as inferred from satellite and ground-based remote sensing," *Geophysical Research Letters*, vol. 28, no. 8, pp. 1479–1482, 2001.

[116] H. E. Redmond, K. D. Dial, and J. E. Thompson, "Light scattering and absorption by wind blown dust: theory, measurement, and recent data," *Aeolian Research*, vol. 2, no. 1, pp. 5–26, 2010.

[117] J. S. Reid, H. H. Jonsson, H. B. Maring et al., "Comparison of size and morphological measurements of coarse mode dust particles from Africa," *Journal of Geophysical Research*, vol. 108, no. D19, 2003.

[118] S. Groß, V. Freudenthaler, M. Wiegner, J. Gasteiger, A. Geiß, and F. Schnell, "Dual-wavelength linear depolarization ratio of volcanic aerosols: lidar measurements of the Eyjafjallajökull plume over Maisach, Germany," *Atmospheric Environment*, vol. 48, pp. 85–96, 2012.

[119] S. E. Bauer, M. I. Mishchenko, A. A. Lacis, S. Zhang, J. Perlwitz, and S. M. Metzger, "Do sulfate and nitrate coatings on mineral dust have important effects on radiative properties and climate modeling?" *Journal of Geophysical Research*, vol. 112, no. D6, 2007.

[120] U. Lohmann and J. Feichter, "Global indirect aerosol effects: a review," *Atmospheric Chemistry and Physics*, vol. 5, no. 3, pp. 715–737, 2005.

[121] S. Twomey, "The influence of pollution on the shortwave albedo of clouds," *Journal of Atmospheric Sciences*, vol. 34, pp. 1149–1152, 1977.

[122] B. A. Albrecht, "Aerosols, cloud microphysics, and fractional cloudiness," *Science*, vol. 245, no. 4923, pp. 1227–1230, 1989.

[123] S. Gasso, "Satellite observations of the impact of weak volcanic activity on marine clouds," *Journal of Geophysical Research*, vol. 113, no. D14, 2008.

[124] F. J. Nober, H.-F. Graf, and D. Rosenfeld, "Sensitivity of the global circulation to the suppression of precipitation by anthropogenic aerosols," *Global and Planetary Change*, vol. 37, no. 1-2, pp. 57–80, 2003.

[125] H. R. Pruppacher and J. D. Klett, *Microphysics of Clouds and Precipitation*, Kluwer, Dordrecht, The Netherlands, 2nd edition, 1997.

[126] W. Szyrmer and I. Zawadzki, "Biogenic and anthropogenic sources of ice-forming nuclei: a review," *Bulletin of the American Meteorological Society*, vol. 78, no. 2, pp. 209–228, 1997.

[127] K. Diehl, C. Quick, S. Matthias-Maser, S. K. Mitra, and R. Jaenicke, "The ice nucleating ability of pollen Part I: laboratory studies in deposition and condensation freezing modes," *Atmospheric Research*, vol. 58, no. 2, pp. 75–87, 2001.

[128] K. Diehl, S. Matthias-Maser, R. Jaenicke, and S. K. Mitra, "The ice nucleating ability of pollen: Part II. Laboratory studies in immersion and contact freezing modes," *Atmospheric Research*, vol. 61, no. 2, pp. 125–133, 2002.

[129] P. J. DeMott, K. Sassen, M. R. Poellot et al., "African dust aerosols as atmospheric ice nuclei," *Geophysical Research Letters*, vol. 30, no. 14, 2003.

[130] P. Seifert, A. Ansmann, S. Groß et al., "Ice formation in ash-influenced clouds after the eruption of the Eyjafjallajökull volcano in April 2010," *Journal of Geophysical Research*, vol. 116, no. D20, 2011.

[131] Y.-S. Choi, R. S. Lindzen, C.-H. Ho, and J. Kim, "Space observations of cold-cloud phase change," *Proceedings of the National Academy of Sciences of the United States of America*, vol. 107, no. 25, pp. 11211–11216, 2010.

[132] P. W. Boyd and M. J. Ellwood, "The biogeochemical cycle of iron in the ocean," *Nature Geoscience*, vol. 3, no. 10, pp. 675–682, 2010.

[133] L. Bopp, K. E. Kohfeld, C. Le Quéré, and O. Aumont, "Dust impact on marine biota and atmospheric CO_2 during glacial periods," *Paleoceanography*, vol. 18, no. 2, 2003.

[134] R. Röthlisberger, M. Bigler, E. W. Wolff, F. Joos, E. Monnin, and M. A. Hutterli, "Ice core evidence for the extent of past atmospheric CO_2 change due to iron fertilisation," *Geophysical Research Letters*, vol. 31, no. 16, 2004.

[135] F. Lambert, B. Delmonte, J. R. Petit et al., "Dust—climate couplings over the past 800,000 years from the EPICA Dome C ice core," *Nature*, vol. 452, no. 7187, pp. 616–619, 2008.

[136] J. K. B. Bishop, R. E. Davis, and J. T. Sherman, "Robotic observations of dust storm enhancement of carbon biomass in the North Pacific," *Science*, vol. 298, no. 5594, pp. 817–821, 2002.

[137] P. W. Boyd, C. S. Wong, J. Merrill et al., "Atmospheric iron supply and enhanced vertical carbon flux in the NE subarctic Pacific: is there a connection?" *Global Biogeochemical Cycles*, vol. 12, no. 3, pp. 429–441, 1998.

[138] A. Martínez-Garcia, A. Rosell-Melé, W. Geibert et al., "Links between iron supply, marine productivity, sea surface temperature, and CO_2 over the last 1.1 Ma," *Paleoceanography*, vol. 24, no. 1, 2009.

[139] B. A. Maher, J. M. Prospero, D. Mackie, D. Gaiero, P. P. Hesse, and Y. Balkanski, "Global connections between aeolian dust, climate and ocean biogeochemistry at the present day and at the last glacial maximum," *Earth-Science Reviews*, vol. 99, no. 1-2, pp. 61–97, 2010.

[140] L. J. Hoffmann, E. Breitbarth, M. V. Ardelan et al., "Influence of trace metal release from volcanic ash on growth of *Thalassiosira pseudonana* and *Emiliania huxleyi*," *Marine Chemistry*, vol. 132-133, pp. 28–33, 2012.

[141] E. P. Achterberg, C. M. Moore, A. Henson et al., "Natural iron fertilization by the Eyjafjallajökull volcanic eruption," *Geophysical Research Letters*, vol. 40, no. 5, pp. 921–926, 2013.

[142] M. T. Jones and S. R. Gislason, "Rapid releases of metal salts and nutrients following the deposition of volcanic ash into aqueous environments," *Geochimica et Cosmochimica Acta*, vol. 72, no. 15, pp. 3661–3680, 2008.

[143] R. C. Hamme, P. W. Webley, W. R. Crawford et al., "Volcanic ash fuels anomalous plankton bloom in subarctic northeast Pacific," *Geophysical Research Letters*, vol. 37, no. 19, Article ID L19604, 2010.

[144] D. Lockwood, P. D. Quay, M. T. Kavanaugh, L. W. Juranek, and R. A. Feely, "High-resolution estimates of net community production and air-sea CO_2 flux in the northeast Pacific," *Global Biogeochemical Cycles*, vol. 26, no. 4, 2012.

[145] A. Lindenthal, B. Langmann, J. Paetsch, I. Lorkorwski, and M. Hort, "The ocean response to volcanic iron fertilization after the eruption of Kasatochi volcano: a regional biogeochemical model study," *Biogeosciences*, vol. 10, pp. 3715–3729, 2013.

[146] T. R. Parsons and F. A. Whitney, "Did volcanic ash from Mt. Kasatochi in 2008 contribute to a phenomenal increase in Fraser River sockeye salmon (Oncohynchus nerka) in 2010?" *Fishery Oceanography*, vol. 21, pp. 374–377, 2012.

[147] S. McKinnell, "Challenges for the Kasatochi volcano hypothesis as the cause of a large return of sockeye salmon (Oncorhynchus nerka) to the Fraser River in 2010," *Fisheries Oceanography*, 2013.

[148] B. R. Jicha, D. W. Scholl, and D. K. Rea, "Circum-Pacific arc flare-ups and global cooling near the Eocene-Oligocene boundary," *Geology*, vol. 37, no. 4, article 303, 2009.

[149] S. M. Cather, N. W. Dunbar, F. W. McDowell, W. C. McIntosh, and P. A. Scholle, "Climate forcing by iron fertilization from repeated ignimbrite eruptions: the icehouse-silicic large igneous province (SLIP) hypothesis," *Geosphere*, vol. 5, no. 3, pp. 315–324, 2009.

[150] R. C. Bay, N. Bramall, and P. B. Price, "Bipolar correlation of volcanism with millennial climate change," *Proceedings of the National Academy of Sciences of the United States of America*, vol. 101, no. 17, pp. 6341–6345, 2004.

[151] R. C. Bay, N. E. Bramall, P. B. Price et al., "Globally synchronous ice core volcanic tracers and abrupt cooling during the last glacial period," *Journal of Geophysical Research*, vol. 111, no. D11, 2006.

[152] S. Bains, R. D. Norris, R. M. Corfield, and K. L. Faul, "Termination of global warmth at the Palaeocene/Eocene boundary through productivity feedback," *Nature*, vol. 407, no. 6801, pp. 171–174, 2000.

[153] P. Censi, L. A. Randazzo, P. Zuddas, F. Saiano, P. Aricò, and S. Andò, "Trace element behaviour in seawater during Etna's pyroclastic activity in 2001: concurrent effects of nutrients and formation of alteration minerals," *Journal of Volcanology and Geothermal Research*, vol. 193, no. 1-2, pp. 106–116, 2010.

[154] J. L. Sarmiento, "Atmospheric CO_2 stalled," *Nature*, vol. 365, no. 6448, pp. 697–698, 1993.

[155] A. J. Watson, "Volcanic iron, CO_2, ocean productivity and climate," *Nature*, vol. 385, no. 6617, pp. 587–588, 1997.

[156] L. M. Mercado, N. Bellouin, S. Sitch et al., "Impact of changes in diffuse radiation on the global land carbon sink," *Nature*, vol. 458, no. 7241, pp. 1014–1017, 2009.

[157] R. A. Scasso, H. Corbella, and P. Tiberi, "Sedimentological analysis of the tephra from the 12–15 August 1991 eruption of Hudson volcano," *Bulletin of Volcanology*, vol. 56, no. 2, pp. 121–132, 1994.

[158] S. L. De Silva and G. A. Zielinski, "Global influence of the AD 1600 eruption of Huaynaputina, Peru," *Nature*, vol. 393, no. 6684, pp. 455–458, 1998.

[159] C. MacFarling Meure, D. Etheridge, C. Trudinger et al., "Law dome CO_2, CH_4 and NCO_2O ice core records extended to 2000 years BP," *Geophysical Research Letters*, vol. 33, no. 14, 2006.

[160] C. D. O'Dowd, M. C. Facchini, F. Cavalli et al., "Biogenically driven organic contribution to marine aerosol," *Nature*, vol. 431, no. 7009, pp. 676–680, 2004.

[161] P. Liss, A. Chuck, D. Bakker, and S. Turner, "Ocean fertilization with iron: effects on climate and air quality," *Tellus B*, vol. 57, no. 3, pp. 269–271, 2005.

[162] C. Bonadonna and A. Foch, "Ash Dispersal Forecast and Civil Aviation Workshop—Consensual Document," 2011, https://vhub.org/resources/503 .

16

Biological and Chemical Diversity of Biogenic Volatile Organic Emissions into the Atmosphere

Alex Guenther

Atmospheric Chemistry Division, NCAR Earth System Laboratory, National Center for Atmospheric Research, 3090 Center Green, Boulder, CO 80301, USA

Correspondence should be addressed to Alex Guenther; guenther@ucar.edu

Academic Editors: P. Massoli, K. Schaefer, and E. Tagaris

Biogenic volatile organic compounds (BVOC) emitted by terrestrial ecosystems into the atmosphere play an important role in determining atmospheric constituents including the oxidants and aerosols that control air quality and climate. Accurate quantitative estimates of BVOC emissions are needed to understand the processes controlling the earth system and to develop effective air quality and climate management strategies. The large uncertainties associated with BVOC emission estimates must be reduced, but this is challenging due to the large number of compounds and biological sources. The information on the immense biological and chemical diversity of BVOC is reviewed with a focus on observations that have been incorporated into the MEGAN2.1 BVOC emission model. Strategies for improving current BVOC emission modeling approaches by better representations of this diversity are presented. The current gaps in the available data for parameterizing emission models and the priorities for future measurements are discussed.

1. Introduction

Terrestrial ecosystems produce and emit many biogenic volatile organic compounds (BVOCs) into the air where they influence the chemistry and composition of the atmosphere including aerosols and oxidants [1–3]. These BVOCs are produced by a variety of sources in terrestrial ecosystems (e.g., flowers, stems, trunks, roots, leaf litter, soil microbes, insects, and animals), but most of the global total emission is from foliage [4–6]. The increasing awareness of the importance of these emissions for earth system modeling has resulted in numerical models of regional air quality and global climate that now routinely include BVOC emissions that are estimated as a function of landcover and environmental driving variables. This is a considerable challenge due to both the hundreds of different BVOC chemical species emitted into the atmosphere [7, 8] and the vast differences in the capacity of various plant species to produce and emit terpenoids and other BVOCs [9, 10]. Furthermore, an individual compound can be emitted by different ecosystem sources that are controlled by a variety of processes. Some compounds are stored in plant tissues that are isolated from the atmosphere and are emitted only if these tissues are damaged, while other compounds are stored in structures that are open to the atmosphere and are continuously being emitted [11]. There are additional compounds that are not stored in tissues but instead are released immediately after production which may happen only in response to stress or specific environmental conditions [12].

Quantitative attempts to account for these BVOC emissions in models must consider all of the processes that control emission variability. Among the greatest of these challenges is characterizing the enormous diversity in BVOC emission types in ecosystems across the world. This paper provides an overview of our current understanding of the chemical and biological diversity of BVOC emissions into the atmosphere. Section 2 describes a compilation of observations in the scientific literature that have been used to quantify BVOC emissions in a widely used numerical model, the Model of Emissions of Gases and Aerosols from Nature (MEGAN) [5], and considers the suitability of these observations for characterizing regional to global BVOC emissions. The known chemical diversity of BVOC emissions is summarized in Section 3, and an approach for improving the representation of this diversity in numerical models is described. BVOC biological diversity is discussed in Section 4, and a framework for better representation of BVOC emission diversity types is presented. Section 5 presents the major conclusions of this summary of our current understanding of the chemical and biological diversity of BVOC emissions.

2. BVOC Emission Observations and Models

Quantitative estimation of global BVOC emissions into the atmosphere began with Went's [13] seminal work that extrapolated measurements of a single group of compounds, monoterpenes, from a single plant species, *Artemisia tridentata*, to the entire earth. Rasmussen [9] recognized the great diversity in BVOC emission capacities of different plants species and introduced an approach for classifying the biosphere into different vegetation groups in order to quantify regional emissions. He noted that at least some vegetation types had "fingerprints" that could be used to represent the emission behavior of those plant species. He combined estimates of USA areas of different forest types (e.g., Loblolly-shortleaf pine forest, oak-gum-cypress forest) with observations of their representative emission rates in order to quantify total BVOC emissions on a USA national and on a global scale. Zimmerman [14] extended this approach using more comprehensive land cover data including broad natural vegetation types (e.g., shrub and brush rangeland, deciduous forest, and mixed forest), agricultural lands (e.g., crops, pasture, and orchards), and a category for residential areas. This approach was limited by the large differences in the emission rates of plant species in landscapes that, for example, are classified as deciduous forest or mixed forest because of the highly variable emission rates of these broad categories of vegetation. Zimmerman made additional progress towards accounting for this by collapsing USA forests into four types: high isoprene (e.g., oak) deciduous forest, low isoprene (e.g., sycamore) deciduous forest, no isoprene deciduous (e.g., maple), and coniferous forest (e.g., loblolly pine). Lamb and colleagues [15] refined this approach using higher resolution (county scale) landcover data that included land area planted with the major crop species. This approach was extended to the global scale [16, 17] by assigning emission factors to ecosystem types in global gridded databases. This was straightforward for categories dominated by a few species (e.g., paddy rice and mangrove) but not for most categories (e.g., farm/city-cool, temperate mixed, and dry evergreen) which did not represent a uniform BVOC emission type. For regions with detailed plant species data, the Biogenic Emission Inventory System 2 (BEIS2) [1] was developed to apply BVOC emission factors for individual tree genera and crop types. However, these data were only available for forests in some regions, and BEIS2 used broad categories for grassland and shrubland ecosystem types.

The MEGAN version 2.1 (MEGAN2.1) [5] BVOC emission model assigns emission factors and parameters to 19 BVOC chemical compound classes for each of the 15 plant functional types (PFTs) used for the Community Land Model (CLM4) [18]. MEGAN2.1 can be run embedded in CLM4 and can also run offline using observations or variables from other models. BVOC emission rate measurements from about 300 studies were synthesized to estimate the emission factors used for MEGAN2.1 including data representative of the major global vegetation types. Measurements representing temperate landscapes are compiled in Table 1 [19–186]. Studies in tropical and boreal landscapes are summarized in Tables 2 [187–229] and 3 [230–268], respectively. Measurements characterizing BVOC emissions from agricultural crops are compiled in Table 4 [70, 269–282]. Terpenoid (e.g., isoprene, MBO, and monoterpene) emission factors were estimated for each of the 15 PFTs. For most of the other compounds, one or a few (e.g., one for woody PFTs and one for herbaceous PFTs) emission factors were used for all PFTs. Terpenoid emission factors are represented with a greater diversity in MEGAN2.1 both because of the greater actual diversity and because more observations have been reported.

Until recently, most BVOC emission measurements were conducted using enclosure techniques, but whole canopy flux measurements using micrometeorological approaches are now becoming more common [215]. Characterizing BVOC emissions with enclosure measurements is challenging due to difficulties in accessing all parts of a mature forest canopy and because of the presence of storage structures which can be disturbed resulting in emissions at rates much higher than for undisturbed conditions [283]. These issues resulted in BVOC emission factors reported by earlier studies that greatly underestimate isoprene emissions, because isoprene emission rates are lower for the shaded leaves in the more easily accessed portion of a forest canopy, and overestimate monoterpene emissions because of disturbances to terpenoid storage structures [284]. The above-canopy flux measurements integrate over the entire canopy and landscape without disturbing emission rates [86]. Capabilities for quantifying biogenic VOC fluxes have steadily improved over the past decades including recent analytical advances such as the time-of-flight proton-transfer reaction mass spectrometer (PTR-TOF-MS) that enables whole canopy measurement of a wide range of BVOC fluxes [285]. Aircraft VOC flux systems have footprints of several km and can characterize fluxes over entire domains of hundreds of km and so are suitable for evaluating fluxes estimated by regional models [286]. Tower-based VOC flux systems typically have a footprint of hundreds of meters and are well suited for quantifying diurnal, seasonal, and interannual variations. Biogenic VOC fluxes have been measured at more than 45 tower locations (Tables 1 to 4 and summarized in [287, 288]), but most of these studies were for a short period (a few weeks or less) of time. The availability of more than 500 above-canopy flux towers constructed for water, carbon, and energy flux studies provides an opportunity to add biogenic VOC measurements without the cost of basic site development [289]. Measurements at a large number of sites can be accomplished with low-cost and low-power relaxed eddy accumulation measurements systems [238].

Figure 1 shows that there were relatively few BVOC emission rate observations reported in the 1960s and 1970s, and all but one of these studies were in temperate regions. Interest in the role of BVOC emissions in regional ozone pollution in the 1970s [9] stimulated publications on this topic by the early 1980s including some investigations of tropical, boreal, and agricultural ecosystems. This interest peaked in the mid-1980s and then declined as some researchers

TABLE 1: Compilation of studies used to estimate temperate vegetation BVOC emission factors for the MEGAN2.1 model [5]. Emission measurement approaches include enclosure (E), canopy micrometeorological (C), and landscape inverse modeling (L) techniques. Compounds include isoprene (Iso), monoterpenes (MT), sesquiterpenes (SQT), and other (Other). PFTs include broadleaf deciduous shrub (BDS), broadleaf evergreen shrub (BES), broadleaf deciduous tree (BDT), C3 grass (C3G), and needleleaf evergreen tree (NET).

Location	Approach	Compounds	PFTs	Reference
MI, USA	C	Iso	BDT	[19]
NC, USA	C	MT	NET	[20]
N T, Australia	L	Iso	BDS, BES	[21]
CO, USA	E	MT, SQT, and Other	BDT	[4]
Inner Mongolia, China	E	Iso, MT	C3G	[22]
CA, USA	C	MBO	NET	[23]
Various, USA	C	MBO, Other	NET	[24]
Austria	C	Other	C3G	[25]
NY, USA	E	MT	BES	[26]
Portugal	L	Iso, MT	BET	[27]
Various, Canada	L	Iso	BDT, NET	[28]
Potted plants	E	MT, SQT, and Other	BDS, BDT, BES, BET, and NET	[29]
FL, USA	E	Iso	BET	[30]
Potted plants	E	Other	NET	[31]
WI, USA	E	Iso	BDT	[32]
UK	E, C	Iso, MT	BES	[33]
Hangzhou, China	E	Iso, MT	BDT, BET, and NET	[34]
Italy	E, C	Iso, MT	BES, BET	[35]
Spain	E, C	MT, SQT	BET	[11]
CA, USA	E	Iso, MT	BDT, BET, and NET	[36]
Italy	C	Iso, MT, and Other	BDS, BES	[37]
Potted plants	E	MT	BES	[38]
CO, USA	C	Other	NET	[39]
Germany	E	MT	BDT	[40]
Mexico	E	Iso, MT	BDT, NET	[41]
MA, USA	L	Iso, Other	BDT	[42]
Potted plants	E	Iso, MT	BDS, BDT, BES, BET, C3G, and NET	[43]
CA	C	MT, SQT, and Other	BET	[44]
Potted plants	E	Other	BDT, BET, and NET	[45]
ON, Canada	C	Iso, MT	BDT	[46–48]
IL, USA	E	Other	C3G	[49]
Zhejiang, China	L	Iso, MT	BDT, NET	[50]
NC, USA	E, C	Iso	BDT	[51]
Various, USA	E	Iso, MT	BDT, BET	[52–54]
CO, USA	L	MBO	NET	[55]
MA, USA	C	Other	BDT	[56]
MA, USA	C	Iso	BDT	[57]
Various, USA	L	Iso, MT	BDT, NET	[58]
CO, USA	E	MT, Other	C3G	[6]
Potted plants	E	Iso, MT	BDT, BET	[59, 60]
NC, USA	C	Iso	BDT	[61]
Various, USA	E, L	Iso, MT	BDS, BDT, BES, and NET	[62–64]
TX, USA	E, C, and L	Iso, MT	BDS, BDT, and BES	[65]
Various, USA	E	Iso, MT	BDT	[66, 67]
CA, USA	E	MBO	NET	[68]
South Africa	E, C	Iso, MT	BDS, BDT, and BES	[69]
Potted plants	E	Other	BDT, C3G, and NET	[70]

TABLE 1: Continued.

Location	Approach	Compounds	PFTs	Reference
Greece	E, L	Iso, MT	NET	[71]
Potted plants	E	MT	BET	[72]
Potted plants	E	Other	NET	[73]
Various, USA	E	MT, SQT	BDT, BES, BET, and NET	[74]
Various, USA	E	MT, SQT	NET	[75, 76]
ID, USA	L	MT	NET	[77]
CA, USA	C	MT	NET	[78]
Germany	E	MT, SQT	NET	[79]
Shenzhen, China	E	Iso, MT	BDS, BDT, BES, BET, C3G, and NET	[80]
WI, USA	E, C, and L	Iso, MT	BDT, C3G, and NET	[81]
Russia	E	Iso, MT, and Other	NDT, NET	[82]
Various, USA	E, C	Other	BDT, NET	[83, 84]
AZ, USA	E, L	Iso, MT, and Other	BES	[85]
CO, USA	C	Iso, MT, Other, and MBO	NET	[86]
MI, USA	C	Iso, MT, and Other	BDT, NET	[87]
CA, USA	C	MT, SQT, and Other	BDT	[88]
CA, USA	E	Iso, MT	BDS, BDT, BES, BET, C3G, and NET	[89]
Italy	E	Iso, MT, and Other	BET	[90, 91]
France	E	Iso, MT	BDT, BET	[92]
Potted plants	E	Other	BDT	[93]
FL, USA	E	MT	NET	[94]
Republic of Korea	E	MT	NET	[95]
MI, USA	C	SQT	BDT	[96]
CO, USA	E, L	MT, SQT, and Other	NET	[97]
MI, USA	E	Iso, MT, SQT, and Other	BDT, NET	[98]
Republic of Korea	L	Iso, MT, and Other	BDT, NET	[99]
VIC, Australia	E	Other	C3G	[100]
Various, China	E	Iso, MT	BDS, BDT, BES, BET, C3G, and NET	[101]
Potted plants	E	Other	BDT, C3G	[102]
Potted plants	E	Other	BDT	[103]
Switzerland	E	Other	BET, BDT	[104]
Various, USA	E, C	Iso, MT	BDT, NET	[105]
WA, USA	E, C	Iso	BDT	[106]
OR, USA	E	MT	NET	[107]
Potted plants	E	MT	NET	[108]
Shenyang, China	E	Iso, MT	NET	[109]
Republic of Korea	E	Iso	BDT, BET	[110]
Potted plants	E	Iso	BDT	[111]
Spain	E	MT	BET	[112]
FL, USA	C	Iso, MT	NET	[113]
Potted plants	E	Iso	BDT	[114]
Various, USA	E	Other	BDS, BDT, and NET	[115, 116]
MI, USA	E	MT	C3G	[117]
PA, USA	L	Iso	BDT	[118]
NM, USA	E	Iso, MT, and Other	BDT, BET, NET	[119]
Various, USA	C	Other	BDS, BDT, and BES	[120]
US, Japan, and Australia	E	SQT	BDS, BDT, BES, and BET	[121]
Japan	E	Iso, Other	NET	[122, 123]
MA, USA	C	Iso, MT, and Other	BDT	[124]
Potted plants	E	Iso	BDT	[125]
CO, USA	E	Iso	BDT	[126]

TABLE 1: Continued.

Location	Approach	Compounds	PFTs	Reference
France	E	Iso	BDT	[127]
Estonia	E	Iso	BDT	[128, 129]
Portugal	E	Iso, MT	BET	[130]
Japan	E	Iso	BET	[131]
Spain	E	MT, SQT	BES, BET, and NET	[132]
MI, USA	E	Iso, MT, and SQT	BDT, NET	[133]
Various, USA	E	Iso, MT, and SQT	BDT, BET, and NET	[134]
Zambia, Botswana	E	Iso, MT	BDS, BDT, BES, and BET	[135]
Italy, France, Spain	E	Iso, MT	BDS, BDT, BES, BET, and NET	[136, 137]
NV, USA	E	Iso, MT, and SQT	BDS, BDT, BES, BET, and NET	[138]
Potted plants	E	Iso	BET	[139]
Spain	E	Iso, MT	BET	[140]
Potted plants	E	Iso	BDT	[141]
AL, USA	E	Iso	BDT	[142, 143]
Portugal	E	MT	NET	[144]
Belgium	E	Iso, MT, and Other	BDT, NET	[145, 146]
WA, USA	E	MT	NET	[147]
MI, USA	C	Iso	BDT	[148]
Austria	L	Other	C3G	[149]
Italy	E	Iso, MT	BDT, BET	[150]
Various, USA	E	Iso	BDT, BET, and NET	[151]
Various, USA	E, L	Iso, MT	BDT, BET, C3G, and NET	[152]
CA, USA	E	Other	C3G	[153]
CO, USA	L	MT	BDT, NET	[154, 155]
Potted plants	E	Iso, MT	BDT	[156]
Potted plants	E	Other	NET	[157]
Potted plants	E	Other	BDT, NET, and C3G	[158]
CA, USA	C	MBO, Other	NET	[159]
Potted plants	E	MT, Other	BET	[160]
NC, USA	E	Iso	BDT	[161]
Various, USA	E	Iso	BDT	[162]
Nepal	L	Iso, MT	NET, BDT, and C3G	[163]
Georgia, USSR	E, L	Iso, MT	BDT, NET	[164]
Belgium	E	MT, Other	BDT	[165]
Various, USA	L	Other	BDT, NET	[166]
Germany	C	Iso, MT, and Other	BDT	[167]
Potted plants	E	MT	BET	[168]
Germany	C	Iso, MT	BDT, NET	[169]
Japan	E	Iso	BDT, BET	[170]
Japan	C	MT	NET	[171]
Potted plants	E	Iso	BET	[172]
Potted plants	E	MT	NET	[173]
Greenhouse	E	Iso, MT	BET, BDT	[174]
Inner Mongolia, China	E	Iso, MT	C3G	[175]
Various, USA	L	Iso, MT	BDT, NET	[176]
France	E	Iso, MT	BDT, BET	[177]
Potted plants	E	MT	BES	[13]
MI, USA	C	Iso	BDT	[178]
TX, USA	L	Iso, MT	NET, BDT	[179]
Various, USA	L	Iso, MT	BDT, C3G	[180]
WA, Australia	E	Iso, MT, and Other	BET	[181]

TABLE 1: Continued.

Location	Approach	Compounds	PFTs	Reference
Beijing, China	E	Iso, MT	BDT	[182]
Potted plants	E	Other	NET	[183]
Japan	L	MT	NET	[184]
Potted plants	E	MT	NET	[185, 186]
Various, USA	E	Iso, MT, and Other	BDS, BDT, BES, BET, C3G, and NET	[14]

concluded that BVOC emissions did not have an important role in regional air quality [290]. An improved understanding of the magnitude of BVOC emissions and the relatively high sensitivity of ozone to BVOC emissions demonstrated that this was not the case [291, 292] and led to a resurgence in BVOC emissions research in boreal, tropical, and agricultural ecosystems in the 1990s. Interest in tropical landscapes was driven by the recognition that the tropics are responsible for 80% of global emissions [17]. There was initially little interest in BVOC from agricultural ecosystems because of the generally low terpenoid emissions from these plant species, but the discovery of substantial amounts of oxygenated VOC emissions from crops [102, 115] led to more studies. The annual publication rate decreased in the mid-2000s, but there has been a recent increase in the number of publications. This has likely been driven by the recognition of the important role of BVOC in secondary organic aerosol production [3, 293].

Figure 2 shows that temperate, tropical, and boreal ecosystems each cover 25 to 35% of the global vegetation-covered land surface with croplands covering the remaining 15%. This figure also shows that although the estimated global BVOC emission is dominated by tropical ecosystems, most studies have focused on temperate ecosystems. Needleleaf trees, broadleaf trees, shrubs, grass and crops each cover 10 to 30% of the global vegetation-covered land area but broadleaf trees are estimated to contribute nearly 80% of the emissions (Figure 3). Investigations of BVOC emission have generally focused on the important emission sources although broadleaf trees are somewhat understudied. Figure 2 shows that isoprene contributes about half of total emissions and was also investigated in about half of these studies. In contrast, other VOC are 36% of the estimated emissions and were examined in only ~20% of the studies. Recent studies provide some balance with relatively more investigations in the tropics and measurements of other VOC (Figure 1).

3. BVOC Chemical Diversity

Terrestrial ecosystems produce thousands of chemical species that can be emitted into the atmosphere [294] but only a few of these compounds are emitted at the rates required to have a significant impact on atmospheric composition [5]. Most of these chemicals are organic compounds including some that contain oxygen, nitrogen, sulfur, or halogens. Biogenic emission models often reflect this dominance by including only a few major compounds, such as isoprene and α-pinene, and omitting the rest or including them as a generic undefined "other" category. More recently, the

MEGAN2.1 biogenic emission model [5] was developed to estimate emissions of 147 compounds that were thought to be significant or potentially significant. This section describes BVOC chemical diversity and potential improvements over the MEGAN2.1 scheme.

3.1. Terpenoid Compounds. Terpenoid compounds have long been considered the dominant global BVOC [9]. This incredibly diverse group includes thousands of chemical species that can be classified as hemiterpenoids (C5), monoterpenoids (C10), sesquiterpenoides (C15), homoterpenes (C11 and C16), diterpenoids (C20), and larger compounds with such low volatility that it is unlikely that they are emitted into the atmosphere in a gaseous form. Terpenoids include oxygenated terpenes such as the hemiterpenoid methylbutenol (MBO), the monoterpenoid linalool, and the sesquiterpenoid cedrol. These oxygenated terpenoids are a small portion of the global total terpenoid emission but may be important in some regions. About half of the 147 BVOC species included in MEGAN2.1 are terpenoid compounds including some that are major contributors to global BVOC emissions (e.g., isoprene, α-pinene) and others that are minor components.

Investigations of BVOC began centuries ago with interest in commercial applications for monoterpenes in the flavor and fragrance industries. These activities led to the development of diverse analytical techniques and a considerable body of the literature describing terpenoid production and distribution in the oleoresins stored within plant tissues. Very little of this information has been incorporated into BVOC emission models because the production of monoterpenes by plants and their release into the atmosphere are not always well correlated, and only a small fraction of the hundreds of monoterpene compounds identified in essential oils have been observed as significant atmospheric BVOC emissions. Some monoterpenoids are oxygenated compounds including some multifunctional oxygenates and acetylated compounds that may make a disproportionate contribution to secondary aerosol production. Early studies indicated that a few monoterpenes (α-pinene, β-pinene, limonene, sabinene, 3-carene, and myrcene) dominated the total monoterpene flux into the atmosphere [284]. However, these studies typically did not attempt to measure all compounds, and some monoterpenes may have been reported more frequently simply because these were the only compounds targeted. It was initially thought that all monoterpenes emanated from storage pools and were controlled only by leaf temperature. The discovery of high emission rates of light-dependent monoterpene emissions, produced from recently synthesized carbon in a manner similar to isoprene,

TABLE 2: Compilation of studies used to estimate tropical vegetation BVOC emission factors for the MEGAN2.1 model [5]. Emission measurement approaches include enclosure (E), canopy micrometeorological (C), and landscape inverse modeling (L) techniques. Compounds include isoprene (Iso), monoterpenes (MT), and other (Other). PFTs include broadleaf deciduous tree (BDT), broadleaf evergreen tree (BET), and warm C4 grass (C4G).

Location	Scale	Compounds	PFTs	Reference
Yunnan, China	C	Iso, MT	BDT	[187]
Malaysia	E, L	Iso, MT	BDT, BET	[188]
AM, Brazil	L	Iso, MT	BDT, BET	[189]
Venezuela	L	Iso	C4G	[190]
Costa Rica	E, C	Iso, MT, and Other	BDT, BET	[191]
AM, Brazil	C	Iso, MT	BDT, BET	[192]
CAR	L	Iso	BDT, C4G	[193]
Botswana	C	MT	BDT	[194]
Various, Brazil	L	Iso, MT	BDT, BET	[195]
Guyana	L	Iso	BDT, BET	[196]
South Africa	E	Iso, MT	BDT, BET	[197]
Various, Brazil	E	Iso, MT	BDT, BET	[198]
Peru	L	Iso, MT	BDT, BET	[199]
Venezuela	L	Iso, MT, and Other	C4G	[200]
Costa Rica	C	Iso, MT, and Other	BDT, BET	[201]
AM, Brazil	C	Iso, MT, and Other	BDT, BET	[202]
AM, Brazil	C	Iso, MT, and Other	BDT, BET	[203]
Panama	E	Iso	BDT, BET	[204]
AM, Brazil	L	Iso, MT	BDT, BET	[205]
RO, Brazil	E, C	Iso, MT, and Other	BDT, BET	[206]
Cameroon, CAR, and Congo	E	Iso, MT	BDT, BET	[207]
Yunnan, China	E	Iso, MT	BDT, BET	[101]
RO, Brazil	E	Iso, MT	BDT	[208]
AM, Brazil	E	Iso, MT	BDT, BET	[209]
AM, Brazil	C	Iso, MT	BDT, BET	[210]
Sabah, Malaysia	C	Iso, MT, and Other	BDT, BET	[211, 212]
Potted plants	E	Iso, MT	BDT, BET	[213]
Sabah, Malaysia	C	Iso, MT, and Other	BET	[214, 215]
PA, Brazil	C	Iso, MT	BDT, BET	[216]
Potted plants	E	Iso	BDT, BET	[217]
India	E	Iso	BDT, BET	[218]
Panama	E, L	Iso	BDT, BET	[151]
AM, Brazil	L	Iso	BDT, BET	[219]
PA, Brazil	C	Iso, MT	BDT, BET	[220]
Malaysia	E, L	Iso, Other	BDT, BET	[221]
Venezuela	L	Iso	C4G	[222]
Benin	E, L	Iso, MT	BDT, BET	[223]
Congo	C	Iso	BDT, BET	[224]
India	E	Iso	BDT, BET	[10]
AM, Brazil	E, L	Other	BDT, BET	[225]
India	E	Iso	BDT, BET	[226]
Surinam	L	Iso, MT, and Other	BDT, BET	[227, 228]
Nigeria, AM, Brazil	L	Iso, MT	BDT, BET	[229]

from European [90] and African [194] savannas, tropical forests [202], and boreal needleleaf trees [237] led to the introduction of multiple emission processes for an individual chemical species in BVOC emission models.

Organic chemists investigating monoterpenes in the late 1800s identified the hemiterpene, isoprene (2-methyl-1,3-butadiene), as the biochemical precursor of monoterpenes, but isoprene was thought to exist only within plant tissues

TABLE 3: Compilation of studies used to estimate boreal vegetation BVOC emission factors for the MEGAN2.1 model [5]. Emission measurement approaches include enclosure (E), canopy micrometeorological (C), and landscape inverse modeling (L) techniques. Compounds include isoprene (Iso), monoterpenes (MT), sesquiterpenes (SQT), and other (Other). PFTs include broadleaf deciduous shrub (BDS), broadleaf deciduous tree (BDT), arctic C3 grass (AC3), needleleaf deciduous tree (NDT), and needleleaf evergreen tree (NET).

Location	Scale	Compounds	PFTs	Reference
Finland	E	MT	NET	[230]
Jilin, China	C	Iso, MT	BDT, NET	[231]
Sweden	E	Iso	AC3	[232, 233]
Sweden	E	Iso	AC3	[234]
Potted plant	E	Iso, MT, and Other	NET	[235]
SK, Canada	C	Iso	BDT	[236]
Finland	C	MT	BDT, NDT, and NET	[237]
Finland	C	Iso, Other	AC3	[238]
Finland	E	MT, SQT	BDS, BDT	[239]
Finland	E, L	Iso, MT, and SQT	BDT	[240, 241]
Finland	E	MT, SQT	NET	[242]
Potted plants	E	Iso	AC3	[243]
Finland	E	Iso, MT, and Other	AC3	[244]
Sweden	C	Iso, MT, and Other	AC3	[245]
Norway	L	MT	NET	[246]
WI, USA	E, C, L	Iso, MT	NDT	[81]
Various, Russia	E	Iso, MT, and Other	BDT, NDT, and NET	[82]
Sweden	L	MT	NET	[247]
Sweden	E	Iso	AC3, NET	[248]
Sweden	E	Iso, MT, and Other	AC3, NET	[249]
ON, Canada	L	Iso	BDT, NET	[250]
Potted plants	E	Iso, MT	NET	[251]
ON, Canada	E	Iso, MT	AC3	[252]
Jilin, China	E	Iso, MT	AC3, BDS, BDT, NDT, and NET	[101]
Potted plants	E	MT, SQT	NET	[253]
SK, Canada	C	Iso	BET	[254]
Sweden	E, L	MT	NET	[255]
AL, USA	C	Iso, MT, and Other	AC3, BDS	[256]
Finland	E, C	MT	NET	[257]
Finland	C	MT	NET	[258]
Finland	E, C	Iso, MT	BDT, NET	[259]
Finland	C	Iso, MT, and Other	NET	[260]
Finland	E	Iso, MT, and SQT	NDT	[261]
Finland	L	Iso, MT	BDT, NET	[262]
Finland	E	MT, SQT	NET	[263]
Sweden	E	Iso	AC3	[264, 265]
Finland	E	MT, SQT, and Other	BDS, BDT	[266]
Sweden	E	MT, SQT, and Other	BDS, BDT	[267]
Various, Canada	C	Iso	AC3, BDS, BDT, and NET	[268]

[151]. The discovery of substantial isoprene emissions from plants into the atmosphere was discovered more than 50 years ago and was initially controversial [152]. Isoprene later became recognized as the dominant global BVOC emission into the atmosphere [17]. Isoprene contributes about half of the total global BVOC flux, and so it is not surprising that it has been investigated more extensively than any other atmospheric BVOC.

Sesquiterpenes (SQTs) are a major component of essential oils stored by some plants, especially broadleaf trees, and can also be directly emitted without being stored [295]. SQTs are emitted from numerous plant species including conifer and broadleaf trees, shrubs, and agricultural crops [296]. While some sesquiterpenes, such as longifolene, have atmospheric oxidation lifetimes on the order of hours, similar to that of the dominant monoterpenes such as α-pinene, some of the

TABLE 4: Compilation of studies used to estimate cropland BVOC emission factors for the MEGAN2.1 model [5]. Emission measurement approaches include enclosure (E), canopy micrometeorological (C), and landscape inverse modeling (L) techniques. Compounds include isoprene (Iso), monoterpenes (MT), sesquiterpenes (SQT), and other (Other).

Location	Scale	Compounds	Crop	Reference
Potted plants	E	MT, SQT, and Other	Potato	[269]
CA, USA	E	Iso, MT, and Other	Various	[270, 271]
CA, USA	E	Iso, MT, and Other	Various	[272]
UK	E	Iso, MT, and Other	Miscanthus, willow coppice	[273]
Potted plants	E	SQT, Other	Tobacco	[274]
Potted plants	E	Iso, MT, SQT, and Other	Switchgrass	[275]
Potted plants	E	SQT	Corn	[276]
Potted plants	E	Other	Sorghum	[70]
Potted plants	E	Iso	Arundo donax	[277]
Spain, Italy	E	Other	Corn, pea, barley, and oat	[93]
VIC, Australia	E	Other	Clover	[100]
Potted plants	E	Iso, MT, SQT, and Other	Wheat, rye, rape, and grape	[102]
Potted plants	E	Iso	Velvet bean, kudzu	[278]
Potted plants	E	Other	Soybean, tomato, bean, and Corn	[115]
Potted plants	E	SQT	Corn	[279]
Potted plants	E	SQT, Other	Sunflower	[280]
Italy	E	Other	Fescue	[281]
CO, USA	C	Other	Alfalfa	[282]
Various, USA	E	Iso, MT, Other	Various	[14]

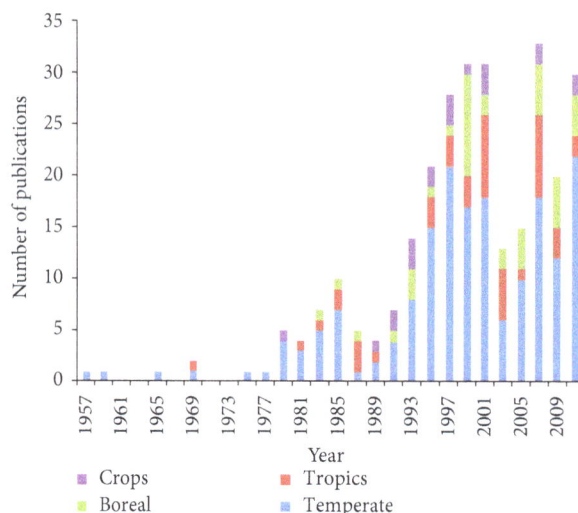

FIGURE 1: Illustration of the number of publications during two-year periods that were used to develop the BVOC emission factors for the MEGAN2.1 model [5].

most dominant sesquiterpenes emitted into the atmosphere (β-caryophyllene and farnesene) are much more reactive and have typical lifetimes of minutes [297]. The low volatility, and in some cases high reactivity, of sesquiterpenes makes them considerably more difficult to detect and quantify. As a result, few studies considered sesquiterpene emission measurements since they were generally thought to be a minor contribution in comparison to monoterpenes. Efforts to quantify sesquiterpene emissions increased in the past decade with the growing interest in atmospheric secondary organic aerosol [76]. Although sesquiterpenes are only a minor

fraction of total BVOCs, they are recognized as important for the atmosphere due to their relatively high SOA yields [298].

Large emissions of an oxygenated hemiterpene, 2-methyl-3-buten-2-ol (referred to here as MBO) were observed from pine trees in the early 1990s although emissions of MBO from insects and flowers had been observed previously [55]. MBO is emitted at high rates from some pine species, such as *Pinus ponderosa*, and low rates from other pines, including most Eurasian pines [299]. The global MBO emission is less than 1% of the global total BVOC, but MBO is the dominant emission in ecosystems dominated by high

(a)

(b)

(c)

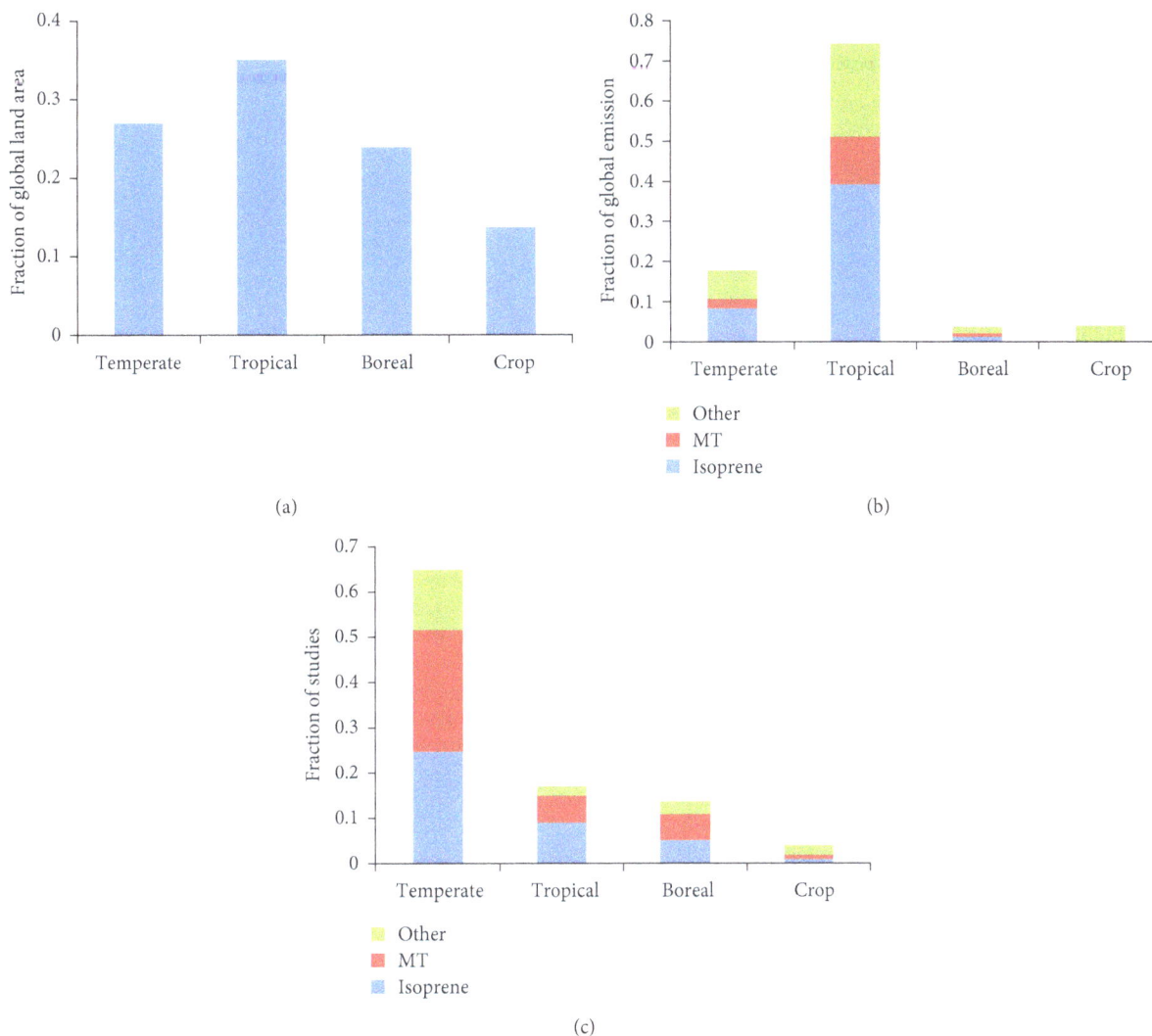

FIGURE 2: Comparison of the global fraction of vegetation-covered land area, global BVOC emissions estimated by [5], and the number of BVOC emission diversity studies (Tables 1–3) for major biomes types (temperate, tropical, boreal, and crop) and compound categories (isoprene, monoterpene, and other).

MBO emitting pines including large areas of western USA forests. Recent studies suggest that MBO may be emitted from most isoprene emitting vegetation at a rate that is ~1% of the isoprene emission rate [300]. This low level emission over a large part of global terrestrial ecosystems could be of the same magnitude as the localized emission from high MBO emitters.

The production of some terpenoid compounds is elevated in response to stress and is often observed as a light dependent, de-novo emission [301]. These include monoterpenes (e.g., ocimene), oxygenated monoterpenes (e.g., linalool), sesquiterpenes (e.g., farnesene), the homoterpenes dimethyl-nonatriene (DMNT), and trimethyl-tridecatetraene (TMTT). Emissions of these compounds are not always present, but when they are observed they can exceed typical monoterpene or sesquiterpene emission rates. The large variability and limited knowledge of factors controlling stress-induced BVOC emissions result in high uncertainties

associated with emissions of these compounds, but they may be a substantial component of total BVOC emissions into the atmosphere, and a better understanding is needed.

3.2. Methanol and Acetone. Methanol and acetone are among the most abundant VOCs in the global atmosphere. High concentrations of atmospheric methanol and acetone observed by investigators in the 1960s were attributed primarily to the atmospheric oxidation of VOC with minor contributions from bacteria, biomass burning, and anthropogenic sources [302]. In the early 1990s, high rates of methanol emissions were observed from vegetation foliage, especially young expanding leaves [115]. Lower rates of acetone emissions were observed from conifer buds [116]. A few years later, decaying leaf litter was found to be a smaller but significant abiotic source of methanol and acetone [303].

Jacob et al. [304] estimated that terrestrial ecosystems (biotic and abiotic) dominate global methanol emissions with

(a)

(b)

(c)

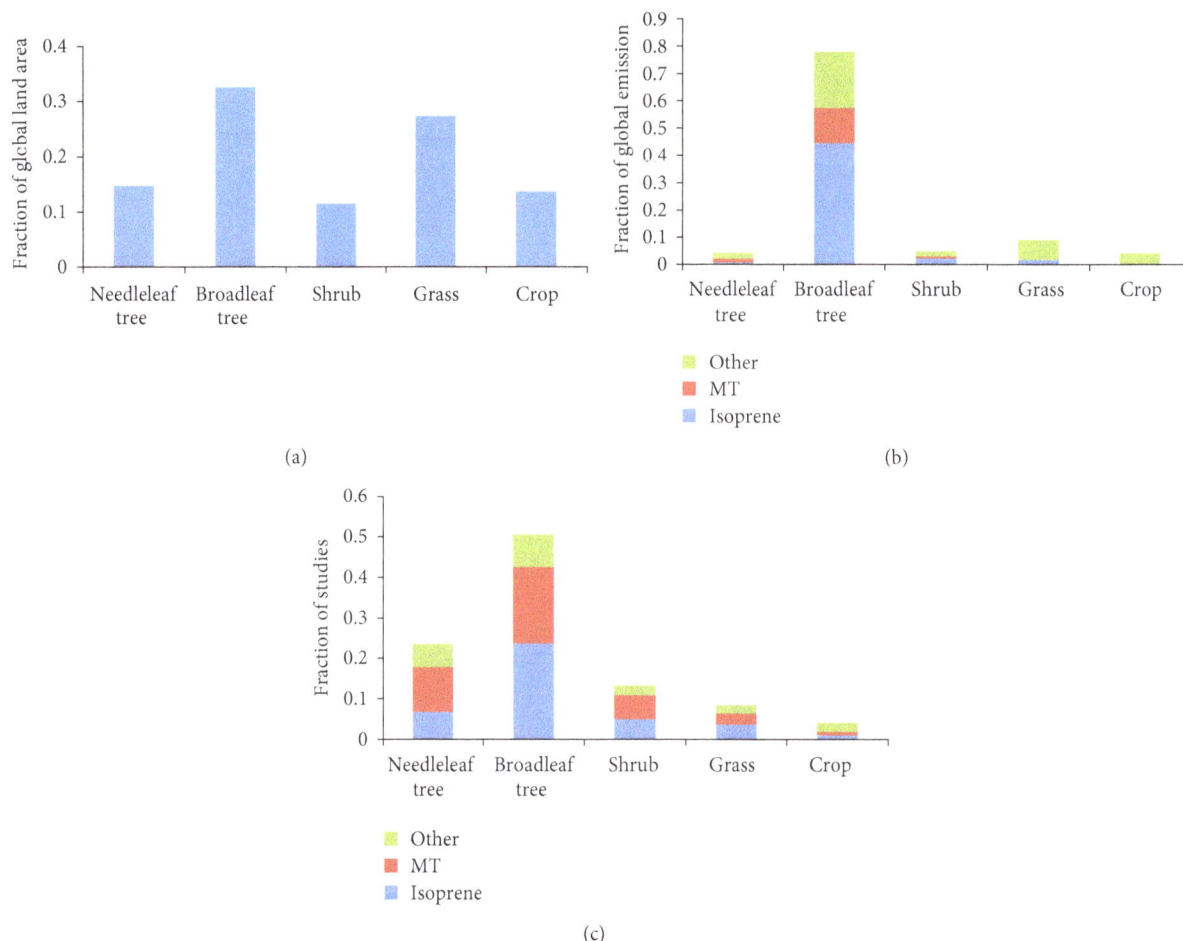

FIGURE 3: Comparison of the global fraction of vegetation-covered land area, global BVOC emissions estimated by [5], and the number of BVOC emission diversity studies (Tables 1–3) for major plant functional types (needleleaf tree, broadleaf tree, shrub, grass, and crop) and compound categories (isoprene, monoterpene, and other).

78% of the global annual production with the remainder being from atmospheric oxidation of VOC (15%), biomass burning (5%), and urban (2%) sources. Millet et al. [305] used additional in situ observations and concluded that oceans were responsible for 35% of the global methanol emission and assigned a contribution of 42% to terrestrial ecosystems. Stavrakou et al. [306] used both satellite and aircraft observations to constrain global methanol distributions and report annual emissions of 187 Tg per year with a contribution of 53% from vegetation. They also identified missing sources in arid and semiarid regions of Central Asia and Western USA.

An analysis of the global acetone budget by Jacob et al. [307] indicated contributions to total emissions from terrestrial ecosystems (37%), atmospheric oxidation of VOC (29%), ocean (28%), biomass burning (5%), and anthropogenic emissions (1%). A more recent analysis concluded that terrestrial ecosystems were responsible for only 22% and oceans contributed 55% [308].

3.3. Acetaldehyde, Formaldehyde, Ethanol, and Organic Acids. Kesselmeier [309] described both the atmospheric importance of short-chained oxygenated VOCs (e.g., acetaldehyde,

formaldehyde, acetic acid, and formic acid) and the challenge of quantifying their atmospheric budgets. This includes the following challenges: (1) there are both natural and anthropogenic sources of these compounds, (2) there are primary and secondary (atmospheric oxidation) sources, (3) these compounds are difficult to measure, and (4) vegetation is both a source and a sink of these compounds. The strong bidirectional exchange exhibited by these compounds requires that both emission and deposition need to be considered. Accurate simulation of land-atmosphere fluxes of these compounds requires estimates of their atmospheric concentrations and the compensation point for each compound.

Alcoholic fermentation in the leaves and roots of plants produces ethanol which is converted to acetaldehyde in a pathway leading to acetate consumption [83]. Millet et al. [310] identified the major sources of atmospheric acetaldehyde as oxidation of VOC (60%), ocean (27%), and terrestrial ecosystems (11%). Biomass burning and anthropogenic emissions contribute the remaining 2%. The introduction of the PTRMS technique has provided an increasing number of measurements of acetaldehyde emissions from vegetation, including whole canopy flux measurements, while there remain relatively few data for ethanol [310].

Substantial emissions of formaldehyde, and lesser amounts of formic and acetic acid, have been reported from studies using enclosure measurements to investigate various tree species [103, 309]. While emissions can be considerable, there is also the potential for a strong uptake of these compounds. These enclosure measurements suggest that the net flux of these compounds is a small emission into the atmosphere. Recent studies using above-canopy measurements have provided evidence that formaldehyde and formic acid emissions could be much larger. An analysis of satellite data suggests that formic acid emissions are two to three times higher than estimated from known sources [311] and that 90% of formic acid has a biogenic origin which includes direct emission and production from terpenoids. The first whole canopy fluxes of formaldehyde measured by eddy covariance have recently been reported [39]. The above-canopy flux, a net emission, was much higher than predicted from enclosure measurements which may be because the flux included both primary emissions and within canopy production. Measurements to better constrain formic acid and formaldehyde fluxes are needed.

3.4. Stress Compounds. Environmental and biotic stresses are important factors controlling BVOC emissions [128]. This includes BVOCs that are emitted at relatively low levels with unstressed conditions and then are elevated under stressed conditions (e.g., α-pinene) and compounds that are typically observed only when plants are stressed (e.g., methyl salicylate). BVOCs associated with pathogen or herbivore-induced stress include ethene, methanol, terpenoids, benzenoids, and green leaf volatiles [73, 128, 312–315]. The biochemical pathways and the defensive roles of these compounds have been the subject of many investigations [316], but there have been few attempts to quantify these emissions and they have not been integrated into regional BVOC emission models. The current limited understanding of the processes controlling stress-induced emissions makes any numerical approach for estimating stress BVOC emissions highly uncertain. Observations that provide an initial assessment of stress-induced emissions provide a first step towards assessing their contribution to total BVOC emissions and the need for accounting for these processes in BVOC emission models.

Ethene is an important phytohormone, and its emission rate from plants has been used as an indicator of stress [317]. Sawada and Totsuka [158] estimated an annual global flux of 18 to 45 Tg of ethene with 74% released from natural sources. This was the first global emission estimate of a nonterpenoid BVOC and was based on an extrapolation of enclosure measurements that indicated widespread ethene production by plants in most landscapes. Canopy scale fluxes measured above a temperate deciduous forest confirmed that substantial amounts of ethene were released into the atmosphere from this landscape [56]. The canopy scale fluxes are in reasonable agreement with the earlier enclosure measurements.

The green leaf volatiles are a major category of BVOC that is associated with plant response to herbivory and other stresses [12]. These compounds are produced in plants from linoleic and linolenic acid which are unsaturated fatty acids. The most prominent of these with respect to emissions into the atmosphere are cis-3-hexenal, trans-2-hexenal, hexanal, 1-hexanol, and cis-3-hexenol [287]. The compound methyl jasmonate is also produced from this pathway and has an important role in plant signaling [318].

3.5. Leaf Surface Compounds. Leaf surfaces are covered by a waxy material that serves as a barrier for keeping water in and pathogens out [319]. Long-chain hydrocarbons, acids, alcohols, and esters are the dominant components of these leaf waxes, but there are a variety of other constituents [320]. While these high molecular weight compounds have low volatility, a small fraction can volatilize into the gas phase, and this may be significant, especially with the high leaf temperatures ($>40°C$) that occur in hot deserts. A study by Matsunaga et al. [120] concluded that some compounds, including homosalate and 2-ethylhexyl salicylate, were emitted at significant rates from a wide variety of plants. These are sunscreen compounds that protect plant tissues from UV solar radiation [120]. The estimated contribution to total emissions from most ecosystems was small, but a large contribution was estimated for desert regions dominated by mesquite (*Prosopis* spp.) which is an important component of large areas in the southwestern USA.

Another source of VOC emissions from vegetation is the oxidation of organics on the surface of leaves and other structures. Fruekilde et al. [45] fumigated leaves with ozone and reported elevated emissions of 6-methyl-5-hepten-2-one, acetone, geranyl acetone, and 4-oxopentanal and suggested that ozonolysis at vegetation surfaces could explain the widespread occurrence of these compounds in ambient air. Karl et al. [321] noted that elevated oxygenated VOC emission from foliage exposed to ozone could also be due to increased production of these compounds in leaves in response to stress or to gas phase oxidation (secondary compounds). They conducted experiments to isolate the mechanisms responsible for oxygenated VOC production and concluded that a substantial amount of oxygenated VOC was primary emissions, originating from reaction of ozone inside of the plant or on plant surfaces, although there were also some secondary products from gas phase reactions.

3.6. Organic Halides. Organic halides including methyl bromide, methyl chloride, and methyl iodide are produced by vegetation and emitted into the atmosphere. Emissions are controlled by environmental conditions including soil moisture and temperature [322]. Even though methyl halide fluxes are small compared to terpenoid emissions, they are an important source of halogens in the stratosphere where they play a role in stratospheric ozone depletion [153]. Quantifying fluxes of methyl halides is challenging because terrestrial ecosystems are both a source and a sink of these compounds [166, 323]. Stable isotopes are now being used to individually quantify gross emission and uptake rates to improve understanding of the processes driving net fluxes [322, 324].

3.7. Organic Sulfur Compounds. Biogenic organic sulfur emissions from marine and terrestrial ecosystems are an important source of atmospheric sulfur compounds in clean environments [325]. Soil microbes and plants are both sources of compounds that include methyl mercaptan, dimethyl sulfide, and dimethyl disulfide. A more recent study [326] estimated that terrestrial ecosystems contribute about 15% of the global dimethyl sulfide flux with the remainder coming from oceans. Higher weight organic sulfur compounds such as diallyl disulfide, methyl propenyl disulfide, and propenylpropyldisulfide can be emitted in substantial amounts from a few plant species [149].

3.8. Alkanes (including Oxygenated Alkanes). Zimmerman [14] reported that a variety of alkanes were a substantial fraction of the biogenic VOCs emitted from vegetation. This was based on gas chromatograph retention times, rather than identification by mass spectrometry, and later studies have found only very low level of emissions of alkanes including ethane [100], propane [249], pentane [82], hexane [136], heptane [157], C6 to C10 saturated aldehydes [327], alcohols [281], ketones [282], pyruvic acid [85], and methane [328]. The potentially large source of methane [328] has been controversial as following studies found either much lower or no methane emission from living plants [329]. Terrestrial ecosystems are, however, a major source of methane emission from soil microbes and termites [328].

3.9. Benzenoid Compounds. The extensive BVOCs emission surveys of Zimmerman [14] also indicated that benzenoid compounds were a substantial fraction of total BVOC emissions. As was the case for alkanes, later studies found much lower benzenoid emissions. However, it is widely recognized that there are many benzenoid compounds (e.g., benzaldehyde, anisole, and benzyl alcohol) emitted as floral scents [330]. These floral benzenoid emissions are thought to make a small contribution to annual regional BVOC emissions [4] but can be a major emission at specific locations [214]. At least some of these compounds (e.g., toluene and methyl salicylate) are associated with plant stress and have been observed at elevated rates from stressed plants [73, 88].

3.10. Other Alkenes (including Oxygenated Alkenes). The terpenoids are not the only alkenes emitted into the atmosphere from terrestrial ecosystems. Propene and butene emissions have been observed in enclosures and confirmed by above-canopy flux measurements [56]. Other longer-chain alkenes have only been observed using enclosure techniques. This includes 1-dodecene and 1-tetradecene [270]. Oxygenated alkenes such as 1,3-octenol [281], neryl acetone [75], terpinyl acetate [183], and nonenal [74] have also been observed but are thought to be minor in comparison to terpenoid emissions.

3.11. Representing BVOC Chemical Diversity in Numerical Models. The first detailed biogenic VOC emission inventory [9] included estimates of just two compounds: isoprene and α-pinene. Several decades later, the USA EPA released the first widely available biogenic emission inventory approach, called BEIS [331]. In addition to emission of isoprene and α-pinene, BEIS included lumped categories for "other monoterpenes" and an "unidentified" category. While this made the emission inventory more comprehensive, the "unidentified" category had limited use in atmospheric chemistry models because BVOCs have such varied atmospheric impacts (e.g., a wide range in aerosol yields and ozone production potential). In addition, some highly reactive BVOC may control the local atmospheric oxidizing capacity, while other less reactive compounds are transported long distances to remote areas or to the stratosphere where they can impact the chemistry of these pristine regions. An initial attempt to account for this was made [17] by using two "other" BVOC categories that included "other reactive VOC," such as 232-MBO and "other VOC" which included less reactive compounds such as methanol and acetone. Emissions of 39 individual BVOCs were later estimated [287] in addition to three other categories: other terpenoids, other reactive NMVOCs, and other NMVOCs. The 39 identified compounds contributed about 94% of the total emission. MEGAN2.1 [5] eliminated the use of any "unidentified" categories and estimated emissions of 149 known compounds.

Most atmospheric chemistry schemes include at most only a few BVOCs and may lump these together with other compounds which limits the advantages of a detailed emissions chemical speciation. The increased number of compounds is a disadvantage if there is a significant increase in the computational resources associated with emissions parameterization, processing inputs, and emission calculations. MEGAN2.1 [5] uses a balanced approach that includes individual representations of 13 major BVOCs along with 5 additional categories for which an emission was calculated, and then the total was speciated into individual BVOC. This approach required the calculation of the emission activity of 18 BVOC types. The emission behavior of a compound, for example, the light dependent response, was treated the same for all vegetation types. This is reasonable for some compounds, such as isoprene, but not for others, such as α-pinene, which have different emission behavior in a tropical forest than in a temperate needleleaf forest [202]. This approach can be improved by using a smaller number of compound types but allowing a different emission behavior for different vegetation types. The 18 BVOC categories used for MEGAN2.1 [5] could be reduced to about half that number. For example, a nine BVOC category scheme could include hemiterpenes, light-dependent monoterpenes, light-independent monoterpenes, sesquiterpenes, methanol, acetone, bidirectional compounds, stress compounds, and other compounds. Each of these nine BVOC emission categories could have a different speciation profile for each vegetation type to simulate differences such as the contributions of individual monoterpenes to the total monoterpene flux from different landscapes.

4. BVOC Biological Diversity

Just as the scent of various flowers can be quite distinct, the total BVOC emission rates of various plants can differ. Some plants have total BVOC emission rates that are less than $0.01\,\mu g$ per gram (dry weight) per hour ($\mu g\,g^{-1}\,h^{-1}$), while others have rates that exceed $100\,\mu\mathrm{g\,g^{-1}\,h^{-1}}$. In addition to the three orders of magnitude variability in total emission, chemical composition can vary greatly with some plants dominated by isoprene, while emissions of other plants are dominated by other compounds such as α-pinene, MBO, or methanol. The BVOC emission rates of different terrestrial ecosystems vary by more than three orders of magnitude because the landscape average emission is determined both by the variability associated with plant-specific BVOC emission rates and the variability in vegetation cover fraction. In order to investigate BVOC emission variations associated with biological diversity, it is useful to define an emission factor for a set of standard conditions such as leaf age, growth environment, light, temperature, CO_2 concentration, soil moisture, and others [5, 129]. While there are clear taxonomic patterns associated with BVOC emissions, with plants of the same species or genus tending to be more similar, there are also many exceptions [332, 333]. This is not unexpected since the taxonomic schemes used to classify plants are not based on their BVOC emissions characteristics. In addition, some BVOC emissions variability is expected within plant species. For example, pine trees emit a variety of monoterpenes that are used for chemical defense against many different pests [334]. If all individuals of a pine species emit the same mix of monoterpenes, then a herbivore that manages to overcome this particular chemical mixture could devastate that pine species. If there are pine populations with different monoterpene emission types, then at least some pine tree individuals will survive.

Welter et al. [177] investigated BVOC emissions of an isoprene emitting oak species (*Quercus canariensis*), a monoterpene emitting oak species (*Q. suber*), and a species that is a hybrid of those two oak species (*Q. afares*). They found that *Q. afares* individuals were monoterpene emitters but at relatively low rates and with high variability. Geron et al. [54] examined isoprene emissions from *Populus* hybrids and found that their emission factors were a factor of two higher than their parents and that the second generation crosses had even higher emission factors. Bäck et al. [230] measured terpenoid emissions of individual Scots Pine (*Pinus sylvestris*) trees in a forest stand in Finland. They found that emissions of some trees were dominated by α-pinene, while others emitted primarily D3-carene, and still others emitted similar amounts of both. These studies demonstrate that there can be substantial within-species variation in terpenoid emissions for at least some plant species. Geron et al. [54] also considered whether there were significant interspecies differences in the isoprene emission factors of isoprene-emitting temperate broadleaf tree species and concluded that there was no clear evidence of this. Variability within and between species was similar suggesting that all temperate broadleaf trees could be divided into just two categories with respect to their isoprene emission: low emitters ($<1\,\mu\mathrm{g\,g^{-1}\,h^{-1}}$) and high emitters (about $90\,\mu\mathrm{g\,g^{-1}\,h^{-1}}$). Isoprene emission factors for high emitting temperate needleleaf trees were much lower than broadleaf trees indicating a need to assign different isoprene emission factors to different PFTs.

Numerical land surface models typically classify terrestrial ecosystems as either a landcover type [335] or a mixture of PFTs [336]. A savanna is an example of an ecosystem that is a mixture of grass and tree PFTs. Models based on a landcover classification have parameterizations that are intended to represent the weighted average for all of the vegetation species found in the biome. Plant functional types represent groups of vegetation species that are similar for at least some physiological and ecological traits. While it is possible for biome schemes to have very detailed classes, those used in global land surface models are simple approaches that provide a limited ability to represent BVOC emission diversity. A scheme with just five vegetation types (e.g., broadleaf forests, needleleaf forests, shrublands, grasslands, and croplands) was able to account for a significant part of BVOC emission diversity [337]. A small to moderate number (5 to 25) of global PFTs provide a reasonable approach for estimating global isoprene emissions at coarse resolution but cannot represent the considerable within-biome emission diversity which results in large errors in local to regional isoprene emission estimates [5].

The MEGAN2.1 [5] approach for simulating BVOC emission diversity is based on the Community Land Model version 4 PFT scheme [338]. The CLM4 approach is typical of the PFT schemes used for the land surface component of global earth system models and includes 6 temperate, 5 boreal/arctic, 3 tropical, and 1 crop PFTs. Table 5 outlines a framework to improve BVOC emission model estimates by expanding the 15 CLM PFTs to 39 PFTs that can better represent the biological BVOC diversity in earth system models. This approach includes a representative "type" species for each of the PFTs listed in Table 5. The first step towards implementing this approach is to conduct an extensive and systematic quantification of the BVOC emission rates of each of these species. This can be accomplished with enclosure measurements [129] or above-canopy flux measurements above monospecific stands [194]. Additional PFTs can be added when it can be demonstrated that their emission characteristics are substantially different from those on this list.

The three needleleaf tree PFTs included in CLM4 are temperate evergreen, boreal evergreen, and boreal deciduous. Figure 3 shows that needleleaf trees cover about 15% of the global vegetation covered land area but are estimated to contribute less than 5% of the total BVOC. Figure 3 also shows that nearly 25% of BVOC studies have targeted needleleaf trees indicating that they are relatively well studied. The studies summarized in Table 4 show that all three PFTs include a monoterpene emitting type. Both temperate and boreal evergreen species also include high isoprene [251] and high MBO [68] emitters, and there is some indication that there should be a low emission category for at least the temperate evergreen trees [284]. It should be noted that the available data for characterizing emissions is limited, and the results of different studies are often conflicting. For example, a literature review [284] indicated that the *Pseudotsuga menziesii* (Douglas fir) monoterpene emission factor is a factor of 8 higher than that of *Tsuga heterophylla* (western hemlock). In contrast, another study [147] found that the

TABLE 5: Plant functional type (PFT) scheme for representing BVOC emission biological diversity. The 15 CLM4 PFTs are subdivided into BVOC emission types (BVOC-PFT). The measured emission characteristics of the indicated representative species can be used to assign BVOC emission factors.

CLM4 PFT description	BVOC-PFT	Representative species
Needleleaf evergreen temperate tree	NETT-IM	*Picea engelmannii* (Engelmann spruce)
Needleleaf evergreen temperate tree	NETT-MT	*Abies grandis* (grand fir)
Needleleaf evergreen temperate tree	NETT-MBO	*Pinus ponderosa* (ponderosa pine)
Needleleaf evergreen temperate tree	NETT-Low	*Thuja plicata* (western redcedar)
Needleleaf evergreen boreal tree	NEBT-MBO	*Pinus contorta* (lodgepole pine)
Needleleaf evergreen boreal tree	NEBT-Iso	*Picea mariana* (black spruce)
Needleleaf evergreen boreal tree	NEBT-MT	*Pinus sylvestris* (Scots pine)
Needleleaf deciduous boreal tree	NDBT-MT	*Larix sibirica* (Siberian larch)
Broadleaf evergreen temperate tree	BETE-Iso	*Quercus virginiana* (southern live oak)
Broadleaf evergreen temperate tree	BETE-IM	*Eucalyptus globulus* (blue gum)
Broadleaf evergreen temperate tree	BETE-MT	*Quercus ilex* (holm oak)
Broadleaf evergreen temperate tree	BETE-Low	*Lithocarpus densiflorus* (tanoak)
Broadleaf evergreen tropical tree	BETR-Iso	*Mangifera indica* (mango)
Broadleaf evergreen tropical tree	BETR-MT	Tropical evergreen MT emitters
Broadleaf evergreen tropical tree	BETR-Low	*Panda oleosa*
Broadleaf deciduous tropical tree	BDTR-Iso	*Hymenaea courbaril* (jatobá)
Broadleaf deciduous tropical tree	BDTR-MT	*Apeiba tibourbou*
Broadleaf deciduous tropical tree	BDTR-Low	*Combretum molle* (velvet bushwillow)
Broadleaf deciduous temperate tree	BDTE-Iso	*Quercus rubra* (red oak)
Broadleaf deciduous temperate tree	BDTE-IM	*Liquidambar styraciflua* (sweetgum)
Broadleaf deciduous temperate tree	BDTE-MT	*Acer saccharum* (sugar maple)
Broadleaf deciduous temperate tree	BDTE-Low	*Sassafras albidum* (sassafras)
Broadleaf deciduous boreal tree	BDBT-Iso	*Populus tremuloides* (aspen)
Broadleaf deciduous boreal tree	BDBT- Low	*Betula pendula* (silver birch)
Broadleaf evergreen temperate shrub	BETS-MT	*Larrea tridentata* (creosote bush)
Broadleaf evergreen temperate shrub	BETS-Iso	*Karwinskia humboldtiana* (coyotillo)
Broadleaf evergreen temperate shrub	BETS-Low	*Atriplex canescens* (four-wing saltbush)
Broadleaf deciduous temperate shrub	BDTS-MT	*Ambrosia dumosa* (white bursage)
Broadleaf deciduous temperate shrub	BDTS-Iso	*Psorothamnus fremontii* (Fremont's dalea)
Broadleaf deciduous temperate shrub	BDTS-Low	*Baccharis texana* (prairie false willow)
Broadleaf deciduous boreal shrub	BDBS-Iso	*Salix arctica* (arctic willow)
Broadleaf deciduous boreal shrub	BDBS-Low	*Alnus crispa* (mountain alder)
C3 grass	C3G-Iso	*Carex appendiculata* (sedge)
C3 grass	C3G-Low	*Koeleria cristata* (June Grass)
C4 Grass	C4G-low	*Bouteloua curtipendula* (sideoats grama)
Arctic C3 grass	AC3G-Iso	*Eriophorum angustifolium* (cottongrass)
Arctic C3 grass	AC3G-Low	*Festuca rubra* (red fescue)
Crop	CRP-MT	*Helianthus annuus* (sunflower)
Crop	CRP-Iso	*Mucuna pruriens* (velvet bean)
Crop	CRP-Low	*Triticum aestivum* (wheat)

western hemlock monoterpene emission factor is more than twice as high as the value for Douglas fir.

The CLM4 PFTs for broadleaf trees include tropical evergreen, tropical deciduous, temperate evergreen, temperate deciduous, and boreal deciduous trees. These broadleaf trees cover about a third of the vegetation-covered earth surface and are estimated to account for almost 80% of the global total

BVOC emission (Figure 3). About half of the BVOC emission diversity studies in Tables 1 to 3 have focused on broadleaf trees resulting in a relatively good characterization of temperate and boreal species, but tropical broadleaf tree emissions have not received enough attention (Figure 2). Each of the five CLM4 broadleaf tree PFTs (Table 5) include a high isoprene emitting type. Some also include a high MT emission type

[241, 339], a high isoprene and high monoterpene emission type [181], and a low emission type [284].

The CLM4 scheme includes just three shrub PFTs: broadleaf deciduous temperate, broadleaf evergreen temperate and broadleaf deciduous boreal. The two temperate shrub PFTs include high monoterpene, high isoprene, and low emitter categories [65, 138, 340, 341]. The boreal shrub PFT includes both high isoprene and low emitters [256].

The three CLM4 grass PFTs are C3 grass, C4 grass, and arctic C3 grass. All three PFTs are dominated by a low terpenoid emitting category, but the temperate and arctic C3 PFTs also include some isoprene emitting species [22, 49, 100, 233, 342]. The crop PFT is dominated by low terpenoid emitters, but there are some examples of high isoprene and high monoterpene emitters [275, 278, 280, 343].

5. Conclusions

This review summarizes the current understanding of BVOC chemical and biological diversity. There are hundreds of BVOCs emitted into the atmosphere, but a relatively few compounds (e.g., isoprene, methanol, α-pinene, acetone, and ethene) dominate the total flux. All BVOCs can influence atmospheric composition, if they are emitted at sufficient rates, but some BVOCs have a relatively high impact due to their reaction rates, products, ozone production potentials, organic aerosol yields, and other properties. As a result, there is a strong need to quantify the chemical diversity of BVOC emissions. On the other hand, a detailed numerical description of BVOC chemical speciation increases computational requirements and the personnel needed to process input variables. In addition, the large uncertainties associated with BVOC emission estimates do not justify an overly detailed parameterization of these compounds. An approach for accurately representing BVOC chemical diversity in emission models requires a balance between providing the appropriate level of details while also minimizing the complexity.

Global land surface models simulate regional variations in ecosystem-atmosphere carbon exchange by assigning values of the photosynthetic parameter $V_{c\,max}$ to each PFT. This parameter describes the maximum rate of carboxylation by the photosynthetic enzyme Rubisco. The values of $V_{c\,max}$ assigned to the 15 PFTs used by CLM4 vary from $52\,\mu mol\,m^{-2}\,s^{-1}$ for grasses to $72\,\mu mol\,m^{-2}\,s^{-1}$ for broadleaf evergreen trees and shrubs [18]. In contrast, the isoprene emission factor, which describes the isoprene emission rate at a set of standard conditions, ranges from $1\,\mu g\,m^{-2}\,h^{-1}$ for boreal deciduous needleleaf trees to $11000\,\mu mol\,m^{-2}\,s^{-1}$ for broadleaf deciduous boreal trees [5]. This comparison illustrates that there is a much greater range in the ability of plants to emit isoprene than there is for photosynthesis. Assigning BVOC emission factors to 15 PFTs is a good initial step towards characterizing BVOC biological diversity, but it is insufficient. A scheme with about 39 PFTs is proposed to improve regional to global BVOC emission estimates.

Reducing uncertainties in BVOC emission estimates will require additional observations. Measurements are especially needed for specific vegetation types (e.g., tropical broadleaf forest and crops) and some nonterpenoid compounds (e.g., ethene, propene, ethanol, ocimene, and hexenal). Leaf-level enclosure measurements are needed to improve representations of the processes controlling emission variations. Tower- and aircraft-based above-canopy flux measurements are also needed to quantify BVOC diversity on landscape to regional scales.

References

[1] T. Pierce, C. Geron, L. Bender, R. Dennis, G. Tonnesen, and A. Guenther, "Influence of increased isoprene emissions on regional ozone modeling," *Journal of Geophysical Research D*, vol. 103, no. 19, pp. 25611–25629, 1998.

[2] A. G. Carlton, R. W. Pinder, P. V. Bhave, and G. A. Pouliot, "To what extent can biogenic SOA be controlled?" *Environmental Science and Technology*, vol. 44, no. 9, pp. 3376–3380, 2010.

[3] D. V. Spracklen, J. L. Jimenez, K. S. Carslaw et al., "Aerosol mass spectrometer constraint on the global secondary organic aerosol budget," *Atmospheric Chemistry and Physics*, vol. 11, no. 23, pp. 12109–12136, 2011.

[4] R. Baghi, D. Helmig, A. Guenther, T. Duhl, and R. Daly, "Contribution of flowering trees to urban atmospheric biogenic volatile organic compound emissions," *Biogeosciences*, vol. 9, pp. 3777–3785, 2012.

[5] A. B. Guenther, X. Jiang, C. L. Heald et al., "The model of emissions of gases and aerosols from nature version 2. 1 (MEGAN2. 1): an extended and updated framework for modeling biogenic emissions," *Geoscientific Model Development*, vol. 5, no. 6, pp. 1471–1492, 2012.

[6] J. P. Greenberg, D. Asensio, A. Turnipseed, A. B. Guenther, T. Karl, and D. Gochis, "Contribution of leaf and needle litter to whole ecosystem BVOC fluxes," *Atmospheric Environment*, vol. 59, pp. 302–311, 2012.

[7] T. E. Graedel, "Terpenoids in the atmosphere," *Reviews of Geophysics & Space Physics*, vol. 17, no. 5, pp. 937–948, 1979.

[8] A. H. Goldstein and I. E. Galbally, "Known and unexplored organic constituents in the earth's atmosphere," *Environmental Science and Technology*, vol. 41, no. 5, pp. 1514–1521, 2007.

[9] R. A. Rasmussen, "What do the hydrocarbons from trees contribute to air pollution?" *Journal of the Air Pollution Control Association*, vol. 22, no. 7, pp. 537–543, 1972.

[10] R. Singh, A. P. Singh, M. P. Singh, A. Kumar, and C. K. Varshney, "Emission of isoprene from common Indian plant species and its implications for regional air quality," *Environmental Monitoring and Assessment*, vol. 144, no. 1–3, pp. 43–51, 2008.

[11] P. Ciccioli, E. Brancaleoni, M. Frattoni et al., "Emission of reactive terpene compounds from orange orchards and their removal by within-canopy processes," *Journal of Geophysical Research D*, vol. 104, no. 7, pp. 8077–8094, 1999.

[12] N. Dudareva, F. Negre, D. A. Nagegowda, and I. Orlova, "Plant volatiles: recent advances and future perspectives," *Critical Reviews in Plant Sciences*, vol. 25, no. 5, pp. 417–440, 2006.

[13] F. W. Went, "Blue hazes in the atmosphere," *Nature*, vol. 187, no. 4738, pp. 641–643, 1960.

[14] P. Zimmerman, "Testing of hydrocarbon emissions from vegetation, leaf litter and aquatic surfaces and devlopment of a method for compiling biogenic emission inventories," Tech. Rep. EPA-450-4-70-004, U.S. Environmental Protection Agency, Research Triangle Park, calif, USA, 1979.

[15] B. Lamb, A. Guenther, D. Gay, and H. Westberg, "A national inventory of biogenic hydrocarbon emissions," *Atmospheric Environment*, vol. 21, no. 8, pp. 1695–1705, 1987.

[16] J.-F. Muller, "Geographical distribution and seasonal variation of surface emissions and deposition velocities of atmospheric trace gases," *Journal of Geophysical Research*, vol. 97, no. 4, pp. 3787–3804, 1992.

[17] A. Guenther, C. N. Hewitt, D. Erickson et al., "A global model of natural volatile organic compound emissions," *Journal of geophysical research*, vol. 100, no. 5, pp. 8873–8892, 1995.

[18] D. M. Lawrence, K. W. Oleson, M. G. Flanner et al., "Parameterization improvements and functional and structural advances in version 4 of the community land model," *Journal of Advances in Modeling Earth Systems*, vol. 3, 27 pages, 2011.

[19] E. C. Apel, D. D. Riemer, A. Hills et al., "Measurement and interpretation of isoprene fluxes isoprene, methacrolein, and methyl vinyl ketone mixing ratios at the PROPHET site during the 1998 intensive," *Journal of Geophysical Research D*, vol. 107, no. 3, pp. 1–15, 2002.

[20] R. R. Arnts, W. B. Petersen, R. L. Seila, and B. W. Gay Jr., "Estimates of α-pinene emissions from a loblolly pine forest using an atmospheric diffusion model," *Atmospheric Environment*, vol. 16, no. 9, pp. 2127–2137, 1982.

[21] G. P. Ayers and R. W. Gillett, "Isoprene emissions from vegetation and hydrocarbon emissions from bushfires in tropical Australia," *Journal of Atmospheric Chemistry*, vol. 7, no. 2, pp. 177–190, 1988.

[22] J. Bai, B. Baker, B. Liang, J. Greenberg, and A. Guenther, "Isoprene and monoterpene emissions from an Inner Mongolia grassland," *Atmospheric Environment*, vol. 40, no. 30, pp. 5753–5758, 2006.

[23] B. Baker, A. Guenther, J. Greenberg, A. Goldstein, and R. Fall, "Canopy fluxes of 2-methyl-3-buten-2-ol over a ponderosa pine forest by relaxed eddy accumulation: field data and model comparison," *Journal of Geophysical Research D*, vol. 104, no. 21, pp. 26107–26114, 1999.

[24] B. Baker, A. Guenther, J. Greenberg, and R. Fall, "Canopy level fluxes of 2-methyl-3-buten-2-ol, acetone, and methanol by a portable relaxed eddy accumulation system," *Environmental Science and Technology*, vol. 35, no. 9, pp. 1701–1708, 2001.

[25] I. Bamberger, L. Hörtnagl, T. M. Ruuskanen et al., "Deposition fluxes of terpenes over grassland," *Journal of Geophysical Research D*, vol. 116, no. 14, Article ID D14305, 2011.

[26] J. N. Barney, J. P. Sparks, J. Greenberg, T. H. Whitlow, and A. Guenther, "Biogenic volatile organic compounds from an invasive species: impacts on plant-plant interactions," *Plant Ecology*, vol. 203, no. 2, pp. 195–205, 2009.

[27] B. Bonsang, G. K. Moortgat, and C. A. Pio, "Overview of the FIELDVOC'94 experiment in a eucalyptus forest of Portugal," *Chemosphere*, vol. 3, no. 3, pp. 211–226, 2001.

[28] J. W. Bottenheim and M. F. Shepherd, "C_2-C_6 hydrocarbon measurements at four rural locations across Canada," *Atmospheric Environment*, vol. 29, no. 6, pp. 647–664, 1995.

[29] A. Bracho-Nunez, S. Welter, M. Staudt, and J. Kesselmeier, "Plant-specific volatile organic compound emission rates from young and mature leaves of Mediterranean vegetation," *Journal of Geophysical Research D*, vol. 116, no. 16, Article ID D16304, 2011.

[30] P. T. Buckley, "Isoprene emissions from a Florida scrub oak species grown in ambient and elevated carbon dioxide," *Atmospheric Environment*, vol. 35, no. 3, pp. 631–634, 2001.

[31] K. E. Burr, S. J. Wallner, and R. W. Tinus, "Ethylene and ethane evolution during cold acclimation and deacclimation of ponderosa pine," *Canadian Journal of Forestry Research*, vol. 21, pp. 601–605, 1991.

[32] C. Calfapietra, G. Scarascia Mugnozza, D. F. Karnosky, F. Loreto, and T. D. Sharkey, "Isoprene emission rates under elevated CO_2 and O_3 in two field-grown aspen clones differing in their sensitivity to O_3," *New Phytologist*, vol. 179, no. 1, pp. 55–61, 2008.

[33] X.-L. Cao, C. Boissard, A. J. Juan, C. N. Hewitt, and M. Gallagher, "Biogenic emissions of volatile organic compounds from gorse (*Ulex europaeus*): diurnal emission fluxes at Kelling Heath, England," *Journal of Geophysical Research D*, vol. 102, no. 15, pp. 18903–18915, 1997.

[34] J. Chang, Y. Ren, Y. Shi et al., "An inventory of biogenic volatile organic compounds for a subtropical urban-rural complex," *Atmospheric Environment*, vol. 56, pp. 115–123, 2012.

[35] P. Ciccioli, C. Fabozzi, E. Brancaleoni et al., "Biogenic emission from the Mediterranean Pseudosteppe ecosystem present in Castelporziano," *Atmospheric Environment*, vol. 31, no. 1, pp. 167–175, 1997.

[36] S. B. Corchnoy, J. Arey, and R. Atkinson, "Hydrocarbon emissions from twelve urban shade trees of the Los Angeles, California, Air Basin," *Atmospheric Environment*, vol. 26, no. 3, pp. 339–348, 1992.

[37] B. Davison, R. Taipale, B. Langford et al., "Concentrations and fluxes of biogenic volatile organic compounds above a Mediterranean Macchia ecosystem in western Italy," *Biogeosciences*, vol. 6, no. 8, pp. 1655–1670, 2009.

[38] W. A. Dement, B. J. Tyson, and H. A. Mooney, "Mechanism of monoterpene volatilization in Salvia mellifera," *Phytochemistry*, vol. 14, no. 12, pp. 2555–2557, 1975.

[39] J. P. Digangi, E. S. Boyle, T. Karl et al., "First direct measurements of formaldehyde flux via eddy covariance: implications for missing in-canopy formaldehyde sources," *Atmospheric Chemistry and Physics*, vol. 11, no. 20, pp. 10565–10578, 2011.

[40] T. Dindorf, U. Kuhn, L. Ganzeveld et al., "Significant light and temperature dependent monoterpene emissions from European beech (*Fagus sylvatica* L.) and their potential impact on the European volatile organic compound budget," *Journal of Geophysical Research D*, vol. 111, no. 16, Article ID D16305, 2006.

[41] P. Dominguez-Taylor, L. G. Ruiz-Suarez, I. Rosas-Perez, J. M. Hernández-Solis, and R. Steinbrecher, "Monoterpene and isoprene emissions from typical tree species in forests around Mexico City," *Atmospheric Environment*, vol. 41, no. 13, pp. 2780–2790, 2007.

[42] P. V. Doskey and W. Gao, "Vertical mixing and chemistry of isoprene in the atmospheric boundary layer: aircraft-based measurements and numerical modeling," *Journal of Geophysical Research D*, vol. 104, no. 17, pp. 21263–21274, 1999.

[43] R. C. Evans, D. T. Tingey, M. L. Gumpertz, and W. F. Burns, "Estimates of isoprene and monoterpene emission rates in plants," *Botanical Gazette*, vol. 143, no. 3, pp. 304–310, 1982.

[44] S. Fares, J. H. Park, D. R. Gentner et al., "Seasonal cycles of biogenic volatile organic compound fluxes and concentrations in a California citrus orchard," *Atmospheric Chemistry and Physics*, vol. 12, no. 20, pp. 9865–9880, 2012.

[45] P. Fruekilde, J. Hjorth, N. R. Jensen, D. Kotzias, and B. Larsen, "Ozonolysis at vegetation surfaces: a source of acetone, 4-oxopentanal, 6-methyl-5-hepten-2-one, and geranyl acetone in the troposphere," *Atmospheric Environment*, vol. 32, no. 11, pp. 1893–1902, 1998.

[46] J. D. Fuentes, D. Wang, G. Den Hartog, H. H. Neumann, T. F. Dann, and K. J. Puckett, "Modelled and field measurements of biogenic hydrocarbon emissions from a Canadian deciduous forest," *Atmospheric Environment*, vol. 29, no. 21, pp. 3003–3017, 1995.

[47] J. D. Fuentes, D. Wang, H. H. Neumann, T. J. Gillespie, G. Den Hartog, and T. F. Dann, "Ambient biogenic hydrocarbons and isoprene emissions from a mixed deciduous forest," *Journal of Atmospheric Chemistry*, vol. 25, no. 1, pp. 67–95, 1996.

[48] J. D. Fuentes and D. Wang, "On the seasonality of isoprene emissions from a mixed temperate forest," *Ecological Applications*, vol. 9, no. 4, pp. 1118–1131, 1999.

[49] Y. Fukui and P. V. Doskey, "Air-surface exchange of nonmethane organic compounds at a grassland site: seasonal variations and stressed emissions," *Journal of Geophysical Research D*, vol. 103, no. 11, pp. 13153–13168, 1998.

[50] F. Geng, X. Tie, A. Guenther, G. Li, J. Cao, and P. Harley, "Effect of isoprene emissions from major forests on ozone formation in the city of Shanghai, China," *Atmospheric Chemistry and Physics*, vol. 11, no. 20, pp. 10449–10459, 2011.

[51] C. D. Geron, D. Nie, R. R. Arnts et al., "Biogenic isoprene emission: model evaluation in a southeastern United States bottomland deciduous forest," *Journal of Geophysical Research D*, vol. 102, no. 15, pp. 18889–18901, 1997.

[52] C. Geron, A. Guenther, T. Sharkey, and R. R. Arnts, "Temporal variability in basal isoprene emission factor," *Tree Physiology*, vol. 20, no. 12, pp. 799–805, 2000.

[53] C. Geron, R. Rasmussen, R. R. Arnts, and A. Guenther, "A review and synthesis of monoterpene speciation from forests in the United States," *Atmospheric Environment*, vol. 34, no. 11, pp. 1761–1781, 2000.

[54] C. Geron, P. Harley, and A. Guenther, "Isoprene emission capacity for US tree species," *Atmospheric Environment*, vol. 35, no. 19, pp. 3341–3352, 2001.

[55] P. D. Goldan, W. C. Kuster, F. C. Fehsenfield, and S. A. Montzka, "The observation of a C_5 alcohol emission in a north American pine forest," *Geophysical Research Letters*, vol. 20, no. 11, pp. 1039–1042, 1993.

[56] A. H. Goldstein, S. M. Fan, M. L. Goulden, J. W. Munger, and S. C. Wofsy, "Emissions of ethene, propene, and 1-butene by a midlatitude forest," *Journal of Geophysical Research D*, vol. 101, no. 4 D, pp. 9149–9157, 1996.

[57] A. H. Goldstein, M. L. Goulden, J. W. Munger, S. C. Wofsy, and C. D. Geron, "Seasonal course of isoprene emissions from a midlatitude deciduous forest," *Journal of Geophysical Research D*, vol. 103, no. 23, pp. 31045–31056, 1998.

[58] J. P. Greenberg, A. Guenther, P. Zimmerman et al., "Tethered balloon measurements of biogenic VOCs in the atmospheric boundary layer," *Atmospheric Environment*, vol. 33, no. 6, pp. 855–867, 1999.

[59] A. B. Guenther, R. K. Monson, and R. Fall, "Isoprene and monoterpene emission rate variability: observations with eucalyptus and emission rate algorithm development," *Journal of Geophysical Research*, vol. 96, no. 6, pp. 10799–10808, 1991.

[60] A. B. Guenther, P. R. Zimmerman, P. C. Harley, R. K. Monson, and R. Fall, "Isoprene and monoterpene emission rate variability: model evaluations and sensitivity analyses," *Journal of Geophysical Research*, vol. 98, no. 7, pp. 12–617, 1993.

[61] A. B. Guenther and A. J. Hills, "Eddy covariance measurement of isoprene fluxes," *Journal of Geophysical Research D*, vol. 103, no. 11, pp. 13145–13152, 1998.

[62] A. Guenther, J. Greenberg, P. Harley et al., "Leaf, branch, stand and landscape scale measurements of volatile organic compound fluxes from U.S. woodlands," *Tree Physiology*, vol. 16, no. 1-2, pp. 17–24, 1996.

[63] A. Guenther, P. Zimmerman, L. Klinger et al., "Estimates of regional natural volatile organic compound fluxes from enclosure and ambient measurements," *Journal of Geophysical Research*, vol. 101, no. 1, pp. 1345–1359, 1966.

[64] A. Guenther, W. Baugh, K. Davis et al., "Isoprene fluxes measured by enclosure, relaxed eddy accumulation, surface layer gradient, mixed layer gradient, and mixed layer mass balance techniques," *Journal of Geophysical Research D*, vol. 101, no. 13, pp. 18555–18567, 1996.

[65] A. Guenther, S. Archer, J. Greenberg et al., "Biogenic hydrocarbon emissions and landcover/climate change in a subtropical savanna," *Physics and Chemistry of the Earth B*, vol. 24, no. 6, pp. 659–667, 1999.

[66] P. Harley, A. Guenther, and P. Zimmerman, "Effects of light, temperature and canopy position on net photosynthesis and isoprene emission from sweetgum (*Liquidambar styraciflua*) leaves," *Tree Physiology*, vol. 16, no. 1-2, pp. 25–32, 1996.

[67] P. Harley, A. Guenther, and P. Zimmerman, "Environmental controls over isoprene emission in deciduous oak canopies," *Tree Physiology*, vol. 17, no. 11, pp. 705–714, 1997.

[68] P. Harley, V. Fridd-Stroud, J. Greenberg, A. Guenther, and P. Vasconcellos, "Emission of 2-methyl-3-buten-2-ol by pines: a potentially large natural source of reactive carbon to the atmosphere," *Journal of Geophysical Research D*, vol. 103, no. 19, pp. 25479–25486, 1998.

[69] P. Harley, L. Otter, A. Guenther, and J. Greenberg, "Micrometeorological and leaf-level measurements of isoprene emissions from a southern African savanna," *Journal of Geophysical Research*, vol. 108, no. 13, 2003.

[70] P. Harley, J. Greenberg, Ü. Niinemets, and A. Guenther, "Environmental controls over methanol emission from leaves," *Biogeosciences*, vol. 4, no. 6, pp. 1083–1099, 2007.

[71] D. Harrison, M. C. Hunter, A. C. Lewis, P. W. Seakins, T. V. Nunes, and C. A. Pio, "Isoprene and monoterpene emission from the coniferous species Abies Borisii-regis: implications for regional air chemistry in Greece," *Atmospheric Environment*, vol. 35, no. 27, pp. 4687–4698, 2001.

[72] C. He, F. Murray, and T. Lyons, "Seasonal variations in monoterpene emissions from Eucalyptus species," *Chemosphere*, vol. 2, no. 1, pp. 65–76, 2000.

[73] A. C. Heiden, T. Hoffmann, J. Kahl et al., "Emission of volatile organic compounds from ozone-exposed plants," *Ecological Applications*, vol. 9, no. 4, pp. 1160–1167, 1999.

[74] D. Helmig, L. F. Klinger, A. Guenther, L. Vierling, C. Geron, and P. Zimmerman, "Biogenic volatile organic compound emissions (BVOCs). I. Identifications from three continental sites in the U.S.," *Chemosphere*, vol. 38, no. 9, pp. 2163–2187, 1999.

[75] D. Helmig, J. Ortega, A. Guenther, J. D. Herrick, and C. Geron, "Sesquiterpene emissions from loblolly pine and their potential contribution to biogenic aerosol formation in the Southeastern US," *Atmospheric Environment*, vol. 40, no. 22, pp. 4150–4157, 2006.

[76] D. Helmig, J. Ortega, T. Duhl et al., "Sesquiterpene emissions from pine trees: identifications, emission rates and flux estimates for the contiguous United States," *Environmental Science and Technology*, vol. 41, no. 5, pp. 1545–1553, 2007.

[77] M. W. Holdren, H. H. Westberg, and P. R. Zimmerman, "Analysis of monoterpene hydrocarbons in rural atmosphere,"

Journal of Geophysical Research, vol. 84, no. 8, pp. 5083–5088, 1979.

[78] R. Holzinger, A. Lee, M. McKay, and A. H. Goldstein, "Seasonal variability of monoterpene emission factors for a Ponderosa pine plantation in California," *Atmospheric Chemistry and Physics*, vol. 6, no. 5, pp. 1267–1274, 2006.

[79] C. Holzke, T. Dindorf, J. Kesselmeier, U. Kuhn, and R. Koppmann, "Terpene emissions from European beech (*Fagus sylvatica* L.): pattern and emission behaviour over two vegetation periods," *Journal of Atmospheric Chemistry*, vol. 55, no. 1, pp. 81–102, 2006.

[80] A.-K. Huang, N. Li, A. Guenther et al., "Investigation on emission properties of biogenic VOCs of landscape plants in Shenzhen," *Huanjing Kexue/Environmental Science*, vol. 32, no. 12, pp. 3555–3559, 2011.

[81] J. G. Isebrands, A. B. Guenther, P. Harley et al., "Volatile organic compound emission rates from mixed deciduous and coniferous forests in Northern Wisconsin, USA," *Atmospheric Environment*, vol. 33, no. 16, pp. 2527–2536, 1999.

[82] V. A. Isidorov, I. G. Zenkevich, and B. V. Ioffe, "Volatile organic compounds in the atmosphere of forests," *Atmospheric Environment*, vol. 19, no. 1, pp. 1–8, 1985.

[83] K. Jardine, T. Karl, M. Lerdau, P. Harley, A. Guenther, and J. E. Mak, "Carbon isotope analysis of acetaldehyde emitted from leaves following mechanical stress and anoxia," *Plant Biology*, vol. 11, no. 4, pp. 591–597, 2009.

[84] K. Jardine, P. Harley, T. Karl, A. Guenther, M. Lerdau, and J. E. Mak, "Plant physiological and environmental controls over the exchange of acetaldehyde between forest canopies and the atmosphere," *Biogeosciences*, vol. 5, no. 6, pp. 1559–1572, 2008.

[85] K. J. Jardine, E. D. Sommer, S. R. Saleska, T. E. Huxman, P. C. Harley, and L. Abrell, "Gas phase measurements of pyruvic acid and its volatile metabolites," *Environmental Science and Technology*, vol. 44, no. 7, pp. 2454–2460, 2010.

[86] T. G. Karl, C. Spirig, J. Rinne et al., "Virtual disjunct eddy covariance measurements of organic compound fluxes from a subalpine forest using proton transfer reaction mass spectrometry," *Atmospheric Chemistry and Physics*, vol. 2, no. 4, pp. 279–291, 2002.

[87] T. Karl, A. Guenther, C. Spirig, A. Hansel, and R. Fall, "Seasonal variation of biogenic VOC emissions above a mixed hardwood forest in northern Michigan," *Geophysical Research Letters*, vol. 30, no. 23, pp. 4–19, 2003.

[88] T. Karl, A. Guenther, A. Turnipseed, E. G. Patton, and K. Jardine, "Chemical sensing of plant stress at the ecosystem scale," *Biogeosciences*, vol. 5, no. 5, pp. 1287–1294, 2008.

[89] J. F. Karlik and A. M. Winer, "Measured isoprene emission rates of plants in California landscapes: comparison to estimates from taxonomic relationships," *Atmospheric Environment*, vol. 35, no. 6, pp. 1123–1131, 2001.

[90] J. Kesselmeier, L. Schäfer, P. Ciccioli et al., "Emission of monoterpenes and isoprene from a Mediterranean oak species *Quercus ilex* L. measured within the BEMA (Biogenic Emissions in the Mediterranean Area) project," *Atmospheric Environment*, vol. 30, no. 10-11, pp. 1841–1850, 1996.

[91] J. Kesselmeier, K. Bode, U. Hofmann et al., "Emission of short chained organic acids, aldehydes and monoterpenes from *Quercus ilex* L. and Pinus pinea L. in relation to physiological activities, carbon budget and emission algorithms," *Atmospheric Environment*, vol. 31, no. 1, pp. 119–133, 1997.

[92] J. Kesselmeier, K. Bode, L. Schafer et al., "Simultaneous field measurements of terpene and isoprene emissions from two

dominant Mediterranean oak species in relation to a North American species," *Atmospheric Environment*, vol. 32, no. 11, pp. 1947–1953, 1998.

[93] J. Kesselmeier, K. Bode, C. Gerlach, and E.-M. Jork, "Exchange of atmospheric formic and acetic acids with trees and crop plants under controlled chamber and purified air conditions," *Atmospheric Environment*, vol. 32, no. 10, pp. 1765–1775, 1998.

[94] J.-C. Kim, "Factors controlling natural VOC emissions in a southeastern US pine forest," *Atmospheric Environment*, vol. 35, no. 19, pp. 3279–3292, 2001.

[95] J.-C. Kim, K.-J. Kim, D.-S. Kim, and J.-S. Han, "Seasonal variations of monoterpene emissions from coniferous trees of different ages in Korea," *Chemosphere*, vol. 59, no. 11, pp. 1685–1696, 2005.

[96] S. Kim, T. Karl, D. Helmig, R. Daly, R. Rasmussen, and A. Guenther, "Measurement of atmospheric sesquiterpenes by proton transfer reaction-mass spectrometry (PTR-MS)," *Atmospheric Measurement Techniques*, vol. 2, no. 1, pp. 99–112, 2009.

[97] S. Kim, T. Karl, A. Guenther et al., "Emissions and ambient distributions of Biogenic Volatile Organic Compounds (BVOC) in a ponderosa pine ecosystem: interpretation of PTR-MS mass spectra," *Atmospheric Chemistry and Physics*, vol. 10, no. 4, pp. 1759–1771, 2010.

[98] S. Kim, A. Guenther, T. Karl, and J. Greenberg, "Contributions of primary and secondary biogenic VOC tototal OH reactivity during the CABINEX (Community Atmosphere-Biosphere INteractions Experiments)-09 field campaign," *Atmospheric Chemistry and Physics*, vol. 11, no. 16, pp. 8613–8623, 2011.

[99] S. Y. Kim, X. Y. Jiang, M. Lee et al., "Impact of biogenic volatile organic compounds on ozone production at the Taehwa Research Forest near Seoul, South Korea," *Atmospheric Environment*, vol. 70, pp. 447–453, 2013.

[100] W. Kirstine, I. Galbally, Y. Ye, and M. Hooper, "Emissions of volatile organic compounds (primarily oxygenated species) from pasture," *Journal of Geophysical Research D*, vol. 103, no. 3339, pp. 10605–10619, 1998.

[101] L. F. Klinger, Q. J. Li, A. B. Guenther, J. P. Greenberg, B. Baker, and J. H. Bai, "Assessment of volatile organic compound emissions from ecosystems of China," *Journal of Geophysical Research*, vol. 107, no. 21, 2002.

[102] G. Konig, M. Brunda, H. Puxbaum, C. N. Hewitt, S. C. Duckham, and J. Rudolph, "Relative contribution of oxygenated hydrocarbons to the total biogenic VOC emissions of selected mid-European agricultural and natural plant species," *Atmospheric Environment*, vol. 29, no. 8, pp. 861–874, 1995.

[103] J. Kreuzwieser, J.-P. Schnitzler, and R. Steinbrecher, "Biosynthesis of organic compounds emitted by plants," *Plant Biology*, vol. 1, no. 2, pp. 149–159, 1999.

[104] J. Kreuzwieser, H. Rennenberg, and R. Steinbrecher, "Impact of short-term and long-term elevated CO_2 on emission of carbonyls from adult Quercus petraea and Carpinus betulus trees," *Environmental Pollution*, vol. 142, no. 2, pp. 246–253, 2006.

[105] B. Lamb, H. Westberg, and G. Allwine, "Biogenic hydrocarbon emissions from deciduous and coniferous trees in the United States," *Journal of Geophysical Research*, vol. 90, no. 1, pp. 2380–2390, 1985.

[106] B. Lamb, H. Westberg, and G. Allwine, "Isoprene emission fluxes determined by an atmospheric tracer technique," *Atmospheric Environment*, vol. 20, no. 1, pp. 1–8, 1986.

[107] M. Lerdau, S. B. Dilts, H. Westberg, B. K. Lamb, and E. J. Allwine, "Monoterpene emission from ponderosa pine," *Journal of Geophysical Research*, vol. 99, no. 8, pp. 16–615, 1994.

[108] M. Lerdau, P. Matson, R. Fall, and R. Monson, "Ecological controls over monoterpene emissions from douglas-fir (Pseudotsuga menziesii)," *Ecology*, vol. 76, no. 8, pp. 2640–2647, 1995.

[109] D. W. Li, Y. Shi, X. Y. He, W. Chen, and X. Chen, "Volatile organic compound emissions from urban trees in Shenyang, China," *Botanical Studies*, vol. 49, no. 1, pp. 67–72, 2008.

[110] Y.-J. Lim, A. Armendariz, Y.-S. Son, and J.-C. Kim, "Seasonal variations of isoprene emissions from five oak tree species in East Asia," *Atmospheric Environment*, vol. 45, no. 13, pp. 2202–2210, 2011.

[111] M. E. Litvak, F. Loreto, P. C. Harley, T. D. Sharkey, and R. K. Monson, "The response of isoprene emission rate and photosynthetic rate to photon flux and nitrogen supply in aspen and white oak trees," *Plant, Cell and Environment*, vol. 19, no. 5, pp. 549–559, 1996.

[112] J. Llusia, J. Penuelas, R. Seco, and I. Filella, "Seasonal changes in the daily emission rates of terpenes by *Quercus ilex* and the atmospheric concentrations of terpenes in the natural park of Montseny, NE Spain," *Journal of Atmospheric Chemistry*, vol. 69, no. 3, pp. 215–230, 2012.

[113] H. W. Loescher, *Non-methane hydrocarbon fluxes from Pinus elliottii and Sereonoa repens: comparing enclosure and above-canopy measurements [Doctoral dissertation]*, University of Florida, 1997.

[114] F. Loreto and T. D. Sharkey, "A gas-exchange study of photosynthesis and isoprene emission in *Quercus rubra* L.," *Planta*, vol. 182, no. 4, pp. 523–531, 1990.

[115] R. C. MacDonald and R. Fall, "Detection of substantial emissions of methanol from plants to the atmosphere," *Atmospheric Environment*, vol. 27, no. 11, pp. 1709–1713, 1993.

[116] R. C. MacDonald and R. Fall, "Acetone emission from conifer buds," *Phytochemistry*, vol. 34, no. 4, pp. 991–994, 1993.

[117] M. B. Madronich, J. P. Greenberg, C. A. Wessman, and A. B. Guenther, "Monoterpene emissions from an understory species, *Pteridium aquilinum*," *Atmospheric Environment*, vol. 54, pp. 308–312, 2012.

[118] R. S. Martin, H. Westberg, E. Allwine, L. Ashman, J. C. Farmer, and B. Lamb, "Measurement of isoprene and its atmospheric oxidation products in a central Pennsylvania deciduous forest," *Journal of Atmospheric Chemistry*, vol. 13, no. 1, pp. 1–32, 1991.

[119] R. S. Martin, I. Villanueva, J. Zhang, and C. J. Popp, "Non-methane hydrocarbon, monocarboxylic acid, and low molecular weight aldehyde and ketone emissions from vegetation in central New Mexico," *Environmental Science and Technology*, vol. 33, no. 13, pp. 2186–2192, 1999.

[120] S. N. Matsunaga, A. B. Guenther, M. J. Potosnak, and E. C. Apel, "Emission of sunscreen salicylic esters from desert vegetation and their contribution to aerosol formation," *Atmospheric Chemistry and Physics*, vol. 8, no. 24, pp. 7367–7371, 2008.

[121] S. N. Matsunaga, A. B. Guenther, J. P. Greenberg et al., "Leaf level emission measurement of sesquiterpenes and oxygenated sesquiterpenes from desert shrubs and temperate forest trees using a liquid extraction technique," *Geochemical Journal*, vol. 43, no. 3, pp. 179–189, 2009.

[122] S. N. Matsunaga, S. Chatani, S. Nakatsuka et al., "Determination and potential importance of diterpene (kaur-16-ene) emitted from dominant coniferous trees in Japan," *Chemosphere*, vol. 87, no. 8, pp. 886–893, 2012.

[123] S. N. Matsunaga, O. Muller, S. Chatani, M. Nakamura, T. Nakaji, and T. Hiura, "Seasonal variation of isoprene basal emission in mature Quercus crispula trees under experimental warming of roots and branches," *Geochemical Journal*, vol. 46, no. 2, pp. 163–167, 2012.

[124] K. A. McKinney, B. H. Lee, A. Vasta, T. V. Pho, and J. W. Munger, "Emissions of isoprenoids and oxygenated biogenic volatile organic compounds from a New England mixed forest," *Atmospheric Chemistry and Physics*, vol. 11, no. 10, pp. 4807–4831, 2011.

[125] R. K. Monson and R. Fall, "Isoprene Emission from Aspen Leaves: influence of Environment and Relation to Photosynthesis and Photorespiration," *Plant Physiology*, vol. 90, no. 1, pp. 267–274, 1989.

[126] R. K. Monson, P. C. Harley, M. E. Litvak et al., "Environmental and developmental controls over the seasonal pattern of isoprene emission from aspen leaves," *Oecologia*, vol. 99, no. 3-4, pp. 260–270, 1994.

[127] S. Moukhtar, B. Bessagnet, L. Rouil, and V. Simon, "Monoterpene emissions from Beech (*Fagus sylvatica*) in a French forest and impact on secondary pollutants formation at regional scale," *Atmospheric Environment*, vol. 39, no. 19, pp. 3535–3547, 2005.

[128] Ü. Niinemets, "Mild versus severe stress and BVOCs: thresholds, priming and consequences," *Trends in Plant Science*, vol. 15, no. 3, pp. 145–153, 2010.

[129] Ü. Niinemets, U. Kuhn, P. C. Harley et al., "Estimations of isoprenoid emission capacity from enclosure studies: measurements, data processing, quality and standardized measurement protocols," *Biogeosciences*, vol. 8, no. 8, pp. 2209–2246, 2011.

[130] T. V. Nunes and C. A. Pio, "Emission of volatile organic compounds from Portuguese eucalyptus forests," *Chemosphere*, vol. 3, no. 3, pp. 239–248, 2001.

[131] K. Ohta, "Diurnal and seasonal variation in isoprene emission from live oak," *Geochemical Journal*, vol. 19, pp. 269–274, 1986.

[132] E. Ormeño, C. Fernandez, A. Bousquet-Mélou et al., "Monoterpene and sesquiterpene emissions of three Mediterranean species through calcareous and siliceous soils in natural conditions," *Atmospheric Environment*, vol. 41, no. 3, pp. 629–639, 2007.

[133] J. Ortega, D. Helmig, A. Guenther, P. Harley, S. Pressley, and C. Vogel, "Flux estimates and OH reaction potential of reactive biogenic volatile organic compounds (BVOCs) from a mixed northern hardwood forest," *Atmospheric Environment*, vol. 41, no. 26, pp. 5479–5495, 2007.

[134] J. Ortega, D. Helmig, R. W. Daly, D. M. Tanner, A. B. Guenther, and J. D. Herrick, "Approaches for quantifying reactive and low-volatility biogenic organic compound emissions by vegetation enclosure techniques. Part B: applications," *Chemosphere*, vol. 72, no. 3, pp. 365–380, 2008.

[135] L. B. Otter, A. Guenther, and J. Greenberg, "Seasonal and spatial variations in biogenic hydrocarbon emissions from southern African savannas and woodlands," *Atmospheric Environment*, vol. 36, no. 26, pp. 4265–4275, 2002.

[136] S. Owen, C. Boissard, R. A. Street, S. C. Duckham, O. Csiky, and C. N. Hewitt, "Screening of 18 Mediterranean plant species for volatile organic compound emissions," *Atmospheric Environment*, vol. 31, no. 1, pp. 101–117, 1997.

[137] S. M. Owen, C. Boissard, B. Hagenlocher, and C. N. Hewitt, "Field studies of isoprene emissions from vegetation in the Northwest Mediterranean region," *Journal of Geophysical Research D*, vol. 103, no. 19, pp. 25499–25511, 1998.

[138] M. R. Papiez, M. J. Potosnak, W. S. Goliff, A. B. Guenther, S. N. Matsunaga, and W. R. Stockwell, "The impacts of reactive terpene emissions from plants on air quality in Las Vegas, Nevada," *Atmospheric Environment*, vol. 43, no. 27, pp. 4109–4123, 2009.

[139] E. Pegoraro, A. Rey, J. Greenberg et al., "Effect of drought on isoprene emission rates from leaves of Quercus virginiana Mill," *Atmospheric Environment*, vol. 38, no. 36, pp. 6149–6156, 2004.

[140] D. Pérez-Rial, J. Peñuelas, P. López-Mahía, and J. Llusià, "Terpenoid emissions from *Quercus robur*. A case study of Galicia (NW Spain)," *Journal of Environmental Monitoring*, vol. 11, no. 6, pp. 1268–1275, 2009.

[141] G. Pétron, P. Harley, J. Greenberg, and A. Guenther, "Seasonal temperature variations influence isoprene emission," *Geophysical Research Letters*, vol. 28, no. 9, pp. 1707–1710, 2001.

[142] P. A. Pier, "Isoprene emission rates from northern red oak using a whole-tree chamber," *Atmospheric Environment*, vol. 29, no. 12, pp. 1347–1353, 1995.

[143] P. A. Pier and C. McDuffie Jr., "Seasonal isoprene emission rates and model comparisons using whole-tree emissions from white oak," *Journal of Geophysical Research D*, vol. 102, no. 20, pp. 23963–23971, 1997.

[144] C. A. Pio and A. A. Valente, "Atmospheric fluxes and concentrations of monoterpenes in resin-tapped pine forests," *Atmospheric Environment*, vol. 32, no. 4, pp. 683–691, 1998.

[145] O. Pokorska, J. Dewulf, C. Amelynck et al., "Isoprene and terpenoid emissions from Abies alba: identification and emission rates under ambient conditions," *Atmospheric Environment*, vol. 59, pp. 501–508, 2012.

[146] O. Pokorska, J. Dewulf, C. Amelynck et al., "Emissions of biogenic volatile organic compounds from *Fraxinus excelsior* and *Quercus robur* under ambient conditions in Flanders (Belgium)," *International Journal of Environmental Analytical Chemistry*, vol. 92, no. 15, pp. 1729–1741, 2012.

[147] S. Pressley, B. Lamb, H. Westberg, A. Guenther, J. Chen, and E. Allwine, "Monoterpene emissions from a Pacific Northwest Old-Growth Forest and impact on regional biogenic VOC emission estimates," *Atmospheric Environment*, vol. 38, no. 19, pp. 3089–3098, 2004.

[148] S. Pressley, B. Lamb, H. Westberg, J. Flaherty, J. Chen, and C. Vogel, "Long-term isoprene flux measurements above a northern hardwood forest," *Journal of Geophysical Research D*, vol. 110, no. 7, Article ID D07301, pp. 1–12, 2005.

[149] H. Puxbaum and G. König, "Observation of dipropenyldisulfide and other organic sulfur compounds in the atmosphere of a beech forest with Allium ursinum ground cover," *Atmospheric Environment*, vol. 31, no. 2, pp. 291–294, 1997.

[150] F. Rapparini, R. Baraldi, F. Miglietta, and F. Loreto, "Isoprenoid emission in trees of Quercus pubescens and Quercus ilex with lifetime exposure to naturally high CO_2 environment," *Plant, Cell and Environment*, vol. 27, no. 4, pp. 381–391, 2004.

[151] R. A. Rasmussen, "Isoprene: identified as a forest-type emission to the atmosphere," *Environmental Science and Technology*, vol. 4, no. 8, pp. 667–671, 1970.

[152] R. A. Rasmussen and F. Went, "Volatile organic material of plant origin in the atmosphere," *Proceedings of the National Academy of Sciences*, vol. 53, pp. 215–220, 1965.

[153] R. C. Rhew, B. R. Miller, and R. F. Weiss, "Natural methyl bromide and methyl chloride emissions from coastal salt marshes," *Nature*, vol. 403, no. 6767, pp. 292–295, 2000.

[154] J. M. Roberts, F. C. Fehsenfeld, D. L. Albritton, and R. E. Sievers, "Measurement of monoterpene hydrocarbons at Niwot Ridge, Colorado," *Journal of Geophysical Research*, vol. 88, no. 15, pp. 10.667–10.678, 1983.

[155] J. M. Roberts, C. J. Hahn, F. C. Fehsenfeld, J. M. Warnock, D. L. Albritton, and R. E. Sievers, "Monoterpene hydrocarbons in the nighttime troposphere," *Environmental Science and Technology*, vol. 19, no. 4, pp. 364–369, 1985.

[156] G. Sanadze, "The nature of gaseous substances emitted by leaves of Robinia pseudoacacia," *Soobshcheniya Akademi Nauk Gruzinskoj*, vol. 27, pp. 747–750, 1957.

[157] T. J. Savage, M. K. Hristova, and R. Croteau, "Evidence for an elongation/reduction/C1-elimination pathway in the biosynthesis of n-heptane in xylem of Jeffrey pine," *Plant Physiology*, vol. 111, no. 4, pp. 1263–1269, 1996.

[158] S. Sawada and T. Totsuka, "Natural and anthropogenic sources and fate of atmospheric ethylene," *Atmospheric Environment*, vol. 20, no. 5, pp. 821–832, 1986.

[159] G. W. Schade and A. H. Goldstein, "Fluxes of oxygenated volatile organic compounds from a ponderosa pine plantation," *Journal of Geophysical Research D*, vol. 106, no. 3, pp. 3111–3123, 2001.

[160] R. Seco, I. Filella, J. Llusià, and J. Peñuelas, "Methanol as a signal triggering isoprenoid emissions and photosynthetic performance in *Quercus ilex*," *Acta Physiologiae Plantarum*, vol. 33, no. 6, pp. 2413–2422, 2011.

[161] T. D. Sharkey, E. L. Singsaas, P. J. Vanderveer, and C. Geron, "Field measurements of isoprene emission from trees in response to temperature and light," *Tree Physiology*, vol. 16, no. 7, pp. 649–654, 1996.

[162] T. D. Sharkey, E. L. Singsaas, M. T. Lerdau, and C. D. Geron, "Weather effects on isoprene emission capacity and applications in emissions algorithms," *Ecological Applications*, vol. 9, no. 4, pp. 1132–1137, 1999.

[163] U. K. Sharma, Y. Kajii, and H. Akimoto, "Characterization of NMHCs in downtown urban center Kathmandu and rural site Nagarkot in Nepal," *Atmospheric Environment*, vol. 34, no. 20, pp. 3297–3307, 2000.

[164] R. W. Shaw Jr., A. L. Crittenden, R. K. Stevens, D. R. Cronn, and V. S. Titov, "Ambient concentrations of hydrocarbons from conifers in atmospheric gases and aerosol particles measured in Soviet Georgia," *Environmental Science and Technology*, vol. 17, no. 7, pp. 389–395, 1983.

[165] M. Šimpraga, H. Verbeeck, M. Demarcke et al., "Clear link between drought stress, photosynthesis and biogenic volatile organic compounds in *Fagus sylvatica* L.," *Atmospheric Environment*, vol. 45, no. 30, pp. 5254–5259, 2011.

[166] B. C. Sive, R. K. Varner, H. Mao, D. R. Blake, O. W. Wingenter, and R. Talbot, "A large terrestrial source of methyl iodide," *Geophysical Research Letters*, vol. 34, no. 17, Article ID L17808, 2007.

[167] C. Spirig, A. Neftel, C. Ammann et al., "Eddy covariance flux measurements of biogenic VOCs during ECHO 2003 using proton transfer reaction mass spectrometry," *Atmospheric Chemistry and Physics*, vol. 5, no. 2, pp. 465–481, 2005.

[168] M. Staudt, A. Ennajah, F. Mouillot, and R. Joffre, "Do volatile organic compound emissions of Tunisian cork oak populations originating from contrasting climatic conditions differ in their responses to summer drought?" *Canadian Journal of Forest Research*, vol. 38, no. 12, pp. 2965–2975, 2008.

[169] R. Steinbrecher, M. Klauer, K. Hauff et al., "Biogenic and anthropogenic fluxes of non-methane hydrocarbons over an urban-impacted forest, Frankfurter Stadtwald, Germany," *Atmospheric Environment*, vol. 34, no. 22, pp. 3779–3788, 2000.

[170] A. Tani and Y. Kawawata, "Isoprene emission from the major native *Quercus* spp. in Japan," *Atmospheric Environment*, vol. 42, no. 19, pp. 4540–4550, 2008.

[171] A. Tani, S. Nozoe, M. Aoki, and C. N. Hewitt, "Monoterpene fluxes measured above a Japanese red pine forest at Oshiba plateau, Japan," *Atmospheric Environment*, vol. 36, no. 21, pp. 3391–3402, 2002.

[172] D. T. Tingey, M. Manning, L. C. Grothaus, and W. F. Burns, "Influence of light and temperature on isoprene emission rates from live Oak," *Physiologia Plantarum*, vol. 47, no. 2, pp. 112–118, 1979.

[173] D. T. Tingey, M. Manning, L. C. Grothaus, and W. F. Burns, "Influence of light and temperature on monoterpene emission rates from slash pine," *Plant Physiology*, vol. 65, no. 5, pp. 797–801, 1980.

[174] J. K.-Y. Tsui, A. Guenther, W.-K. Yip, and F. Chen, "A biogenic volatile organic compound emission inventory for Hong Kong," *Atmospheric Environment*, vol. 43, no. 40, pp. 6442–6448, 2009.

[175] H. J. Wang, J. Y. Xia, Y. J. Mu, L. Nie, X. G. Han, and S. Q. Wan, "BVOCs emission in a semi-arid grassland under climate warming and nitrogen deposition," *Atmospheric Chemistry and Physics*, vol. 12, no. 8, pp. 3809–3819, 2012.

[176] C. Warneke, J. A. de Gouw, L. Del Negro et al., "Biogenic emission measurement and inventories determination of biogenic emissions in the eastern United States and Texas and comparison with biogenic emission inventories," *Journal of Geophysical Research D*, vol. 115, no. 5, Article ID D00F18, 2010.

[177] S. Welter, A. Bracho-Nunez, C. Mir et al., "The diversification of terpene emissions in Mediterranean oaks: iessons from a study of *Quercus suber* , *Quercus canariensis* and its hybrid *Quercus afares*," *Tree Physiology*, vol. 32, no. 9, pp. 1082–1091, 2012.

[178] H. Westberg, B. Lamb, R. Hafer, A. Hills, P. Shepson, and C. Vogel, "Measurement of isoprene fluxes at the PROPHET site," *Journal of Geophysical Research D*, vol. 106, no. 20, pp. 24347–24358, 2001.

[179] C. Wiedinmyer, S. Friedfeld, W. Baugh et al., "Measurement and analysis of atmospheric concentrations of isoprene and its reaction products in central Texas," *Atmospheric Environment*, vol. 35, no. 6, pp. 1001–1013, 2001.

[180] C. Wiedinmyer, J. Greenberg, A. Guenther et al., "Ozarks Isoprene Experiment (OZIE): measurements and modeling of the 'isoprene volcano," *Journal of Geophysical Research D*, vol. 110, no. 18, Article ID D18307, pp. 1–17, 2005.

[181] A. J. Winters, M. A. Adams, T. M. Bleby et al., "Emissions of isoprene, monoterpene and short-chained carbonyl compounds from Eucalyptus spp. in southern Australia," *Atmospheric Environment*, vol. 43, no. 19, pp. 3035–3043, 2009.

[182] Z. Xiaoshan, M. Yujing, S. Wenzhi, and Z. Yahui, "Seasonal variations of isoprene emissions from deciduous trees," *Atmospheric Environment*, vol. 34, no. 18, pp. 3027–3032, 2000.

[183] A. Yani, G. Pauly, M. Faye, F. Salin, and M. Gleizes, "The effect of a long-term water stress on the metabolism and emission of terpenes of the foliage of Cupressus sempervirens," *Plant, Cell and Environment*, vol. 16, no. 8, pp. 975–981, 1993.

[184] Y. Yokouchi, M. Okaniwa, Y. Ambe, and K. Fuwa, "Seasonal variation of monoterpenes in the atmosphere of a pine forest," *Atmospheric Environment*, vol. 17, no. 4, pp. 743–750, 1983.

[185] Y. Yokouchi, A. Hijikata, and Y. Ambe, "Seasonal variation of monoterpene emission rate in a pine forest," *Chemosphere*, vol. 13, no. 2, pp. 255–259, 1984.

[186] Y. Yokouchi and Y. Ambe, "Factors affecting the emission of monoterpenes from red pine (*Pinus densiflora*)," *Plant Physiology*, vol. 75, no. 4, pp. 1009–1012, 1984.

[187] B. Baker, J.-H. Bai, C. Johnson et al., "Wet and dry season ecosystem level fluxes of isoprene and monoterpenes from a southeast Asian secondary forest and rubber tree plantation," *Atmospheric Environment*, vol. 39, no. 2, pp. 381–390, 2005.

[188] D. R. Cronn and W. Nutmagul, "Analysis of atmospheric hydrocarbons during winter MONEX (Borneo)," *Tellus*, vol. 34, no. 2, pp. 159–165, 1982.

[189] P. Crutzen, M. Coffey, A. Delany et al., "Observations of air composition in Brazil between the equator and 20˚S during the dry season," *Acta Amazonica*, vol. 15, pp. 77–119, 1985.

[190] L. Donoso, R. Romero, A. Rondón, E. Fernandez, P. Oyola, and E. Sanhueza, "Natural and anthropogenic C_2 to C_6 hydrocarbons in the Central-Eastern Venezuelan atmosphere during the rainy season," *Journal of Atmospheric Chemistry*, vol. 25, no. 2, pp. 201–214, 1996.

[191] C. Geron, A. Guenther, J. Greenberg, H. W. Loescher, D. Clark, and B. Baker, "Biogenic volatile organic compound emissions from a lowland tropical wet forest in Costa Rica," *Atmospheric Environment*, vol. 36, no. 23, pp. 3793–3802, 2002.

[192] J. P. Greenberg, P. R. Zimmerman, L. Heidt, and W. Pollock, "Hydrocarbon and carbon monoxide emissions from biomass burning in Brazil," *Journal of Geophysical Research*, vol. 89, no. 1, pp. 1350–1354, 1984.

[193] J. P. Greenberg, "Biogenic volatile organic compound emissions in central Africa during the Experiment for the Regional Sources and Sinks of Oxidants (EXPRESSO) biomass burning season," *Journal of Geophysical Research D*, vol. 104, no. 23, pp. 30659–30671, 1999.

[194] J. P. Greenberg, A. Guenther, P. Harley et al., "Eddy flux and leaf-level measurements of biogenic VOC emissions from mopane woodland of Botswana," *Journal of Geophysical Research D*, vol. 108, no. 13, pp. 2–9, 2003.

[195] J. P. Greenberg, A. B. Guenther, G. Pétron et al., "Biogenic VOC emissions from forested Amazonian landscapes," *Global Change Biology*, vol. 10, no. 5, pp. 651–662, 2004.

[196] G. Gregory, R. Harriss, R. Talbot et al., "Air chemistry over the tropical forest of Guyana," *Journal of Geophysical Research*, vol. 91, pp. 8603–8612, 1986.

[197] A. Guenther, L. Otter, P. Zimmerman, J. Greenberg, R. Scholes, and M. Scholes, "Biogenic hydrocarbon emissions from southern African savannas," *Journal of Geophysical Research D*, vol. 101, no. 20, pp. 25859–25865, 1996.

[198] P. Harley, P. Vasconcellos, L. Vierling et al., "Variation in potential for isoprene emissions among Neotropical forest sites," *Global Change Biology*, vol. 10, no. 5, pp. 630–650, 2004.

[199] D. Helmig, B. Balsley, K. Davis et al., "Vertical profiling and determination of landscape fluxes of biogenic nonmethane hydrocarbons within the planetary boundary layer in the Peruvian Amazon," *Journal of Geophysical Research D*, vol. 103, no. 19, pp. 25519–25532, 1998.

[200] R. Holzinger, E. Sanhueza, R. von Kuhlmann, B. Kleiss, L. Donoso, and P. J. Crutzen, "Diurnal cycles and seasonal variation of isoprene and its oxidation products in the tropical savanna atmosphere," *Global Biogeochemical Cycles*, vol. 16, no. 4, pp. 22-1, 2002.

[201] T. Karl, M. Potosnak, A. Guenther et al., "Exchange processes of volatile organic compounds above a tropical rain forest: implications for modeling tropospheric chemistry above dense vegetation," *Journal of Geophysical Research D*, vol. 109, no. 18, pp. D18306–19, 2004.

[202] T. Karl, A. Guenther, R. J. Yokelson et al., "The tropical forest and fire emissions experiment: emission, chemistry, and transport of biogenic volatile organic compounds in the lower atmosphere over Amazonia," *Journal of Geophysical Research D*, vol. 112, no. 18, Article ID D18302, 2007.

[203] T. Karl, A. Guenther, A. Turnipseed, G. Tyndall, P. Artaxo, and S. Martin, "Rapid formation of isoprene photo-oxidation products observed in Amazonia," *Atmospheric Chemistry and Physics*, vol. 9, no. 20, pp. 7753–7767, 2009.

[204] M. Keller and M. Lerdau, "Isoprene emission from tropical forest canopy leaves," *Global Biogeochemical Cycles*, vol. 13, no. 1, pp. 19–29, 1999.

[205] J. Kesselmeier, U. Kuhn, A. Wolf et al., "Atmospheric volatile organic compounds (VOC) at a remote tropical forest site in central Amazonia," *Atmospheric Environment*, vol. 34, no. 24, pp. 4063–4072, 2000.

[206] J. Kesselmeier, U. Kuhn, S. Rottenberger et al., "Concentrations and species composition of atmospheric volatile organic compounds (VOCs) as observed during the wet and dry season in Rondônia (Amazonia)," *Journal of Geophysical Research D*, vol. 107, no. 20, pp. 1–20, 2002.

[207] L. F. Klinger, "Patterns in volatile organic compound emissions along a savanna-rainforest gradient in central Africa," *Journal of Geophysical Research D*, vol. 103, no. 1, pp. 1443–1454, 1998.

[208] U. Kuhn, S. Rottenberger, T. Biesenthal et al., "Isoprene and monoterpene emissions of Amazonian tree species during the wet season: direct and indirect investigations on controlling environmental functions," *Journal of Geophysical Research D*, vol. 107, no. 20, pp. XCXLIII–XCXLIV, 2002.

[209] U. Kuhn, S. Rottenberger, T. Biesenthal et al., "Seasonal differences in isoprene and light-dependent monoterpene emission by Amazonian tree species," *Global Change Biology*, vol. 10, no. 5, pp. 663–682, 2004.

[210] U. Kuhn, M. O. Andreae, C. Ammann et al., "Isoprene and monoterpene fluxes from Central Amazonian rainforest inferred from tower-based and airborne measurements, and implications on the atmospheric chemistry and the local carbon budget," *Atmospheric Chemistry and Physics*, vol. 7, no. 11, pp. 2855–2879, 2007.

[211] C. E. Jones, J. R. Hopkins, and A. C. Lewis, "In situ measurements of isoprene and monoterpenes within a south-east Asian tropical rainforest," *Atmospheric Chemistry and Physics*, vol. 11, no. 14, pp. 6971–6984, 2011.

[212] B. Langford, P. K. Misztal, E. Nemitz et al., "Fluxes and concentrations of volatile organic compounds from a South-East Asian tropical rainforest," *Atmospheric Chemistry and Physics*, vol. 10, no. 17, pp. 8391–8412, 2010.

[213] D. Y. C. Leung, P. Wong, B. K. H. Cheung, and A. Guenther, "Improved land cover and emission factors for modeling biogenic volatile organic compounds emissions from Hong Kong," *Atmospheric Environment*, vol. 44, no. 11, pp. 1456–1468, 2010.

[214] P. K. Misztal, S. M. Owen, A. B. Guenther et al., "Large estragole fluxes from oil palms in Borneo," *Atmospheric Chemistry and Physics*, vol. 10, no. 9, pp. 4343–4358, 2010.

[215] P. K. Misztal, E. Nemitz, B. Langford et al., "Direct ecosystem fluxes of volatile organic compounds from oil palms in South-East Asia," *Atmospheric Chemistry and Physics*, vol. 11, no. 17, pp. 8995–9017, 2011.

[216] J.-F. Müller, T. Stavrakou, S. Wallens et al., "Global isoprene emissions estimated using MEGAN, ECMWF analyses and a detailed canopy environment model," *Atmospheric Chemistry and Physics*, vol. 8, no. 5, pp. 1329–1341, 2008.

[217] H. Oku, M. Fukuta, H. Iwasaki, P. Tambunan, and S. Baba, "Modification of the isoprene emission model G93 for tropical tree *Ficus virgata*," *Atmospheric Environment*, vol. 42, no. 38, pp. 8747–8754, 2008.

[218] P. K. Padhy and C. K. Varshney, "Isoprene emission from tropical tree species," *Environmental Pollution*, vol. 135, no. 1, pp. 101–109, 2005.

[219] R. A. Rasmussen and M. A. K. Khalil, "Isoprene over the Amazon Basin," *Journal of Geophysical Research*, vol. 93, no. 2, pp. 1417–1421, 1988.

[220] H. J. I. Rinne, A. B. Guenther, J. P. Greenberg, and P. C. Harley, "Isoprene and monoterpene fluxes measured above Amazonian rainforest and their dependence on light and temperature," *Atmospheric Environment*, vol. 36, no. 14, pp. 2421–2426, 2002.

[221] T. Saito, Y. Yokouchi, Y. Kosugi, M. Tani, E. Philip, and T. Okuda, "Methyl chloride and isoprene emissions from tropical rain forest in Southeast Asia," *Geophysical Research Letters*, vol. 35, no. 19, Article ID L19812, 2008.

[222] E. Sanhueza, M. Santana, D. Trapp et al., "Field measurement evidence for an atmospheric chemical source of formic and acetic acids in the tropic," *Geophysical Research Letters*, vol. 23, no. 9, pp. 1045–1048, 1996.

[223] J. E. Saxton, A. C. Lewis, J. H. Kettlewell et al., "Isoprene and monoterpene measurements in a secondary forest in northern Benin," *Atmospheric Chemistry and Physics*, vol. 7, no. 15, pp. 4095–4106, 2007.

[224] D. Sercanda, A. Guenther, L. Klinger et al., "EXPRESSO flux measurements at upland and lowland Congo tropical forest site," *Tellus B*, vol. 53, no. 3, pp. 220–234, 2001.

[225] R. W. Talbot, M. O. Andreae, H. Berresheim, D. J. Jacob, and K. M. Beecher, "Sources and sinks of formic, acetic, and pyruvic acids over central Amazonia. 2. Wet season," *Journal of Geophysical Research*, vol. 95, no. 10, pp. 16–811, 1990.

[226] C. K. Varshney and A. P. Singh, "Isoprene emission from Indian trees," *Journal of Geophysical Research D*, vol. 108, no. 24, pp. 24–7, 2003.

[227] C. Warneke, S. L. Luxembourg, J. A. de Gouw, H. J. I. Rinne, A. B. Guenther, and R. Fall, "Disjunct eddy covariance measurements of oxygenated volatile organic compounds fluxes from an alfalfa field before and after cutting," *Journal of Geophysical Research D*, vol. 107, no. 7-8, pp. 6–1, 2002.

[228] J. Williams, U. Pöschl, P. J. Crutzen et al., "An atmospheric chemistry interpretation of mass scans obtained from a proton transfer mass spectrometer flown over the tropical rainforest of Surinam," *Journal of Atmospheric Chemistry*, vol. 38, no. 2, pp. 133–166, 2001.

[229] P. R. Zimmerman, J. P. Greenberg, and C. E. Westberg, "Measurements of atmospheric hydrocarbons and biogenic emission fluxes in the Amazon Boundary Layer," *Journal of Geophysical Research*, vol. 93, no. 2, pp. 1407–1416, 1988.

[230] J. Bäck, J. Aalto, M. Henriksson, H. Hakola, Q. He, and M. Boy, "Chemodiversity in terpene emissions at a boreal Scots pine stand," *Biogeosciences Discussions*, vol. 8, no. 5, pp. 10577–10615, 2011.

[231] J. Bai, F. Lin, X. Wan, A. Guenther, A. Turnipseed, and T. Duhl, "Volatile organic compound emission fluxes from a temperate forest in Changbai Mountain," *Acta Scientiae Circumstantiae*, vol. 32, no. 3, pp. 545–554, 2012.

[232] A. Ekberg, A. Arneth, H. Hakola, S. Hayward, and T. Holst, "Isoprene emission from wetland sedges," *Biogeosciences*, vol. 6, no. 4, pp. 601–613, 2009.

[233] A. Ekberg, A. Arneth, and T. Holst, "Isoprene emission from Sphagnum species occupying different growth positions above the water table," *Boreal Environment Research*, vol. 16, no. 1, pp. 47–59, 2011.

[234] P. Faubert, P. Tiiva, Å. Rinnan, A. Michelsen, J. K. Holopainen, and R. Rinnan, "Doubled volatile organic compound emissions from subarctic tundra under simulated climate warming," *New Phytologist*, vol. 187, no. 1, pp. 199–208, 2010.

[235] I. Filella, M. J. Wilkinson, J. Llusià, C. N. Hewitt, and J. Peñuelas, "Volatile organic compounds emissions in Norway spruce (*Picea abies*) in response to temperature changes," *Physiologia Plantarum*, vol. 130, no. 1, pp. 58–66, 2007.

[236] J. D. Fuentes, D. Wang, and L. Gu, "Seasonal variations in isoprene emissions from a boreal aspen forest," *Journal of Applied Meteorology*, vol. 38, no. 7, pp. 855–869, 1999.

[237] A. Ghirardo, K. Koch, R. Taipale, I. Zimmer, J.-P. Schnitzler, and J. Rinne, "Determination of de novo and pool emissions of terpenes from four common boreal/alpine trees by 13CO$_2$ labelling and PTR-MS analysis," *Plant, Cell and Environment*, vol. 33, no. 5, pp. 781–792, 2010.

[238] S. Haapanala, J. Rinne, K.-H. Pystynen, H. Hellén, H. Hakola, and T. Riutta, "Measurements of hydrocarbon emissions from a boreal fen using the REA technique," *Biogeosciences*, vol. 3, no. 1, pp. 103–112, 2006.

[239] H. Hakola, J. Rinne, and T. Laurila, "The hydrocarbon emission rates of tea-leafed willow (*Salix phylicifolia*), silver birch (*Betula pendula*) and European aspen (*Populus tremula*)," *Atmospheric Environment*, vol. 32, no. 10, pp. 1825–1833, 1998.

[240] H. Hakola, T. Laurila, J. Rinne, and K. Puhto, "The ambient concentrations of biogenic hydrocarbons at a northern European, boreal site," *Atmospheric Environment*, vol. 34, no. 29-30, pp. 4971–4982, 2000.

[241] H. Hakola, T. Laurila, V. Lindfors, H. Hellén, A. Gaman, and J. Rinne, "Variation of the VOC emission rates of birch species during the growing season," *Boreal Environment Research*, vol. 6, no. 3, pp. 237–249, 2001.

[242] H. Hakola, V. Tarvainen, J. Bäck et al., "Seasonal variation of mono- and sesquiterpene emission rates of Scots pine," *Biogeosciences*, vol. 3, no. 1, pp. 93–101, 2006.

[243] D. T. Hanson, S. Swanson, L. E. Graham, and T. D. Sharkey, "Evolutionary significance of isoprene emission from mosses," *The American Journal of Botany*, vol. 86, no. 5, pp. 634–639, 1999.

[244] H. Hellén, H. Hakola, K.-H. Pystynen, J. Rinne, and S. Haapanala, "C$_2$-C$_{10}$ hydrocarbon emissions from a boreal wetland and forest floor," *Biogeosciences*, vol. 3, no. 2, pp. 167–174, 2006.

[245] T. Holst, A. Arneth, S. Hayward et al., "BVOC ecosystem flux measurements at a high latitude wetland site," *Atmospheric Chemistry and Physics*, vol. 10, no. 4, pp. 1617–1634, 2010.

[246] O. Hov, J. Schjoldager, and B. M. Wathne, "Measurement and modeling of the concentrations of terpenes in coniferous forest air (Norway)," *Journal of Geophysical Research*, vol. 88, no. 15, pp. 10679–10688, 1983.

[247] R. Janson, "Monoterpene concentrations in and above a forest of Scots pine," *Journal of Atmospheric Chemistry*, vol. 14, no. 1-4, pp. 385–394, 1992.

[248] R. Janson and C. de Serves, "Isoprene emissions from boreal wetlands in Scandinavia," *Journal of Geophysical Research D*, vol. 103, no. 19, pp. 25513–25517, 1998.

[249] R. Janson, C. de Serves, and R. Romero, "Emission of isoprene and carbonyl compounds from a boreal forest and wetland in Sweden," *Agricultural and Forest Meteorology*, vol. 98-99, pp. 671–681, 1999.

[250] B. T. Jobson, Z. Wu, H. Niki, and L. A. Barrie, "Seasonal trends of isoprene, C$_2$-C$_5$ alkanes, and acetylene at a remote boreal site in Canada," *Journal of Geophysical Research*, vol. 99, pp. 1589–1599, 1994.

[251] K. Kempf, E. Allwine, H. Westberg, C. Claiborn, and B. Lamb, "Hydrocarbon emissions from spruce species using environmental chamber and branch enclosure methods," *Atmospheric Environment*, vol. 30, no. 9, pp. 1381–1389, 1996.

[252] L. F. Klinger, P. R. Zimmerman, J. P. Greenberg, L. E. Heidt, and A. B. Guenther, "Carbon trace gas fluxes along a successional gradient in the Hudson-Bay Lowland," *Journal of Geophysical Research*, vol. 99, no. 1, pp. 1469–1494, 1994.

[253] D. M. Martin, J. Gershenzon, and J. Bohlmann, "Induction of volatile terpene biosynthesis and diurnal emission by methyl jasmonate in foliage of Norway spruce," *Plant Physiology*, vol. 132, no. 3, pp. 1586–1599, 2003.

[254] E. Pattey, R. L. Desjardins, H. Westberg, B. Lamb, and T. Zhu, "Measurement of isoprene emissions over a black spruce stand using a tower-based relaxed eddy-accumulation system," *Journal of Applied Meteorology*, vol. 38, no. 7, pp. 870–877, 1999.

[255] G. Petersson, "High ambient concentrations of monoterpenes in a Scandinavian pine forest," *Atmospheric Environment*, vol. 22, no. 11, pp. 2617–2619, 1988.

[256] M. J. Potosnak, B. Baker, L. LeStourgeon et al., "Isoprene emissions from a tundra ecosystem," *Biogeosciences*, vol. 10, pp. 871–889, 2013.

[257] T. Räisänen, A. Ryyppö, and S. Kellomäki, "Monoterpene emission of a boreal Scots pine (*Pinus sylvestris* L.) forest," *Agricultural and Forest Meteorology*, vol. 149, no. 5, pp. 808–819, 2009.

[258] J. Rinne, H. Hakola, and T. Laurila, "Vertical fluxes of monoterpenes above a Scots pine stand in the boreal vegetation zone," *Physics and Chemistry of the Earth B*, vol. 24, no. 6, pp. 711–715, 1999.

[259] J. Rinne, H. Hakola, T. Laurila, and Ü. Rannik, "Canopy scale monoterpene emissions of *Pinus sylvestris* dominated forests," *Atmospheric Environment*, vol. 34, no. 7, pp. 1099–1107, 2000.

[260] J. Rinne, R. Taipale, T. Markkanen et al., "Hydrocarbon fluxes above a Scots pine forest canopy: measurements and modeling," *Atmospheric Chemistry and Physics*, vol. 7, no. 12, pp. 3361–3372, 2007.

[261] T. M. Ruuskanen, H. Hakola, M. K. Kajos, H. Hellén, V. Tarvainen, and J. Rinne, "Volatile organic compound emissions from Siberian larch," *Atmospheric Environment*, vol. 41, no. 27, pp. 5807–5812, 2007.

[262] C. Spirig, A. Guenther, J. P. Greenberg, P. Calanca, and V. Tarvainen, "Tethered balloon measurements of biogenic volatile organic compounds at a Boreal forest site," *Atmospheric Chemistry and Physics*, vol. 4, no. 1, pp. 215–229, 2004.

[263] V. Tarvainen, H. Hakola, H. Hellén, J. Bäck, P. Hari, and M. Kulmala, "Temperature and light dependence of the VOC emissions of Scots pine," *Atmospheric Chemistry and Physics*, vol. 5, no. 4, pp. 989–998, 2005.

[264] P. Tiiva, R. Rinnan, T. Holopainen, S. K. Mörsky, and J. K. Holopainen, "Isoprene emissions from boreal peatland microcosms; effects of elevated ozone concentration in an open field experiment," *Atmospheric Environment*, vol. 41, no. 18, pp. 3819–3828, 2007.

[265] P. Tiiva, P. Faubert, A. Michelsen, T. Holopainen, J. K. Holopainen, and R. Rinnan, "Climatic warming increases isoprene emission from a subarctic heath," *New Phytologist*, vol. 180, no. 4, pp. 853–863, 2008.

[266] T. Vuorinen, A.-M. Nerg, E. Vapaavuori, and J. K. Holopainen, "Emission of volatile organic compounds from two silver birch (*Betula pendula* Roth) clones grown under ambient and elevated CO_2 and different O_3 concentrations," *Atmospheric Environment*, vol. 39, no. 7, pp. 1185–1197, 2005.

[267] Q.-H. Zhang, F. Schlyter, and P. Anderson, "Green leaf volatiles interrupt pheromone response of spruce bark beetle, Ips typographus," *Journal of Chemical Ecology*, vol. 25, no. 12, pp. 2847–2861, 1999.

[268] T. Zhu, D. Wang, R. L. Desjardins, and J. I. Macpherson, "Aircraft-based volatile organic compounds flux measurements with relaxed eddy accumulation," *Atmospheric Environment*, vol. 33, no. 12, pp. 1969–1979, 1999.

[269] N. G. Agelopoulos, K. Chamberlain, and J. A. Pickett, "Factors affecting volatile emissions of intact potato plants, Solanum tuberosum: variability of quantities and stability of ratios," *Journal of Chemical Ecology*, vol. 26, no. 2, pp. 497–511, 2000.

[270] J. Arey, "Terpenes emitted from agricultural species found in California's Central Valley," *Journal of Geophysical Research*, vol. 96, no. 5, pp. 9329–9336, 1991.

[271] J. Arey, A. M. Winer, R. Atkinson, S. M. Aschmann, W. D. Long, and C. L. Morrison, "The emission of (Z)-3-hexen-1-ol, (Z)-3-hexenylacetate and other oxygenated hydrocarbons from agricultural plant species," *Atmospheric Environment*, vol. 25, no. 5-6, pp. 1063–1075, 1991.

[272] J. Arey, D. E. Crowley, M. Crowley, M. Resketo, and J. Lester, "Hydrocarbon emissions from natural vegetation in California's South Coast Air Basin," *Atmospheric Environment*, vol. 29, no. 21, pp. 2977–2988, 1995.

[273] N. Copeland, J. N. Cape, and M. R. Heal, "Volatile organic compound emissions from Miscanthus and short rotation coppice willow bioenergy crops," *Atmospheric Environment*, vol. 60, pp. 327–335, 2012.

[274] C. M. de Moraes, M. C. Mescher, and J. H. Tumlinson, "Caterpillar-induced nocturnal plant volatiles repel conspecific females," *Nature*, vol. 410, no. 6828, pp. 577–579, 2001.

[275] A. S. D. Eller, K. Sekimoto, J. B. Gilman et al., "Volatile organic compound emissions from switchgrass cultivars used as biofuel crops," *Atmospheric Environment*, vol. 45, no. 19, pp. 3333–3337, 2011.

[276] S. P. Gouinguené and T. C. J. Turlings, "The effects of abiotic factors on induced volatile emissions in corn plants," *Plant Physiology*, vol. 129, no. 3, pp. 1296–1307, 2002.

[277] C. N. Hewitt, R. K. Monson, and R. Fall, "Isoprene emissions from the grass Arundo donax L. are not linked to photorespiration," *Plant Science*, vol. 66, no. 2, pp. 139–144, 1990.

[278] F. Loreto and T. D. Sharkey, "Isoprene emission by plants is affected by transmissible wound signals," *Plant Cell and Environment*, vol. 16, no. 5, pp. 563–570, 1993.

[279] J. Ruther and S. Kleier, "Plant-plant signaling: ethylene synergizes volatile emission in Zea mays induced by exposure to (Z)-3-hexen-1-ol," *Journal of Chemical Ecology*, vol. 31, no. 9, pp. 2217–2222, 2005.

[280] G. Schuh, A. C. Heiden, T. Hoffmann et al., "Emissions of volatile organic compounds from sunflower and beech: dependence on temperature and light intensity," *Journal of Atmospheric Chemistry*, vol. 27, no. 3, pp. 291–318, 1997.

[281] A. Tava, N. Berardo, C. Cunico, M. Romani, and M. Odoardi, "Cultivar differences and seasonal changes of primary metabolites and flavor constituents in tall fescue in relation to palatability," *Journal of Agricultural and Food Chemistry*, vol. 43, no. 1, pp. 98–101, 1995.

[282] C. Warneke, S. L. Luxembourg, J. A. de Gouw, H. J. I. Rinne, A. B. Guenther, and R. Fall, "Disjunct eddy covariance measurements of oxygenated volatile organic compounds fluxes from an alfalfa field before and after cutting," *Journal of Geophysical Research D*, vol. 107, no. 7-8, pp. 6–1, 2002.

[283] S. Juuti, J. Arey, and R. Atkinson, "Monoterpene emission rate measurements from a monterey pine," *Journal of Geophysical Research*, vol. 95, no. 6, pp. 7515–7519, 1990.

[284] A. Guenther, P. Zimmerman, and M. Wildermuth, "Natural volatile organic compound emission rate estimates for U.S. woodland landscapes," *Atmospheric Environment*, vol. 28, no. 6, pp. 1197–1210, 1994.

[285] S. Kim, A. Guenther, and E. Apel, "Quantitative and qualitative sensing techniques for biogenic volatile organic compounds and their oxidation products," *Environmental Science*, 2013.

[286] T. Karl, E. Apell, A. Hodzic, D. D. Riemer, D. R. Blake, and C. Wiedinmyer, "Emissions of volatile organic compounds inferred from airborne flux measurements over a megacity," *Atmospheric Chemistry and Physics*, vol. 9, no. 1, pp. 271–285, 2009.

[287] A. Guenther, C. Geron, T. Pierce, B. Lamb, P. Harley, and R. Fall, "Natural emissions of non-methane volatile organic compounds, carbon monoxide, and oxides of nitrogen from North America," *Atmospheric Environment*, vol. 34, no. 12–14, pp. 2205–2230, 2000.

[288] J. Kesselmeier, A. Guenther, T. Hoffmann, M. T. Piedade, and J. Warnke, "Natural volatile organic compound emissions from plants and their roles in oxidant balance and particle formation," in *Amazonia and Global Change*, M. Keller, Ed., Geophysical Monograph Series, 2009.

[289] A. Guenther, M. Kulmala, A. Turnipseed, J. Rinne, T. Suni, and A. Reissell, "Integrated land ecosystem-atmosphere processes study (iLEAPS) assessment of global observational networks," *Boreal Environment Research*, vol. 16, no. 4, pp. 321–336, 2011.

[290] A. P. Altshuller, "Review: natural volatile organic substances and their effect on air quality in the United States," *Atmospheric Environment*, vol. 17, no. 11, pp. 2131–2165, 1983.

[291] M. Trainer, "Models and observations of the impact of natural hydrocarbons on rural ozone," *Nature*, vol. 329, no. 6141, pp. 705–707, 1987.

[292] W. L. Chameides, R. W. Lindsay, J. Richardson, and C. S. Kiang, "The role of biogenic hydrocarbons in urban photochemical smog: atlanta as a case study," *Science*, vol. 241, no. 4872, pp. 1473–1475, 1988.

[293] C. R. Hoyle, M. Boy, N. M. Donahue et al., "A review of the anthropogenic influence on biogenic secondary organic aerosol," *Atmospheric Chemistry and Physics*, vol. 11, no. 1, pp. 321–343, 2011.

[294] J. T. Knudsen, R. Eriksson, J. Gershenzon, and B. Ståhl, "Diversity and distribution of floral scent," *Botanical Review*, vol. 72, no. 1, pp. 1–120, 2006.

[295] J. H. Langenheim, "Higher plant terpenoids: a phytocentric overview of their ecological roles," *Journal of Chemical Ecology*, vol. 20, no. 6, pp. 1223–1280, 1994.

[296] T. R. Duhl, D. Helmig, and A. Guenther, "Sesquiterpene emissions from vegetation: a review," *Biogeosciences*, vol. 5, no. 3, pp. 761–777, 2008.

[297] N. C. Bouvier-Brown, A. H. Goldstein, J. B. Gilman, W. C. Kuster, and J. A. de Gouw, "In-situ ambient quantification of monoterpenes, sesquiterpenes and related oxygenated compounds during BEARPEX 2007: implications for gas- and particle-phase chemistry," *Atmospheric Chemistry and Physics*, vol. 9, no. 15, pp. 5505–5518, 2009.

[298] T. Sakulyanontvittaya, A. Guenther, D. Helmig, J. Milford, and C. Wiedinmyer, "Secondary organic aerosol from sesquiterpene and monoterpene emissions in the United States," *Environmental Science and Technology*, vol. 42, no. 23, pp. 8784–8790, 2008.

[299] D. W. Gray, M. T. Lerdau, and A. H. Goldstein, "Influences of temperature history, water stress, and needle age on methylbutenol emissions," *Ecology*, vol. 84, no. 3, pp. 765–776, 2003.

[300] D. W. Gray, S. R. Breneman, L. A. Topper, and T. D. Sharkey, "Biochemical characterization and homology modeling of methylbutenol synthase and implications for understanding hemiterpene synthase evolution in plants," *Journal of Biological Chemistry*, vol. 286, no. 23, pp. 20582–20590, 2011.

[301] G.-I. Arimura, K. Matsui, and J. Takabayashi, "Chemical and molecular ecology of herbivore-induced plant volatiles: proximate factors and their ultimate functions," *Plant and Cell Physiology*, vol. 50, no. 5, pp. 911–923, 2009.

[302] J. R. Snider and G. A. Dawson, "Tropospheric light alcohols, carbonyls, and acetonitrile: concentrations in the southwestern United States and Henry's law data," *Journal of Geophysical Research*, vol. 90, pp. 3797–3805, 1985.

[303] C. Warneke, T. Karl, H. Judmaier et al., "Acetone, methanol, and other partially oxidized volatile organic emissions from dead plant matter by abiological processes: significance for atmospheric HO(X) chemistry," *Global Biogeochemical Cycles*, vol. 13, no. 1, pp. 9–17, 1999.

[304] D. J. Jacob, B. D. Field, Q. Li et al., "Global budget of methanol: constraints from atmospheric observations," *Journal of Geophysical Research D*, vol. 110, no. 8, pp. 1–17, 2005.

[305] D. B. Millet, D. J. Jacob, T. G. Custer et al., "New constraints on terrestrial and oceanic sources of atmospheric methanol," *Atmospheric Chemistry and Physics*, vol. 8, no. 23, pp. 6887–6905, 2008.

[306] T. Stavrakou, A. Guenther, A. Razavi et al., "First space-based derivation of the global atmospheric methanol emission fluxes," *Atmospheric Chemistry and Physics*, vol. 11, no. 10, pp. 4873–4898, 2011.

[307] D. J. Jacob, B. D. Field, E. M. Jin et al., "Atmospheric budget of acetone," *Journal of Geophysical Research D*, vol. 107, no. 9-10, pp. 5–1, 2002.

[308] E. V. Fischer, D. J. Jacob, D. B. Millet, R. M. Yantosca, and J. Mao, "The role of the ocean in the global atmospheric budget of acetone," *Geophysical Research Letters*, vol. 39, no. 1, Article ID L01807, 2012.

[309] J. Kesselmeier, "Exchange of short-chain oxygenated volatile organic compounds (VOCs) between plants and the atmosphere: a compilation of field and laboratory studies," *Journal of Atmospheric Chemistry*, vol. 39, no. 3, pp. 219–233, 2001.

[310] D. B. Millet, A. Guenther, D. Siegel et al., "Global atmospheric budget of acetaldehyde: 3-D model analysis and constraints from in-situ and satellite observations," *Atmospheric Chemistry and Physics*, vol. 10, no. 7, pp. 3405–3425, 2010.

[311] T. Stavrakou, J.-F. Müller, J. Peeters et al., "Satellite evidence for a large source of formic acid from boreal and tropical forests," *Nature Geoscience*, vol. 5, no. 1, pp. 26–30, 2012.

[312] J. Engelberth, H. T. Alborn, E. A. Schmelz, and J. H. Tumlinson, "Airborne signals prime plants against insect herbivore attack," *Proceedings of the National Academy of Sciences of the United States of America*, vol. 101, no. 6, pp. 1781–1785, 2004.

[313] T. C. Turlings and J. Ton, "Exploiting scents of distress: the prospect of manipulating herbivore-induced plant odours to enhance the control of agricultural pests," *Current Opinion in Plant Biology*, vol. 9, no. 4, pp. 421–427, 2006.

[314] C. C. Von Dahl, M. Hävecker, R. Schlögl, and I. T. Baldwin, "Caterpillar-elicited methanol emission: a new signal in plant-herbivore interactions?" *Plant Journal*, vol. 46, no. 6, pp. 948–960, 2006.

[315] K. Hüve, M. M. Christ, E. Kleist et al., "Simultaneous growth and emission measurements demonstrate an interactive control of methanol release by leaf expansion and stomata," *Journal of Experimental Botany*, vol. 58, no. 7, pp. 1783–1793, 2007.

[316] M. R. Kant, P. M. Bleeker, M. V. Wijk, R. C. Schuurink, and M. A. Haring, "Plant volatiles in defence," *Advances in Botanical Research*, vol. 51, pp. 613–666, 2009.

[317] F. A. M. Wellburn and A. R. Wellburn, "Variable patterns of antioxidant protection but similar ethene emission differences in several ozone-sensitive and ozone-tolerant plant selections," *Plant, Cell and Environment*, vol. 19, no. 6, pp. 754–760, 1996.

[318] J. Browse and G. A. Howe, "New weapons and a rapid response against insect attack," *Plant physiology*, vol. 146, no. 3, pp. 832–838, 2008.

[319] A. Hansjakob, M. Riederer, and U. Hildebrandt, "Wax matters: absence of very-long-chain aldehydes from the leaf cuticular wax of the glossy11 mutant of maize compromises the prepenetration processes of *Blumeria graminis*," *Plant Pathology*, vol. 60, no. 6, pp. 1151–1161, 2011.

[320] D. Chachalis, K. N. Reddy, and C. D. Elmore, "Characterization of leaf surface, wax composition, and control of redvine and trumpetcreeper with glyphosate," *Weed Science*, vol. 49, no. 2, pp. 156–163, 2001.

[321] T. Karl, P. Harley, A. Guenther et al., "The bi-directional exchange of oxygenated VOCs between a loblolly pine (*Pinus taeda*) plantation and the atmosphere," *Atmospheric Chemistry and Physics*, vol. 5, no. 11, pp. 3015–3031, 2005.

[322] M. A. H. Khan, M. E. Whelan, and R. C. Rhew, "Effects of temperature and soil moisture on methyl halide and chloroform fluxes from drained peatland pasture soils," *Journal of Environmental Monitoring*, vol. 14, no. 1, pp. 241–249, 2012.

[323] Y. Yoshida, Y. Wang, C. Shim, D. Cunnold, D. R. Blake, and G. S. Dutton, "Inverse modeling of the global methyl chloride sources," *Journal of Geophysical Research D*, vol. 111, no. 16, Article ID D16307, 2006.

[324] R. C. Rhew, "Sources and sinks of methyl bromide and methyl chloride in the tallgrass prairie: applying a stable isotope tracer technique over highly variable gross fluxes," *Journal of Geophysical Research G*, vol. 116, no. 3, Article ID G03026, 2011.

[325] T. S. Bates, B. K. Lamb, A. Guenther, J. Dignon, and R. E. Stoiber, "Sulfur emissions to the atmosphere from natural sources," *Journal of Atmospheric Chemistry*, vol. 14, no. 1–4, pp. 315–337, 1992.

[326] S. F. Watts, "The mass budgets of carbonyl sulfide, dimethyl sulfide, carbon disulfide and hydrogen sulfide," *Atmospheric Environment*, vol. 34, no. 5, pp. 761–779, 2000.

[327] J. Wildt, K. Kobel, G. Schuh-Thomas, and A. C. Heiden, "Emissions of oxygenated volatile organic compounds from plants part II: emissions of saturated aldehydes," *Journal of Atmospheric Chemistry*, vol. 45, no. 2, pp. 173–196, 2003.

[328] F. Keppler, J. T. G. Hamilton, M. Braß, and T. Röckmann, "Methane emissions from terrestrial plants under aerobic conditions," *Nature*, vol. 439, no. 7073, pp. 187–191, 2006.

[329] T. A. Dueck, R. de Visser, H. Poorter et al., "No evidence for substantial aerobic methane emission by terrestrial plants: a 13C-labelling approach," *New Phytologist*, vol. 175, no. 1, pp. 29–35, 2007.

[330] S.-L. Steenhuisen, R. A. Raguso, A. Jürgens, and S. D. Johnson, "Variation in scent emission among floral parts and inflorescence developmental stages in beetle-pollinated Protea species (Proteaceae)," *South African Journal of Botany*, vol. 76, no. 4, pp. 779–787, 2010.

[331] T. E. Pierce and P. S. Waldruff, "PC-BEIS: a personal computer version of the Biogenic Emissions Inventory System," *Journal of the Air and Waste Management Association*, vol. 41, no. 7, pp. 937–941, 1991.

[332] M. T. Benjamin and A. M. Winer, "Estimating the ozone-forming potential of urban trees and shrubs," *Atmospheric Environment*, vol. 32, no. 1, pp. 53–68, 1998.

[333] P. C. Harley, R. K. Monson, and M. T. Lerdau, "Ecological and evolutionary aspects of isoprene emission from plants," *Oecologia*, vol. 118, no. 2, pp. 109–123, 1999.

[334] R. G. Latta, Y. B. Linhart, M. A. Snyder, and L. Lundquist, "Patterns of variation and correlation in the monoterpene composition of xylem oleoresin within populations of ponderosa pine," *Biochemical Systematics and Ecology*, vol. 31, no. 5, pp. 451–465, 2003.

[335] E. Sertel, A. Robock, and C. Ormeci, "Impacts of land cover data quality on regional climate simulations," *International Journal of Climatology*, vol. 30, no. 13, pp. 1942–1953, 2010.

[336] G. B. Bonan, S. Levis, L. Kergoat, and K. W. Oleson, "Landscapes as patches of plant functional types: an integrating concept for climate and ecosystem models," *Global Biogeochemical Cycles*, vol. 16, no. 2, pp. 5–1, 2002.

[337] A. Guenther, T. Karl, P. Harley, C. Wiedinmyer, P. I. Palmer, and C. Geron, "Estimates of global terrestrial isoprene emissions using MEGAN (Model of Emissions of Gases and Aerosols from Nature)," *Atmospheric Chemistry and Physics*, vol. 6, no. 11, pp. 3181–3210, 2006.

[338] G. B. Bonan, P. J. Lawrence, K. W. Oleson et al., "Improving canopy processes in the Community Land Model version 4 (CLM4) using global flux fields empirically inferred from FLUXNET data," *Journal of Geophysical Research*, vol. 116, no. G02, 2011.

[339] B. Clement, M. L. Riba, R. Leduc, M. Haziza, and L. Torres, "Concentration of monoterpenes in a maple forest in Quebec," *Atmospheric Environment*, vol. 24, no. 9, pp. 2513–2516, 1990.

[340] C. Geron, A. Guenther, J. Greenberg, T. Karl, and R. Rasmussen, "Biogenic volatile organic compound emissions from desert vegetation of the southwestern US," *Atmospheric Environment*, vol. 40, no. 9, pp. 1645–1660, 2006.

[341] K. Jardine, L. Abrell, S. A. Kurc, T. Huxman, J. Ortega, and A. Guenther, "Volatile organic compound emissions from Larrea tridentata (creosotebush)," *Atmospheric Chemistry and Physics*, vol. 10, no. 24, pp. 12191–12206, 2010.

[342] D. A. Exton, D. J. Suggett, M. Steinke, and T. J. McGenity, "Spatial and temporal variability of biogenic isoprene emissions from a temperate estuary," *Global Biogeochemical Cycles*, vol. 26, 2012.

[343] E. Ormeño, D. R. Gentner, S. Fares, J. Karlik, J. H. Park, and A. H. Goldstein, "Sesquiterpenoid emissions from agricultural crops: correlations to monoterpenoid emissions and leaf terpene content," *Environmental Science and Technology*, vol. 44, no. 10, pp. 3758–3764, 2010.

17

Rain Height Statistics Based on 0°C Isotherm Height Using TRMM Precipitation Data for Earth-Space Satellite Links in Nigeria

Joseph Sunday Ojo

Department of Physics, The Federal University of Technology, Akure PMB 704, Akure 340213, Nigeria

Correspondence should be addressed to Joseph Sunday Ojo; josnno@yahoo.com

Academic Editors: L. Mona, I. A. Pérez, and C. Serio

In the prediction of attenuation due to precipitation related phenomena, the 0°C isotherm height plays a vital role. In this paper, 2 years of precipitation data obtained from the Tropical Rain Measuring Mission (TRMM) satellite had been analyzed to establish the distribution of rain height based on 0°C isotherm heights over six locations in Nigeria. Probability of exceedance of rain heights in each of the locations was compared between the two seasons in Nigeria. Rain heights distribution was also compared with the ITU-R P.839 recommendation. The overall results show seasonal, rainfall type's dependence and overestimation of the rain height predicted by the ITU for Nigeria.

1. Introduction

As the satellite communication link continues to expand its bandwidths for higher data rates, among others, system performances can be marred by the increased propagation challenges at these higher frequency ranges. Rain induced attenuation of microwaves poses a serious challenge to system/signal availability at frequencies above 10 GHz. For good system design, the designer must provide adequate fade margins to ensure system availability. There are a number of propagation mechanisms affecting Earth-space satellite communications that are a major concern in the system design. These include gaseous attenuation, cloud and fog attenuation, and rain and ice attenuation [1, 2]. However, rain-induced attenuation has been identified to cause the most serious degradation to system performances of both terrestrial and satellite links operating at frequencies above 10 GHz [3–5]. In order to estimate the level of the degradation, most researchers normally used prediction methods when actual measurements are not available. Among the meteorological parameters needed for prediction method is the annual average of the 0°C isotherm height. This 0°C isotherm height is the height in the stratiform type of rain where the frozen hydrometeors begin to change state into

liquid rain due to temperature difference [3, 6]. The ITU-R [7] gave a global map of 0°C isotherm height above mean sea level (km) on a resolution of 1.5° by 1.5° in both latitude and longitude to be used in regions of the world where no location specific information is available. However, study revealed that this height strongly depends on local weather [8, 9].

In this paper, the annual average, year-to-year variability, probability distribution levels, and seasonal variation of 0°C isotherm height derived from TRMM-PR data for some locations in Nigeria are presented.

2. Climatology of Nigeria and Source of Data

Nigeria lies between latitudes 4° and 14°N and longitudes 2° and 15°E with a varied landscape and climate. The country is in the western part of Africa. The seasonal northward and southward oscillatory movement of the intertropical discontinuity (ITD) largely dictates the weather pattern of Nigeria. The moist southwesterly winds from the South Atlantic Ocean, which is the source of moisture needed for rainfall and thunderstorms to occur, prevail over the country during the rainy season (April–October). In reverse, northeasterly winds which raise and transport dust particles from the Sahara Desert prevail all over the country during

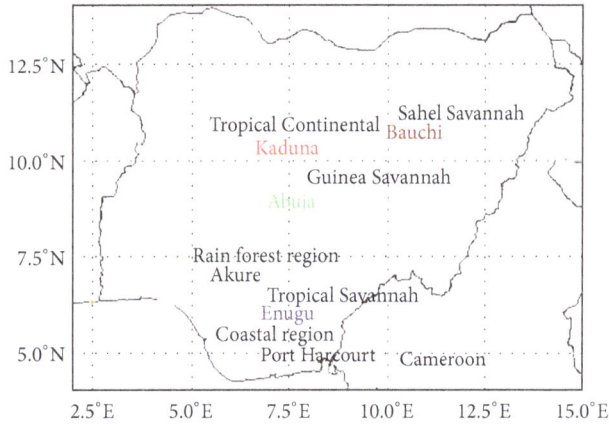

FIGURE 1: Map of Nigeria showing the characteristics of the observation sites.

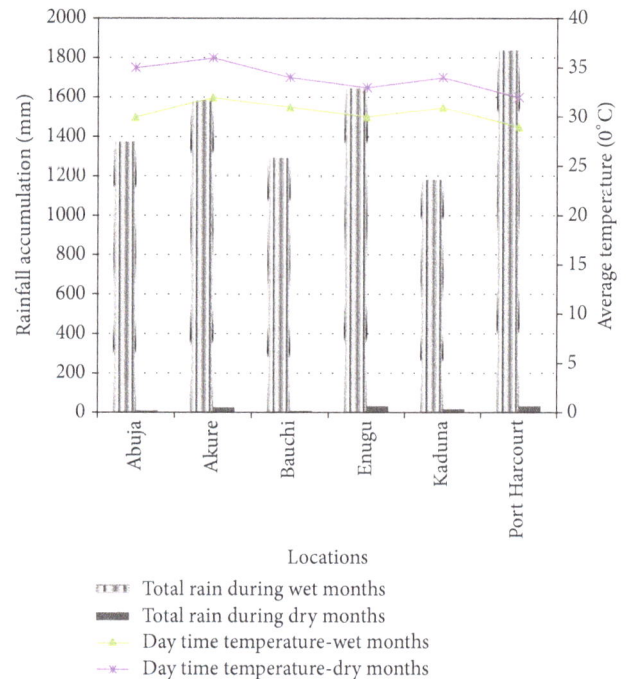

FIGURE 2: Local weather observed during the period of study.

the dry/Harmattan period (November–March). The overall changes in temperature, rainfall, and other meteorological parameters determine the changes in climate in the country each year. Nigeria being a tropical country experiences abundant sunshine all through the year.

The $0°C$ isotherm height is derived from TRMM-PR observations taken over some locations in Nigeria: Abuja, Akure, Bauchi, Enugu, Kaduna, and Port Harcourt which are located in 344, 358, 800, 223, 645, and 80 m above sea level, respectively. Figure 1 presents a map showing the characteristics of the sites while Figure 2 shows the local weather observed in each of the locations during the period of study. Each of the locations has been chosen to represent the different climatic region of Nigeria (Figure 1). The TRMM-PR is the first space borne rain radar and the only instrument on TRMM that can directly observe vertical distributions of rain. The satellite was launched in November 1997 with an altitude of 350 km and frequency of 13.8 GHz. Although the design lifetime of the satellite was 3 years, the radar still remains in good condition and continues to send high-quality data but presently at a higher altitude [3]. Further details of the radar are available in the works of Thurai et al. [3], Awaka et al. [10], and Iguchi et al. [11]. Event classification in TRMM radar data is assigned with a routine algorithm, namely, 2A23. The algorithm identifies each of measured profile data based on vertical and horizontal distribution of radar reflectivity factor. They are classified as stratiform, convective, or other precipitation on a 4.3 km by 4.3 km pixel basis. The bright-band height is obtained from the peak value of the algorithm outputs at the point when the bright band is detected.

Two years (January 2010 to December 2011) of TRMM PR 2A23 data had been processed to obtain the bright-band heights. The algorithm is tested whether a bright band exists in rain echoes and determines the bright-band height when it exists [3, 10, 12]. It also detects isolated warm rain whose height is below the $0°C$ height. Averaging of the data on a grid basis was not performed in areas where there is no PR data as well as areas where all grid points with bright band-heights are less than 2 km. This is because some of the range gates below 2 km are considered to have significant clutter

contamination and the radar reflectivity is assumed to be constant below this height. Hence, the algorithm could not identify the event as stratiform precipitation as pointed out in the work of Thurai et al., [3]. The bright-band height almost lies below the $0°C$ isotherm height.

The $0°C$ isotherm height or freezing height level height is recommended by the ITU-R, in particular ITU-R [13] for calculating rain-induced attenuation statistics meant for planning and design of Earth-space telecommunication systems.

3. Distribution of $0°C$ Isotherm Heights over Some Stations in Nigeria

Figure 3 presents the annual distribution of monthly average $0°C$ isotherm heights over the six locations considered as well as the value recommended by the ITU-R. It could be observed that $0°C$ isotherm heights are higher during the wet months of the year (April–October) than the rest of the calendar months (dry months—November–March). This is in agreement with the observation made by [9] that during the period of enhanced shower activities, $0°C$ isotherm heights are higher than during the dry months. In overall, $0°C$ isotherm heights vary from 4.160 to 4.457 km, 4.109 to 4.458 km, 4.170 to 4.460 km, 4.146 to 4.451 km, 4.162 to 4.455 km and 4.117 to 4.462 km over Abuja, Akure, Bauchi, Enugu, Kaduna and PortHarcourt respectively. However, ITU-R [7] predicted an annual average value of 4.86 km for this region with an average percentage difference of about 4% of the measured values. The implication is that rain attenuation estimated using the value predicted by the ITU will be overestimated in this region due to the annual average estimated value of the $0°C$ isotherm height used.

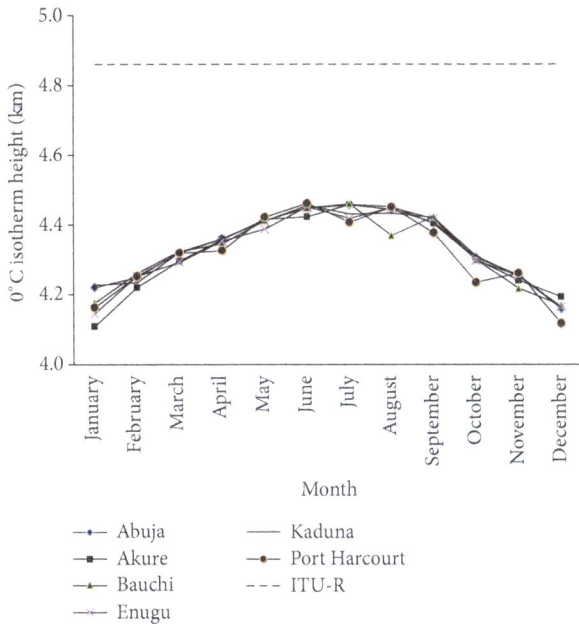

FIGURE 3: Average monthly variation of the 0°C isotherm height.

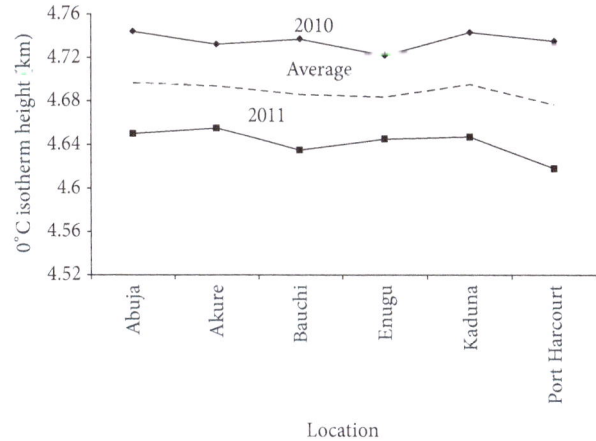

FIGURE 4: Year-to-year variability of 0°C isotherm height over all the stations.

Figure 4 shows the year-to-year variation of 0°C isotherm height for the period of study as well as overall average as a function of location of study. The difference is noticeable in the 2010 case, which shows higher values when compared with 2011. For example, in 2010, the mean values of 0°C isotherm height are found to be 4.744, 4.743, 4.737, 4.735, 4.732, and 4.722 km over Abuja, Kaduna, Bauchi, Port Harcourt, Akure, and Enugu, respectively, while in 2011, the values are 4.655, 4.650, 4.647, 4.645, 4.635, and 4.618 km over Akure, Abuja, Kaduna, Enugu, Bauchi, and Port Harcourt, respectively. Judging from these results, we observed variation in 0°C isotherm height at different location and in different year. For example, the percentage difference between 2010 and 2011 when compared with the mean height is 2, 1.6, 2.2, 1.6, 2, and 2.5% over Abuja, Akure, Bauchi, Enugu, Kaduna, and Port Harcourt, respectively. This may be due to the activities of the intertropical discontinuity over the locations. Further observation shows that Abuja with latitude 9.04°N recorded the highest value of 4.744 km while Enugu with latitude 6.24°N recorded the least value of 4.722 km. However, in 2011, Akure with latitude 7.18°N recorded the highest value of 4.655 km while Port Harcourt with latitude 4.43°N recorded the least value of 4.618 km. These results do not depend very much on latitude of the stations as earlier observed by [8].

Figures 5(a), 5(b), 5(c), 5(d), 5(e), and 5(f) show the averaged distribution of 0°C isotherm height for the two seasons in Nigeria as well as the average years. As earlier stated, the two seasons considered are wet/rainy season (April–October) and dry season (November–February of the following year) for all stations. Generally, the variation shows the expected seasonal trend of a wet season showing the higher rain heights and the dry months showing the lowest rain heights. For example, Figure 5(a) presents the result of 0°C isotherm height over Abuja, Nigeria. It could be observed that, during the wet months, the heights vary between 3.48 and 4.85 km

over probability range 0.01–99.99% of time the 0°C isotherm height is exceeded, while, during the dry months, the heights vary between 3 and 4.76 km over probability range 0.01–99.99% of the time the ordinate is exceeded. The same trend could be observed in other locations (Figures 5(b), 5(c), 5(d), 5(e), and 5(f)) although with different variation in the 0°C isotherm heights values. This is imputable to the fact that Nigeria located in the tropical region has closely related climatic conditions and rainfall pattern throughout the country. The variation of ground temperature over Abuja, Akure, Bauchi, Enugu, Kaduna, and Port Harcourt is 30–35°C, 32–36°C, 31–34°C, 30–23°C, 31–34°C, and 29–32°C, respectively, during wet and dry seasons. The total rainfall during the dry seasons over Abuja, Akure, Bauchi, Enugu, Kaduna, and Port Harcourt is 5.2 mm, 21.1 mm, 3.2 mm, 28 mm, 12.1 mm, and 26.85 mm, respectively, and the wet season is 1378.0 mm, 1588.5 mm, 1297.3 mm, 1649.2 mm, 1186.6 mm, and 1842.9 mm, respectively. It could be further observed that, for all the months, the heights vary from 3.21 to 4.81, 3.43 to 4.83, 3.49 to 4.7, 3.46 to 4.72, 3.62 to 4.94, and 3.55 to 4.64 over probability range 0.01–99.99% of time the ordinate is exceeded over Abuja, Akure, Bauchi, Enugu, Kaduna, and Port Harcourt, respectively.

It is worth mentioning here that the 0°C isotherm heights over Kaduna in different seasons are appreciable, but the variation of 0°C isotherm height over Abuja is insignificant at low probability levels. The appreciable values observed over Kaduna might be due to the nature of the rain in the region which is mostly of thunderstorm rain during the months of observation. It has been reported in the work of Ajayi and Barbaliscia [6] that, in the tropics, during the months when thunderstorm rain is more prevalent, 0°C isotherm heights during rainy conditions increased. The overall results show distinct seasonal dependence and no variation of latitudinal dependence is observed.

4. Conclusion

Two years of data from the TRMM-PR satellite had been employed to study the 0°C isotherm height over some

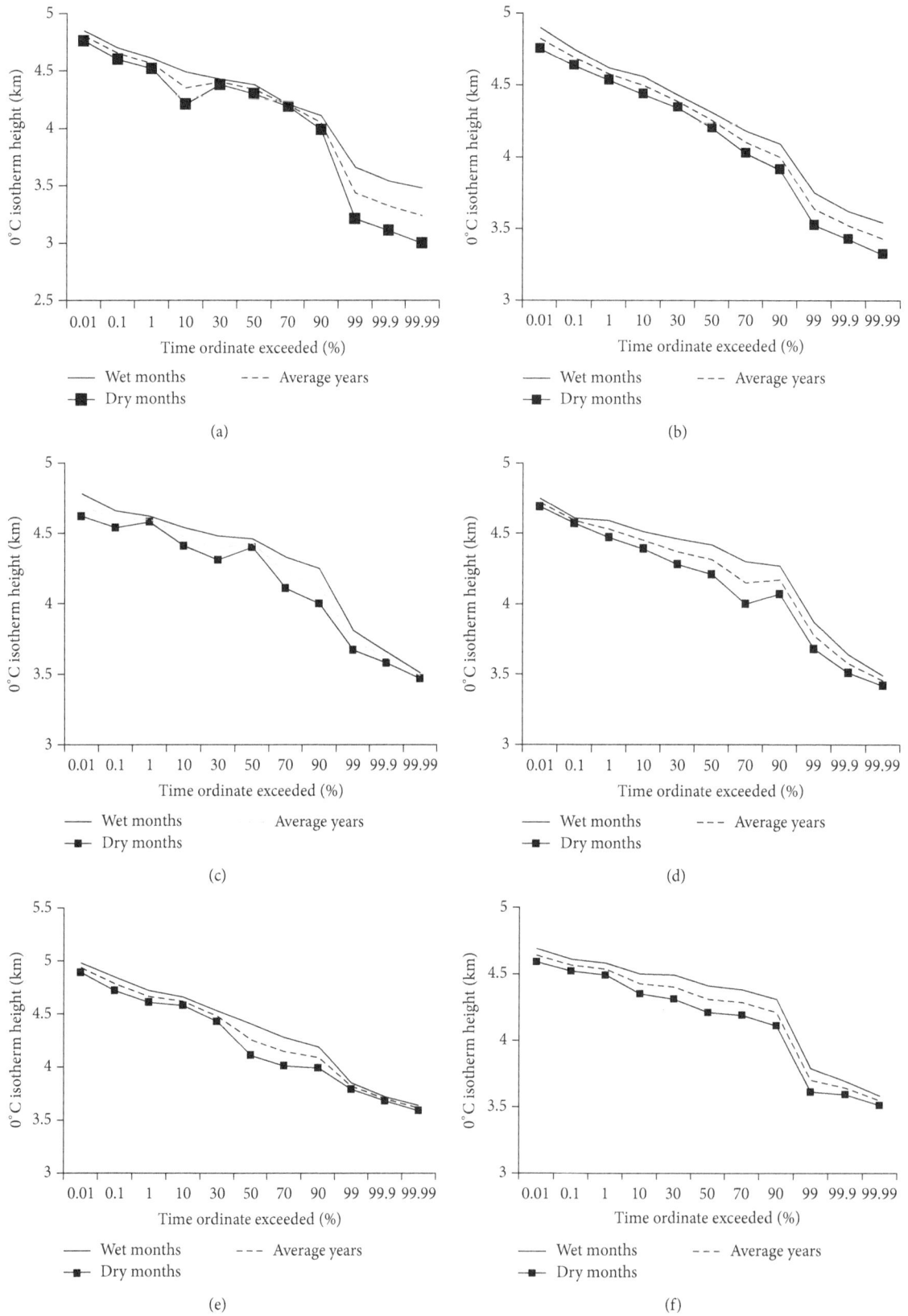

FIGURE 5: Distribution of 0°C isotherm heights during different seasons over (a) Abuja (b) Akure, (d) Enugu, (e) Kaduna, and (f) Port Harcourt.

selected locations in Nigeria and from the result, it could be observed that the 0°C isotherm height shows seasonal type dependence. Year-to-year variation shows slight differences with 2010 showing higher values than 2011. The result also shows that the 0°C isotherm height is location dependent with no two stations studied having the same value. Comparison with the predicted value by the ITU-R shows overestimation of about 4% as compared to the measured value over the stations. The result will be useful in estimating location dependent fade margins required for user availability of satellite and terrestrial line of sight communication links for this region.

Conflict of Interests

The authors declare that there is no conflict of interests regarding the publication of this paper.

Acknowledgment

The author acknowledges the support of NASA Goddard Space Flight, Centre Earth Enterprise Distributed Active Archive Centre (GES DAAC), for providing data used for this work.

References

[1] L. J. Ippolito, "Propagation effects handbook for satellite systems designs. A summary of propagation impairments on 10 to 100 GHz satellite links with techniques for system design," in *National Aeronautics and Space Administration (NASA) Doc*, vol. 1082, no. 2, Scientific and Technical Information Branch, 1989.

[2] A. Kumar and S. K. Sarkar, "Cloud attenuation and cloud noise temperature over some Indian eastern stations for satellite communication," *Indian Journal of Radio and Space Physics*, vol. 36, no. 5, pp. 375–379, 2007.

[3] M. Thurai, E. Deguchi, T. Iguchi, and K. Okamoto, "Freezing height distribution in the tropics," *International Journal of Satellite Communications and Networking*, vol. 21, no. 6, pp. 533–545, 2003.

[4] J. S. Mandeep, "Rain height statistics for satellite communication in Malaysia," *Journal of Atmospheric and Solar-Terrestrial Physics*, vol. 70, no. 13, pp. 1617–1620, 2008.

[5] J. S. Ojo and T. V. Omotosho, "Comparison of rain rate derived from TRMM satellite data and raingauge data for microwave applications in Nigeria," *Journal of Atmospheric and Solar-Terrestrial Physics*, vol. 7, no. 27, pp. 5026–5035, 2013.

[6] G. O. Ajayi and F. Barbaliscia, "Prediction of attenuation due to rain. Characteristics of the 0°C isotherm in temperate and tropical climates," *International Journal of Satellite Communications*, vol. 8, no. 3, pp. 187–198, 1990.

[7] International TelecommunicationUnion Recommendation 839-3: Rain height for prediction methods, International Telecommunication Union, Geneva, Switzerland, 2003.

[8] N. C. Mondal and S. K. Sarkar, "Rain height in relation to 0°C isotherm height for satellite communication over the Indian subcontinent," *Theoretical and Applied Climatology*, vol. 76, no. 1-2, pp. 89–104, 2003.

[9] J. S. Mandeep, "0°C isotherm height for satellite communication in Malaysia," *Advances in Space Research*, vol. 43, no. 6, pp. 984–989, 2009.

[10] J. Awaka, T. Iguchi, and T. Okamoto, "Early results on rain type classification by the Tropical Rainfall measuring Mission (TRMM) precipitation radar," in *Proceedings of the 8th URSI Commission F Open Symposium*, pp. 143–146, Aveiro, Portugal.

[11] T. Iguchi, T. Kozu, R. Meneghini, J. Awaka, and K. I. Okamoto, "Rain-profiling algorithm for the TRMM precipitation radar," *Journal of Applied Meteorology*, vol. 39, no. 12, pp. 2038–2052, 2000.

[12] C. Kummerow, W. Barnes, T. Kozu, J. Shiue, and J. Simpson, "The Tropical Rainfall Measuring Mission (TRMM) sensor package," *Journal of Atmospheric and Oceanic Technology*, vol. 15, no. 3, pp. 809–817, 1998.

[13] International Telecommunication Union Recommendation 618-10, *Propagation Data and Prediction Methods Required for the Design of Earth-Space Telecommunication Systems (ITUR)*, Geneva, Switzerland, 2009.

18

Simulation of Water Vapor Condensation in a Partly Closed Structure: The Influence of the External Conditions of Temperature and Humidity

Jean Batina[1] and René Peyrous[2]

[1] *Laboratoire des Sciences de l'Ingénieur Appliquées à la Mécanique et à l'Electricité (SIAME), Université de Pau et des Pays de l'Adour, BP 1155, 64013 Pau, France*
[2] *Laboratoire d'Electronique des Gaz et des Plasmas, Université de Pau et des Pays de l'Adour, 64000 Pau, France*

Correspondence should be addressed to Jean Batina; jean.batina@univ-pau.fr

Academic Editors: H. A. Flocas and A. Luhar

Our aim is to determine the more significant parameters acting on the water vapor condensation in a partly closed structure, submitted to external constraints (temperature and humidity) which induce convective movements and thermal variations inside. These constraints locally lead to condensation of the water vapor, initially contained in the air of the volume and/or on the walls. The inside bottom wall is remained dry. Condensed water quantities depend on: (1) dimensions of the structure, (2) the air renewing and its hygrometry, and (3) the phase between thermal and hydrometric conditions. Peculiar conditions are needed to obtain a maximum of condensation.

1. Introduction

A lot of experimental and fundamental works have shown the possibility to condense the water vapor contained in ambient air with a view to recover it. We can mention the work of Nikolayev et al. [1] who highlighted the importance to study the basic physical phenomena in the formation of dew and the possibility to improve its recovering. In the same way, Muselli et al. [2] studied a dew water collector for potable water in Ajaccio. In their investigations concerning the mechanism of soil water vapor adsorption in arid regions, Beysens et al. [3] concluded that "there are areas in which, during the dry season, the dominant process is vapor adsorption, and dew formation is a rare occurrence." This study and several others on the same subject shows the importance to characterize the main parameters intervening in the formation of water vapor condensation, particularly in arid regions. All these scientific works demonstrate that successes concerning water recovering possibilities by the mean of water vapor contained in ambient air remain limited and the subject causes scientific debates. The numerical simulation,

more and more current in engineering sciences, is not very useful to describe, understand, and explain this complex process in which heat and mass transfer and thermodynamic of the mixture occur in a strong interactive manner at the same time, Gandhidasan and Abualhamayel [4], Beysens et al. [5], Caltagirone and Breil [6].

The aim of this study is to determine the more significant parameters acting on the conditions of the water vapor condensation present in air, inside a closed or partly closed structure (Batina et al. [7], and Batina et al. [8]). An illustration of such phenomena can be found in the case of the Arles-sur-Tech (France) sarcophagus (nonwatertight) Beysens et al. [9], Perard and Leborgne [10], the marble walls dimensions of which result in an about $0.33 \, \text{m}^3$ internal volume. This structure is submitted to external atmospheric conditions (temperature and humidity) which generate convective movements and thermal variations inside this space. These ones are then able to generate locally, in the structure or/and on the walls, the condensation of the water vapor of the air initially contained in the volume or renewed by introducing through an orifice. The approach of the external

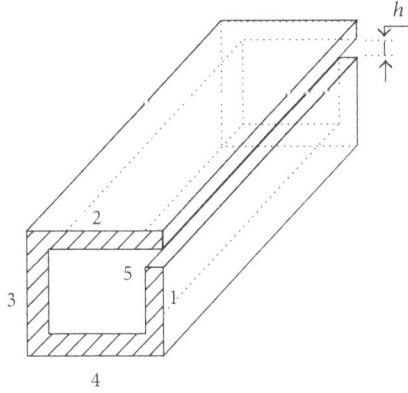

FIGURE 1: Transverse section showing the boundary conditions—(1) and (2): walls submitted to periodic external constraints; (3) and (4): adiabatic walls; (5): orifice (renewing internal air).

atmospheric conditions (temperature and hygrometry) was, in a first time and in a simplified way, made by representing the daily thermal variations (T) by a sinusoidal function and the relative hydrometric variations (RH) by crenels with a same period. These two parameters are linked by a temporal phase φ (T/RH) which is varied by steps of 3 h, from 0 to 24 h. The thermal and hydrometric evolutions and the resulting condensed water quantities, inside the considered volume, are represented, for a transverse section, at various moments of the day as a function of the phase φ (T/RH) for $\varphi = 0$ h and $\varphi = 12$ h. Taking into account the extent of the inquiry field, the numerical simulation permits to consider a lot of variable parameters which can be modified. Our study is essentially founded on the Arles-sur-Tech sarcophagus case for which we have some experimental results running on some months. In a first time, despite the importance of their role, the volume of the structure, the dimension of the orifice, and the nature of the walls material (marble) remain unchanged. The study, as a function of φ, on the dynamical and thermal behaviour and on the condensation effects, shows the existence of an optimum of condensation linked with this parameter φ.

2. Modelisation

In the parallelepipedic structure of internal dimensions: $0.47\,\text{m} \times 0.40\,\text{m} \times 1.76\,\text{m}$, we will consider only the transverse section ($0.47\,\text{m} \times 0.40\,\text{m}$), named "width section" (Figure 1), in which the convective movements take place. The marble walls, of $0.1\,\text{m}$ thick, are assumed as watertight and smooth. The frontal side 1 and the top side 2 (Figure 1) are submitted to the external atmospheric conditions (temperature and humidity) varying periodically in time, when all the other sides (3, 4, etc.) are considered as adiabatic (null thermal flux conditions). The orifice 5, of which the thickness h can be varied from $0.02\,\text{m}$ up to $0.30\,\text{m}$, and from which the inside air can be renewed only by the outside-inside thermal gradient, is situated on the upper and right part of the width section, on the side of the walls submitted to the constraints (Figure 1). The external air is without initial speed on the level of the orifice, at the initial temperature of $15°\text{C}$ with

a relative hygrometry in a 60% to 90% range at 6 h a.m. (reference values). The humid air contained in the structure is consequently submitted to a natural convective phenomenon and its renewing is only obtained by the mean of the orifice.

Atmospheric pressure, relative humidity, and temperatures involved are issued from mean values coming from the experimental results obtained by Perard and Leborgne [10]. Moreover, the temperature variations are sufficiently reduced to justify the Boussinesq approximation. In these conditions, the equations system to solve contains the movement quantity and energy conservation and continuity (Navier-Stokes). Their classical vectorial forms are the following

$$\rho \left[\frac{\partial \vec{V}}{\partial t} + \left(\vec{V} \cdot \vec{\nabla} \right) \vec{V} \right] = -\vec{\nabla} p + \mu \vec{\nabla} \cdot \left(\nabla \vec{V} + \nabla^t \vec{V} \right) + \rho \vec{g}, \quad (1)$$

$$\vec{\nabla} \cdot \vec{V} = 0, \quad (2)$$

$$\rho c_p \left[\frac{\partial T}{\partial t} + \vec{V} \cdot \vec{\nabla} T \right] = \lambda \Delta T + S. \quad (3)$$

To these dynamic and thermal equations, we must add the transport equation of the water vapor given by:

$$\frac{\partial c}{\partial t} + \vec{V} \cdot \vec{\nabla} c = \vec{\nabla} \left(D \vec{\nabla} c \right) + S', \quad (4)$$

where S and S', respectively, represent the source terms (heat for (3), and mass for (4)), coming from the external air, and representing the balance between evaporation and condensation of the water vapor contained in air.

The state equations of the humid air, given by the Mollier diagram, enable to relate dry temperature and relative humidity with the other characteristic variables of the air: more particularly absolute (or specific) humidity and dew-point temperature.

Starting from a defined configuration, we vary in time the thermal and hydrometric conditions of the external air (boundary conditions on the constrained walls), with a 24 h period: sinusoidal variations of the temperature ($T = 15 \pm 5°\text{C}$), $15°\text{C}$ at 6 h in the morning, maximum at 12 h (noon), and minimum at 24 h (midnight); crenelled periodical variations for the relative hygrometry (RH = 60% during 10 h from 7 h a.m. up to 5 h p.m., centered at noon; RH = 90% during 14 h from 5 h p.m. up to 7 h a.m. the next morning, centered at midnight).

Our initial phase is $\varphi = 0$ h (Figure 2). Though nonrealistic, these crenelled variations of RH have permitted to check the stability of the numerical model. At the moment $t = 0$ (6 h in the morning), we assume that the fluid (wet air), inside the considered volume, is in a steady state with uniform mean relative humidity and temperature. The thermal constraint is applied over two adjacent external sides, the other sides remaining adiabatic.

The modelisation of the process is in two dimensions, according to a transverse section, and the mesh grid is structured, strongly refined in the orifice wake and in the vicinity of the walls.

The grid includes 5940 (66×90) meshes in the totality.

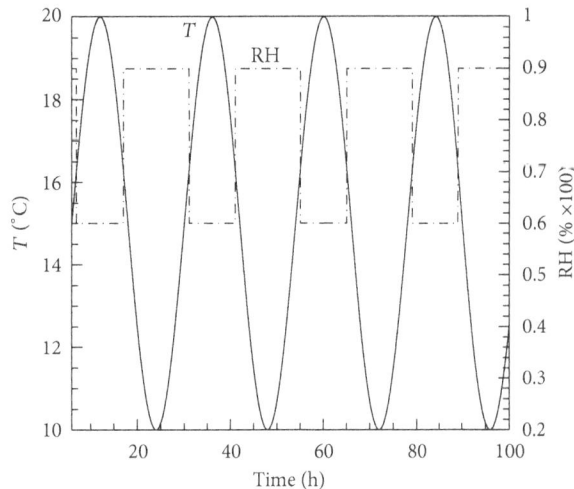

FIGURE 2: Temperature (T) and relative humidity (RH) temporal evolutions of the external air (boundary conditions on the constrained walls): phase $\varphi = 0$ h.

Equations are solved according to the finite volumes method particularly developed by Caltagirone and Breil [6], using the Gear time scheme of order 2, the time-step of which is fixed at 5 s.

Starting from these stable initial conditions, we show that the considered system reaches a steady state regime (no more variations at the same time from day to day) after 3 days (see Figure 6).

This study on the dynamical behaviour as well as the condensation effects, as a function of φ and of the orifice thickness h, shows the existence of an optimum of condensation linked with these parameters.

We will present our results by firstly taking into consideration the simplified hypothesis of a remaining dry inside bottom wall.

In a future paper, we will present the results obtained with a wet bottom wall and a small water layer remaining level-constant in time. This second hypothesis will permit to work with constant inner volume of air.

3. Results and Discussion

We have chosen to show only the more significant results in a fixed configuration: width section, 0.08 m orifice, for two phases ($\varphi = 0$ h and $\varphi = 12$ h) at various moments.

3.1. Phase $\varphi = 0$ h. In Figures 3(a) and 3(b) we can see the main role played by the thermal gradient between the outside and the inside of the structure and, in the same way, the role of the thermal inertia of the walls on the fluid circulation inside the studied volume: mainly clockwise circulation in the upper part at about 0 h (midnight), double cell circulation at 6 h (am), and mainly anticlock wise at 12 h (noon) and 18 h (6 h pm). The more important velocities are about the size order of $5.10{-}2$ m/s. We can note a "buffer effect" (very small speeds of exchanges) at the structure entrance at 0 h (midnight) and more important at 18 h (6 h p.m.). At 6 h

(a.m.) and at noon, the circulations at the structure entrance are very looking like (coming into the structure from the top, going out by the down).

This circulation will act on the isothermal and isohumidity distributions inside the structure. The behaviour of the bi-dimensional section of the local heat-fields is represented Figure 4.

Isotherms (difference value from 15°C) show a natural convective activity, fairly important, indicated by temperature difference values of about 5°C in some occurrences (0 h), able to produce condensation phenomena or drying effects inside the structure.

In Figure 4, at midnight (0 h) and at 6 pm (18 h), the numbered isotherms 1 represent the external air temperature (gap value compared with 15°C: $\delta T = 0$°C initial, see Figure 2; −5°C at midnight (0 h) and 0°C at 6 pm (18 h). In the same Figure 4, at 6 a.m. and at noon (12 h) the isotherms, respectively, numbered 10 and 15 play the same role: 0°C at 6 am and 5°C at noon (12 h). The used temperature scales are function of the obtained thermal variations for the considered moment.

It can be noted that the upper temperature variation compared with the external one is obtained at noon (12 h) ($\delta T = -3$°C), inside the structure, all along the inner walls.

About the relative iso humidity (Figure 5), we point out a condensation possibility relatively high in the vicinity of the walls under constraints at 0 h and 6 h (am) when, at 6 pm (18 h), the imposed external hygrometry is of 90%; there is not a possibility of condensation due to the too low negative temperature gap.

The best possibilities of condensation will be between midnight (0 h) and 6 h (am), when δT is the more important. Drying effects can be noted in some zones, particularly at noon (12 h), but in this case the external humidity is of only 60%. At 18 h (6pm), the imposed external hygrometry is 90%, but the inside mean temperature gap δT is always positive.

The hygrometry dispatch outside of these zones remains relatively homogeneous. Actually, the water vapor diffusion phenomenon inside the structure has a more important kinetic than the convection one and participates more actively to the humidity homogenisation.

As for the isothermal curves, the relative hygrometry variations scales depend on the considered moment. The following isohumidity curves (Figure 5), associated with the previous isothermal figures (Figure 4), permit a local view of the dynamical, thermal, and hydrometric behaviour at each moment of the studied case.

Nevertheless, a global view of this behaviour (by numerical integration of the variables in space) is better illustrated by the evaporation/condensation phenomena taking place inside the structure. This behaviour points out the phases due to the presence of the marble walls which have higher thermal inertia than the inner air. Starting from the synchronous boundary conditions $\varphi = 0$ h (Figure 2), we can see in Figure 6(a) that mean temperature and hygrometry values follow a similar behaviour to the ones outside the structure with RH upper values higher than 90% and lower than 60%.

We can notice a dephasing of about 2 hours between temperature and relative hygrometry due to the walls effect.

Simulation of Water Vapor Condensation in a Partly Closed Structure: The Influence of the External
Conditions of Temperature and Humidity

205

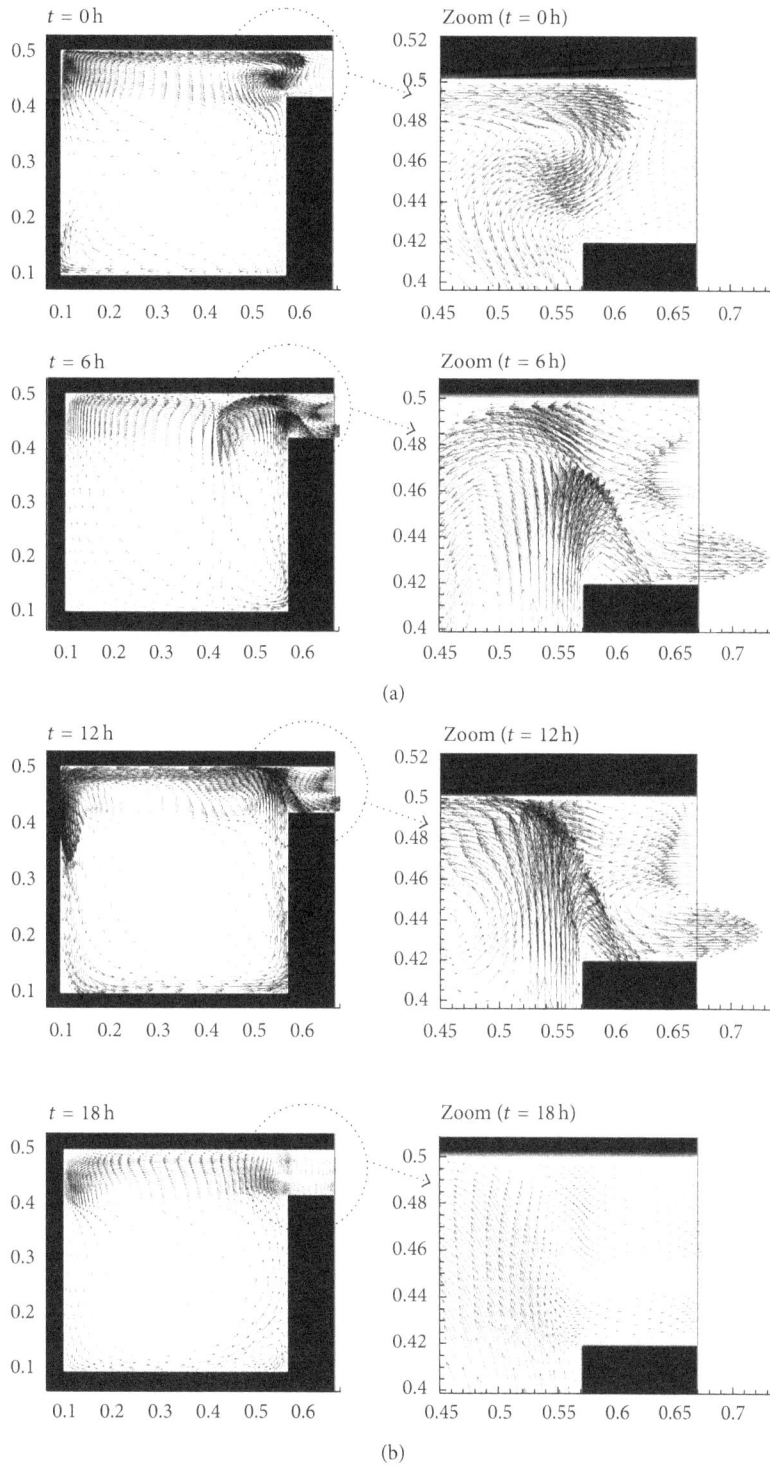

FIGURE 3: (a) Examples of the inside circulation at midnight (0 h) and 6 h (6 a.m.) with zoom on the orifice entrance. Scale (m/s) for the vector length: (Grid Units/Magnitude) = 1. (b) Examples of the inside circulation at noon (12 h) and 18 h (6 p.m.) with zoom on the orifice entrance. Scale (m/s) for the vector length: (Grid Units/Magnitude) = 1.

Figure 6(b) shows the evolution consequences through the in-time cumulated water quantities over 4 days (ratio between the condensed water and the water vapor quantity contained in the volume at the initial moment). This figure shows also the steady-state effect obtained in 3 days.

Even if low, the condensation phenomenon, however, is not negligible, and its temporal variation starts by condensing at about 7 pm (19 h), with a first maximum around midnight (24 h) and an end of the phenomenon (transition from condensation to evaporation) at 6 am the next day (30 h). This

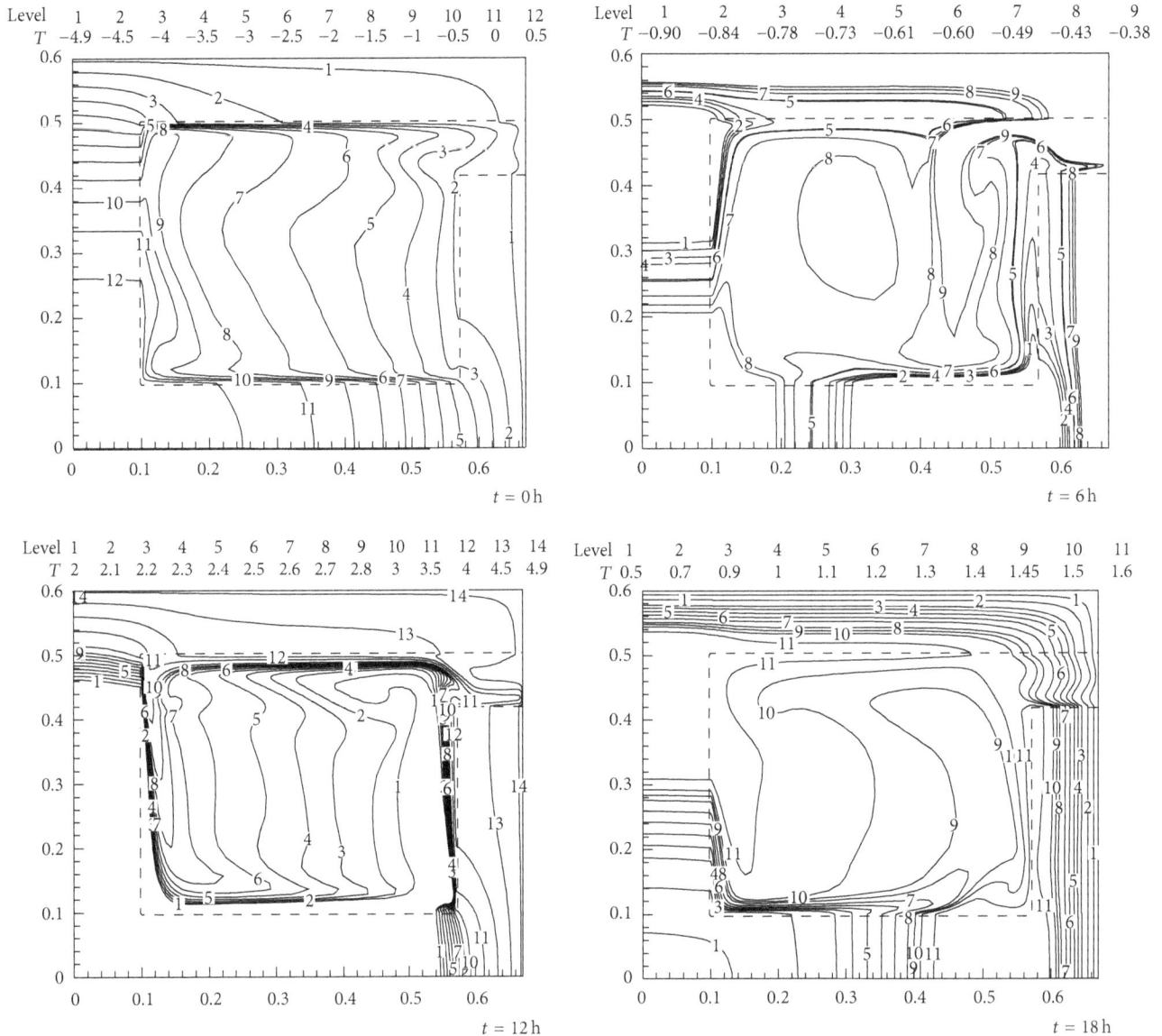

FIGURE 4: Isotherms (difference temperatures from 15°C) evolution inside the structure at various moments. Phase $\varphi = 0\,h$: $t = 0\,h$ with $T_{ext} = 10°C$; $t = 6\,h$ with $T_{ext} = 15°C$; $t = 12\,h$ with $T_{ext} = 20°C$; $t = 18\,h$ with $T_{ext} = 15°C$. T_{ext} is the external temperature.

first step matches with the internal temperature decrease. A second step starts immediately after and ends at 7 am (31 h) the second day, with a maximum at around 6.30 a.m. This second step, more upper than the first one, matches with a water vapor providing by renewing, that is to say, by hydrometric exchanges between outside and inside of the structure.

These figures show in an amazing way the importance of the thermal hydrometric exchanges with the external medium compared with the pure convective phenomenon inside the structure.

To better quantify the hydrometric exchanges between the outside and inside of the volume at each instant t_1, we have defined the percentage of condensed water vapor by the following equation as a function of phase, the bottom wall of the structure being assumed remaining dry at each moment:

$$\eta = \frac{\int_0^{t_1} \left(q_e\left(t\right) - q_s\left(t\right)\right) dt}{\int_0^{t_1} q_e\left(t\right) dt} \times 100\%. \tag{5}$$

In Figure 7 which follows, we can see that the mean level of condensation is situated in a 5% and 6% range at the end of the third day and then is stabilized.

3.2. *Phase* $\varphi = 12\,h$. Now we show the more significant results obtained with a phase $\varphi = 12\,h$ (Figure 8).

Simulation of Water Vapor Condensation in a Partly Closed Structure: The Influence of the External
Conditions of Temperature and Humidity

207

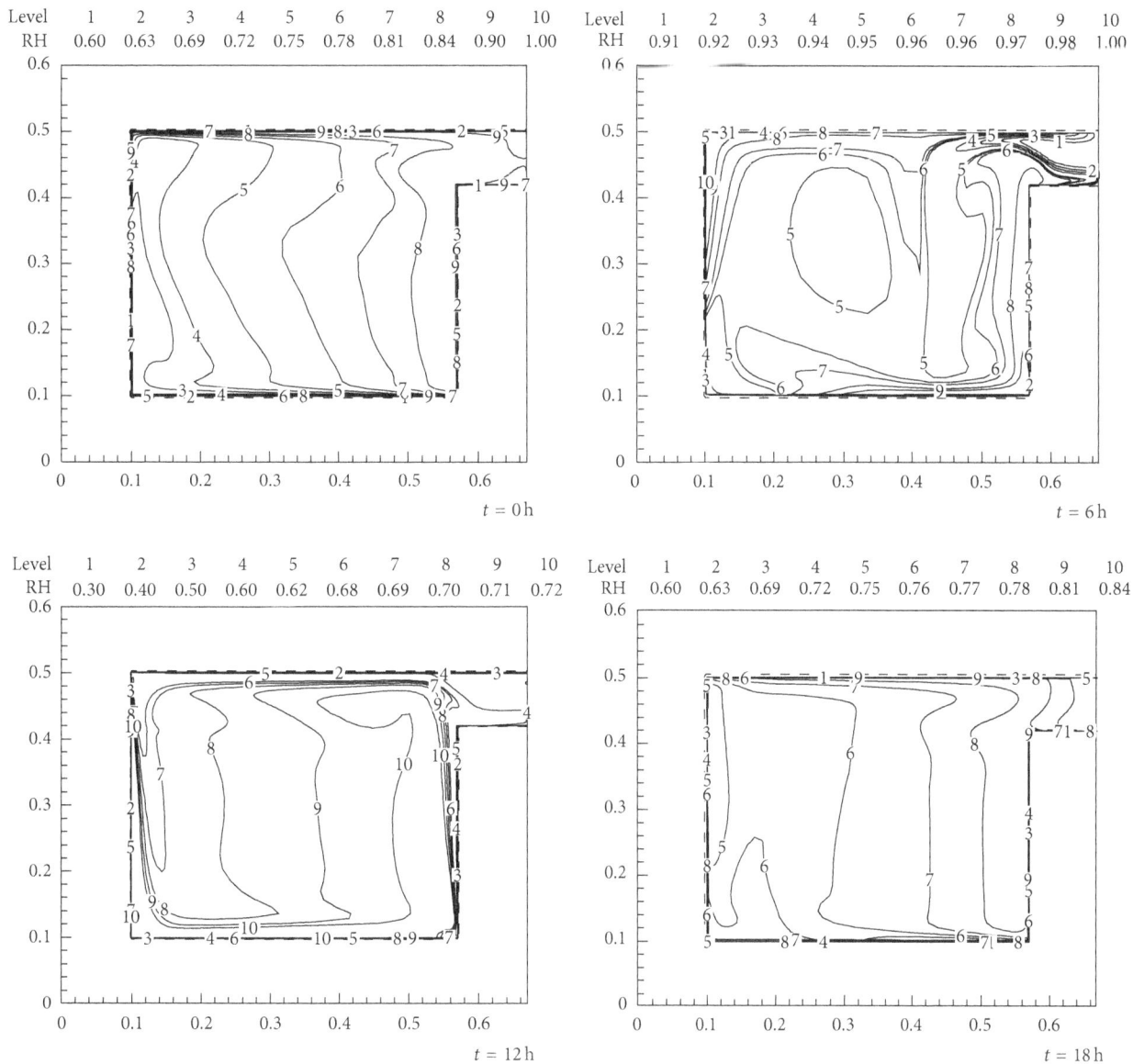

Level	1	2	3	4	5	6	7	8	9	10
RH	0.60	0.63	0.69	0.72	0.75	0.78	0.81	0.84	0.90	1.00

$t = 0\,$h

Level	1	2	3	4	5	6	7	8	9	10
RH	0.91	0.92	0.93	0.94	0.95	0.96	0.96	0.97	0.98	1.00

$t = 6\,$h

Level	1	2	3	4	5	6	7	8	9	10
RH	0.30	0.40	0.50	0.60	0.62	0.68	0.69	0.70	0.71	0.72

$t = 12\,$h

Level	1	2	3	4	5	6	7	8	9	10
RH	0.60	0.63	0.69	0.72	0.75	0.76	0.77	0.78	0.81	0.84

$t = 18\,$h

FIGURE 5: Relative isohumidity evolutions inside the structure at various moments t. Phase $\varphi = 0\,$h: $t = 0\,$h with RH = 90%; $t = 6\,$h with RH = 90%; $t = 12\,$h with RH = 60%; $t = 18\,$h, RH = 90%.

A primary observation is that dynamics of the circulation inside the structure are the same, at about some details, at the same times as the ones presented in the previous Figure 3. This demonstrates the main role of the "thermal driving force" on the internal circulation.

The width section of local temperature and relative hygrometry are represented in Figures 9 and 10. Isotherms, gap values from 15°C (Figure 9), show convective movements due to temperature steps relatively important in some cases, up to about 5°C at 0 h (midnight) and 3°C at noon (12 h).

Dealing with the relative iso humidity (Figure 10), we can see a humidity concentration relatively high in the vicinity of the orifice and on the walls submitted to the constraints at 6 h (am), 12 h (noon), and 18 h (6 pm), where the external relative hygrometry is the highest (90%). That causes an internal condensation all along the walls, essentially at 6 h (am).

The hygrometry dispatch remains relatively homogeneous out of these zones.

It also appears that the diffusion effect of the water vapor inside the structure has more important kinetics than the convection one and more actively participates in the humidity homogenisation.

As for the phase $\varphi = 0\,$h, starting from the synchronized boundary conditions, that is to say, coincidence between maxima and minima of temperature and relative hygrometry ($\varphi = 12\,$h, Figure 8), we can see, in Figure 11(a), that the temperature and hygrometry mean values have a similar behaviour to the one at the outside of the structure. However, we can note a phase displacement of about 2 hours time between T and RH essentially depending on the nature of the walls of the structure.

The time during which mean internal RH > 90% is identical to the external hydrometric constraint when this one is

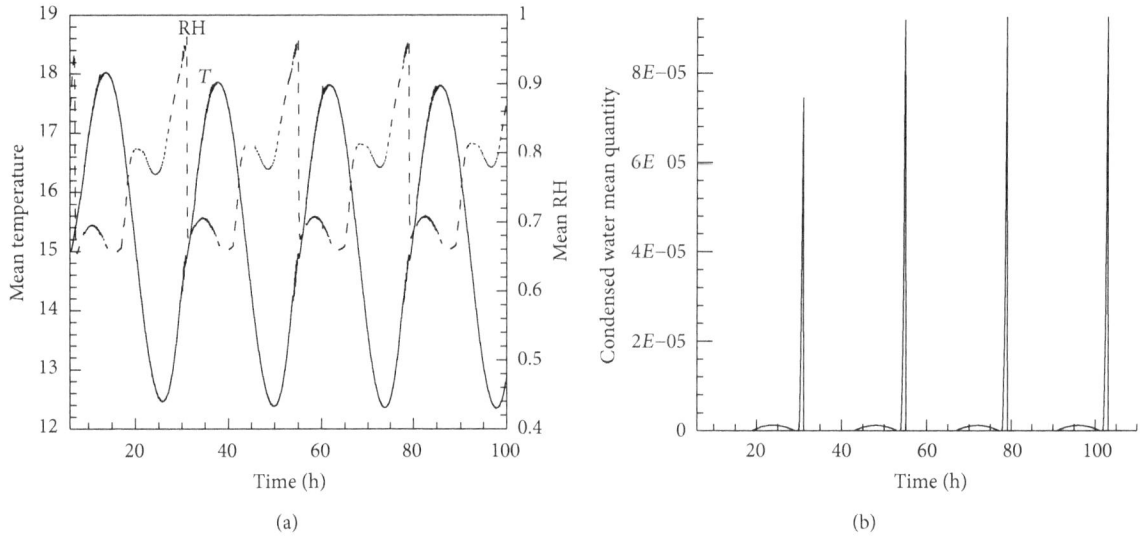

FIGURE 6: (a) Mean temperature and relative humidity evolutions of the inner air. (b) Evolutions of the mean quantity of condensed water vapour inside the structure.

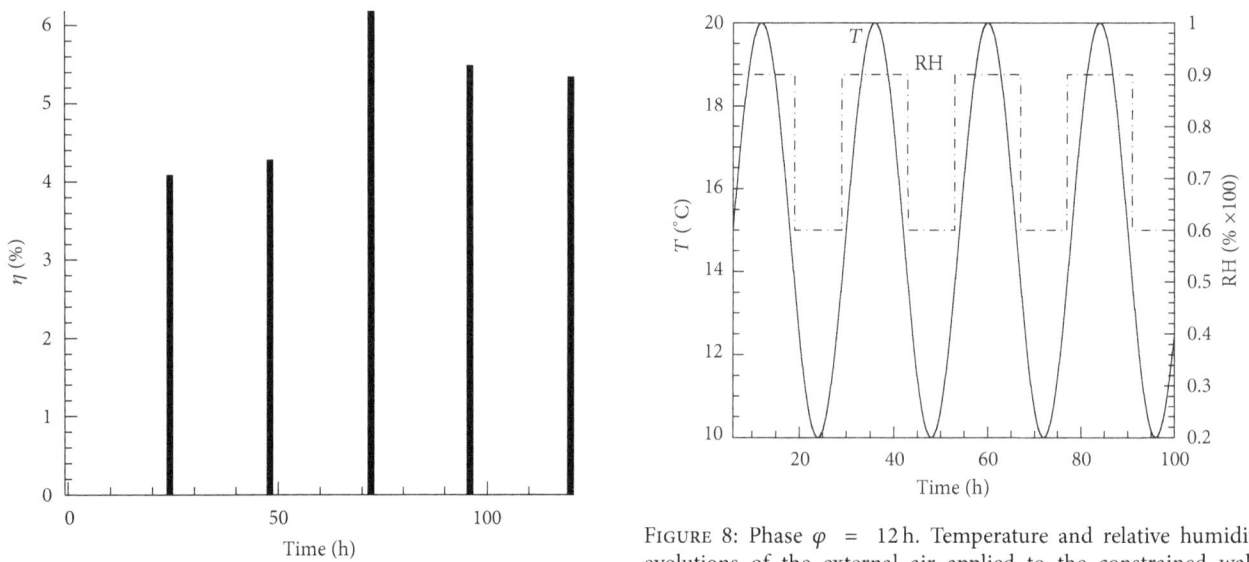

FIGURE 7: Day after day evolution of the condensed water vapour ratio. Phase $\varphi = 0$ h.

FIGURE 8: Phase $\varphi = 12$ h. Temperature and relative humidity evolutions of the external air applied to the constrained walls (boundary conditions of the model).

equal to 90%. But there are some times during which internal RH is lower than 60% pointing out drying periods of the air inside the volume. This fact can be particularly observed at 0 h (midnight), when the imposed outside relative hygrometry is equal to 60%, but when the internal gap of temperature is positive in comparison with external ones.

The spatio temporal evolutions of the relative humidity (Figure 10) lead to hold in time water quantities, that is to say, ratio between the condensed water vapor quantity and the water vapor contained in the air inside the volume at the initial time. In Figure 11(b) showing the condensed water evolutions over 4 days, we can again observe the stabilization effect of the process in about 3 days. The temporal variation

of the condensation starts at 7 h (am), reaches a maximum at about 12 h (noon), and ends around 15 h (3 pm). This first condensation step is linked with a water vapor renewing, that is to say, with the external-internal hydrometric exchanges through the orifice of the structure, and could be used to recover some atmospheric water. The amplitude of the phenomenon is about 10 times upper than the one observed in the case of $\varphi = 0$ h.

We must point out a second condensation step which starts at 19 h (7 pm) and ends at 5am the next day (29 h) with a maximum at about 1 h30 the following day. This second step, lower than the first one, cannot be seen on the figure because of its very small amplitude. It corresponds with an internal falling temperature. We can still verify the importance of the

Simulation of Water Vapor Condensation in a Partly Closed Structure: The Influence of the External
Conditions of Temperature and Humidity

209

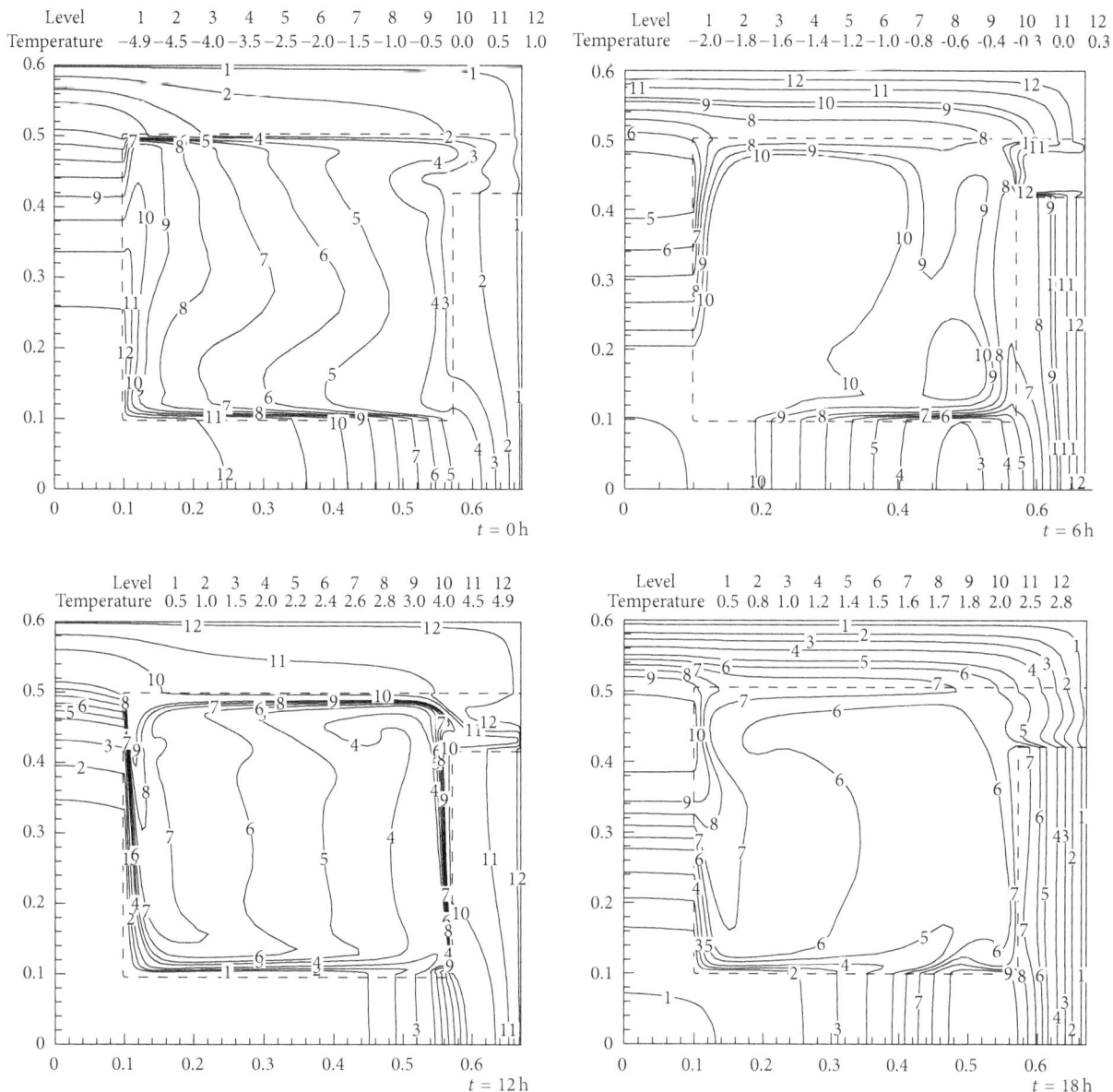

FIGURE 9: Isotherms (difference temperatures from 15°C) evolution inside the structure at various moments.

thermal and hydrometric exchanges with the outside medium compared to the pure convective phenomenon inside the structure.

By comparing Figures 6 and 11 we can point out the various possibilities resulting from the phase between external temperature and hygrometry.

The mean condensation level (Figure 12) takes place above 12% at the end of the third day and then is stabilized. By comparing Figure 12 with Figure 7 ($\varphi = 0$ h) it can be seen that the efficiency is 2 times higher, which demonstrates the important role of the phase between T and RH.

As in the case of $\varphi = 0$ h, the previous curves have permitted to display the mass and thermal dynamics of the air at each moment inside the structure. All the same numerical

integration of the variables in the space, the evaporation-condensation phenomena were displayed and quantified.

4. Conclusion/Prospects

We have studied the influence of parameters such as the thermal and hydrometric external boundary conditions and the phase between temperature and hygrometry on the condensed water vapor quantities, in a fixed parallelepiped structure the wall thickness of which and nature (marble) and the orifice dimension are fixed. We have assumed that the inside bottom wall remains dry all along the computation. The influence of the other parameters such as the residence time of the water vapor inside the structure, depending on the

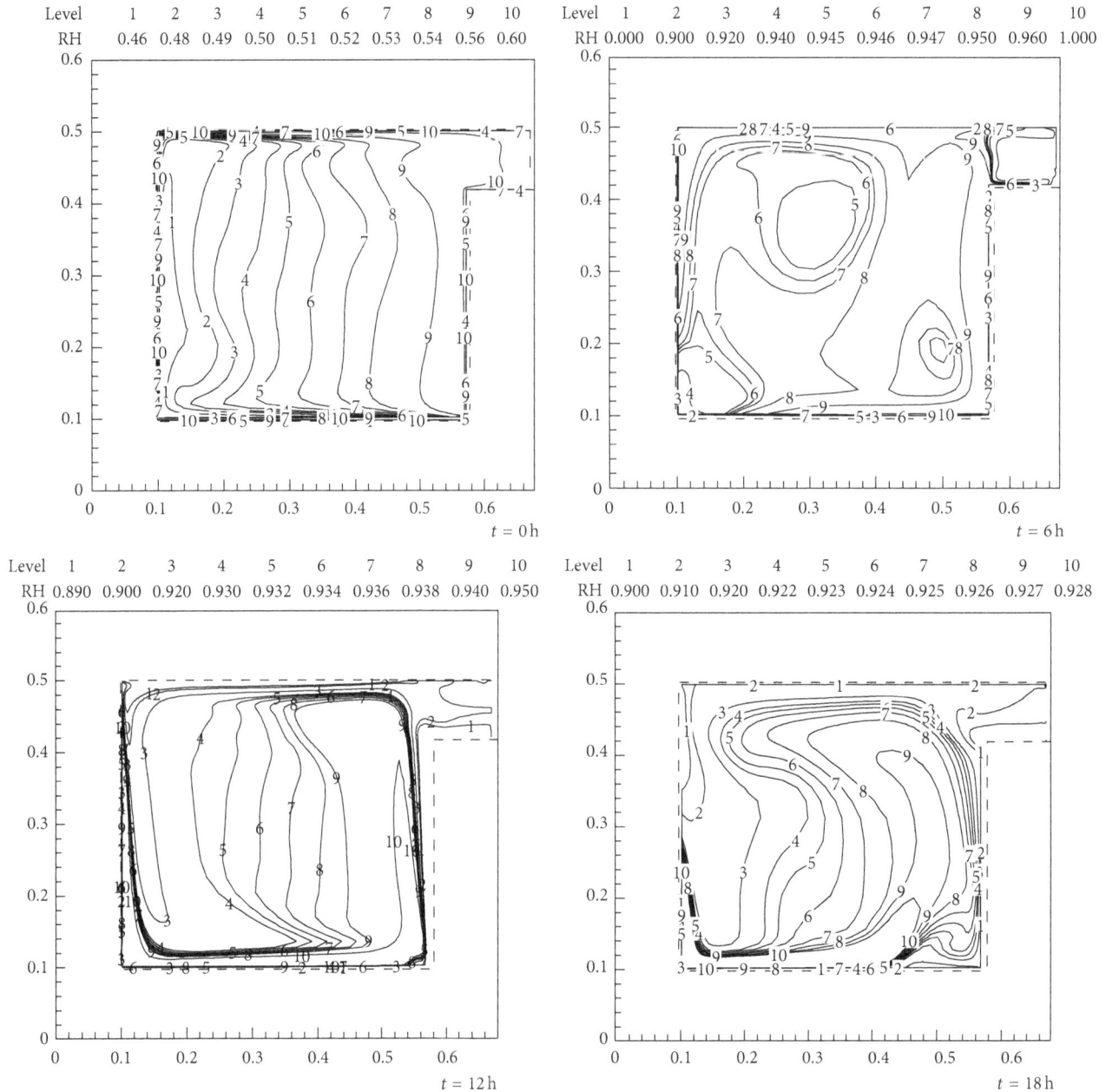

FIGURE 10: Relative isohumidity evolutions inside the structure at various moments. Phase $\varphi = 12\,\text{h}$ $t = 0\,\text{h}$ (midnight) with RH = 60%; $t = 6\,\text{h}$ (am) with RH = 90%; $t = 12\,\text{h}$ with RH = 90%; $t = 18\,\text{h}$ with RH = 90%.

orifice dimension and the longitudinal or transversal section, nature and thickness of the walls, and the presence or not of a thin layer of water on the inside bottom wall, and so on, will be the aim of other publications.

The present study points out an important dependency on the condensed water vapor quantities as a function of the considered parameters. We have noted, for the width section only, that the more significant parameter is the phase between temperature and hygrometry. The external-internal temperature gap acts as a "thermal driving force" on the internal circulation and the renewing of the water vapor inside the structure.

In Figure 13, we can see that the condensed water vapor quantities are very depending on the phase φ (T/RH) and reach a maximal value when φ is about 15 h.

The actual running computations using other conditions as length section and variable orifice size, acting on the residence time and renewing of the water vapor inside the volume, equally show the existence of condensed water vapor maximal values depending on the fixed conditions Batina et al. [8]. One of these studies taking into consideration the inside bottom wall covered by a thin layer of water remaining at a constant level will be presented in a future paper.

Simulation of Water Vapor Condensation in a Partly Closed Structure: The Influence of the External
Conditions of Temperature and Humidity

211

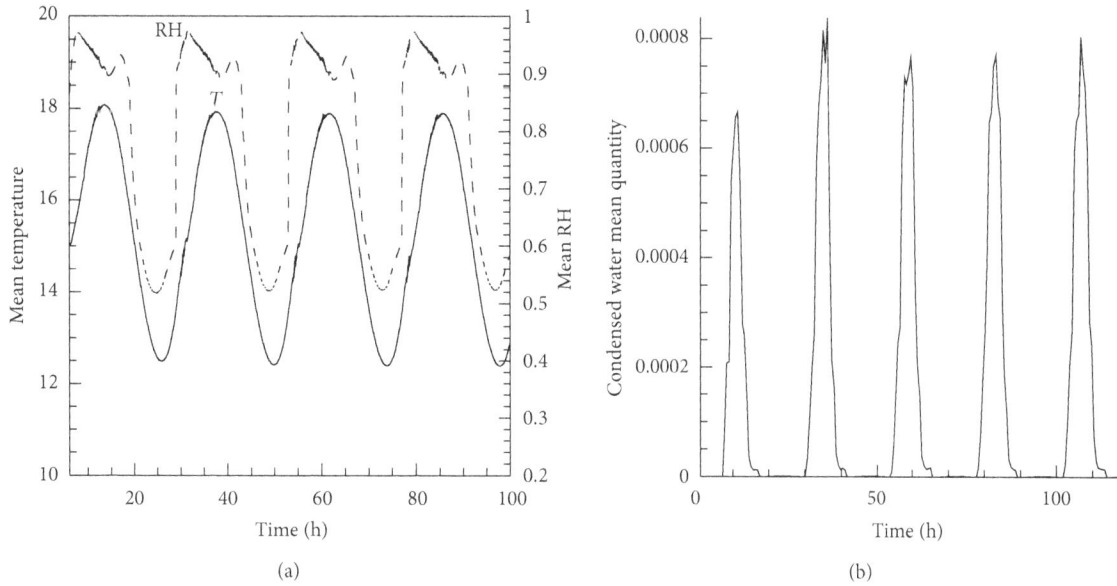

FIGURE 11: (a) Mean temperature and relative hygrometry of the internal air. (b) Mean condensed water vapour evolution inside the structure
over four days.

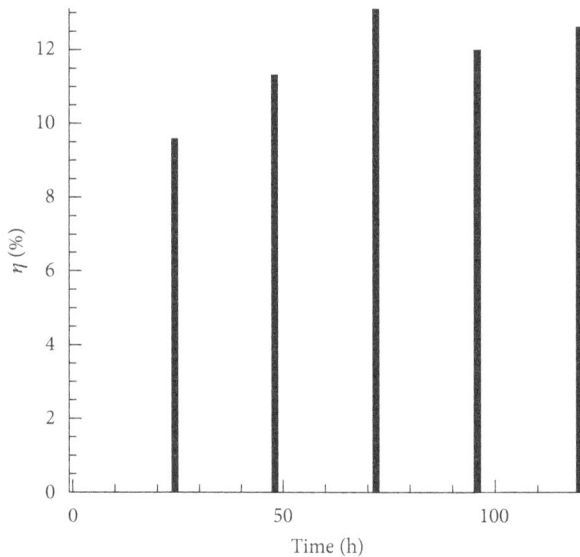

FIGURE 12: Day after day evolution of the condensed water vapour
ratio. Phase $\varphi = 12$ h.

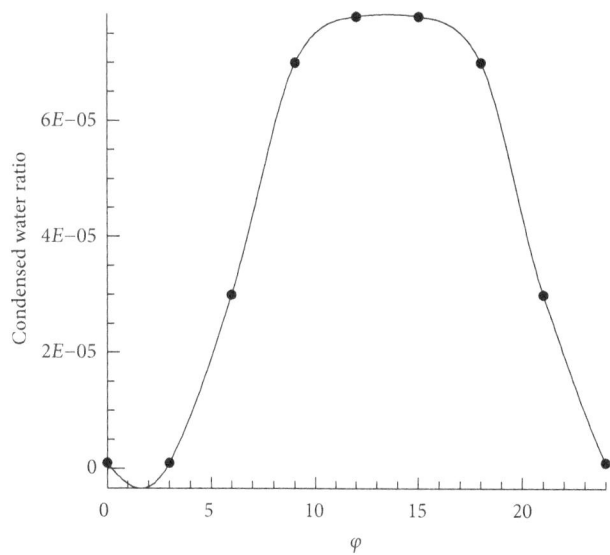

FIGURE 13: Ratio of condensed water vapour as a function of the
phase (T/RH) for an 8 cm orifice.

The systematic study of the influence of all these parameters and the results already obtained open some new promising prospects able to enhance the condensation phenomena inside this type of structure.

In a near future, the software offers numerous possibilities whose results concern water recovery inside partly closed structure where very high humidity can take place (maturing cellars, mushroom beds, etc.). This software will permit to determine a better hygrometry control system and improvement ways.

Nomenclature and Units

C: Vapor concentration

c_p: Heat capacity, $\text{J·kg}^{-1}\text{·K}^{-1}$

D: Mass diffusivity, $\text{m}^2\text{·s}^{-1}$

g: Gravitational acceleration, m·s^{-2}

La: Latent heat, J·kg^{-1}

p: Pressure, Pa

q_e: Entering quantity of water vapor, kg

q_s: Outgoing quantity of water vapor, kg

RH: Relative Humidity, in %
t: Time, s
T: Temperature, K
T_{ext}:   External temperature, K
$\vec{V} = (u, v)$: Speed field, m·s^{-1}.

Greek Symbols

δT: Temperature difference from 15°C (0 initial value), K
φ: Phase between T and
ρ: Volumic mass, kg·m^{-3}
μ: Dynamic viscosity, Pa·s
λ: Thermal conductivity, $\text{W·m}^{-1}\text{·K}^{-1}$
η: Condensed water ratio.

Acknowledgment

The authors thank Mr. Stéphane Glockner of the TREFLE Laboratory - UMR 8508, for his assistance in using the finite volumes method.

References

[1] V. S. Nikolayev, D. Beysens, A. Gioda, I. Milimouk, E. Katiushin, and J.-P. Morel, "Water recovery from dew," *Journal of Hydrology*, vol. 182, no. 1–4, pp. 19–35, 1996.

[2] M. Muselli, D. Beysens, J. Marcillat, I. Milimouk, T. Nilsson, and A. Louche, "Dew water collector for potable water in Ajaccio (Corsica Island, France)," *Atmospheric Research*, vol. 64, no. 1–4, pp. 297–312, 2002.

[3] D. Beysens, I. Milimouk, V. S. Nikolayev et al., "Comment on "The moisture from the air as water resource in arid region: hopes, doubt and facts" by Kogan and Trahtman," *Journal of Arid Environments*, vol. 67, no. 2, pp. 343–352, 2006.

[4] P. Gandhidasan and H. I. Abualhamayel, "Modeling and testing of a dew collection system," *Desalination*, vol. 180, no. 1–3, pp. 47–51, 2005.

[5] D. Beysens, M. Muselli, V. Nikolayev, R. Narhe, and I. Milimouk, "Measurement and modelling of dew in island, coastal and alpine areas," *Atmospheric Research*, vol. 73, no. 1-2, pp. 1–22, 2005.

[6] J. P. Caltagirone and J. Breil, "A vector projection method for solving the Navier-Stokes equations," *Comptes Rendus de l'Académie des Sciences—Séries IIB—Mechanics-Physics-Astronomy*, vol. 327, no. 11, pp. 1179–1184, 1999.

[7] J. Batina, R. Peyrous, and J. Castaing-Lasvinottes, "Condensation de vapeur d'eau en milieu partiellement clos soumis à des conditions périodiques de température et d'humidité : simulationet influence des conditions opératoires," in *Proceedings of the Society for Theriogenology (SFT) Congress*, pp. 345–350, Toulouse, France, June 2008.

[8] J. Batina, R. Peyrous, and J. Castaing-Lasvinottes, "Conditions opératoires optimales favorisant la condensation de vapeur d'eau en milieu partiellement clos soumis à des variations de température et d'humidité," in *Proceedings of the JITH Congress*, pp. 345–350, Djerba, Tunisia, March 2009.

[9] D. Beysens, M. Muselli, J.-P. Ferrari, and A. Junca, "Water production in an ancient sarcophagus at Arles-sur-Tech (France)," *Atmospheric Research*, vol. 57, no. 3, pp. 201–212, 2001.

[10] G. Perard and C. Leborgne, "Sarcophage d'Arles-sur-Tech, rapport technique," *Houille Blanche*, vol. 6, pp. 873–881, 1961.

Permissions

The contributors of this book come from diverse backgrounds, making this book a truly international effort. This book will bring forth new frontiers with its revolutionizing research information and detailed analysis of the nascent developments around the world.

We would like to thank all the contributing authors for lending their expertise to make the book truly unique. They have played a crucial role in the development of this book. Without their invaluable contributions this book wouldn't have been possible. They have made vital efforts to compile up to date information on the varied aspects of this subject to make this book a valuable addition to the collection of many professionals and students.

This book was conceptualized with the vision of imparting up-to-date information and advanced data in this field. To ensure the same, a matchless editorial board was set up. Every individual on the board went through rigorous rounds of assessment to prove their worth. After which they invested a large part of their time researching and compiling the most relevant data for our readers. Conferences and sessions were held from time to time between the editorial board and the contributing authors to present the data in the most comprehensible form. The editorial team has worked tirelessly to provide valuable and valid information to help people across the globe.

Every chapter published in this book has been scrutinized by our experts. Their significance has been extensively debated. The topics covered herein carry significant findings which will fuel the growth of the discipline. They may even be implemented as practical applications or may be referred to as a beginning point for another development. Chapters in this book were first published by Hindawi Publishing Corporation; hereby published with permission under the Creative Commons Attribution License or equivalent.

The editorial board has been involved in producing this book since its inception. They have spent rigorous hours researching and exploring the diverse topics which have resulted in the successful publishing of this book. They have passed on their knowledge of decades through this book. To expedite this challenging task, the publisher supported the team at every step. A small team of assistant editors was also appointed to further simplify the editing procedure and attain best results for the readers.

Our editorial team has been hand-picked from every corner of the world. Their multi-ethnicity adds dynamic inputs to the discussions which result in innovative outcomes. These outcomes are then further discussed with the researchers and contributors who give their valuable feedback and opinion regarding the same. The feedback is then collaborated with the researches and they are edited in a comprehensive manner to aid the understanding of the subject.

Apart from the editorial board, the designing team has also invested a significant amount of their time in understanding the subject and creating the most relevant covers. They scrutinized every image to scout for the most suitable representation of the subject and create an appropriate cover for the book.

The publishing team has been involved in this book since its early stages. They were actively engaged in every process, be it collecting the data, connecting with the contributors or procuring relevant information. The team has been an ardent support to the editorial, designing and production team. Their endless efforts to recruit the best for this project, has resulted in the accomplishment of this book. They are a veteran in the field of academics and their pool of knowledge is as vast as their experience in printing. Their expertise and guidance has proved useful at every step. Their uncompromising quality standards have made this book an exceptional effort. Their encouragement from time to time has been an inspiration for everyone.

The publisher and the editorial board hope that this book will prove to be a valuable piece of knowledge for researchers, students, practitioners and scholars across the globe.

List of Contributors

P. A. Sedykh and I. Yu. Lobycheva
Institute of Solar-Terrestrial Physics SB RAS, Lermontov Street, 126 a, P.O. Box 291, Irkutsk 664033, Russia

Daniel Carbunaru, Monica Sasu and Victor Stefanescu
Department of Atmospheric Physics, Faculty of Physics, University of Bucharest, P.O. Box MG-11, 077125 Bucharest, Romania
National Meteorological Administration, Bucuresti-Ploiesti Avenue, No. 97, 013686 Bucharest, Romania

Sabina Stefan
Department of Atmospheric Physics, Faculty of Physics, University of Bucharest, P.O. Box MG-11, 077125 Bucharest, Romania

Alexander Mishev
Institute for Nuclear Research and Nuclear Energy, Bulgarian Academy of Sciences, 1784 Sofia, Bulgaria
Sodankyla Geophysical Observatory (Oulu Unit), University of Oulu, 90014 Oulu, Finland

Gamal El Afandi
Division of International Research and Development, College of Agriculture, Environment and Nutrition Sciences and College of Engineering, Tuskegee University, Tuskegee, AL 36088, USA
Astronomy and Meteorology Department, Faculty of Science, Al Azhar University, Cairo 11884, Egypt

Mostafa Morsy and Fathy El Hussieny
Astronomy and Meteorology Department, Faculty of Science, Al Azhar University, Cairo 11884, Egypt

Thiago R. Rodrigues, Sérgio R. de Paulo, Jonathan W. Z. Novais, Leone F. A. Curado, José S. Nogueira and Renan G. de Oliveira
Instituto de Fisica, Universidade Federal de Mato Grosso, 78060-900 Cuiaba-MT, Brazil

Francisco de A. Lobo
Departamento de Agronomia e Medicina Veterinaria, Universidade Federal de Mato Grosso, 78060-900 Cuiaba-MT, Brazil

George L. Vourlitis
Department of Biological Sciences, California State University, San Marcos, San Diego, CA 92096, USA

Roman Corobov
Eco-TIRAS International Environmental Association, 9/1 Independentii Street, Apartment, 133, 2060 Chisinau, Moldova

Scott Sheridan
Department of Geography, Kent State University, Kent, OH 44242, USA

Kristie Ebi
Department of Global Ecology, Carnegie Institution for Science, Stanford, CA 94305, USA

Nicolae Opopol
Hygiene and Epidemiology Department, State Medical and Pharmaceutical University, 67aGh, Asachi Street, 2028Chisinau, Moldova

Shi Kai and Liu Chun-Qiong
Key Laboratory of Hunan Ecotourism, Jishou University, Jishou, Hunan 416000, China
College of Biology and Environmental Sciences, Jishou University, Jishou, Hunan 416000, China

Li Si-Chuan
College of Biology and Environmental Sciences, Jishou University, Jishou, Hunan 416000, China

B. H. Vaid
TMSI, National University of Singapore, Singapore
Department of Applied Sciences, Raj Kumar Goel Institute of Technology forWomen, Near Jain Tube, Delhi Meerut Road, Ghaziabad 200203, India

Adil Rasheed
SINTEF ICT, Applied Mathematics, Strindveien 4, 7050 Trondheim, Norway

Darren Robinson
LESO-PB, EPFL, CH-1015 Lausanne, Switzerland

R. Shrivastava and R. B. Oza
Radiation Safety Systems Division, Bhabha Atomic Research Centre, Mumbai 400 085, India

S. K. Dash
Centre for Atmospheric Sciences, Indian Institute of Technology Delhi, New Delhi 110 016, India

D. N. Sharma
Health Safety and Environment Group, Bhabha Atomic Research Centre, Mumbai 400 085, India

Churchill Okonkwo
Beltsville Center for Climate System Observation, Atmospheric Science Program, Howard University, Washington, DC 20059, USA

Andre Lenouo
Department of Physics, Faculty of Science, University of Douala, P.O. Box 24157, Douala, Cameroon

Francois Kamga Nkankam
Laboratory for Environmental Modelling and Atmospheric Physics, Department of Physics, University of Yaounde 1, Yaounde, Cameroon

Avit Kumar Bhowmik
Institute for Environmental Sciences, Quantitative Landscape Ecology, University of Koblenz-Landau, Fortstraße 7, 76829 Landau (Pfalz), Germany

M. Tesfaye
Department of Geography, Geoinformatics and Meteorology, University of Pretoria, Lynwood Road, Pretoria 0002, South Africa
National Laser Centre, Council for Scientific and Industrial Research, P.O. Box 395, Pretoria 0001, South Africa

J. Botai
Department of Geography, Geoinformatics and Meteorology, University of Pretoria, Lynwood Road, Pretoria 0002, South Africa

V. Sivakumar
Department of Geography, Geoinformatics and Meteorology, University of Pretoria, Lynwood Road, Pretoria 0002, South Africa
Discipline of Physics, School of Chemistry and Physics, University of KwaZulu Natal, Westville, Durban 4000, South Africa

G. Mengistu Tsidu
Department of Physics, Addis Ababa University, P.O. Box 1176, Addis Ababa, Ethiopia

Baerbel Langmann
Institute of Geophysics, University of Hamburg, Geomatikum, Office 1411, Bundesstraße 55, 20146 Hamburg, Germany

Alex Guenther
Atmospheric Chemistry Division, NCAR Earth System Laboratory, National Center for Atmospheric Research, 3090 Center Green, Boulder, CO 80301, USA

Joseph Sunday Ojo
Department of Physics, The Federal University of Technology, Akure PMB 704, Akure 340213, Nigeria

Jean Batina
Laboratoire des Sciences de l Ingenieur Appliquees a la Mecanique et a l Electricite (SIAME), Universite de Pau et des Pays de l Adour, BP 1155, 64013 Pau, France

René Peyrous
Laboratoire d Electronique des Gaz et des Plasmas, Universite de Pau et des Pays de lAdour, 64000 Pau, France